AF598017

METHODS IN MOLECULAR BIOLOGY™

Series Editor
John M. Walker
School of Life Sciences
University of Hertfordshire
Hatfield, Hertfordshire, AL10 9AB, UK

For other titles published in this series, go to
www.springer.com/series/7651

Protein Secretion

Methods and Protocols

Edited by

Anastassios Economou

Institute of Molecular Biology and Biotechnology, Foundation for Research and Technology - Hellas, Heraklion, Greece

Editor
Anastassios Economou
Foundation for Research & Technology – Hellas (FORTH)
Institute of Molecular Biology & Biotechnology and Department of Biology
University of Crete
Nikolaou Plastira 100
700 13 Iraklion
Greece
aeconomo@imbb.forth.gr

ISSN 1064-3745 e-ISSN 1940-6029
ISBN 978-1-60327-167-7 e-ISBN 978-1-60327-412-8
DOI 10.1007/978-1-60327-412-8
Springer New York Dordrecht Heidelberg London

Library of Congress Control Number: 2010920086

Printed on acid-free paper

Humana Press is part of Springer Science+Business Media (www.springer.com)

Preface

The elucidation of the complete information content in hundreds of genomes has brought with it a surprising realization. More than a third of all the proteins in any given proteome are comprised of non-cytoplasmic polypeptides. These can be resident membrane proteins such as channels and receptors or secretory proteins such as hydrolytic enzymes and toxins. Membrane biogenesis and protein trafficking and secretion are central to the biology and pathology of the cell. Optimal protein trafficking is essential for cell viability, communication, and programmed death, for cells to modulate and yield metabolic goods from their environment, for pathogens to attack, and for hosts to fend them off.

Since all polypeptides in prokaryotes and most in eukaryotes are synthesized by cytoplasmic ribosomes, the cell has acquired tools that enable it to accurately and efficiently sort exported proteins from the cytoplasmic residents. Various specialized chaperones, pilots, and ushers have evolved to correctly recognize secretory and membrane proteins and in several instances this recognition prevents or stalls folding reactions. Moreover, this chaperone-mediated "face-control" effectively sorts extra-cytoplasmic from cytoplasmic proteins and then delivers them to complex membrane-associated cellular nanomachines. These catalyze the transmembrane crossing of the targeted polypeptides. Exported proteins come in different functional and structural flavors and are destined for residency of different subcellular compartments or the outside world. Some of them are even savvy enough to cross several prokaryotic and eukaryotic membranes before they reach their final destination. Hence we are now aware that various secretory proteins carry different export signals that act as specific address tags. In many instances, the features of these export signals are well understood and have hence predictive value when a new genome is deciphered through the use of biocomputing.

The study of protein secretion comes with some challenging biochemistry since a large part of the reactions take place at or in the membrane. Elegant genetic and biochemical approaches have been combined over the past 30 years in order to dissect the ways by which the membranes are negotiated so that the exported protein lands on the other side. Overexpression systems have allowed the purification of the subunits involved in large amounts, and this in turn facilitated structural studies. Such is the progress in this approach that, in many of the newer secretion systems, the structures of the components precede the description of biochemical functions. This is less true for the structural elucidation of membrane proteins. Despite recent progress, <200 membrane protein structures are known. In recent years, other powerful cell biology tools that can even offer real-time observation of the secretion process have become available. Finally, organism-wide proteomics is providing insight into how protein secretion is incorporated in the whole metabolic reaction network of the cell and in many instances has revealed interesting links with the rest of the cell's physiology.

The purpose of *Protein Secretion: Methods and Protocols* is to provide some examples of the multiplicity of tools that have been developed to study protein sorting, membrane targeting, transmembrane crossing, and secretion across multiple membranes. A wide variety of methods are covered that range from bioinformatics and proteomics to fundamental enzymology and genetics to cell biology, structural analyses, and biophysics. This only reflects the highly multidisciplinary nature one expects from a mature field. It is hoped that the study of the various systems and the tools developed to decipher their secrets will provide users with inspiration in finding ways to tackle problems encountered in their research. The multiplicity of protein export systems discovered to date suggests that we are nowhere near a complete inventory. I chose to focus on well-characterized paradigms so that the reader can benefit from robust, well-established protocols in which many of the experimental wrinkles have been ironed out. Several systems have been chosen from both prokaryotic and eukaryotic organisms. The book is aimed at the biochemist, geneticist, or biologist (cell, molecular, or structural) who is a protein secretion novice and also at seasoned protein secretion experts who wish to incorporate new experimental tools in their studies. The book is also aimed at researchers who want to explore the immense biotechnological potential of secretion systems in the manipulation of protein export pathways for the production of heterologous proteins (be they biopharmaceuticals or industrial enzymes) as well as their use to develop vaccines and anti-microbials. The reader may gain insight from the difficulties encountered in the more established systems and use this rationally in the dissection of less characterized protein secretion machines, in less characterized organisms, or other cell biology and membrane-related questions.

I would like to thank the authors who have contributed to this work for their enthusiastic response and efforts; to John Walker, the series editor, for his constant help, encouragement, and vigilant eye; to Georgia Houlaki for her expert secretarial help and exceptional organizational skills; and to the staff at Humana Press who helped produce this volume.

Tassos Economou

Contents

Contributors

SUHILA APPATHURAI • *Cell Biology and Metabolism Program, National Institute of Child Health and Human Development, National Institutes of Health, Bethesda, MD, USA*

LAURENT AUDRY • *Institut Pasteur, Groupe "Dynamique des interactions hôte-pathogène", Paris, France*

DIBYENDU BHATTACHARYYA • *Department of Molecular Genetics and Cell Biology, The University of Chicago, Chicago, IL, USA*

DOMINIQUE BELIN • *Department of Pathology and Immunology, University Medical Center, University of Geneva, Geneva, Switzerland*

BEN C. BERKS • *Department of Biochemistry, Oxford University, Oxford, UK*

GEERT VAN DER BOGAART • *Department of Membrane Enzymology, Groningen Biomolecular Sciences and Biotechnology Institute and Zernike Institute for Advanced Materials, University of Groningen, Groningen, The Netherlands*

MIKHAIL BOGDANOV • *Department of Biochemistry and Molecular Biology, University of Texas Medical School, Houston, TX, USA*

BETTINA BÖLTER • *Department Biologie I-Botanik, Ludwig-Maximilians-Universität, Planegg-Martinsried and Munich Center for Integrated Protein Science, CiPSM, Ludwig-Maximilians- Universität, Munich, Germany*

JEFFREY L. BRODSKY • *Department of Biological Sciences, University of Pittsburgh, Pittsburgh, PA, USA*

NERMIN CELIK • *Department of Biochemistry and Molecular Biology, Monash University, Clayton, Victoria, Australia*

AGNIESZKA CHACINSKA • *Institut für Biochemie und Molekularbiologie, ZBMZ, and Centre for Biological Signalling Studies (bioss), Universität Freiburg, Freiburg, Germany*

JENNINE M. CRANE • *Department of Biochemistry, University of Missouri, Columbia, MO, USA*

MICHAEL DAGLEY • *Department of Biochemistry and Molecular Biology, Monash University, Clayton, Victoria, Australia*

KUSH DALAL • *Department of Biochemistry and Molecular Biology, Life Sciences Institute, Faculty of Medicine, University of British Columbia, Vancouver, BC, Canada*

ROSS E. DALBEY • *Department of Chemistry, The Ohio State University, Columbus, OH, USA*

PAVEL DOLEZAL • *Department of Parasitology, Faculty of Science, Charles University, Prague, Czech Republic*

WILLIAM DOWHAN • *Department of Biochemistry and Molecular Biology, University of Texas Medical School, Houston, TX, USA*

ARNOLD DRIESSEN • *Department of Molecular Microbiology, Groningen Biomolecular Sciences and Biotechnology Institute and Zernike Institute for Advanced Materials, University of Groningen, Groningen, The Netherlands*

FRANCK DUONG • *Department of Biochemistry and Molecular Biology, Life Sciences Institute, Faculty of Medicine, University of British Columbia, Vancouver, BC, Canada*

ANASTASSIOS ECONOMOU • *Department of Biology, University of Crete and Institute of Molecular Biology and Biotechnology-Foundation for Research and Technology Hellas, Heraklion, Crete, Greece*

JOST ENNINGA • *Institut Pasteur, Groupe "Dynamique des interactions hôte-pathogène", Paris, France*

JAN-WILLEM DE GIER • *Center for Biomembrane Research, Department of Biochemistry and Biophysics, Stockholm University, Stockholm, Sweden*

BENJAMIN S. GLICK • *Department of Molecular Genetics and Cell Biology, The University of Chicago, Chicago, IL, USA*

GIORGOS GOURIDIS • *Department of Biology, University of Crete and Institute of Molecular Biology and Biotechnology-Foundation for Research and Technology Hellas, Heraklion, Crete, Greece*

ADAM T. HAMMOND • *Institute for Biophysical Dynamics, The University of Chicago, Chicago, IL, USA*

PHILIP N. HEACOCK • *Department of Biochemistry and Molecular Biology, University of Texas Medical School, Houston, TX, USA*

RAMANUJAN S. HEGDE • *Cell Biology and Metabolism Program, National Institute of Child Health and Human Development, National Institutes of Health, Bethesda, MD, USA*

STEPHEN HIGH • *Faculty of Life Sciences, University of Manchester, Manchester, UK*

I. BARRY HOLLAND • *Institut de Genetique et Microbiologie, UMR 8621 CNRS, Universite de Paris-Sud, Orsay, France*

SPYRIDOULA KARAMANOU • *Institute of Molecular Biology and Biotechnology-Foundation for Research and Technology Hellas, Heraklion, Crete, Greece*

MIRJAM KLEPSCH • *Center for Biomembrane Research, Department of Biochemistry and Biophysics, Stockholm University, Stockholm, Sweden*

MARINA KOUKAKI • *Institute of Molecular Biology and Biotechnology-Foundation for Research and Technology Hellas, Heraklion, Crete, Greece*

VIKTOR KRASNIKOV • *Department of Optical Condensed Matter Physics, Zernike Institute for Advanced Materials, University of Groningen, Groningen, The Netherlands*

ANDREAS KUHN • *Institute of Microbiology and Molecular Biology, University of Hohenheim, Stuttgart, Germany*

ILJA KUSTERS • *Department of Molecular Microbiology, Groningen Biomolecular Sciences and Biotechnology Institute and Zernike Institute for Advanced Materials, University of Groningen, Groningen, The Netherlands*

VLADIMIR A. LIKIC • *Bio21 Molecular Science and Biotechnology Institute, University of Melbourne, Parkville, Victoria, Australia*

ANGELA A. LILLY • *Department of Biochemistry, University of Missouri, Columbia, MO, USA*

TREVOR LITHGOW • *Department of Biochemistry and Molecular Biology, Monash University, Clayton, Victoria, Australia*

MALAIYALAM MARIAPPAN • *Cell Biology and Metabolism Program, National Institute of Child Health and Human Development, National Institutes of Health, Bethesda, MD, USA*

CHRIS MEISINGER • *Institut für Biochemie und Molekularbiologie, ZBMZ, and Centre for Biological Signalling Studies (bioss), Universität Freiburg, Freiburg, Germany*

MATTHIAS MÜLLER • *Institute of Biochemistry and Molecular Biology, ZBMZ, University of Freiburg, Freiburg, Germany*

KUNIO NAKATSUKASA • *Department of Biological Sciences, University of Pittsburgh, Pittsburgh, PA, USA*

SHIN-ICHIRO NARITA • *Institute of Molecular and Cellular Biosciences, University of Tokyo, Tokyo, Japan*

TRACY PALMER • *Division of Molecular Microbiology, College of Life Sciences, University of Dundee, Dundee, Scotland*

SASCHA PANAHANDEH • *Institute of Biochemistry and Molecular Biology, ZBMZ, University of Freiburg, Freiburg, Germany*

NIKOLAUS PFANNER • *Institut für Biochemie und Molekularbiologie, ZBMZ, and Centre for Biological Signalling Studies (bioss), Universität Freiburg, Freiburg, Germany*

BERT POOLMAN • *Department of Membrane Enzymology, Groningen Biomolecular Sciences and Biotechnology Institute and Zernike Institute for Advanced Materials, University of Groningen, Groningen, The Netherlands*

LINDA L. RANDALL • *Department of Biochemistry, University of Missouri, Columbia, MO, USA*

SHRUTI RASTOGI • *Department of Biochemistry and Molecular Biophysics, Columbia University and Columbia University Center for Computational Biology and Bioinformatics (C2B2), New York, NY, USA*

FERNANDA RODRIGUEZ • *Biozentrum, Growth and Development, University of Basel, Basel, Switzerland*

BURKHARD ROST • *Department of Biochemistry and Molecular Biophysics, Columbia University and Columbia University Center for Computational Biology and Bioinformatics (C2B2) and NorthEast Structural Genomics Consortium (NESG) & New York Consortium on Membrane Protein Structure (NYCOMPS), New York, NY, USA*

FRANK SARGENT • *Division of Molecular Microbiology, College of Life Sciences, University of Dundee, Dundee, Scotland*

ANNE-KATHRIN SCHUBERT • *Institute of Microbiology and Molecular Biology, University of Hohenheim, Stuttgart, Germany*

AJAY SHARMA • *Cell Biology and Metabolism Program, National Institute of Child Health and Human Development, National Institutes of Health, Bethesda, MD, USA*

SUSAN SCHLEGEL • *Center for Biomembrane Research, Department of Biochemistry and Biophysics, Stockholm University, Stockholm, Sweden*

DIONISIA P. SIDERIS • *Department of Biology, University of Crete and Institute of Molecular Biology and Biotechnology-Foundation for Research and Technology Hellas, Heraklion, Crete, Greece*

NANDI SIMPSON • *Institut Cochin, Universite Paris Descartes, CNRS (UMR 8104) Inserm, U567, Paris, France*

OLIVER SCHMIDT • *Institut für Biochemie und Molekularbiologie, ZBMZ, and Centre for Biological Signalling Studies (bioss), Universität Freiburg, Freiburg, Germany*

JÜRGEN SOLL • *Department Biologie I-Botanik, Ludwig-Maximilians-Universität, Planegg-Martinsried and Munich Center for Integrated Protein Science, CiPSM, Ludwig-Maximilians-Universität, Munich, Germany*

ANNE SPANG • *Biozentrum, Growth and Development, University of Basel, Basel, Switzerland*

NATALIE STIEGLER • *Institute of Microbiology and Molecular Biology, University of Hohenheim, Stuttgart, Germany*

PENELOPE STRITTMATTER • *Department Biologie I-Botanik, Ludwig-Maximilians-Universität, Planegg-Martinsried and Munich Center for Integrated Protein Science, CiPSM, Ludwig-Maximilians-Universität, Munich, Germany*

STEVEN M. THEG • *Department of Plant Biology, University of California, Davis, CA, USA*

KOSTAS TOKATLIDIS • *Department of Materials Science and Technology, University of Crete and Institute of Molecular Biology and Biotechnology-Foundation for Research and Technology Hellas, Heraklion, Crete, Greece*

HAJIME TOKUDA • *Institute of Molecular and Cellular Biosciences, University of Tokyo, Tokyo, Japan*

F.-NORA VÖGTLE • *Institut für Biochemie und Molekularbiologie, ZBMZ, and Centre for Biological Signalling Studies (bioss), Universität Freiburg, Freiburg, Germany*

SAMUEL WAGNER • *Center for Biomembrane Research, Department of Biochemistry and Biophysics, Stockholm University, Stockholm, Sweden*

PENG WANG • *Department of Chemistry, The Ohio State University, Columbus, OH, USA*

DAVID WICKSTRÖM • *Center for Biomembrane Research, Department of Biochemistry and Biophysics, Stockholm University, Stockholm, Sweden*

CORNELIA M. WILSON • *Faculty of Medicine, University of Limoges, Limoges, France*

JANNY DE WIT • *Department of Molecular Microbiology, Groningen Biomolecular Sciences and Biotechnology Institute and Zernike Institute for Advanced Materials, University of Groningen, Groningen, The Netherlands*

JIJUN YUAN • *Department of Chemistry, The Ohio State University, Columbus, OH, USA*

Chapter 1

The Extraordinary Diversity of Bacterial Protein Secretion Mechanisms

I. Barry Holland

Abstract

I have tried to cover the minimal properties of the prolific number of protein secretion systems identified presently, particularly in Gram negative bacteria. New systems, however, are being reported almost by the month and certainly I have missed some. With the accumulating evidence one remains in awe of the complexity of some pathways, with the Type III, IV and VI especially fearsome and impressive. These systems illustrate that protein secretion from bacteria is not only about passage of large polypeptides across a bilayer but also through long tunnels, raising quite different questions concerning mechanisms. The mechanism of transport via the Sec-translocase–translocon is well on the way to full understanding, although a structure of a stuck intermediate would be very helpful. The understanding of the precise details of the mechanism of targeting specificity, and actual polypeptide translocation in other systems is, however, far behind. Groups willing to do the difficult (and risky) work to understand mechanism should therefore be more actively encouraged, perhaps to pursue multidisciplinary, collaborative studies. In writing this review I have become fascinated by the cellular regulatory mechanisms that must be necessary to orchestrate the complex flow of so many polypeptides, targeted by different signals to such a wide variety of transporters. I have tried to raise questions about how this might be managed but much more needs to be done in this area. Clearly, this field is very much alive and the future will be full of revealing and surprising twists in the story.

Key words: Protein secretion pathways, translocon, translocase, bacteria, transport tunnels, insertases, SRP, SecA, Y.

1. Introduction

1.1. A Myriad of Protein Translocation Pathways in Bacteria

Bacterial and other membranes are relatively well packed with proteins, with a high proportion of total cell protein targeted to this crucially important compartment. As noted by Engelman (1),

A. Economou (ed.), *Protein Secretion*, Methods in Molecular Biology 619,
DOI 10.1007/978-1-60327-412-8_1, © Springer Science+Business Media, LLC 2010

the membrane is now considered to have a structure more mosaic than fluid as proposed in the nevertheless farsighted model put forward by Singer and Nicholson in 1972 (2). Most estimates have indicated that at least 20% of the coding capacity of the genomes of Gram negative bacteria, corresponding to approximately 1000 distinct polypeptides in bacteria like *Escherichia coli*, are localized to the cytoplasmic membrane. Moreover, 25–30% of total cell protein is associated with the cell envelope (inner + outer membrane) together with proteins that are secreted to the medium under laboratory conditions. Surprisingly, Yamane et al. (3) have estimated from the genome of the Gram positive *Bacillus subtilis* that at least 2000 genes (over 40% of the coding capacity) encode proteins with predicted transmembrane spanning regions. Moreover, this study in *B. subtilis* noted a minimum additional 260 genes encoding proteins carrying *recognisable* secretion signals – certainly an underestimate since some secreted proteins have no easily recognisable signal sequence. Notably, genome studies show that several of the secretion systems to be described below occur in the same species and are often encoded in multiple copies.

Gram negative bacteria utilise many systems to translocate proteins into (insertion) or through (translocation) one or both membranes. Gram positive bacteria appear at the moment to be restricted to 6 or 7: the Sec-system, YidC and Tat insertase/translocases. More specialised translocators of Type I (ABC), Type IV (conjugation-like system), and a recently identified apparently novel mechanism (4–7), together with the specialised flagellar and pili organelle assembly systems, complete the set (8). This contrast between Gram positive and negative organisms is testimony to the importance of the additional outer membrane barrier that has to be negotiated in the Gram negatives. In *E. coli*, the three protein translocase/translocon complexes are also present in the inner membrane. The SecA ATPase and the core SecYEG translocon channel (corresponding to three Sec61 subunits in eukaryotes) transport proteins targeted by the chaperone SecB in an unfolded form to the periplasm (and outwards where appropriate). In addition, targeted by the signal recognition particle (SRP), integral membrane proteins are co-translationally inserted (assembled) into the bilayer via SecYEG assisted by YidC (*see* 9, 10). YidC interestingly, although its precise function is not yet well understood, has homologues in both mitochondria and chloroplasts. YidC facilitates the lateral release of TM helices from the SecY-translocon during assembly of integral membrane proteins. Thus, YidC fulfils a vital auxiliary role enabling the translocon to exert its dual function of directing proteins to the periplasm or the bilayer, as appropriate (11). The YidC protein also acts as an independent insertase in some way for certain integral membrane proteins. The Tat pathway is involved in translocating fully folded proteins across the inner membrane, including many proteins of the redox system and some integral membrane proteins (12).

In addition to the assembly machines for Type IV pili and for flagellar assembly, at least four systems in Gram negative organisms (Types I, III, IV plus now Type VI) translocate proteins directly through 'tunnels' from cytoplasm to the exterior, while at least a dozen systems have been discovered in different bacteria that transport proteins from the periplasm across the outer membrane. Finally, the elusive translocase for insertion of proteins into the outer membrane was discovered 5 years ago and turns out to involve a protein, Omp85, in *Neisseria menigitidis* (13) and its *E. coli* homologue, YaeT (14), distantly related to chloroplast and mitochondrial protein translocases (13,14).

2. Evolutionary Power: Multiple Strategies to Cross the Outer Membrane in Gram Negative Bacteria

2.1. Crossing via Tunnels

Gram negative bacteria possess an extraordinary array of one step and two step mechanisms to secrete proteins to the exterior. So far, four one step systems have been characterised that effectively form transenvelope, multi-subunit 'tunnels', connecting the cytosol to the exterior, bypassing the periplasm. The Type one tunnel (15), composed of three proteins is a relatively simple affair, while remarkably the complex Type III and Type IV tunnels continue well beyond the cell surface, constituting a needle that penetrates the surface of a mammalian or bacterial cell, for injection of protein effectors or nucleoprotein complexes, respectively (16,17). The ultimate 'tunnel', however, is provided for the flagellin subunits (largely unfolded) that must travel 15–20 microns to the tip of the flagellum, whose basal body in the inner membrane and overall assembly and structure (18) are strikingly similar to those of the Type III protein translocator (16).

2.2. Translocation from the Periplasm

The evolution of a second membrane barrier in Gram negative bacteria, rather than the thick cell wall of Gram negative organisms, while providing additional protection for the cell, raised the problem of how, on the one hand, to supply this barrier with proteins necessary for controlling cell permeability, and on the other hand, how to secrete important polypeptides to the exterior. The power of the evolutionary process has solved this by producing a zoo of alternative systems. These so-called two step pathways constitute a variety of outer membrane translocators, with substrates supplied from the periplasm, dependent on Tat or, in the majority of cases, the Sec-machinery. A recent exception, demonstrating the extraordinary variety of mechanisms that are possible for protein secretion, has shown that some proteins can be translocated from the periplasm via the SecY translocon and then somehow exit through the outer membrane TolC portal, by a Type I system with an ABC-ATPase translocator (19), although classically Type I substrates enter the pathway from the cytosol.

From the discovery in the mid-1980s of the Type II, two step secretion pathway (20, 21), several variations on this theme have been characterised – two to three even in the last year. These new pathways also include what can be described as a hybrid pathway in the case of pertussis toxin. The first stage is Sec-dependent transport to the periplasm, while a variation on the Type IV pathway, with a specialised secretin, in someway carries the toxin to the exterior. However, of course, unlike Type IV substrates, which normally enter the transport tunnel from the cytoplasm, the PT toxin accesses from the periplasm (22). A particularly ingenious hybrid system (23), as seen from a human perspective, though hardly surprising now in this field, involves the initial Sec-dependent translocation of an autotransporter to the periplasm, followed by transport of the passenger domain to the medium, apparently utilising the YaeT outer membrane protein assembly machinery (discussed below). In this way, the functions of two insertase/translocases are combined in order to facilitate the sequential crossing of the two membranes. Another intriguing variation on the hybrid theme is the discovery of the 'P-usher' system in *Pseudomonas aeruginosa* (24) in which a 'classical' usher (as found in Type P pilin transport) carries a PORTRA domain that otherwise is conserved in a subfamily of the Type V pathway (autotransporters and two partner systems), where it is necessary for recognition of the passenger domain, prior to its translocation on to the surface. Moreover, Ruer et al. (24) showed that the hybrid usher translocated not only pilin subunits but at a lower frequency also an orphan TpsA adhesin, producing a mixed pilus.

3. A More Detailed Look at the Fundamentals of Different Systems

Two comprehensive overviews of bacterial protein transport systems (25, 26) and an excellent coverage of the structure and function of SecA and the SecYEG translocon (9) are recommended. In addition, several excellent recent reviews concerning the detailed structure of individual translocases and translocons of the many different protein transport systems in bacteria are indicated throughout the text. I shall simply note here some important general features distinguishing different pathways, including striking advances in the last 2–3 years. I would like first to clarify a point of nomenclature. A protein or protein complex that is integrated into a membrane and acts as a specific gate *through* which other proteins pass across a membrane should be considered a *translocon*. Long oligomeric structures that traverse membrane and non-membranous regions, involving presumably a quite different molecular mechanism of translocation of proteins, should be considered a special case, qualifying as tunnels or conduits. An enzyme or complex of proteins that directly energise transport through a translocon is a *translocase*. In some complex transenvelope machineries a distinction has to be made between the energy necessary for assembly of the machine and the actual translocase that drives the translocation of the ultimate transport substrate. In most cases this is not clear as yet. SecA and SecYEG are easily defined as translocase and translocon respectively. In Type I secretion, passage across the inner membrane, at least, is carried out by ABC transporters that usually constitute a fused translocase and translocon. Concerning the integral membrane protein YidC and its independent insertase activity, as in most other systems for crossing the inner membrane, what is translocase and what is translocon remains to be defined precisely.

3.1. Post-translational Transport via SecA and SecY

Many proteins are translocated post-translationally in unfolded form across the cytoplasmic membrane to the periplasm by the general secretion pathway. Such proteins, recognised by their characteristic N-terminal signal sequences, are chaperoned by SecB, passed to SecA and inserted by the translocase into the channel of the SecYEG translocon. This is followed by segmental translocation of short peptides coinciding with insertion-deinsertion of SecA into the bilayer, accompanied by hydrolysis of ATP. Additional input of energy is provided from the PMF (proton-motive force) at some later stage until transport is completed and finally proteins are released by cleavage of the signal sequence. The nature of the catalytic cycle for SecA and the interaction between the two key players SecA and SecY, involving several essential contact sites, is now understood in considerable detail (27). Enormous progress has been made therefore in understanding the mechanism of translocation of polypeptides by SecA–SecY, since in particular the pioneering work of William Wickner with in vitro systems initiated in the 1980s. Nevertheless, detailed knowledge of many aspects still remain unknown; not least the precise mechanism of segmental threading of polypeptides through the SecYEG translocon by SecA. However, in October 2008, three major papers appeared in *Nature* (28–30) to provide us, although still just glimpses of this complex dynamic process, with some exciting new insights. These separate studies from the laboratories of Tom Rapoport, and Osamu Nureki and Koreaki Ito, involved a combination of new crystal structures of SecY and a SecA–SecY complex (capturing different conformations in SecA) and cysteine cross-linking of preprotein–SecA–SecY complexes (i.e. with a translocation intermediate). From the combined results, and building on other recent studies, we now have a clearer picture of the conformational changes, induced by binding to the signal sequence (and presumably ATP) that, produces movement of specific small domains in SecA. These movements directly result in the insertion of a segment of the preprotein into the translocon, concomitant with displacement of the plug that normally blocks the channel of SecY.

3.2. Co-translational Transport

In contrast to translocation of proteins to the periplasm, the assembly/insertion of many integral membrane proteins involves co-translational targeting of unfolded polypeptides bound to the SRP, through recognition of N-terminal hydrophobic sequences, as they emerge from the ribosome. In most cases, these appear to be targeted directly to the SecYEG machinery and/or the YidC insertase. With the help of YidC, successive transmembrane spanning regions (TMS) are then partitioned laterally from the SecYEG translocon into the membrane (10). It is still not clear precisely how future transmembrane helices are recognised by either SecY or the YidC machinery for partitioning but progressively the nature of the specific topogenic motifs that signal domains for lateral transfer from the appropriate translocon are being elucidated. The indications are that the same 'rules' may apply in bacteria and higher organisms (31). Since assembly of many integral membrane proteins is not apparently directly dependent on SecA, energy must be supplied by the translation process itself or the PMF. However, membrane proteins with extensive hydrophilic loops require and somehow utilise SecA to drive translocation of these regions.

3.3. Transport of Folded Proteins Across the Cytoplasmic Membrane

The Tat system in Gram negative organisms, in complete contrast to the SecY–dependent pathway, translocates fully folded proteins, often in complex with their essential cofactors. The translocon is composed of a large complex containing three different subunits. TatC apparently binds the polypeptide substrate, while a complex of TatA, B, C may form the large translocation channel across the inner membrane. Curiously, while a large number of A-subunits are present in the Tat complex in *E. coli*, in *Pseudomonas* and in *B. subtilis*, for example, the system completely lacks the TatA protein. On the other hand, in *B. subtilis*, two types of Tat complexes are present that show distinct transport substrate specificity (32). Despite important recent progress in elucidating the mechanism of action of the Tat machinery, in reality much needs to be discovered about the nature of the translocon, targeting specificity, and above all how the PMF functions as an energy source and how folded polypeptides can be squeezed through the bilayer.

3.4. Type I, III and IV: Tunnels from Cytoplasm to the Exterior in Gram Negative Bacteria

Three pathways, Type I, III and IV, that transport individual proteins directly to the exterior have been studied in detail. Type I requires three proteins (15,33), a cytoplasmic, ABC ATPase (e.g. HlyB), forming a complex with a membrane-anchored MFP (Membrane Fusion Protein) protein (HlyD) spanning the periplasm, which completes the transenvelope tunnel for the secreted haemolysin (HlyA) by recruiting on 'demand', the outer membrane protein, TolC (34). TolC forms a trimeric structure (35) and provides a mechanism whereby the channel is opened to facilitate translocation by a transport substrate induced iris-like movement of the long periplasmic helices of TolC (36). Unfolded HlyA molecules contact the HlyB-HlyD complex via a C-terminal signal sequence (37) and are translocated dependent upon ATP hydrolysis and apparently a contribution from the PMF (38). With the exception of SecA, the structure of the HlyB dimer and details of the catalytic cycle and therefore its possible coupling to transport are the best understood of any translocase ATPase (39,40). Nevertheless, the nature of the early stages of initiation of translocation involving HlyB,D and the final folding of HlyA as it emerges on the cell surface are unclear.

The transport apparatus for Types III and IV is much more complex, being composed of up to 20 different proteins, many with unknown or poorly understood functions. Transport substrates are recognised in association with specific chaperones by distinct non-cleaved N-terminal signals and transport is associated with dedicated cytoplasmic ATPases at the base of the transenvelope structures (16,41), although precisely how energy for transport is generated and used, certainly in the case of Type III, is not clear (42). In most cases a clearly characterised secretin or secretin-like oligomeric structure in the outer membrane has been identified but the inner membrane translocon is not so well defined. Proteins are likely translocated across the outer membrane in an unfolded or partially folded form and certainly in the case of the type III 'effectors' are refolded within the host cell after injection from the distal terminus of the needle-like structure that protrudes from the bacterial surface. This last step involves a third translocon encoded by the bacterium inserted into the target membrane (43). Remarkably, the general features of the Type III transport apparatus and the flagellum assembly are quite similar, with the exception that the flagellin and flagellum tip proteins remain an integral component of the structure and no proteins are released to the outside (18). Type IV secretion (41,17) is involved in many Gram negative bacteria in transfer of proteins, DNA or DNA-protein complexes in conjugal transfer of plasmids to other bacteria, plants or animal cells. The transenvelope tunnel-structure requires at least 12 distinct polypeptides, including a terminal section composed of pilin-like subunits that extends through the periplasm well beyond the surface to penetrate target cells in some way. However, none of the translocons for either the inner or outer membranes or that inserted into the target cell membrane appear to be clearly established in this system. Three apparent ATPases are located at the base of the conduit, two in the cytoplasm, but their respective functions are unclear.

3.5. Type V Secretion, Autotransporters

One of the largest groups of protein translocation systems in Gram negative bacteria constitutes the Type V group that includes autotransporter polypeptides (ATs) and the functionally similar two partner systems (composed of TpsA and TpsB proteins). Following transport to the periplasm by the Sec system, the C-terminal (passenger) domain of an autotransporter, or its TpsB analogue, spontaneously insert into the outer membrane as a β-barrel in order, in some way to promote translocation of the N-terminal domain of the autotransporter, or of TpsA (23,44). Surprisingly, recent studies have shown that contrary to the 'simple' idea that the passenger is transported through the AT's own membrane domain, translocation was shown to be dependent on the outer membrane protein assembly (insertase) BamA protein. Indeed it has been speculated that BamA (Omp85 complex) may be directly implicated in translocation of the AT passenger domain rather than simply the insertase for the AT (45). AT substrates appear to partially fold prior to translocation while the TpsA proteins have a remarkably elongated β-helical structure, forming a thin multi-loop structure that may facilitate easy translocation (46). The transmembrane domain in both AT and TPS systems forms channels in membranes, nevertheless it remains unclear whether a monomer or a multimer or even another outer membrane protein, like BamA, constitutes the actual translocon. Interestingly, TpsB proteins contain an N-terminal periplasmic POTRA domain (*po*lypeptide *tr*ansport *a*ssociated) that is also found in the outer membrane protein assembly (insertase) BamA protein. The POTRA domain is therefore likely to be required for docking recognition with the TpsA protein. Following transport, the AT subdomain proteins may be released by autocatalysis or remain bound to the surface but in both AT and TPS systems, the mechanism for final folding of the translocated proteins is apparently not clear.

3.6. Type II Secretion, and Pilus Assembly

The Type II secretion system is the best studied of the two step mechanisms (21), all of which involve a periplasmic intermediate with the transport substrates being provided by the Sec or Tat machines. Surprisingly, however, recently some type II transport substrates have been shown to be targeted to the SecYEG translocon by the SRP, implying co-translational transport across the cytoplasmic membrane (47,48). However, from my point of view other aspects of Type II transport are even more curious. In particular, what precisely is the role of the transenvelope structure and how is it accessed by an apparently already folded transport substrate. The transenvelope structure itself commences with a cytoplasmic ATPase at its base, while the exit to the outside involves a well-conserved, multimeric translocon or secretin in the outer membrane. Related secretin family members apparently fulfil the same role in Type III secretion and in secretion of Type IV pilins (49). A secretin-like protein is also involved in secretion of the pectin lyase in *Dickeya dadantii* PlnA (50), using a pathway which has characteristics of a hybrid between Tat-transport across the inner membrane and then to the exterior in as yet unknown way, via a Type II-like mechanism.

The transenvelope structure in the Type II pathway includes pilin like subunits, clearly with the implication that these form a channel or conduit for translocation. Recognition of the secretion path across the outer membrane appears to involve tertiary structural motifs in the Type II protein substrates. But the puzzle is how does the largely folded Type II (and presumably Type VI) transport substrate gain access from the periplasm into the transenvelope transport channel or tunnel for ultimate passage through the outer membrane via the secretin. An intriguing possibility, given the presence of the pilus-like structure incorporated into the transenvelope tunnel, is that this might provide, as in Type IV 'twitching pili' (51), a retraction-expansion mechanism intermittently expelling Type II proteins to the exterior (52). However, as far as I am aware, no evidence to support this has yet appeared.

A recent structural study of a related member of the secretin family, Wza, which in fact is involved in polysaccharide transport, has produced another novelty – an octameric α-helical barrel (53), in contrast to the conventional β-barrel outer membrane proteins. Interestingly, in the Type III, IV and VI injectisome structures, the secretin or its analogues most likely plays a dual role, initially as a translocation channel for subunits of the structure and finally as a gasket, stabilising the needle structure as well as maintaining the integrity of the outer membrane.

Finally, the Type 1/P pilin apparatus, the so-called usher/chaperone pathway, transporting pilin subunits from periplasm to the external surface, is a relatively simple two protein complex. First, the specialised chaperone delivers already folded pilin subunits from the periplasm to the usher and inserts them into the base of the growing pilus as it emerges from the surface. The dimerized usher forms two potential channels in the membrane, however, one is closed and acts as the initial docking point for new pilin subunits, while only the second molecule functions as the actual translocation channel (54). This is intriguingly reminiscent of the division of labour now proposed for the operation of the double-barrelled SecYEG translocon.

4. Avoiding Traffic Problems

Faced with a complex array of protein transport systems how does the cell avoid traffic problems? In fact, a number of possible mechanisms to avoid snarl ups are beginning to emerge. First, in contrast to shake flask cultures, we may presume that in 'natural' conditions non-essential secretion systems will be shut down or substantially down regulated until an appropriate environmental stimulus triggers activation. Moreover, there are clear indications in the literature that, for example, different Type III 'effectors' are programmed for injection into host cells according to a temporal programme during the course of an infection (43). This would echo the elegant mechanism of the sequential regulation of the production of components of the flagellum, whereby new components for the growing structure are only produced and translocated as required (55,18).

4.1. Spatial Localization of Translocation Machineries

Evidence has also recently accumulated indicating that the translocation machinery for Type III (56), IV (57), V (58) and for Type II, at least in *Pseudomonas* (59), is specifically localised to one or both poles. This presumably permits spatial separation from the Sec system. In contrast, the Type 1 translocators appear to be dispersed throughout the cell (A. Pimenta, J, Young and I. B. Holland, unpublished). On the other hand, Campo et al. (60) have shown that SecA and Y are distributed in a spiral pattern in *B. subtilis* and such an organisation is apparently also found in *E. coli*, as shown by the tracing of the distribution of SecG, an important component of the translocon, throughout the cell (61). Joumouille et al. (56) have suggested the possible advantage of the restricted, polar localization of 'injectisomes' as a means to increase the local concentration of proteins to be transported. Interestingly, in the case of the Type III substrate IpaC, Joumouille et al. (56) have also shown that prior to transport, this protein is found as a cytoplasmic pool, close to a pole, in proximity to the translocator. This indicates that in this system, and conceivably for other translocators, it may be advantageous to compartmentalise the production of the transport substrate to avoid queuing and increase efficiency of delivery.

4.2. Subpopulations of Dedicated Secretors

It is also reasonable to assume that regulation of secretion activity for some systems will prove to be further examples of population regulation through 'bistability', that is activated in only a subpopulation of cells (62). A likely example is the secretion of the TasA protein, necessary for normal biofilm formation, as a component of the extracellular matrix in biofilms in *B. subtilis.* Another component of the matrix, co-regulated with the operon that includes *tasA*, is exopolysaccharide (EPS), which indeed is now known to be expressed only in a subpopulation of cells in which the antirepressor for the repressor SinR is switched on (63).

4.3. Can the Translocon SecYEG Cope with the Traffic?

What about the handling of the large flux of the varied proteins, which constitute the transport substrates for the SecYEG-machinery. These are known to include integral membrane proteins, translocated co-translationally, in close association with the SRP and a ribosome, together with YidC. This plays a critical role sequential lateral partitioning of one or more transmembrane spanning regions. In addition, the SecA,Y system has to deal with post-translational transport of SecB/SecA associated proteins that are targeted to the periplasm and outer membrane, plus a wide range of other proteins that are ultimately secreted to the exterior. We can estimate that in total the flux might constitute up to 1000 distinct polypeptides. Importantly, these may vary enormously in abundance – from, for example, in *E. coli*, a basal level of a few Lac-permease molecules per cell, 50 or so penicillin-binding proteins (e.g. PBP 3), to at least 100,000 molecules of outer membrane porins, and even that is dwarfed by the 7×10^5 copies of the envelope bound Braun lipoprotein (64). All these proteins with differing requirements have to be identified, sorted and translocated each generation by apparently only 500 molecules of SecY (65), that is, 250 dimers in which only one apparently functions as a translocation channel at any one time (66). A simple calculation indicates that for a 60 min generation time, with a very conservative minimum of 3×10^5 molecules to be translocated via SecY, the average residence time with the translocon should be a second or so for each molecule. On the other hand the co-translation insertion of one 30 kDa membrane protein at 37°C could tie up a translocon for 60s at least. One might say that the numbers do not add up.

This simplistic analysis suggests the possibility that some of our fundamental suppositions may not be correct, for example, that the load on the Sec system is less than we have believed and that YidC, or even an additional so far unknown mechanism of insertion of membrane proteins, play greater roles than thought. A recent depletion study of the essential translocon component SecE indeed surprisingly showed that the effect on the formation of inner membrane proteins was minimal although outer membrane protein production was affected (67). In contrast, we showed that the assembly of both inner and outer membrane proteins was reduced 70% when Ts mutants of either SecA or SecY were shifted to the restrictive temperature (68). On the other hand, the apparent distribution of Sec translocators throughout the bacterial cell as some form of 'spiral structure' sits well with the idea that this is essential to distribute the load and minimise queuing of quite different proteins at individual translocon complexes. Finally, curiously, although evidence for the up-regulation of *secA* expression in relation to demand has been obtained both in *E. coli* and *B. subtilis* (69), no such evidence to my knowledge is available for *secYEG* regulation.

5. Final Stages of Transport of Proteins to the Gram Negative Outer Membrane: A New Translocase

5.1. Periplasmic Chaperones/Foldase

The biogenesis of outer membrane proteins is an interesting case of how we all underestimated for so long the need for a specialised mechanism for final insertion into the outer membrane. Following the identification of major outer membrane proteins in Gram negative bacteria in the mid-1970s, carrying characteristic Sec-dependent N-terminal secretion signals, there was a tendency to assume a spontaneous folding mechanism in the periplasm, an equally spontaneous partitioning mechanism into the outer membrane.

With all post-transport steps driven by physico-chemistry. This is an example, not unknown in other areas of recent microbiological research that when confronted with a complex phenomenon, with no obvious precedents, we have tended to 'leave it to physics' rather than looking for a specialised hard wired mechanism through a genetical approach. On the first question, the folding of outer membrane proteins in the periplasm, progress was slow due to the preconception that since classical chaperones (and ATP) were absent from the periplasm then *no* chaperones could be involved. Ulf Henning (70) in fact provided the first evidence for a periplasmic chaperone Skp but the finding was greeted with some early scepticism. Despite the accumulation of subsequent evidence identifying several proteins with unusual properties that helped to fold proteins, the idea was still slow to be accepted that outer membrane proteins required some help to fold correctly and quickly. Now indeed it is well established that the *E. coli* periplasm contains a variety of chaperones quite distinct from those found in the cytoplasm (71).

5.2. An Insertase for the Assembly of Outer Membrane Proteins

It took much longer to settle the second question that not only chaperones but a dedicated machine was also finally required to assemble proteins into the outer membrane. In 2003, Voulhoux et al. (72, *see* also 13) identified the protein, Omp85, in *N. mengingitidis*, followed by identification of its homologue in *E. coli*, YaeT (73). These proteins, required for assembly of proteins into the outer membrane are sometimes referred to as BamA (beta *b*arrel *a*ssembly *m*achinery). Omp85/YaeT forms a large complex of unknown composition and its mechanism of action is poorly understood. However, there is general agreement that this is the missing insertase. Interestingly, as indicated above, the YaeT complex was also suggested to be the translocase for the passenger domain of autotransporters (74), which would indicate a dual function similar in principle to that of SecYEG in the inner membrane.

From this story and from the recent dramatic discovery of cytoskeletal proteins in bacteria we should learn the lesson that the enormous success of bacteria and other micro-organisms is due to evolutionary pressure over the last 4 billion years ensuring that these organisms *harness* physical and chemical forces, through evolution of appropriate proteins, based on a genetic programme, rather than leaving such forces to work unsupervised.

6. The Varied Functions of Secreted Proteins

6.1. Not Only Toxins

Production of toxins or effectors of host functions associated with bacterial pathogenicity is clearly one of the most important processes dependent upon secretion, and the revelation of the exquisite details of the Type III needle injection system over the last few years has illustrated this. However, it is worth reminding ourselves that secretion of proteins (and peptides) from bacteria encompasses many other important functions. These include interestingly, many examples of bacteria expressing Type III systems, whose secreted proteins can also apparently facilitate establishment of mutalistic symbiotic relationships with plants, insects and vertebrates, with results that may include protection of these hosts from other pathogens (75). The Type I and II secretion systems frequently involve secretion of hydrolysing enzymes necessary to break down large molecules for subsequent uptake and metabolism. Secretion of proteins to form organelles involved in motility – pili and flagella – are vital for processes in the natural environment such as biofilm formation and swarming to colonise a desirable niche. Equally, many bacteria secrete or express surface adhesins, while Gram positive bacteria, in particular, secrete intercelluar signalling peptides also involved in colonisation of many habitats not only in human or plant hosts. Bacteria secrete building blocks for biofilm formation, capsular proteins and antibiotic degrading enzymes to provide protection. Finally, in many environments, certainly in soils or the human gut, bacterial species are continually in competition with other micro-organisms and secreted bacteriocins are vital weapons in the struggle for survival.

7. Moonlighting Proteins and Other Novelties

7.1. Cytoplasmic Proteins with No Obvious Secretion Signals Appear to be Secreted

A curious group of proteins, with as yet no indication of functional significance, are constituted by so-called 'moonlighting' proteins (76). At first appearance these are seemingly perfectly legitimate cytoplasmic proteins with well-established physiological roles. For example, enolase in *B. subtilis* (carbon metabolism) or elongation factor EF-Tu in *Listeria monocytogenes* appear in significant amounts in culture supernatants. It is difficult to envisage that a particular subset of proteins could regularly appear outside the cell as the result of a non-specific accident, and they deserve more study. In fact, although these proteins display no known secretion signals, Brunak and co workers (76) have identified a set of properties characteristic for this family of proteins that conceivably could include secretion 'signals'. Indeed, some of these proteins may be secreted by dedicated mechanisms yet to be discovered. An attractive possibility could be secretion via, for example, an exocytosis mechanism. The recently discovered mechanism of secretion of the *E. coli* ClyA toxin (with no identifiable secretion signal) involving its release associated with outer membrane vesicles could be a interesting precedent (77), and again we look forward to hearing more about this intriguing system. In fact, at least some strains of *E. coli* were found to release substantial amounts of haemolysin A, the prototype Type 1 transport substrate, associated with similar vesicles (78). This is in contrast to the previously clearly established mechanism of direct release from the cells via the Type I, ABC-dependent mechanism culminating in passage through TolC (79,36,33). In our view it still remains to be seen whether the reported association with vesicles is a primary secretion mechanism or simply represents free HlyA molecules that have been reabsorbed on to the cell surface as we have observed can occur (unpublished, Orsay laboratory), before subsequent release as vesicles. Nevertheless, whatever the mechanism, if the vesicle associated toxin, which has an apparently enhanced specific activity (77), is used as a specific delivery system to target cells in vivo, as Wai and co-workers have suggested, this could also have important implications for the action of other secreted toxins.

7.2. Newcomers

The recently discovered Type VI system is an important virulence determinant in organisms such as Cholera and Yersinia with five distinct copies of Type VI genes in the latter. Type VI constitutes a fourth example of a one step, transenvelope system – cytosol to exterior. This is also likely to incorporate a needle structure for an injection mechanism although the organization of the 15 or so components is poorly understood so far (80). Initially Type VI was compared with the Type IV pathway, in that the inner membrane channel probably involves two conserved Type IV-like proteins. Unfortunately, as in many other bacterial secretion systems, we are burdened with incomprehensible, impossible to remember names, with these two being extreme examples – DotU and IcmF! – DotU is an integral membrane protein, while IcmF is likely a membrane-linked ATPase. IcmF may be involved in assembly of the secretion conduit rather than directly in translocation of secreted proteins (80). On the other hand, the Type VI machine in fact has several interesting features distinguishing it from other pathways. These include a member of the AAA^{+} (e.g. ClpB) family of hexameric ATPases located at the cytoplasmic base of the conduit. This ATP-motor may participate in presenting unfolded proteins to the transport channel. The transport tunnel itself may be stacked with hexameric rings of Hcp subunits (79). However, it is not clear how the Hcp structure is precisely organised and even less clear how it penetrates the outer membrane of the producer cell. In fact, the Hcp protein is often found in culture supernatants, but this may reflect laboratory conditions where the secretion machinery is perhaps incompletely assembled. In this context it is also unclear how the presumed transport conduit might be capped to form a needle to enter a susceptible host. In *Vibrio cholera*, the Vrg1 protein, although found to be secreted by the Type VI system, contains on the one hand a conserved region similar to the phage T4–tail spike, and a C-terminal domain capable of cross-linking actin (81). These characteristics suggest that the protein may be both part of a membrane puncturing device and a toxic effector that is released into host cells. Finally, a truly novel feature of this pathway in *P. aeruginosa* at least is the presence of a ser/thr kinase-phosphatase couple located at the base of the machine. This controls the phosphorylation state of at least one protein component of the transenvelope structure, thereby affecting the activity/assembly of the transport tunnel (82).

Finally, other recently discovered examples of protein transporters appearing to mix elements from already known pathways and two in particular may be noted. The PlnH pectin lyase curiously somehow escapes from the Tat translocator without cleavage of the N-terminal signal region and then is secreted via a modified Type II pathway with a specialised secretin (50). In a really bizarre example, the assembly of subunit II (cyo) of cytochrome bo (3) oxidase involves the N-terminal domain being inserted by YidC while the C-terminal is inserted via SecYEG (83).

8. Some Perspectives

In summary, by my count there are now more than 20 different protein translocation systems identified in different bacteria – the majority in Gram negative bacteria but at least 6 in Gram positive organisms. Remarkably, these vary in complexity from a minimum of three protein components, to some transenvelope systems in Gram negative bacteria requiring at least 20 polypeptides to build the conduit to cross two bacterial membranes and beyond. In the Type II, IV and VI systems, indeed further complexity is required for inserting into the host membrane. For the future, much needs to be understood concerning the actual mechanism of protein movement through these structures, precisely how this is energised and whether function involves repeated expansion–retraction accompanying insertion de-insertion into the host membrane. Alternatively, are these needles used once only in vivo, leaving the attacking bacterium fixed to the host cell, reflecting a suicide mission?

In terms of understanding the detailed assembly and function of these complex systems, the prospects unfortunately appear bleak for constructing an in vitro system for analysis. Ideally, progress would be quicker if we concentrated efforts on establishing the basic principles of a few prototype systems rather than *describing* any more new ones. This should enable us, for example, to settle the nature of initial recognition/specificity of transport substrates, the source and mechanism of action of the energy motor for transport, how proteins, on the one hand wiggle or snake through a 'simple' translocon in the inner or outer membrane, or, presumably, involving quite different principles, how proteins 'crawl' through long multi-subunit tunnels. A telling example to follow in this field is the great progress made with the understanding of the function of the Sec-translocase/translocon, where many laboratories have concentrated on two organisms, *E. coli* and *B. subtilis*. As a result we have a very good idea how the SecB chaperone/SecA translocase interacts with the SecYEG translocon. In addition, from the most recent spectacular biochemical and crystal structure data there is a plausible model of how the translocase docks to the translocon and how insertion of the signal sequence may open the periplasmic gate to allow translocation to the other side of the membrane. Nevertheless, despite this great progress I am still impatient to see the first crystal structure of a translocation intermediate stuck in the SecYEG channel in order to have a visual idea of how polypeptides thread their way through the translocon at the atomic level.

References

1. Engelman, D.M. (2005) Membranes are more mosaic than fluid. *Nature.* **438**, 578–580.
2. Singer, S.J. and Nicolson, G.L. (1972) The fluid mosaic model of the structure of cell membranes. *Science.* **175**, 720–723.
3. Yamane, K., Bunai, K. and Kakeshita, H. (2007) Protein traffic for secretion and related machinery of *Bacillus subtilis. Biosci. Biotech. Biochem.* **68**, 2007–2023.
4. Fu, L.L. (2007) Protein secretion pathways in *Bacillus subtilis*: Implication for optimization of heterologous protein secretion. *Biotechnol. Advances.* **25**, 1–12.
5. Economou, A., Christie, P. J., Fernandez, R. C., Palmer, T., Plano, G. V. and Pugsley, A. T. (2006) Secretion by numbers: protein traffic in prokaryotes. *Mol. Microbiol.* **62**, 308–319.
6. McLaughlin, B., Chon, J.S., MacGurn J.A., Carlsson, F. Cheng, T.L. and Cox, J.S. (2007) A Mycobacterium ESX-1-secreted virulence factor with unique requirements for export. *PloS Pathogens.* **3**(8), e105.
7. Abdallah, A.M., Gey van Pittius, N. C., Champion, P. A., Cox, J., Luirink, J., Vandenbroucke-Grauls, C. M., *Appelmelk*, B. J. and Bitter, W. (2007). Type VII secretion – mycobacteria show the way. *Nat. Rev. Microbiol.* **5**, 883–891.
8. Mandlik, A., Swierczynski, A., Das, A. and Ton-That, H. (2007) Pili in Gram positive bacteria: assembly, involvment in colonization and biofilm development. *Trends Microbiol.* **16**, 33–40.
9. Papanikou, E., Karamanou, S. and Economou, A. (2007) Bacterial protein secretion through the translocase nanomachine. *Nat. Rev. Microbiol.* **5**, 839–851.
10. Driessen. A. J. and Nouwen. N. (2008) Protein translocation across the bacterial membrane. *Annu. Rev. Biochem.* **77**, 643–667.
11. Eijlander, R. T., Jongbloed J. D. and Kuipers, O. P. (2009) Relaxed specificity of the *Bacillus subtilis* TatAdCd translocase in Tat dependent protein secretion. *J. Bacteriol.* **191**, 196–202.
12. Sargent, F., Berks, B.C. and Palmer, T. (2003) Pathfinders and trailblazers: a prokaryotic targeting system for transport of folded proteins. *FEMS Microbial. Lett.* **254**, 198–207.
13. Gentle, I., Gabriel, K., Beech, P., Waller, R. and Lithgow, T. (2004) The Omp85 family of outer membrane proteins is essential for outer membrane biogenesis in mitochondria and bacteria. *J. Cell. Biol.* **164**, 19–24.
14. Wu, T., Malinvemi, J., Ruiz, M., Kim, S., Silhavy, T. J. and Kahne, D. (2005) Identification of a multicompnent complex required for outer membrane biogenesis in *Escherichia coli. Cell.* **121**, 235–245.
15. Holland, I.B., Schmitt, L. and Young, J. (2005) Type I protein secretion in bacteria, the ABC transporter dependent pathway. *Molec. Mem. Biol.* **22**, 29–39.
16. Mueller, C.A., Broz, P. and Coenelis, G.R. (2008) The Type III secretion system tip complex and translocon. *Molec. Microbiol.* **68**, 1085–1095.
17. Backert. S., Fronzes, R. and Waksman, G. (2008) VirB2 and VirB5 proteins: specialised adhesins in bacterial type-IV secretion systems? *Trends Microbiol.* **16**, 409–413.
18. Apel, D. and Surette. M.G. (2008) Bringing order to a complex machine: The assembly of the bacterial flagella. *Biochim. Biophys. Acta.* **1778**, 1851–1858.
19. Yamanaka, H., Kobayashi, H., Takahashi, E. and Keinosuke, O. (2008) MacAB is involved in the secretion of *Escherichia coli* heat stable enterotoxin II. *J. Bacteriol.* **190**, 7693–7698.
20. D'Enfert. C., Ryter, A and Pugsley, A. P. (1987) Cloning and expression in *Escherichia coli* of the *Klebsiella pneumoniae* genes for production, surface localisation and secretion of the lipoprotein pullulanase. *EMBO J.* **6**, 3531–3538.
21. Johnson, T. L., Abendroth, J., Wim, G. J. Hol. and Sandkvist, M. (206) Type II secretion: from structure to function. *FEMS Microbiol. Lett.* **255**, 175–186.
22. Verma, A. and Burns, D. L. (2007) Requirements for assembly of PtlH with the pertussis toxin transporter apparatus of *Bordetella pertussis. Infection Immun.* **75**, 2297–2306.
23. Hodak, H. and Jacob-Dubuisson, F. (2007) Current challenges in autotransport and two-partner protein secretion pathways. *Res. Microbiol.* 158, 631–637.
24. Ruer, S., Ball, G., Filloux, A. and de Bentzmann, S. (2008) The 'P-usher', a novel protein transporter involved in fimbrial assembly and TpsA secretion. *EMBO J.* **7**, 2669–2680.
25. Holland, I. B. (2005) Translocation of bacterial proteins. *Biochim. Biophys. Acta.* **1694**, 5–16.
26. Saier, M. H. (2006) Protein secretion and membrane insertion systems in Gram negative bacteria. *J. Membrane. Biol.* **214**, 75–90.
27. Karamanou, S., Bariami, V., Papanikou, E., Kalodimos, C. G. and Economou, A. (2008) Assembly of the translocase motor onto the

preprotein-conducting channel. *Mol. Microbiol.* **70**, 311–322.
28. Erlandsen, K.J., Miller, S. B. M., Nam, Y., Osborne, A. R., Zimmer, J. and Rapoport, T.A. (2008) A role for the two-helix finger of the SecA ATPase in protein translocation. *Nature.* **455**, 984–987.
29. Zimmer, J., Nam, Y. and Rapoport, T.A. (2008) Structure of a complex of the ATPase SecA and the protein-translocation channel. *Nature.* **455**, 936–943.
30. Tsukazaki, T., Mori, H., Fukai, S., Ishitani, R., Mori, T., Dohmae, N., Perederina, A., Sugita, Y., Vassylyev, D. G., Ito, K. and Nureki, O. (2008) Conformational transition of Sec machinery inferred from bacterial SecYE structures. *Nature.* **455**, 988–991.
31. Xie, K., Hessa, T., Seppala, S., Rapp, M., von Heijne G, and Dalbey, R. E. (2007) Features of transmembrane segments that promote lateral release from the translocase into the lipid phase. *Biochemistry.* **46**, 15153–15161.
32. Eijlander, R.T. Jongbloed, J.D. and Kuipers, O.P. (2009) Relaxed specificity of the Bacillus subtilis TatAdCd translocase in Tat-dependent protein secretion. *J. Bacteriol.* **191**, 196–202.
33. Blight. M. A., Chervaux, C. and Holland. (1994) Protein secretion pathway in *E. coli.* *Curr. Opin. Biotechnol.* **5**, 468–474.
34. Balakrishnan, L., Hughes, C. and Koronakis, V. (2001) Substrate-triggered recruitment of the TolC channel-tunnel during type I export of hemolysin in *Escherchia coli. J. Mol. Biol.* **313**, 501–510.
35. Koronakis, V., Sharff, A; Koronakis, E., Luisi, B. and Hughes, C. (2000) Crystal structure of the bacterial membrane protein TolC, central to multidrug efflux and protein export. *Nature.* **405**, 4914–919.
36. Eswaren, J., Hughes, C. and Koronakis, V. (2003) lockingTolC entrance helices to prevent protein translocation by the bacterial type I export apparatus. *J. Mol. Biol.* **327**, 309–315.
37. Chervaux, C. and Holland, I. B. (1996) Random and directed mutagenesis to elucidate the functional importance of helix II and F-989 in the C-terminal secretion signal of *Escherichia coli. J. Bacteriol.* **178**, 1232–1236.
38. Koronakis, V., Hughes, C. and Koronakis, V. (1991) Energetically distinct early and late stages of HlyB/HlyD-dependent secretion across both *Escherichia coli* membranes. *EMBO J.* **10**, 3263–3272.
39. Zaitseva, J., Jenewein, S., Wiedenmann, A., Holland, I. B. and Schmitt, L. (2005) H662 is the linchpin of ATP hydrolysis in the nucleotide-binding domain of the ABC transporter HlyB. *EMBO J.* **24**, 1901–1910.
40. Zaitseva, J., Oswald, C., Jumpertz, T., Wiedenmann, A., Holland, I. B. and Schmitt, L. (2006) A structural analysis of symmetry required for catalytic activity of an ABC-ATPase domain dimer. *EMBO J.* **25**, 3432–3443.
41. Baron, C. (2006) VirB8: a conserved type IV secretion system assembly factor and drug target. *Biochem. Cell Biol.* **84**, 890–899.
42. Paul, K. Erhardt, M, Hirano, T., Blair, D. F. and Hughes, K.T. (2009) Energy source and type III flagella secretion. *Nature.* **451**, 489–492.
43. Galan, J. E. and Wolf-Watz, H. (2006) Protein delivery into eukaryoteic cells by type III secretion machines. *Nature.* **444**, 567–573.
44. Hendersen, I.R., Navarro-Garcia, F., Desvaux, M., Fernandez, R. C. and Ala'Aldeen, D. (2004) Type V secretion pathway: the autotransporter story. *Microbiol. Molec. Biol. Rev.* **68**, 692–744.
45. Oomen, C. J., van Ulsen, P., Van Gelder, P., Feijen, M., Tommassen, J. and Gros, P. (2004) Structure of the translocator domain of a bacterial autotransporter. *EMBO J.* **23**, 1257–1266.
46. Wells, T. J., Tree, J. J., Ulett, G. C. and Schembri, M. A. (2007) Autotransporter proteins: novel targets at the bacterial cell surface. *FEMS Microbiol. Lett.* **274**, 163–172.
47. Francetic, O., Buddeimeijer, N., Lewenza, S., Kumamoto, C. A. and Pugsley A. P. (2007) Signal recognition particle-dependent inner membrane targeting of the PulG pseudoplin of a type II secretion system.*J. Bacteriol.* **189**, 1783–1793.
48. Arts, J., van Boxtel, R., Filloux, A., Tommassen J. and Koster, M. (2007) Expression of the pseudopilin XcpT of the *P. aeruginosa* type II secretion system via the signal recognition particle – Sec pathway. *J. Bacteriol.* **189**, 2069–2076.
49. Craig L. and Li, J. (2008) Type IV pili: paradoxes in form and function. *Curr. Opin. Struct. Biol.* **18**, 267–277.
50. Ferrandez, Y. and Condemine, G. (2008) Novel mechanism of outer membrane targeting of proteins in Gram negative bacteria. *Mol. Microbiol.* **69**, 1349–1357.
51. Proft, T. and Baker, E. N. (2008) Pili in Gram-negative and Gram-positive bacteria - structure, assembly and their role in disease. *Cell Mol. Life Sci.* Oct. 27 E-Pub
52. Vignon, G., Kohler, R., Larquet, E., Giroux, S., Prevost, M. C., Roux, P. and Pugsley, A. P. (2003) Type IV-like pili formed by

the type II secreton: specificity, composition, bundling, polar localization, and surface presentation of peptides. *J. Bacteriol.* **185**, 3416–3428.
53. Collins, R. F. (2007) Wza: a new structural paradigm for outer membrane secretory proteins? *Trends Microbiol.* **15**, 96–100.
54. Remaut., H., Tang, C., Henderson, N. S., Pinkner, J. S., Wang, T., Hultgren, S. J., Thanassi, D. G., Waksman, G. and Li, H. (2008) Fiber formation across the outer membrane by the chaperone/usher pathway. *Cell.* **133**, 640–652.
55. Chevance, F. F. and Hughes, K. T. (2008) Coordinating assembly of a bacterial macromolecular machine. *Nat. Rev. Microbiol.* **6**, 445–455.
56. Jaumouille, V., Francertic, O., Sansonetti, P. J. and Nhieu, G. T. V. (2008) Cytoplasmic targeting of IpaC to the bacterial pole directs polar type III secretion in *Shigella. EMBO J.* **27**, 447–457.
57. Judd, P. K., Kumar, R. B. and Das, A. (2005) The type IV apparatus protein VirB6 of *Agrobacterium tumefaciens* localizes to a cell pole. *Mol. Microbiol.* **55**, 115–124.
58. Jain, S., van Ulsen, P., Benz, I., Schmidt, M. A., Fernandez, R., Tommassen, J. and Goldberg, M. B. (2006) Polar localization of the autotransporter family of large bacterial virulence proteins. *J. Bacteriol.* **188**, 4841–4850.
59. Senf, F., Tommassen, J. and Koster, M. (2008) Polar secretion of proteins via the Xcp type II secretion system in *Pseudomonas aeruginosa. Microbiology.* **154**, 3025–3032.
60. Campo, N., Tjalsma, H., Buist, G., Stepniak, D., Meijer, M., Veenhuis, M., Westermann, M., Muller, J. P., Bron, S., Kok, J., Kuipers, O. P. and Jongbloed, J. D. H. (2004) Subcellular secretion sites for bacterial export. *Mol. Microbiol.* **53**, 1583–1599.
61. Shiomi, D., Yoshimoto, M., Homma, M. and Kawagishi, I. (2006) Helical distribution of the bacterial chemorecptor via colocalization with the Sec protein translocation machinery. *Mol. Microbiol.* **60**, 894–906.
62. Dubnau, D. and Losick, R. (2006) Bistability in bacteria. *Mol. Microbiol.* **61**, 564–572.
63. Chai, Y., Chu, F., Kolter, R. and Losick, R. (2008) Bistability and biofilm formation in *Bacillus subtilis. Mol. Microbiol.* **67**, 254–263.
64. Nikaido, H. (1996) The outer membrane. In F. C. Neidhardt (Ed.) *Escherichia coli and Salmonella*: cellular and molecular biology, Second edition, volume 1, pp. 29–47, ASM Press, Washington D.C.
65. Schatz, P. J., Bieker, K. L., Ottamann, K. M., Silhavy, T. J. and Beckwith J. (1991) One of three transmembrane stretches is sufficient for the functioning of the SecE protein, a membrane component of the *E. coli* secretion machinery. *EMBO J.* **10**, 1749–1757.
66. Osborne, A. R. and Rapoport. T. A. (2007) Protein translocation is mediated by oligomers of the SecY complex with one SecY copy forming the channel. *Cell.* **129**, 97–110.
67. Baars, L., Wagner, S., Wickstrom, D., Klepsch, M., Ytterberg, A. J., van Vijk, K. J. and de Gier, J. W. (2008) Effects of SecE depletion on the inner and outer membrane proteome of *Escherichia coli. J. Bacteriol.* **190**, 3505–3525.
68. Baker, K., Mackman, N., Jackson, M. and Holland, I. B. (1987) Role of SecA and SecY in protein export as revealed by studies of TonA assembly into the outer membrane of *Escherchia coli. J. Mol. Biol.* **198**, 693–703.
69. Oliver, D. B. (1993) SecA protein: autoregulated ATPase catalysing preprotein insertion and translocation across the *Escherichia coli* inner membrane. *Mol. Microbiol.* 7, 159–165.
70. Chen, R. and Henning, U. (1996) A periplasmic protein (Skp) of *Escherichia coli* selectively binds a class of outer membrane proteins. *Mol. Microbiol.* **19**, 1287–1294.
71. Bos, M. P., Robert, V. and Tommassen J. (2007) Biogenesis of the Gram negative bacterial outer membrane. *Annu. Rev. Microbiol.* **61**, 191–214.
72. Voulhoux, R., Bos, M. P., Mols, M. and Tommassen J. (2003) Role of a highly conserved bacterial protein in outer membrane assembly. *Science.* **299**, 262–265.
73. Doerrier, W. T. and Raetz, R. H. (2005) Loss of outer membrane proteins without inhibition of lipid export in an *Escherichia coli* YaeT mutant. *J. Biol. Chem.* **280**, 27679–27687.
74. Bernstein H.D. (2007) Are bacterial 'autotransporters' really transporters? *Trends Microbiol.* **15**, 441–447.
75. Preston, G. M. (2007) Metropolitan microbes: type III secretion in multihost symbionts. *Cell Host Microbe.* **2**, 291–294.
76. Bendtsen, J. D., Kiemer, L., Fausboll, A. and Brunak, S. (2005) Non-classical secretion in bacteria. *BMC Microbiol.* Oct 7, 5, 38.
77. Wai, S. N., Lindmark, B., Soderblom, T., Takadi, A., Westermark, M., Oscarsson, J., Jass, J., Richter-Dahlfors, A., Mizunoe, Y. and Uhlin, B. E. (2003) Vesicle-mediated export and assembly of pore forming oligomers of the enterobacterial ClyA cytotoxin. *Cell.* **115**, 25–35.

78. Balsalobre, C., Silvan, J. M., Berglund, S., Mizunoe, Y., Uhlin, B. E. and Wai, S. N. (2006) Release of the type 1 secreted α-haemolysin via outer membrane vesicles from *Escherichia coli. Mol. Microbiol.* **59**, 99–112.
79. Mackman, N., Baker, K., Gray, L., Haigh, R., Nicaud, J. M. and Holland I.B. (1987) Release of a chimeric protein into the medium from *Escherichia coli* using the C-terminal secretion signal of haemolysin. *EMBO J.* **6**, 2835–2841.
80. Filloux, A., Hachani, A. and Bleves, S. (2008) The bacterial type VI secretion machine: yet another player for protein transport across membranes. *Microbiology.* **154**, 1570–1583.
81. Pukatzki, S., Ms, A. T., Revel, A. T., Sturtevant, D. and Mekalanos, J. J. (2007) Type IV secretion system translocates a phage tail spike-like protein into target cells where it cross-links actin. *Proc. Acad. Sci. USA.* **103**, 1528–1533.
82. Mougous, J. D., Gifford, C. A., Ramsdell, T.L. and Mekalanos, J. J. (2007) Threonine phosphorylation postranslationally regulates protein secretion in *Pseudomonas aeruginosa. Nat. Cell. Biol.* **9**, 797–803.
83. Celebi, N., Dalbey, R. E. and Yuan, J. (2008) Mechanism and hydrophobic forces driving membrane protein insertion of subunit II of cytochrome bo 3 oxidase. *J. Mol. Biol.* **375**, 1282–1292.

Chapter 2

In Vitro and In Vivo Approaches to Studying the Bacterial Signal Peptide Processing

Peng Wang and Ross E. Dalbey

Abstract

Protein targeting in both eukaryotic and prokaryotic cells is often directed by a signal sequence located at the amino-terminus of the protein. In eukaryotes, proteins that are sorted into different compartments of the cell, such as endoplasmic reticulum, mitochondria, and chloroplast, require different signal sequences. In bacteria, proteins which are exported to the outer membrane or the periplasmic space are also guided by signal peptides. After the protein is translocated across the cytoplasmic membrane, the signal peptide is proteolytically removed by signal peptide cleavage. Here, in this chapter, we describe methods to study signal peptide processing in bacteria, including purification of signal peptidase and its substrates. We also describe the measurement of the catalytic constants of signal peptidases using an in vitro assay. In addition, we will present an in vivo assay using a temperature sensitive signal peptidase strain to determine which preproteins are processed by Signal peptidase 1.

Key words: Signal peptide, signal peptidase, purification, activity assay, preprotein.

1. Introduction

Signal peptides located at the amino terminus of preproteins are typically 20 to 25 residues long. While these signal peptides do not show strong sequence homology, they do have three conserved domains (1): an amino-terminal basic region, a central hydrophobic region, and a carboxyl-terminal region containing residues important for signal peptide processing. After signal peptides initiate membrane translocation of the mature region of the protein, they are cleaved off typically by a type I signal peptidase

A. Economou (ed.), *Protein Secretion*, Methods in Molecular Biology 619,
DOI 10.1007/978-1-60327-412-8_2,

(SPase I). This enzyme was first purified by Zwizinski and Wickner in 1980 (2).

Purified SPase I can cleave a number of different precursors of membrane and secreted proteins. M13 procoat, the precursor of the leucine binding protein, pre-β-lactamase, pre-maltose binding protein, and pro-OmpA are all examples of substrates (3). Substrate recognition typically requires Ala-X-Ala at the −1 to −3 positions of the signal peptide (relative to the cleavage site) where X can be any residue (4). Substrate cleavage occurs after the −1 Ala.

SPase I family of proteases utilize a serine/lysine catalytic mechanism to perform its cleavage reaction (5). This is accomplished by a nucleophilic attack by the Ser90 Oγ atom on the preprotein substrate scissile peptide bond of the preprotein substrate located between the signal sequence and the mature protein. Early studies to examine SPase processing were rather qualitative and used a radioactively labeled substrate, such as M13 procoat (6). Later studies used chemical amounts of peptide substrates (7), fluorescently labeled peptide substrates (8) or purified preprotein substrates (9) where it was possible to measure the kinetic constants. The SPase and the preprotein substrates can be easily over-expressed and purified to homogeneity in large amount for in vitro assays. Various methods have been developed to test the activity of SPase I both in vivo and in vitro. Details will be discussed below.

There is another family of bacterial SPases, SPase II (for review see (10)), which processes lipoprotein substrates. These lipoproteins also contain the so-called lipobox that is located in the signal sequence and the mature region that has the consensus sequence Leu-Ala/Ser-Gly/Ala-Cys at the −3 to +1 position. The cysteine at position +1 needs to be modified with a diacylglycerol before cleavage can occur. This chapter will focus only on the *Escherichia coli* SPase I.

2. Materials

2.1. Purification of 6-His-Tagged SPase I Proteins

1. Cell strain: BLR(DE3) (Novagen).
2. Plasmid: pET23 (Novagen).
3. Luria-Bertani (LB) broth MILLER (EMD). Dissolve 25 g powder per 1 L water and autoclave.
4. IPTG: Isopropyl-β-D-thiogalactopyranoside, Analytical Grade (Anatrace). Prepare 1 M stock solution and store at −20°C.
5. Lysis buffer: 50 mM Tris-HCl, pH 8.0, 20% sucrose.

6. Lysozyme, from chicken egg white (Sigma).
7. DNase, RNase-free (Promega).
8. Solubilization buffer: 10 mM triethanolamine (TEA)-HCl, 10% glycerol, 1% Triton X-100, pH 7.9.
9. Q-Sepharose (Amersham Pharmacia Biotech).
10. Ni^{2+}-nitrilotriacetic acid agarose (Qiagen).
11. 6-His buffer: 10 mM Tris-HCl, pH 8.5, 100 mM KCl, 20 mM imidazole, 10 mM β-mercaptoethanol (BME), 1% detergent (either Triton X-100 or β-D-octylglucopyranoside).
12. Wash buffer: 10 mM Tris-HCl, pH 8.5, 100 mM KCl, 20 mM imidazole, 10 mM BME, 1% detergent (either Triton X-100 or β -D-octylglucopyranoside), 900 mM KCl.
13. Dialysis buffer: 20 mM Tris-HCl, pH 8.0, 0.5% Triton X-100.
14. Column buffer:10 mM TEA HCl, pH 7.9, 10% glycerol, 1% Triton X-100.

2.2. Purification of ProOmpA Nuclease A

1. Cell strain: BL21(DE3) (Novagen).
2. Plasmid: pET-21a (Novagen).
3. Luria-Bertani (LB) broth MILLER (EMD). Dissolve 25 g powder per 1 L water and autoclave.
4. IPTG: prepare 1 M stock solution and store at −20°C.
5. PMSF: phenylmethylsulfonyl fluoride (Thermo Scientific). Prepare 100 mM stock solution in isopropanol.
6. Ammonium sulfate.
7. Ni^{2+}-nitrilotriacetic acid agarose (Qiagen).
8. Imidazole (Sigma). Prepare 1 M stock solution and store at 4°C.
9. FRENCH pressure cell press (SLM Instrument, Inc.).
10. Sonic Dismembrator, Model 500 (Fisher Science).
11. Sephacryl S-300 gel column (Pharmacia).
12. TEP buffer: 25 mM Tris-HCl, pH 8.0, 5 mM EDTA, 1 mM PMSF.
13. MEB buffer: 50 mM Tris-HCl, pH 8.0, 1 mM EDTA, 1 M KCl, 2 M guanidinium-HCl, 3 mM BME.
14. Dialysis buffer 1: 50 mM Tris-HCl, pH 8.8, 1 M KCl, 10 mM $CaCl_2$, 20% glycerol.
15. Dialysis buffer 2: 25 mM Tris-HCl, pH 8.8, 1 M KCl, 10 mM $CaCl_2$.
16. Dialysis buffer 3: 25 mM Tris-HCl, pH 8.8, 10 mM $CaCl_2$.

17. Guanidine buffer: 7.2 M guanidinium-HCl, pH 3.0, 50 mM Na citrate, 5 mM BME.

2.3. In Vitro SPase Activity Assay and Kinetic Analyses

1. BCA™ protein assay kit: Pierce
2. 5X sample buffer: 250 mM Tris-HCl pH 6.8, 10% SDS, 50% glycerol, 0.02% bromophenol blue. Add 10% BME prior to use.

2.4. In Vivo SPase Activity Assay

1. Cell strain: IT41(DE3).
2. Plasmid: pET23a.
3. IPTG. Prepare a 1 M solution and store at −20°C.
4. [^{35}S]methionine: Trans [^{35}S]-LABEL Metabolic Labeling Reagent (MP Biomedicals, Inc.).
5. Trichloroacetic acid (TCA), 20% solution.
6. Acetone. Chill on ice prior to use.
7. Staph A (Calbiochem): Pansorbin cells.
8. M9 minimal medium (for 100 mL): 90 mL of M9 salt solution (made from M9 minimal salts, 5X, Sigma), 10 mL of 19 amino acids (500 μg/mL each except methionine), 2.5 mL of 20% fructose, 100 μL of thiamine (1 mg/mL), 100 μL of 1 M $MgSO_4$, adjust pH to 7.0. Store at 4°C.
9. 1X Triton buffer: 10 mM Tris-HCl, pH 8.0, 5 mM EDTA, 150 mM NaCl, 2.5% Triton X-100.
10. Fixing buffer (for 1 L): 500 mL CH_3OH, 400 mL H_2O, 100 mL acetic acid.
11. Gel dyer (Bio-Rad).
12. Phosphorimaging screen and exposure cassette (Molecular Dynamics).
13. Imager and image analysis software (Typhoon Imager, GE Healthcare).

2.5. Sodium Dodecyl Sulfate-Polyacrylamide Gel Electrophoresis (SDS-PAGE)

1. Separating gel buffer: 1.5 M Tris-HCl, pH 8.8, filter and store at room temperature.
2. Stacking gel buffer: 0.5 M Tris-HCl, pH 6.8, filter and store at room temperature.
3. 10% (w/v) SDS.
4. 29% (w/v) acrylamide/1% (w/v) bisacrylamide solution (Bio-Rad).
5. N,N,N,N’-tetramethylethylene diamine (TEMED) (Sigma).
6. Ammonium persulfate (AP) (Sigma). Prepare 10% (w/v) solution in water. Prepare fresh solution before use.

7. SDS-PAGE Running buffer: 25 mM Tris, 192 mM glycine, 0.1% (w/v) SDS.
8. 10X SDS-PAGE running buffer can be prepared as follow: Tris base (30.3 g), glycine (144 g), SDS (10 g), dissolve in 1 L water. Store at room temperature.
9. Separating gel (12%) : prepare 10 mL of 12% separating gel by mixing 3.3 mL water, 4.0 mL of 29% acrylamide/1% bisacrylamide solution, 2.5 mL separating gel buffer, 0.1 mL of 10% SDS, 0.1 mL of 10% AP and 5 μL TEMED.
10. Separating gel (17.2%): prepare 10 mL of 17.2% separating gel by mixing 1.57 mL water, 5.73 mL of 29% acrylamide/1% bisacrylamide solution, 2.5 mL separating gel buffer, 0.1 mL of 10% SDS, 0.1 mL of 10% AP, 5 μL TEMED.
11. Stacking gel: prepare 5 mL stacking gel by mixing 3.1 mL water, 0.7 mL of 29% acrylamide/1% bisacrylamide solution, 1.3 mL stacking gel buffer, 0.05 mL of 10% SDS, 0.05 mL of 10% AP, 3 μL TEMED.
12. Prestained molecular weight marker: Precision Plus Protein Standards All Blue (Bio-Rad).
13. 2X SDS sample loading buffer: 100 mM Tris-HCl, pH 6.8, 4% (w/v) SDS (electrophoresis grade), 0.2% (w/v) bromophenol blue, 20% (v/v) glycerol, 200 mM dithiothreitol (DTT). Add DTT prior to using the solution.
14. Gel Code Blue staining kit (Pierce).
15. Gel dryer: Bio-Rad Model 583.
16. Filter paper: Fisher Pure Cellulose Chromatography Paper (0.35 mm thick).
17. Phosphorimager screen (Molecular Dynamics).

3. Methods

Almost all proteins exported to the outer membrane and periplasmic space of Gram negative bacteria are synthesized with a signal peptide that is cleaved off by signal peptidases. SPases function to release exported proteins from the membrane so they can go on to the outer membrane or periplasmic space of the cell; uncleaved exported proteins are translocated across the membrane but remain bound to the inner membrane by their uncleaved hydrophobic signal peptides (11).

To measure the in vitro activity of SPase I in detergent both SPase and the substrates are over-expressed in *E. coli* and are purified to homogeneity. Early methods for purification of SPase employed the isolation of membranes, Triton X-100 extraction and multiple chromatography steps (2). However, taking advantage of His-tags and affinity chromatography these days, the protease can be purified in a much simpler way, as described below. Substrates can be prepared by in vitro translation or from over-expression strains where export by the Sec machinery is impaired (12). In this chapter, we will describe the over-expression and purification of pro-OmpA nuclease A with a His-tag which is used as a substrate. To assay the activity of SPase in vitro, SPase and its preprotein substrate are mixed together in detergent for various times and the processing of the precursor substrate to the mature protein can be determined (**Fig. 2.1**). Typically, the intensities of the precursor and mature bands are measured on a Coomassie blue stained SDS-PAGE gel (13). In this fashion, the V_{max}, k_{cat}, and K_m values of SPase can be calculated from in vitro studies.

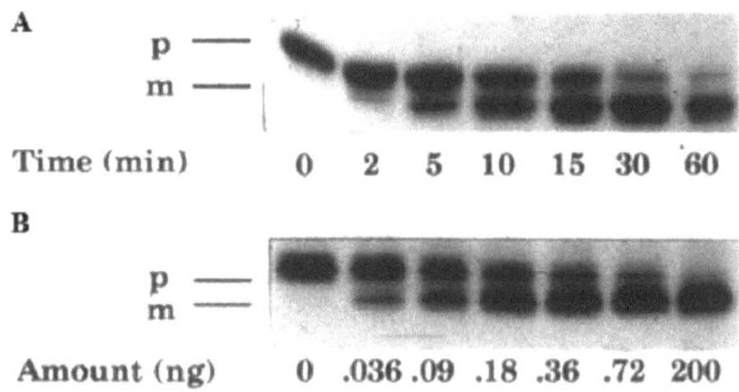

Fig. 2.1. SPase 1 cleavage of pro-OmpA nuclease A is time and enzyme dependent. (**A**) Time dependence of pro-OmpA nuclease A cleavage. (**B**) Pro-OmpA nuclease A cleavage as a function of the amount of SPase I present. p: pro-OmpA nuclease A. m: mature form of OmpA nuclease A.

Signal peptide processing can be measured in vivo either by utilizing a temperature sensitive signal peptidase strain or by using a strain whose SPase I expression is regulated. There are several strains that are available. First, the strain H560/pRD9 can be used to decrease the SPase I level (11,14). This strain has a SPase I gene under the control of the *araBAD* promoter integrated into chromosome and the strain growth is arabinose-dependent. Second, in the temperature sensitive IT41 strain, the SPase activity is strongly impaired at the non-permissive temperature (42°C) (15). IT41 has an amber mutation in the DNA at the position that normally encodes Gln39 in the SPase I protein located upstream to the proteolytic domain (16) and is temperature sensitive even in the absence of a temperature sensitive suppressor (15). This suggests that at 30°C, but not at 42°C, there is read through of the amber codon to generate sufficient amounts of SPase to support growth (16). The advantage of IT41 is that the switch of temperature can cause the chromosomally encoded SPase I to inactivate much more rapidly than in H560/pRD9 strain where the cell needs to be grown for a long time in minimal medium supplemented with glucose, typically 6–7 h to decrease the SPase to limiting amounts in the cell (11,14). IT41 was modified in order to facilitate the induction of a plasmid-encoded protein, by using λDE3 prophage to incorporate the T7 RNA polymerase gene into the chromosome of IT41 to produce IT41(DE3). The protein used as a substrate in IT41 or H560/pRD9 is usually radiolabeled during the experiment (17). SDS-PAGE is employed to separate the precursor from the mature protein (**Fig. 2.2**). To estimate the activity of SPase in vivo, the ratio of precursor and mature forms is calculated by scanning the autoradiograms of the gel.

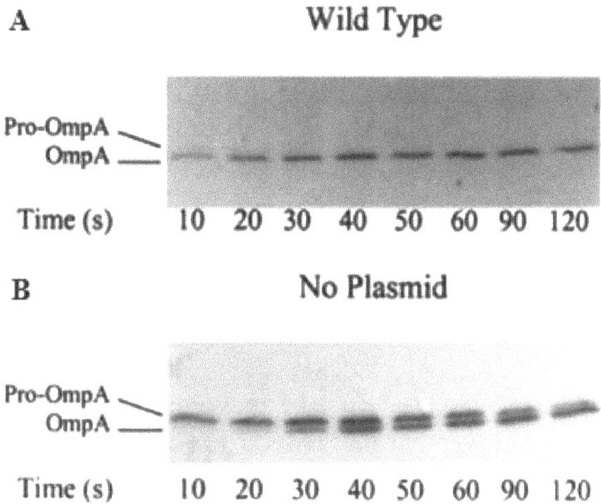

Fig. 2.2. In vivo SPase 1 activity assay. IT41 was labeled at the non-permissive temperature with [^{35}S]-translabel for 15 s and chased with non-radioactive methionine for the indicated times. (**A**) IT41 with a plasmid encoding wild type SPase I. (**B**) IT41 with no plasmid as a negative control.

3.1. Purification of 6-His-Tagged SPase Proteins

SPase I with a 6-His tag is purified by ion-exchange and nickel affinity chromatography. The first ion-exchange step gives a semi-pure, diluted SPase I preparation. The second nickel affinity chromatography step serves to concentrate and further purify the enzyme. Typically, 3 mg of pure protein (>95%) can be isolated from 3 L of culture (18).

1. Overnight cultures of *E. coli* BLR(DE3) cells harboring the pET23Lep vector are back-diluted 1:40 in 3 L of LB media (*see* **Note 1**) supplemented with 100 μg/mL ampicillin and 12.5 μg/mL tetracycline. The cell culture is grown at 37°C to an absorbance of 0.6 at 600 nm.
2. Expression of SPase 1 is induced by the addition of 0.5 mM IPTG. Growth of the culture is continued for an additional 4 h.

3. Cells are harvested by centrifugation and resuspended in 25 mL of lysis buffer (*see* **Section 2.1.5**). Lysozyme (6 mg) and RNase-free DNase (60 μL at 10 mg/mL) are added, and the solution is stirred for 10 min followed (*see* **Note 2**) by freezing at −80°C and thawing to lyse the cells (this freeze/thaw step is repeated several times) (*see* **Note 3**).
4. The frozen lysed cells are thawed and 200 μL of 1 M magnesium acetate is added and mixed by stirring for 10 min at room temperature.
5. The solution is then centrifuged at 40,000 *g* (Ti-70 rotor, Beckman centrifuge) for 30 min at 4°C, and the pellet is resuspended in 25 mL of 10 mM TEA-HCl, 10% glycerol, pH 7.9.
6. After centrifugation again at 40,000 *g* for 30 min, the pellet is resuspended by douncing in solubilization buffer containing 1% Triton X-100 (*see* **Section 2.1.8**) to extract the SPase from the membrane and re-centrifuged a third time.
7. The SPase I-rich supernatant is loaded onto a 15 mL Q-Sepharose column previously equilibrated in solubilization buffer (*see* **Note 4**).
8. The column is washed with 20 mL solubilization buffer plus 5 mM magnesium sulfate, pH 7.9.
9. SPase I is eluted with a continuous gradient of 0–0.1 M KCl in column buffer. 2 mL fractions are collected and assayed for SPase I protein by SDS-PAGE and Gel Code Blue staining.
10. Fractions containing the enzyme are pooled and loaded onto a 1 mL Ni^{2+} nitrilotriacetic acid-agarose (*see* **Note 5**) column equilibrated with 6-His buffer.
11. The column is then washed with 7 mL of 6-His buffer followed by 1 mL of wash buffer (*see* **Section 2.1.12**).
12. SPase I is then eluted using a 100–300 mM imidazole step gradient.
13. Eluted fractions are assayed for SPase I protein by SDS-PAGE followed by Gel Code Blue staining.
14. To remove the imidazole, pooled proteins are dialyzed against dialysis buffer (*see* **Section 2.1.13**) or washed with 20 mM phosphate, 1% β-D-octylglucopyranoside, pH 8.0 buffer and centrifugation using a Centricon-10 centrifugation tube (Amicon).

3.2. Purification of ProOmpA Nuclease A

ProOmpA nuclease A containing a 6-His tag is purified by ion-exchange and nickel chelate chromatography (19). The procedure is a modification of the method described in (9) that involves refolding of a denatured protein. In this new procedure, a

6His-tag is added to proOmpA nuclease A at the N-terminus. This allows an additional purification step involving Ni^{2+}chelate chromatography.

1. Use the glycerol stock BL21 pET-21a bearing the proOmpA nuclease A gene to make the starter culture. Grow overnight by shaking at 37°C with vigorous aeration.
2. Add 25 mL of overnight culture to 1 L LB media and grow to $OD_{600} \sim 0.8$. Induce with 1 mM IPTG for 3 h.
3. Centrifuge cells at 6,500 *g* (JA-10 rotor, Beckman centrifuge) for 10 min.
4. Resuspend cells in TEP buffer (*see* **Section 2.2.12**) with 5 mL per gram of cells.
5. Add 5μL 100 mM PMSF (*see* **Note 6**) to the resuspended cells and French press at 10,000 psi once. Then add another 500 μL of 100 mM PMSF and French press two more times (*see* **Note 7**).
6. Centrifuge sample to pellet the membranes at 120,000 *g* in Ti-70 (Beckman) for 1 h. Discard supernatant.
7. Resuspend pellet containing proOmpA nuclease A in 20 mL MEB guanidinium buffer (*see* **Section 2.2.13**) by sonicating it 2 or 3 times and stir for 30 min at 4°C. The sonication conditions are 60 s per sonication, 60% pulse time, 70% output, a minimum of 60 s cooling on ice between each sonication (*see* **Notes 8**, **9**, and **10**).
8. Centrifuge sample at 230,000 *g* (Ti-70 rotor, Beckman ultracentrifuge) for 30 min to pellet the membranes. Keep supernatant containing the proOmpA nuclease A.
9. Dialyze supernatant in 2 L dialysis buffer 1 overnight (*see* **Note 11**).
10. Dialyze again in 1 L dialysis buffer 2 for 4 h.
11. Dialyze a third time in 1 L dialysis buffer 3 for 4 h.
12. Centrifuge the dialyzed sample to remove any residual aggregates at 230,000 *g* in Ti-70 (Beckman) for 30 min.
13. Add 2.3 g of ammonium sulfate per 20 mL to make a solution of 20% saturation of ammonium sulfate. Incubate at 4°C for 30 min.
14. Spin down to remove precipitated proteins at 23,000 *g* for 30 min.
15. Add 4.5 g ammonium sulfate per 20 mL to the proOmpA nuclease A containing supernatant to increase percent ammonium sulfate to 55% saturation. Incubate at 4°C for 30 min.

16. Collect the precipitated proOmpA nuclease A by centrifugation at 23,000 *g* for 30 min.
17. Resuspend the sample containing proOmpA nuclease A in 5 mL of guanidinium buffer (*see* **Section 2.2.17**). The denatured proOmpA nuclease should become soluble.
18. Centrifuge at 230,000 *g* (TI-70 rotor, Beckman ultracentrifuge) for 30 min to remove non-soluble protein.
19. Apply supernatant containing the denatured proOmpA nuclease A to a S-300 sepharose column (120 × 3 cm) equilibrated with 7.2 M guanidinium HCl. Collect 8 mL fractions. Pool fractions in the second peak and then analyze by SDS-PAGE. Protein should be approximately 90% pure at this step.
20. Refold denatured proOmpA nuclease A by dialyzing 3 times against 2 L of 25 mM HEPES pH 7.5 (*see* **Note 12**). Each dialysis step should be for at least 4 h.
21. Load refolded proteins onto an Ni^{2+}-nitrilotriacetic acid agarose column and elute with a stepwise 100–500 mM imidazole gradient.
22. Eluted fractions are assayed for proOmpA nuclease A by SDS-PAGE followed by GelCode Blue staining.

3.3. In Vitro SPase Activity Assay and Kinetic Assay

3.3.1. Activity Assay

1. Pro-OmpA nuclease A prepared in **Section 3.2** is used as the substrate. The Pierce BCA protein assay kit is used to determine the concentration of purified SPase (*see* **Section 3.1**). An $E^{1\%}$ at 280 nm equal to 8.3 is used to determine the concentration of purified pro-OmpA nuclease A. An absorbance at 280 nm equal to 8.3 equals a concentration of 1 mg/mL protein.
2. SPase I is added to substrate (15 μM final concentration) in 50 mM Tris HCl, pH 8.0, 1% Triton X-100. Typically, the reaction volume is 15 μL.
3. The reaction is incubated at 37°C (*see* **Note 13**) for various times (**Fig. 2.1A**). The reaction is stopped by the addition of 4μL of 5X sample buffer and placing the sample in a dry-ice/ethanol bath.
4. Processing of the proOmpA nuclease A substrate to its mature form is monitored on a 17.2% SDS-PAGE gel.
5. Enzyme dependence of the reaction can be determined by examining processing with dilutions of SPase I (**Fig. 2.1B**).
6. A typical starting concentration of SPase for the dilution study is 0.1 mg/mL.
7. Aliquots of the enzyme are added to 15 μM substrate in 50 mM Tris-HCl, pH 8.0, 1% Triton X-100 (reaction volume is 15 μL).

8. The reaction is incubated at 37°C for 1 h and then stopped by the addition of 4 μL of 5X sample buffer followed by incubating sample in a dry-ice/ethanol bath.
9. Preprotein processing is examined by using a 17.2% SDS-PAGE gel.

3.3.2. Kinetic Analyses

1. Pro-OmpA nuclease A is used as substrate. Its concentration is determined by using an $E^{1\%}$ at OD_{280} of 8.3 (*see* **Section 3.3.1.1**).
2. The cleavage reaction is performed in 50 mM Tris-HCl, pH 8.0, 10 mM $CaCl_2$, 1% Triton X-100 buffer, with the substrate at five different concentrations (37.3, 24.9, 18.7, 12.4, and 6.2 μM).
3. The reaction is initiated by the addition of SPase I. The concentration of SPase is determined by the Pierce BCA protein assay kit.
4. The reaction is carried out at 37°C, and aliquots of the reaction are removed at various times. The last time point should have less than 10% processing of the substrate.
5. The reaction is stopped by the addition of 5 μL of 5X sample buffer, and the samples are frozen immediately in a dry-ice/ethanol bath.
6. The amount of pro-OmpA nuclease A that is converted to mature nuclease A by SPase I is analyzed by SDS-PAGE on a 17.2% gel, followed by staining with Gel Code Blue staining kit.
7. The precursor and mature proteins are quantified by scanning the gels using a scanning densitometer (Technology Resources, Inc. Line Tamer PCLT 300). Percentage processing is determined by dividing the area of the mature protein band by the sum of the mature and precursor band areas.
8. The initial rates are determined by plotting the amount of product versus time. The V_{max}, k_{cat}, and K_m values are extracted from a $1/V_i$ versus $1/[S]$ plot where V_i is the initial velocity. A computer program Microcal Origin is used to plot the data and for linear regression analysis of the data.
9. At least three independent experiments are necessary for obtaining reliable values.

3.4. In Vivo SPase Activity Assay

1. IT41(DE3) (*see* **Note 14**) is grown at 30°C in M9 minimal medium with the SPase I gene on the IPTG-inducible pET23 plasmid. No plasmid IT41 (DE3) is grown in this fashion and analyzed following the same protocol below.

2. At the mid-log phase at an OD_{600} ~0.3, cells are shifted to 42°C and grown at this temperature for 1 h to impair SPase I activity.
3. IPTG (1 mM final concentration) is added to induce synthesis of the plasmid-encoded SPase I and incubated at 42°C for an additional 30 min.
4. Cell cultures (1 mL) are labeled with 200 μCi of [^{35}S]methionine for 15 s and chased with non-radioactive methionine (500 μg/mL) (*see* **Note 15**).
5. At indicated times (such as 10, 30, or 120 s), aliquots (100 μL) are removed and quenched with an equal volume of ice-cold 20% TCA (**Fig. 2.2**).
6. The mixtures are kept on ice for 1 h, followed by centrifugation at 16,000 *g*. Supernatants are removed and the precipitated proteins are first washed by an equal volume of ice-cold acetone and the sample centrifuged. The pellet is dissolved in 100 μL 10 mM Tris-HCl, pH 8.0, 2% SDS by heating at 95°C for 5 min (*see* **Note 16**).
7. Add 1X Triton buffer (900 μL) (*see* **Section 2.4.9**) and Staph A (30 μL) (*see* **Section 2.4.7**), mix and incubate on ice for 15 min.
8. Centrifuge sample at 13, 000 rpm for 15 s to pellet Staph A to remove the non-specifically bound proteins. Then transfer the supernatant to new tubes, add 3 μL of antibody (*see* **Note 17**) that recognizes the tested SPase substrates such as proOmpA, mix and incubate on ice for 1 h.
9. Add 30 μL Staph A, mix and incubate on ice for 1 h.
10. Centrifuge sample for 15 s to pellet Staph A-antibody protein complex and discard supernatant. 1 mL 1X Triton buffer is added and the pellet is resuspended by vortexing. Repeat the wash and centrifugation two times.
11. Dissolve the pellets in 40 μL 2X SDS sample loading buffer and analyze the protein sample by SDS-PAGE (*see* **Note 18**). Typically, 15 μL of sample is loaded onto the gel.
12. The proteins in the gel are fixed by incubation of the gel for 15 min in fixing buffer at room temperature and dried using a Bio-Rad gel dryer. The dried radioactive gel is exposed on a phosphorimaging screen for at least 3 h. The image is obtained by Typhoon Variable Mode Imager and analyzed with the ImageQuant software.
13. Typically, processing of the proOmpA to the mature OmpA is used to determine the in vivo activity of SPase I (**Fig. 2.2**).

3.5. SDS-PAGE

Procedure according to Bio-Rad.

1. Make sure the glass plates and the gel casting apparatus are clean and dry before use.
2. Prepare 5 mL separating gel mix for each gel. AP and TEMED should be added last. Pour the gel mix in between the gel casting plates.
3. Cover the separating gel with isopropanol and leave the gel at room temperature for 30 min to polymerize (*see* **Note 19**).
4. Remove the isopropanol.
5. Prepare 2.5 mL of stacking gel per gel. Pour the gel mix in between the gel casting plates over the separating gel until it reaches the top. Insert the desired comb. Leave the gel at room temperature for 30 min to polymerize.
6. Carefully remove the comb and rinse the wells with SDS-PAGE running buffer (*see* **Section 2.5.7**).
7. Assemble the gel apparatus and fill the chambers with SDS-PAGE running buffer.
8. Load 2–10 μL of sample into the well. Apply power (200 volts) to the gel running apparatus and start electrophoresis.
9. When the electrophoresis is complete, remove the buffer and disassemble the gel apparatus.
10. For staining the proteins with Gel Code Blue staining kit, first soak the separating gel in distilled water (20–50 mL) and microwave for 1 min.
11. Remove the water and place the separating gel in the same amount of distilled water. Then microwave for 1 min again.
12. Pour off the water and soak the separating gel in the same amount of distilled water. Shake for 10 min at room temperature.
13. Pour off the water and soak the separating gel in the same amount of Gel Code Blue staining buffer. Microwave for 1 min.
14. Shake the gel in Gel Code Blue staining buffer at room temperature until the gel in the buffer cools down. Protein bands are visualized soon after the stain.
15. Once the protein bands show up on the gel, pour off the Gel Code Blue staining solution and destain the gel using distilled water by shaking.

To analyze the radioactive proteins on the gel by phosphorimaging, follow steps 16 to 20.

16. For analyzing radioactive protein samples, soak the separating gel for 5–10 min in 50 mL gel fixing buffer (*see* **Section 2.4.9**). Shake at room temperature.
17. Transfer the gel to a filter paper (size should be slightly larger than the gel) and cover it with a plastic wrap film. The gel is left in the gel dryer with the filter paper side facing down.
18. Dry the gel at 70°C while vacuum is on for 1 h.
19. Transfer the dried gel to the exposure cassette. Cover it with the phosphorimager screen. Close the cassette. Expose the gel for at least 3 h.
20. Scan the phosphorimager screen using the phosphorimager.

4. Notes

1. All media (LB, M9) should be pre-warmed before used for culturing the cells.
2. When stirring the pellet to resuspend proOmpA nuclease A or SPase I, minimize the amount of foaming corresponding to denatured protein.
3. When purifying SPase I, the freeze-thaw steps might not be sufficient to break the cell. Then, a sonication step can always be added before loading the samples onto the Q-Sepharose column. Follow the sonication settings described in **Section 3.2**. To check if the cells are completely broken, centrifuge a small amount of the cells at 6,000 *g* for 10 min. Broken cells are not pelleted at this condition.
4. Steady flow speed through the columns (Q-sepharose and S-300) in the gel filtration and ion exchange chromatography steps used to purify SPase I and proOmpA nuclease A, respectively, is important for getting better separation.
5. Cobalt beads can be used instead of Nickel beads for affinity purification of the His-tagged SPase I and proOmpA nuclease A proteins. If the binding efficiency of the His-tagged proteins is low when running the sample through the column, seal the column and incubate it with the entire sample while gently shaking in a cold room overnight.
6. During purification of proOmpA nuclease A, PMSF is toxic, and very unstable in water. Handle with care.
7. When disrupting the cells in the French press, keep the flow speed low. Cell lysate should come out slowly drop

by drop. The cell suspension that comes out at the end of French press step should be a deep bronze color and almost transparent.

8. During sonication, the sonication tip should be maintained between $^{1}/_{2}$ and 2/3 depth inside the liquid to achieve better efficiency.
9. The sonication step in **Section 3.2.7** in the purification of proOmpA nuclease A is used to better resuspend the membrane pellet isolated from the previous ultracentrifugation step, which is usually very tightly packed. Dounce homogenization can be performed for the same purpose here.
10. The duration of sonication should be experimentally determined by the operator. Normally, at the end of the sonication step, the cell suspension should be a deep bronze color and almost clear.
11. The dialysis bag should be sealed carefully to prevent loss of samples during this step. If possible, leave the open end(s) of the dialysis bag out of the buffer.
12. While refolding the denatured proOmpA nuclease A, watch out for precipitates that may occur during the dialysis. Increasing amount of precipitates means the concentration of the denatured protein solution is too high or the dialysis is going too quickly.
13. In the activity and kinetic assays, when incubating a reaction at a certain temperature, a water bath is always preferred.
14. To test if an IT41(DE3) colony is temperature-sensitive, prepare LB media with a concentration of 2.5 g NaCl/L (LS2.5 media). Grow IT41(DE3) on LS2.5 plate at 30°C until colonies reached 1–2 mm in size. Single colonies are streaked in duplicate on LS2.5 plates for incubation at 30°C and 42°C. No growth should be seen at 42°C, the non-permissive temperature. To test if a IT41(DE3) strain can express T7 RNA polymerase, use the T7 tester phage to infect the cell and induce with IPTG. λDE3 lysogens should give very large plaques compared to non-induced condition.
15. Trans-[^{35}S] label has a half-life of 87.4 days. Use [^{35}S]-label that has been recently purchased.
16. While doing the in vivo signal peptidase activity assay, cell pellets after the acetone wash step can be hard to dissolve. In this case, either heat it using a heat block and vortex or leave it at room temperature for overnight.
17. The amount of antibody needed for immunoprecipitating OmpA should be adjusted for each different batch of

antibodies to achieve the best results. Too much antibody can cause background signal to occur.

18. The difference of molecular weights between pro-OmpA and the mature OmpA (or the preprotein and the mature protein of other SPase substrates) is not large. Carefully control the length of time the SDS-PAGE gel is run for better separation.
19. If the SDS-PAGE gel does not polymerize after 1 h, do not continue waiting. Prepare fresh AP solution and repeat the polymerization of the gel.

Acknowledgment

The work was supported by National Institute of Health Grant (GM63862-05) to R.E.D

References

1. von Heijne, G. (1983) Patterns of amino acids near signal-sequence cleavage sites *Eur J Biochem* **133**, 17–21.
2. Zwizinski, C. and Wickner, W. (1980) Purification and characterization of leader (signal) peptidase from Escherichia coli *J Biol Chem* **255**, 7973–7977.
3. Watts, C., Wickner, W. and Zimmermann, R. (1983) M13 procoat and a pre-immunoglobulin share processing specificity but use different membrane receptor mechanisms *Proc Natl Acad Sci USA* **80**, 2809–2813.
4. Shen, L.M., Lee, J.I., Cheng, S.Y., Jutte, H., Kuhn, A. and Dalbey, R.E. (1991) Use of site-directed mutagenesis to define the limits of sequence variation tolerated for processing of the M13 procoat protein by the Escherichia coli leader peptidase *Biochemistry* **30**, 11775–11781.
5. Tschantz, W.R., Sung, M., Delgado-Partin, V.M. and Dalbey, R.E. (1993) A serine and a lysine residue implicated in the catalytic mechanism of the Escherichia coli leader peptidase *J Biol Chem* **268**, 27349–27354.
6. Chang, C.N., Blobel, G. and Model, P. (1978) Detection of prokaryotic signal peptidase in an Escherichia coli membrane fraction: endoproteolytic cleavage of nascent fl pre-coat protein *Proc Natl Acad Sci USA* **75**, 361–365.
7. Dev, I.K., Ray, P.H. and Novak, P. (1990) Minimum substrate sequence for signal peptidase I of Escherichia coli *J Biol Chem* **265**, 20069–20072.
8. Stein, R.L., Barbosa, M.D. and Bruckner, R. (2000) Kinetic and mechanistic studies of signal peptidase I from Escherichia coli *Biochemistry* **39**, 7973–7983.
9. Chatterjee, S., Suciu, D., Dalbey, R.E., Kahn, P.C. and Inouye, M. (1995) Determination of Km and kcat for signal peptidase I using a full length secretory precursor, pro-OmpA-nuclease A *J Mol Biol* **245**, 311–314.
10. Paetzel, M., Karla, A., Strynadka, N.C. and Dalbey, R.E. (2002) Signal peptidases *Chem Rev* **102**, 4549–4580.
11. Dalbey, R.E. and Wickner, W. (1985) Leader peptidase catalyzes the release of exported proteins from the outer surface of the *Escherichia coli* plasma membrane *J Biol Chem* **260**, 15925–15931.
12. Ito, K. (1982) Purification of the precursor form of maltose-binding protein, a periplasmic protein of Escherichia coli *J Biol Chem* **257**, 9895–9897.
13. Paetzel, M., Strynadka, N.C., Tschantz, W.R., Casareno, R., Bullinger, P.R. and Dalbey, R.E. (1997) Use of site-directed

chemical modification to study an essential lysine in Escherichia coli leader peptidase *J Biol Chem* **272**, 9994–10003.

14. Dalbey, R.E. (1991) In vivo protein translocation into or across the bacterial plasma membrane *Methods Cell Biol* **34**, 39–60.
15. Inada, T., Court, D.L., Ito, K. and Nakamura, Y. (1989) Conditionally lethal amber mutations in the leader peptidase gene of Escherichia coli *J Bacteriol* **171**, 585–587.
16. Cregg, K.M., Wilding, I. and Black, M.T. (1996) Molecular cloning and expression of the spsB gene encoding an essential type I signal peptidase from Staphylococcus aureus *J Bacteriol* **178**, 5712–5718.
17. Bilgin, N., Lee, J.I., Zhu, H.Y., Dalbey, R. and von Heijne, G. (1990) Mapping of catalytically important domains in Escherichia coli leader peptidase *EMBO J* **9**, 2717–2722.
18. Klenotic, P.A., Carlos, J.L., Samuelson, J.C., Schuenemann, T.A., Tschantz, W.R., Paetzel, M. et al. (2000) The role of the conserved box E residues in the active site of the Escherichia coli type I signal peptidase *J Biol Chem* **275**, 6490–6498.
19. Carlos, J.L., Paetzel, M., Brubaker, G., Karla, A., Ashwell, C.M., Lively, M.O. et al. (2000) The role of the membrane-spanning domain of type I signal peptidases in substrate cleavage site selection *J Biol Chem* **275**, 38813–38822.

Chapter 3

Membrane Insertion of Small Proteins

Andreas Kuhn, Natalie Stiegler, and Anne-Kathrin Schubert

Abstract

Proteins that are less than 10 kDa in size are easily purified under denaturing conditions and can often be refolded by removal of the denaturing agents. The purified small membrane proteins are competent for membrane insertion when the denaturing agent is diluted out and a membranous system like liposomes or proteoliposomes is added. This system allows the characterization of the membrane insertion process at the molecular level. The insertion of the protein into proteoliposomes can be followed by protease digestion and Western blot analysis. Only if the antigenic region of the protein has translocated into the lumen of the proteoliposome it is protected from the protease. When combining this approach with fluorophores that are placed within the membrane protein, membrane insertion can also be followed by fluorescence correlation spectroscopy.

Key words: Pf3 coat protein, M13 procoat protein, YidC membrane insertase, liposomes, fluorescence labelling.

1. Introduction

The membrane insertion of small proteins that contain at most two transmembrane regions differs in the requirements for translocases and insertases compared to multispanning membrane proteins and proteins that contain large periplasmic domains. Insertion of most of these small proteins is Sec-independent, as they show efficient membrane insertion in the absence of the Sec translocase. This was first discovered with the M13 procoat protein that inserts normally in strains that are deficient in SecA or SecY (1), later with Pf3 coat protein (2), the F_1F_o ATP synthase subunit c (3) and the sensory protein MscL (4).

Small proteins cannot insert by a cotranslational mechanism since the ribosome accommodates amino acid residues in its exit

A. Economou (ed.), *Protein Secretion*, Methods in Molecular Biology 619,
DOI 10.1007/978-1-60327-412-8_3, © Springer Science+Business Media, LLC 2010

tunnel. The exit tunnel measures about 8 nm in length; extended protein chains of 23 amino acid residues are buried inside until they reach the surface of the ribosome. However, the width of the exit tunnel measures 1.37 nm which should allow the formation of an alpha helix (5). In this case, the exit tunnel could accommodate about 55 amino acid residues making it difficult for small proteins to reach the cytosol during translation.

To analyse membrane insertion in vivo external proteases are added that cleave the periplasmic region of the protein (6) or a single cysteinyl residue in the periplasmic region is used that can be modified resulting in an electrophoretic shift (4). These methods are discussed in Chapter 4 of this book. Membrane insertion is also analysed in vitro into inverted membrane vesicles (7) or into liposomes (8) by externally added protease. Fluorescently labelled proteins can be monitored for translocation by quenching either in liposomes or proteoliposomes.

The results from the in vivo and in vitro experiments show that most small membrane proteins use the membrane insertase YidC to traverse the bilayer (8, 9). YidC is a 62 kDa protein of *Escherichia coli* that spans the membrane 6 times. Most likely, the hydrophobic transmembrane regions of YidC interact with the substrate protein and support the formation of a transmembrane α-helix. After the substrate protein has achieved the transmembrane configuration it is released from YidC. YidC can be easily purified from *E. coli* cells by affinity chromatography and reconstituted using *E. coli* phospholipids into unilamellar proteoliposomes using the extrusion technique (8).

2. Materials

2.1. Protein Purification

2.1.1. Expression and Purification of YidC Protein

1. LB-Medium: 10 g tryptone, 5 g yeast extract, 5 g NaCl in 1 L. Autoclave.
2. Prepare 200 mg/mL ampicillin. Store at 4°C.
3. For induction of the plasmid-encoded YidC prepare 1 M isopropyl-thio-β-D-galactoside (IPTG). Store at −20°C.
4. Prepare 100 mM phenylmethylsulfonyl fluoride (PMSF) in ethanol. Store at 4°C.
5. TSB buffer: 20 mM Tris-HCl pH 8.0, 300 mM NaCl, 10% (v/v) glycerol.
6. Extraction buffer: 1% (w/v) Fos-choline-12 (Anatrace), 20 mM Tris-HCl pH 8.0, 300 mM NaCl, 10% glycerol. Store at room temperature.
7. Buffer S: 0.1% (w/v) Fos-choline-12, 20 mM Tris-HCl pH 8.0, 300 mM NaCl, 10% glycerol, 30 mM imidazole. Store at room temperature.

8. Ni^{2+}-NTA-agarose 1:1 in absolute ethanol: equilibrate with buffer S.
9. Buffer A: 0.2% (w/v) Fos-choline-12, 20 mM Tris-HCl pH 8.0, 300 mM NaCl, 10% glycerol, 40 mM imidazole. Store at room temperature.
10. Buffer B: 0.2% (w/v) Fos-choline-12, 20 mM Tris-HCl pH 8.0, 300 mM NaCl, 10% glycerol, 300 mM imidazole. Store at room temperature.

2.1.2. Expression and Purification of Small Membrane Proteins

2.1.2.1. Expression and Purification of M13 Procoat H5 Protein

1. LB-Medium: 10 g tryptone, 5 g yeast extract, 5 g NaCl in 1 L. Autoclave.
2. Prepare 200 mg/mL ampicillin. Store at 4°C.
3. For induction of synthesis of plasmid-encoded M13 procoat protein prepare 1 M IPTG. Store at −20°C.
4. Lysis buffer: 50 mM Tris-HCl pH 7.6, 10% sucrose.
5. Prepare 20 mg/mL lysozyme. Store at −20°C.
6. Prepare 20 mg/mL DNase II. Store at −20°C.
7. Prepare 1 M $MgCl_2$.
8. Buffer T: 10 mM triethanolamine-HCl pH 7.5, 10% (v/v) glycerol.
9. Extraction buffer: 20 mM Tris-HCl pH 7.9, 500 mM NaCl, 1% (w/v) lauroyl-sarcosine, 10% (v/v) glycerol.
10. Ni^{2+}-NTA-agarose 1:1 in absolute ethanol: equilibrate with buffer A.
11. Buffer A: 20 mM Tris-HCl pH 7.9, 500 mM NaCl, 1% (w/v) lauroyl-sarcosine, 5 mM imidazole.
12. Buffer A1: 20 mM Tris-HCl pH 7.9, 500 mM NaCl, 1% (w/v) lauroyl-sarcosine, 100 mM imidazole.
13. Buffer A2: 20 mM Tris-HCl pH 7.9, 500 mM NaCl, 1% (w/v) lauroyl-sarcosine, 200 mM imidazole.
14. Buffer A3: 20 mM Tris-HCl pH 7.9, 500 mM NaCl, 1% (w/v) lauroyl-sarcosine, 500 mM imidazole.
15. Dialysis buffer: 50 mM Tris-HCl pH 7.9, 1% (w/v) lauroyl-sarcosine.
16. Denaturation buffer: 2 M guanidium-HCl, 50 mM Tris-HCl pH 8.0. Store at 4°C.

2.1.2.2. Expression and Purification of Pf3 Coat Protein

1. LB-Medium: 10 g tryptone, 5 g yeast extract, 5 g NaCl in 1 L. Autoclave. Store at 4°C.
2. 100 mg/mL ampicillin stock solution is prepared and stored at 4°C.
3. For induction of the plasmid-encoded Pf3 coat protein prepare 1 M IPTG. Store at −20°C.

4. Lysis buffer: 100 mM Tris-HCl pH 8.0, 20% (w/v) sucrose.
5. Resuspension buffer: 100 mM Tris-HCl pH 8.5.
6. Solubilization buffer: 100 mM Tris-HCl pH 8.5, 8 M urea. Store above 20°C
7. Dialyse buffer: 100 mM Tris-HCl pH 8.5, 10% (v/v) 2-propanol.
8. Gel filtration buffer: 100 mM Tris-HCl pH 7.5, 10% (v/v) 2-propanol.

2.2. Preparation of Different Gel Systems

2.2.1. SDS-PAGE (Sodium Dodecyl Sulfate–Polyacrylamide Gel Electrophoresis)

For 4 mini gels (glass plate 7 × 9 cm):

1. 12% SDS-PAGE Separating gel: prepare the gel by adding 12 mL of 30% acrylamide (Roth, Germany), 5.62 mL of 2.0 M Tris-HCl pH 8.8, 12.1 mL H_2O, 120 μL of 25% (w/v) SDS, 150 μL of 10% (w/v) ammonium persulfate (APS), 10 μL N, N, N′, N′-tetramethylethylenediamine (TEMED).
2. SDS-PAGE 7.5% Stacking gel: prepare the gel by adding 2.66 mL of 30% acrylamide, 2.5 mL of 1.0 M Tris-HCl pH 6.8, 14.6 mL H_2O, 80 μL of 25% (w/v) SDS, 100 μL of 10% (w/v) APS, 10 μL TEMED.
3. SDS-PAGE running buffer: 25 mM Tris, 192 mM glycine, 0.1% (w/v) SDS. 10x SDS-buffer stock can be prepared and stored at 4°C.
4. SDS-PAGE loading dye is freshly prepared: 5 volumes of solution 1, 4 volumes of solution 2, 1 volume of 1 M dithiothreitol (DTT). Solution 1: 1 mL of 1 M Tris-HCl pH 7.0, 0.5 mL of 0.2 M EDTA pH 7.0, 3.5 mL H_2O. Solution 2: 4 mL of 25% (w/v) SDS, 1 mL of 1 M Tris-HCl pH 7.0, 3.5 mL glycerol, 3.5 mL of 0.5% (w/v) bromophenol blue.

2.2.2. Tricine-SDS-PAGE

For 4 mini gels (glass plate 7 × 9 cm):

1. 15% Separating gel: prepare the gel by adding 15 mL of 30% acrylamide (Roth), 10 mL of 3.0 M Tris-HCl pH 8.45 with 0.3% (w/v) SDS, 4 g glycerol, 1 mL H_2O, 100 μL of 10% (w/v) APS, 10 μL TEMED.
2. 10% Spacer gel: prepare the gel by adding 3.3 mL of 30% acrylamide, 4 mL of 3.0 M Tris-HCl pH 8.45 with 0.3% (w/v) SDS, 4.8 mL H_2O, 40 μL of 10% (w/v) APS, 4 μL TEMED.
3. 4% Stacking gel: prepare the gel by adding 1.1 mL of 30% acrylamide, 2.5 mL of 3.0 M Tris-HCl pH 8.45 with 0.3% (w/v) SDS, 6.5 mL H_2O, 80 μL of 10% (w/v) APS, 8 μL TEMED.

4. Tricine-SDS-PAGE anode buffer: 0.2 M Tris-HCl, pH 8.9; tricine-SDS-PAGE cathode buffer: 0.1 M Tris-HCl pH 8.25, 0.1 M tricine, 0.1% (w/v) SDS. 10x buffer stocks can be prepared and stored at 4°C.
5. SDS-PAGE loading dye as in step 4 of **Section 2.2.1**.

2.2.3. SDS-Urea-PAGE

SDS-gel containing 22% urea. For 4 mini gels (glass plate 7 × 9 cm) or 2 normal gels (glass plate 14 × 18 cm):

1. Separating gel containing 22% (w/v) urea: prepare the gel by adding 10.8 g urea, 3.75 mL of 3.3 M Tris-HCl pH 8.7, 14.7 mL of 45% acrylamide, 1.2 mL of 2% bis-acrylamide, 150 μL of 25% (w/v) SDS, 105 μL of 10% (w/v) APS, 10 μL TEMED.
2. Stacking gel 22% (v/v) urea: prepare the stacking gel by adding 3.6 g urea, 0.6 mL of 1 M Tris-HCl pH 6.8, 1.1 mL of 45% acrylamide, 0.65 mL of 2% bis-acrylamide, 4.85 mL H_2O, 40 μL of 25% (w/v) SDS, 100 μL of 10%(w/v) APS, 8 μL TEMED.
3. SDS-PAGE running buffer: 25 mM Tris, 192 mM glycine, 0.1% (w/v) SDS. 10x running buffer stock can be prepared and stored at room temperature.
4. Loading buffer is prepared by adding 0.3 M Tris-HCl pH 6.8, 50% (v/v) glycerol, 5% (w/v) SDS, 0.05% (w/v) bromophenol blue. 900 μL of loading buffer is added to 100 μL of 1 M DTT. The buffer is stored at room temperature.

2.3. Preparation of Proteins for SDS-PAGE Systems

1. Prepare a 30% (w/v) solution of trichloroacetic acid (TCA) and store at 4°C.
2. Acetone. Store at −20°C
3. Loading dye (*see* step 4 of **Section 2.2.1**) and loading buffer (*see* step 4 of **Section 2.2.3**).

2.4. Detection of Proteins in Gel Systems

2.4.1. Silver Staining

1. Fixing solution: 200 mL absolute ethanol, 50 mL acetic acid, 250 mL H_2O.
2. Sensitizing solution: 15 mL absolute ethanol, 62.5 μL glutaraldehyde, 0.1 g sodium thiosulfate, 3.4 g sodium acetate, fill to 50 mL with H_2O.
3. Staining solution: 1.25 g silver nitrate, fill to 500 mL with H_2O. Store at 4°C.
4. Developing solution: 12.5 g sodium carbonate, fill to 500 mL with H_2O. Store at 4°C.
5. Stopping solution: 7.3 g sodium ethylenediaminotetraacetic acid (EDTA) fill to 500 mL with H_2O. Store at room temperature.

2.4.2. Western Blotting

1. Transfer buffer (10x): 1 L of 10x transfer buffer is prepared with 50 mM Tris and 200 mM glycine. The buffer is stored at room temperature. For preparing 1x transfer buffer, add 5% methanol.
2. Hybond^{TM}-ECL^{TM} Nitrocellulose membrane (GE Healthcare).
3. Whatman paper (Schleicher & Schuell).
4. Ponceau S solution: 0.1% (w/v) Ponceau S, 5% (v/v) acetic acid.
5. Tris buffered saline (TBS buffer): 1 L of 10x TBS buffer is prepared with 248 mM Tris-HCl pH 7.4, 730 mM NaCl, 27 mM KCl.
6. Phosphate buffered saline (PBS buffer): 1 L of 10x PBS buffer is prepared with 1.5 mM NaCl, 30 mM KCl, 15 mM KH_2PO_4, 80 mM Na_2HPO_4, pH 7.4. PBS-Tween: 1x PBS with 0.005% Tween-20. The buffer is stored at 4°C.
7. Blocking buffer: 5% Non-fat milk in TBS buffer or PBS-Tween.
8. ECL-Kit (GE Healthcare)
9. Developing solution (Tetenal), prepare 120 mL by 1:10 dilution in H_2O.
10. Stopping solution is prepared by adding 2 drops of acetic acid to 500 mL H_2O.
11. Fixing solution (Tetenal), is prepared by adding 25 mL fixer to 125 mL H_2O.

2.5. Preparation of Multilamellar Vesicles

1. *E. coli* total lipid extract (Avanti Lipids) as a lipid powder. Store at −20°C.
2. Diacylglycerol (DAG) is purchased from Lipid Products (South Nutfield, Great Britain). DAG is purchased as a solution in chloroform at a final concentration of 20 mg/mL. Store at −20°C.
3. Chloroform (Roth). Store at room temperature.
4. Methanol (pure). Store at room temperature.
5. Acetone. Store at −20°C.
6. 250 mL glass plunger for 100 mg lipid powder.
7. Nitrogen gas (4.6 quality).
8. Glass beads (Baack, Germany) with a mean diameter of 5 mm.

2.6. Thin Layer Chromatography

1. Running solvent: prepare by adding 30 mL methanol, 4 mL H_2O and 85 mL chloroform. Prepare immediately before equilibration of the chamber.

2. Chromatography chamber.
3. TLC plates: silica gel 60 F_{254}(Merck).
4. Glass capillary.
5. Elementary iodine (Merck).
6. Ninhydrin solution: 0.2 g ninhydrin, 0.5 mL acetic acid, 96 mL acetone, add H_2O to 100 mL.

2.7. Preparation of Unilamellar Liposomes and YidC-Proteoliposomes

1. Buffer N: 100 mM Na_2SO_4, 10 mM Hepes pH 8.0. Store at −20°C.
2. 2x buffer N: 200 mM Na_2SO_4, 20 mM Hepes pH 8.0. Store at −20°C.
3. Purified YidC protein with a concentration of 3 mg/mL in 0.2% (w/v) Fos-choline-12, 20 mM Tris-HCl pH 8.0, 300 mM NaCl, 10% (v/v) glycerol, 213 mM imidazole. Store at −80°C.
4. The Sub-Micron Particle Analyzer Model N4SD (Beckman/Coulter) is used to measure the mean diameter of the vesicles by adding 10 μL of extruded liposomes to 990 μL buffer N. The measurement is carried out with a 1 × 1 cm^2 glass cuvette (Helma), 1 mL total volume.

2.8. Reconstitution of Purified YidC Protein

1. Preparation of YidC-proteoliposomes as described in **Section 2.7**, steps 1–4.
2. Prepare a 1 mg/mL stock solution of trypsin (Sigma). Store in aliquots at −20°C.
3. Prepare a 1 mg/mL stock solution of trypsin inhibitor (Sigma). Store at −20°C.
4. Triton X-100: 10% (v/v). Store at room temperature and in the dark.
5. Two antibodies that recognize either the C-terminus or the periplasmic loop of YidC are used (our laboratory collection). Store at −80°C.

2.9. Generation of a Membrane Potential

1. Buffer K: 100 mM K_2SO_4, 10 mM Hepes pH 8.0. Store at −20°C.
2. Valinomycin solution: 1 μM valinomycin in absolute ethanol. Store at −20°C.

2.10. The Measurement of Membrane Potential

1. Oxonol VI (Fluka): 1 mM oxonol VI in H_2O. Store at −20°C.
2. 1×1 cm^2 quartz cuvette (Helma), 1 mL total volume.

3. *Jasco* FP 750 Spectrofluorometer: Measurement at 25°C. Settings are as follows:

Excitation wavelength	599 nm
Emission wavelength	634 nm
Band width (Ex)	5 nm
Band width (Em)	5 nm
Response	1 s
Sensitivity	Medium
Measurement range	0–1200 s
Data pitch	1 s

4. Thermostat: Julabo VC is used to keep the sample in the cuvette at 25°C.
5. To clean the cuvette after usage wash twice with 2-propanol and H_2O.

2.11. Membrane Insertion of M 13 Procoat H5 into Proteoliposomes

1. Purified YidC protein with a concentration of 3 mg/mL in 0.2% (w/v) Fos-choline-12, 20 mM Tris-HCl pH 8.0, 300 mM NaCl, 10% (v/v) glycerol, 213 mM imidazole. Store at −80°C.
2. Valinomycin (Sigma): 230 μM valinomycin in absolute ethanol. Store at −20°C.
3. Denaturation buffer: 2 M guanidium-HCl, 50 mM Tris-HCl pH 8.0. Store at 4°C.
4. Proteinase K (Invitrogen) stock is stored at −20°C at 10 mg/mL.
5. Triton X-100: 10% (v/v). Store at room temperature, in the dark.
6. Multilamellar vesicles with a concentration of 20 mg/mL. Store at −80°C.
7. 30% TCA solution (*see* **Section 2.3.1**).

2.12. Labelling of Proteins at Unique Cysteine Residues with Fluorescent Dyes

1. Atto520-maleimide (ATTO-TEC, Germany). Store at −20°C.
2. Bodipy® FL *N*-(2-aminoethyl)maleimide (Molecular probes). Store at −20°C.
3. Labelling buffer: 100 mM Tris-HCl pH 7.25, 100 mM KCl, 10% (v/v) 2-propanol.
4. Tris-(2-carboxyethyl)-phosphine-hydrochloride (TCEP, T-2556, Molecular Probes) stock solution in H_2O is stored at 4°C with the concentration of 30 mg/mL.

5. Nitrogen gas.
6. Dimetylsulfoxide (DMSO) is stored at 4°C.
7. 0.3 g Dowex ion exchange resin (Serva 1 × 8, mesh size: 200–400) is washed with 1.5 mL of 1 M NaOH, with H_2O, with 4 M acetic acid and with H_2O. It is stored as 50% slurry in 1 M Tris-HCl pH 7.25 at 4°C. The washing steps are prepared for 15 min using a rotating wheel at room temperature. The matrix is pelleted (500 *g*, 1 min) and the supernatant is removed.
8. The Typhoon confocal scanner (GE Healthcare) is used to visualize the labelled proteins separated on a SDS-urea gel.

2.13. Binding of Fluorescently Labelled Proteins to the Membrane Surface

1. For preparing unilamellar vesicles, the Mini-Extruder (Avanti Inc.) with 1 mL syringes (Avanti no: 610017), filter supports (Avanti no: 610014) and 19 mm Nucleopore track-etch membrane with a pore size of 0.4 μm (Schleicher & Schuell) is used.
2. The vesicles are made from total lipid extract (Avanti 100500P *E. coli* total lipid extract as powder) and diacylglycerol (DAG; Lipid Products, Great Britain).
3. Buffer N: 100 mM Na_2SO_4, 10 mM Hepes pH 8.0.
4. Proteinase K (Invitrogen) stock is stored at −20°C with the concentration of 10 mg/mL.
5. Prepare a 30% (w/v) solution of TCA and store at 4°C.

3. Methods

The membrane insertion of small phage proteins like M13 procoat and Pf3 coat are strictly dependent on the membrane protein YidC in vivo (9). To test if YidC is sufficient for membrane insertion, the insertion of these proteins can be tested in vitro under different conditions (8). YidC is purified and reconstituted into liposomes, and the orientation of YidC is tested after reconstitution by protease digestion. The actual orientation of YidC is a critical factor for further experiments since only one orientation is functional. The quality of the liposomes is an additional factor for the success of the membrane insertion experiments. The tightness of the bilayer of the liposomes can be determined by measuring the stability of the electrochemical membrane potential over an extended time frame (**Fig. 3.1**). The antibiotic valinomycin is used to generate a potential since it induces an influx of K^+ ions resulting in a positive inside potential after the addition to

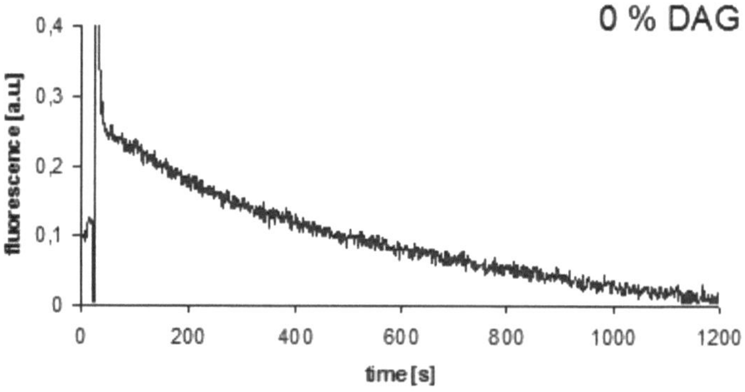

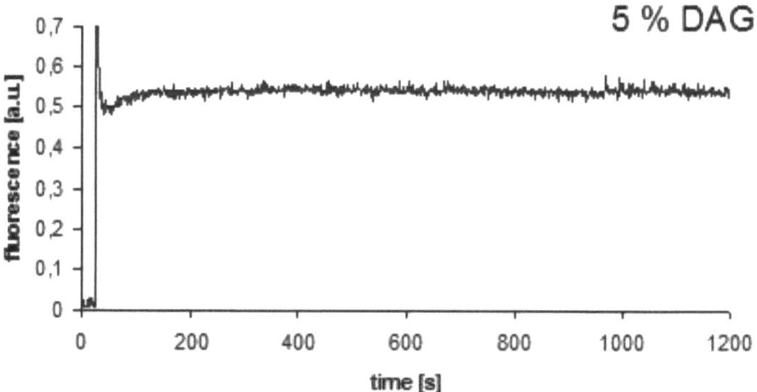

Fig. 3.1. The membrane potential of liposomes containing different amounts of diacylglycerol (DAG). The fluorescence at an excitation/emission of 599 nm/634 nm was recorded for 20 min.

the liposomes. The presence of a membrane potential is indicated by the fluorescence of oxonol VI that is negatively charged and membrane permeable. It incorporates into the membrane when a transmembrane potential is present (inside positively charged) and the increase of the fluorescence can be measured by a fluorometer.

Thiol-reactive reagents will react with cysteine residues on proteins to give thioether-coupled products. To label the Pf3 coat protein with thiol-reactive fluorescent dye, N- or C-terminal single-cysteine mutants of Pf3 coat can be made and the proteins are then purified. The fluorescent dye Atto520 is 589 Da and membrane permeable (**Fig. 3.2**). It has a quantum yield of 90%. The absorption maximum is at 525 nm and the emission maximum is at 545 nm. The fluorescent dye BodipyFL is 414 Da and membrane permeable. It has a quantum yield of 99%. The absorption maximum is at 504 nm and the emission maximum is at 511 nm. The fluorescently labelled proteins can be tested for insertion into proteoliposomes or membrane vesicles.

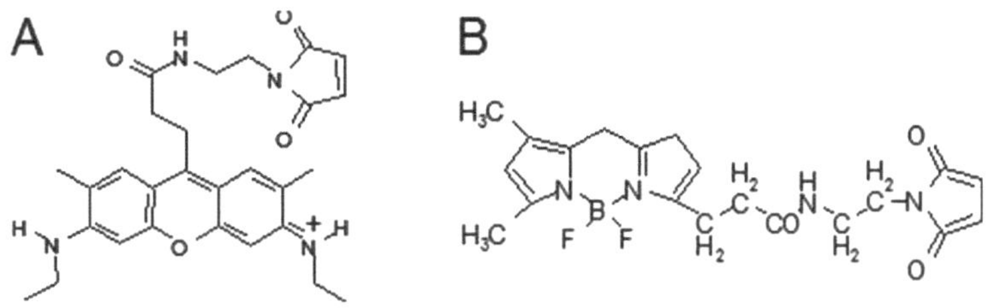

Fig. 3.2. The fluorescent dyes Atto520-maleimide (**A**) and BodipyFL-maleimide (**B**) were used to label the coat proteins.

3.1. Protein Purification

3.1.1. Expression and Purification of YidC Protein

1. To optimize purification of YidC, ten histidines are attached to the N-terminus of YidC.
2. The *E. coli* strain C43 (DE3) is transformed with the plasmid encoding YidC.
3. The cells are grown in 2 L of LB growth medium containing 200 μg/mL ampicillin at 37°C with shaking.
4. The expression of YidC is induced with 1 mM of isopropylthiogalactoside (IPTG) at an $OD_{600\ nm}$= 0.5.
5. After 3 h of growth, the cells are harvested by centrifugation (10,000 *g*, 15 min, 4°C).
6. The pellet is resuspended in 1 mL of TSB buffer per gram cells.
7. The cells are lysed by two passages through a French pressure cell (8,000 psi). PMSF is added to a final concentration of 1 mM to prevent proteolysis.
8. Total membranes are collected by centrifugation (200,000 *g*, 50 min, 4°C).
9. The membranes are washed by resuspension in 20 mL of extraction buffer.
10. The membranes are solubilized with a pre-chilled Dounce-homogenizer and kept on ice (*see* **Note 1**).
11. Non-solubilized proteins and aggregates are removed by centrifugation as in step 8 above.
12. The YidC protein is then purified by immobilized metal affinity chromatography. The supernatant is incubated with equilibrated Ni-NTA-agarose for 2 h at 4°C under rotation.
13. After binding, the column is washed with 20 bed volumes of buffer A to remove the majority of contaminants.
14. Bound protein is eluted with 10 bed volumes of buffer B and collected in 1 mL fractions.
15. The purified protein YidC is snap frozen in liquid nitrogen and stored at −80°C.

3.1.2. Expression and Purification of Small Membrane Proteins

3.1.2.1. Expression and Purification of M13 Procoat H5 Protein

1. The C-terminal His-tag modified protein M13 procoat H5 (6) is overproduced in *E. coli* strain HB101 and purified by Ni^{2+}-chelating chromatography.
2. The cells are grown at 37°C in 2 L of LB growth medium containing 200 μg/mL of ampicillin.
3. The protein expression is induced with 1 mM IPTG at $OD_{600\ nm}$= 0.5 and growth of the culture is continued for 2.5 h.
4. The cells are then harvested by centrifugation (10,000 *g*, 15 min, 4°C).
5. The pellet is resuspended in 1 mL of lysis buffer per gram cells.
6. The cells are broken by addition of 0.5 mg/mL lysozyme, 0.2 mg/mL DNase II and 5 mM $MgCl_2$.
7. The cells are stirred for 1 h at 4°C and the membranes are sedimented by centrifugation (40,000 *g*, 30 min, 4°C).
8. The membranes are washed with 40 mL of buffer T, homogenized with a Dounce-homogenizer and centrifuged as in step 8 of **Section 3.1.1**.
9. To extract the M13 procoat protein, the membrane fraction is solubilized with a Dounce-homogenizer in 20 mL of extraction buffer.
10. Non-solubilized protein is removed by ultracentrifugation (250,000 *g*, 20 min, 4°C).
11. The supernatant containing the procoat protein is incubated with equilibrated Ni^{2+}-NTA-agarose for 2 h at 4°C using a rotating wheel.
12. After binding to the column it is washed with 20 bed volumes of buffer A.
13. The bound protein is then eluted in a stepwise gradient of increasing concentrations of imidazole.
14. Firstly, with 10 bed volumes of buffer A1.
15. Followed by 10 bed volumes of buffer A2.
16. Finally with 20 bed volumes of buffer A3.
17. 2 mL fractions are collected in every step.
18. The protein is dialysed against dialysis buffer.
19. The sample is snap frozen with liquid nitrogen and stored at −80°C.
20. Before starting the membrane insertion experiment, the purified M13 procoat H5 is diluted 1:1 in denaturation buffer.

3.1.2.2. Expression and Purification of Pf3 Coat Protein

1. The plasmid pT7-7 encoding the Pf3 coat protein was used to overexpress the protein in *E. coli* BL21(DE3)pLysS that has the T7 polymerase gene under the inducible lacUV 5 promoter (10).
2. Cells are grown in LB medium with 100 μg/mL ampicillin at 37°C until an OD_{600} of 0.6 was reached. The culture is induced with 1 mM IPTG, and grown for additional 3 h at 37°C. The cells are harvested by centrifugation (10,000 *g*, 15 min, 4°C).
3. Cells are resuspended in lysis buffer and frozen in liquid nitrogen as cell nuggets and stored at −20°C.
4. 5 g of the cell nuggets are resuspended on ice while stirring in 10 mL resuspension buffer and the cells are broken by 5 passes with a French pressure cell at 8,000 psi.
5. The solution is centrifuged for 15 min at 3,020 *g* to remove the cell debris.
6. To pellet the membrane the supernatant is centrifuged for 1 h at 200,000 *g*.
7. The pellet is resuspended in 5 mL solubilization buffer using a Dounce-homogenizer and stirred for 1 h at room temperature.
8. The solution is then centrifuged for 1 h at 170,000 *g* and the solubilized coat protein is in the supernatant.
9. 2.5 mL of the crude extract is diluted with 2.5 mL of dialysis buffer and dialysed against 2 × 1 L of the dialysis buffer overnight at 4°C.
10. If the solution appears turbid after dialysis, it is filtered through a sterile filter with a pore size of 0.2 μm to remove the precipitated material.
11. The dialysed soluble extract was applied onto a Superdex 200 column and the protein containing fractions were pooled and further purified by a second size exclusion chromatography (Superdex 75 column) (**Fig. 3.3**).
12. The protein was stored in a glass tube and the purity was analysed by Tricine-SDS-PAGE (*see* **Note 2**).

3.2. Description of Different Gel Systems

3.2.1. SDS-PAGE Analysis

1. Clean the glass plates with 70% ethanol and allow the plates to dry for 10 min.
2. For preparing 4 mini gels a multiple mini-gel casting chamber (Sigma) is assembled (glass plates, spacers and plastic plates). The chamber has to be closed properly to prevent leakage of the gel solution.

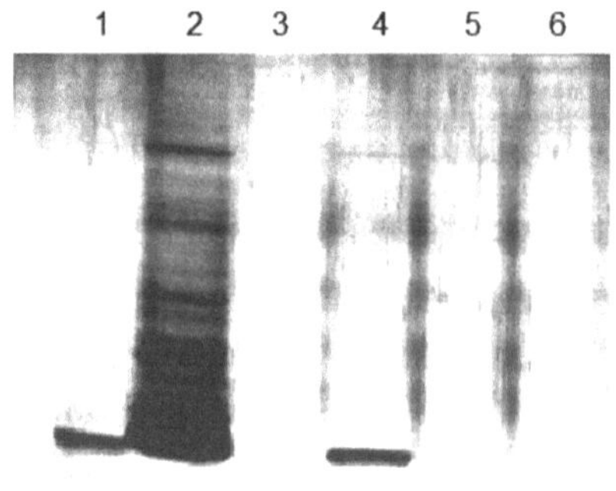

Fig. 3.3. Purification of the NC-Pf3 coat protein mutant. The protein was collected after a Superdex 200 column (*lane 2*) followed by Superdex 75 chromatography (*lanes 3–6*: fractions 11 to 14). The protein concentration in fraction 12 was about 1 mg/mL. *Lane 1* shows the purified wild-type protein as a marker.

3. Prepare the separating gel. Pour at least 4 cm (in height) separating gel and overlay with 2-propanol. The gel should polymerize within 30 min (*see* **Note 3**).
4. The 2-propanol is removed after the polymerization.
5. Prepare the stacking gel. Fill in the stacking gel to the top and quickly insert the comb. The gel should polymerize within 30 min. Let it stand for another hour.
6. Load samples and run at 17 mA per mini gel for 1 h and 15 min.

3.2.2. Tricine-SDS-PAGE Analysis

1. Clean the glass plates with 70% ethanol and allow the plates to dry for 10 min.
2. For preparing 4 mini gels, see step 2 of **Section 3.2.1**.
3. Prepare the separating gel. Pour separating gel at 3 cm (in height) and overlay with 2-propanol. The gel should polymerize within 30 min (*see* **Note 3**).
4. The 2-propanol is removed after the polymerization.
5. Prepare the spacer gel. Pour at least 1 cm spacer gel and overlay with 2-propanol. The gel should polymerize within 30 min. Let it stand for another 30 min.
6. The 2-propanol is removed after the polymerization.
7. Prepare the stacking gel. Fill in the stacking gel to the top and quickly insert the comb. The gel should polymerize within 30 min. Let it stand for another hour.
8. Load samples and run at 17 mA per mini gel for about 3 h.

3.2.3. SDS-Urea-PAGE Analysis

1. Clean the glass plates with 70% ethanol and allow the plates to dry for 10 min.

2. Fix the assembled glass plates with clamps and seal them up with 1% (w/v) agarose. The agarose should harden within 10 min.
3. Prepare the separating gel. See **Note 3**. Urea can be dissolved by shaking the mixture for several minutes in a 37°C water bath. Air bubbles have to be removed from the gel solution by a vacuum aspirator. Pour at least a 7 cm (in height) separating gel and overlay with 2-propanol. The gel should polymerize within 30 min. Let it sit for 3 h for complete polymerization overlayed with water (*see* **Note 4**).
4. Prepare the stacking gel. Urea can be dissolved by shaking the mixture for several minutes in a 37°C water bath. Fill the stacking gel to the top and quickly insert the comb.
5. Use 20–25 mA current per gel for about 3 h.

3.3. Preparation of Proteins for SDS-PAGE Systems

1. The protein samples are precipitated by adding 15% TCA (*see* **Section 2.3.1**). Keep the sample on ice for at least 1 h.
2. The samples are centrifuged for 15 min at 15,000 *g* and 4°C.
3. The supernatant is removed and the pellet is washed twice with acetone. After every washing step, the samples are centrifuged for 15 min at 15,000 × *g* and 4°C and the supernatant is removed.
4. The samples are dried in the Speed Vac for 5 min.
5. The pellets are resuspended in loading dye (*see* step 4 of **Section 2.2.1** or **2.2.3**).
6. At maximum, 20 μL sample is loaded onto a mini gel.

3.4. Detection of Proteins in Gel Systems

3.4.1. Silver Staining

1. To fix the proteins in the gel, it is soaked in 50 mL of fixing solution for 30 min.
2. The sensitizing solution has to be made fresh every time it is used. The gel has to be incubated for 30 min.
3. Wash the gel three times with 50 mL H_2O for 5 min.
4. To stain the gel, incubate the gel at least 20 min with 50 mL staining solution containing 20.5 μL of 37% formaldehyde. It is preferable to perform this step in the dark.
5. Wash the gel twice with 50 mL H_2O for 1 min.
6. Developing lasts at least 2 min (up to 45 min) with 50 ml developing solution containing 10.5 μL of 37% formaldehyde. This step should be performed in the dark (*see* **Note 5**).
7. Stop the developing by adding 50 mL of stopping solution (*see* **Note 6**).

3.4.2. Western Blotting

1. For immunoblotting of YidC or M13 procoat samples, TBS buffer is used. For Pf3 coat protein, PBS-Tween is used.
2. The gel is moved to transfer buffer for immunoblotting analysis and left in buffer for 5–20 min.
3. A transfer system is set up in a blotting apparatus made in a sandwich-like manner. Therefore, from anode to cathode the following is assembled: two Whatman sheets, nitrocellulose membrane, the gel, and two Whatman sheets. All pieces have to be soaked in transfer buffer.
4. The membrane transfer is accomplished at 10 V. The transfer of YidC, M13 and Pf3 takes 60, 35 and 45 min, respectively.
5. The nitrocellulose membrane is removed and the transferred proteins are stained by shaking the membrane in a Ponceau S solution for 3 min. The membrane is then washed with H_2O until the purple staining of the membrane fades.
6. Non-specific binding is reduced by shaking the membrane in blocking buffer for at least 1 h at room temperature or overnight at 4°C.
7. The nitrocellulose membrane is washed 4 times in either TBS or PBS-Tween buffer, each time for 5 min.
8. The nitrocellulose membrane is incubated for at least 3 h at room temperature (or overnight at 4°C) with 40 mL of the same buffer with added 1:10,000 diluted first antibody. The membrane is washed 4 times as in step 7.
9. The nitrocellulose membrane is incubated with 40 mL of the same buffer with 1:10,000 diluted secondary antibody for 1 h at room temperature. The membrane is washed 4 times as in step 7.
10. The nitrocellulose membrane is incubated for 3 min with ECL reagent that is prepared by dilution of solution 1 and solution 2.
11. The membrane is first transferred into a Saran wrap and then put into an exposure cassette. A blue light sensitive x-ray film is laid on top of the membrane and the cassette is tightly closed.
12. The film is exposed for a suitable short time.
13. The film is developed by incubation in developing solution until clear bands are visible.
14. The film is washed in stopping solution for a few seconds.
15. The film is fixed in fixing solution, washed with H_2O and dried.

3.5. Preparation of Multilamellar Vesicles (MLV)

1. To minimize oxidation of the lipids all following steps (3 to 17) are done under nitrogen atmosphere and under green light (*see* **Note 7**).
2. 100 mg of the lipid powder is dissolved in chloroform to a final concentration of 10 mg/mL under nitrogen (*see* **Note 8**).
3. Two drops of methanol are added.
4. Since the *E. coli* total lipid extract contains 17% non-lipid material an acetone precipitation step is performed.
5. The lipid solution is transferred into a 45 mL Teflon centrifugation tube.
6. 35 mL ice cold acetone is added to the Teflon tube and the lipids are precipitated for 2 h at −20°C.
7. The lipids are sedimented for 30 min at 5,000 *g*, 4°C.
8. The supernatant is discarded and the pelleted lipids are resuspended in chloroform and methanol as described above (steps 3 and 4). See **Note 8**.
9. The lipid solution is transferred to a 250 mL glass plunger.
10. The lipids are protected from light and oxygen by wrapping the glass flask in aluminium foil and adding nitrogen gas, respectively (*see* **Note 7**).
11. The appropriate amount of diacylglycerol (DAG) is added at a concentration of 1–5% depending on the lipid preparation.
12. The solvent is removed by rotary evaporation at a pressure of 50 mbar for about 15 min at room temperature.
13. The lipid film is dried under vacuum conditions for 6 h at room temperature until all solvent has disappeared.
14. The dry lipid film is resuspended in degassed double distilled H_2O at a concentration of 20 mg/mL. The resuspension is improved by adding glass beads.
15. The lipid suspension is distributed in aliquots, snap frozen in liquid nitrogen and stored at −80°C (*see* **Note 9**).

3.6. Thin Layer Chromatography

1. Equilibrate the chamber with freshly prepared running solvent for 1 h at room temperature.
2. The thin layer chromatography (TLC) plate has to be dried for at least 4 h at 100°C prior to using.
3. Prepare samples of the lipid powder either directly from Avanti or after an acetone precipitation. Then, test by TLC the acetone supernatant and the lipid suspension.
4. The lipid samples are diluted into chloroform.

5. The samples are applied on the silica plate using a glass capillary tube.
6. The plate is placed into the chamber.
7. The chromatography is run until the solvent front is 2 cm below the end of the plate.
8. Possible lipid spots should be marked.
9. After drying for 10 min at room temperature the plate is placed into an iodine atmosphere for at least 30 min at room temperature to stain the lipids.
10. The positions of the lipids are marked with a pencil.
11. The iodine is removed overnight by leaving the TLC plate under the hood.
12. The plate is sprayed with a solution of ninhydrin to visualize amino groups of the lipids (*see* **Note 10**).
13. The plate is heated for 45 min at 100°C.
14. The purple spots are marked with a pencil.

3.7. Preparation of Unilamellar Liposomes and YidC-Proteoliposomes

1. The MLVs stored at −80°C are thawed at room temperature and kept on ice.
2. To prepare unilamellar vesicles with a mean diameter of about 250–350 nm, the MLV suspension (*see* **Section 3.5**) (100–400 μL) is forced through an extruder by using a pore size of 400 nm (steps 3–7).
3. The extruder consisting of two syringes and a Teflon cylinder is assembled with two filters and a nitrocellulose membrane in between.
4. The syringes and the interior of the Teflon cylinder are equilibrated with buffer N.
5. Before loading the lipid suspension into the gas-tight syringes of the extruder, it is diluted 1:1 with 2x buffer N.
6. To yield YidC-containing proteoliposomes, the purified YidC protein is added with a YidC:lipid ratio of 1:25,000 to the suspension of multilamellar vesicles immediately prior to extrusion.
7. The suspension is slowly extruded 10 times.
8. The extruded vesicles are stored on ice and kept in the dark until used.

3.8. Orientation of the Reconstituted YidC Protein

1. YidC-proteoliposomes are prepared as described in **Section 3.7**, steps 1–8. Step 6 is not performed. Instead, the YidC:lipid ratio is increased to 1:5,000 for a better detection of the reconstituted YidC protein by SDS-PAGE. This is achieved by the addition of the purified YidC protein to the suspension of multilamellar vesicles (*see* **Note 11**).

2. For the reconstitution assay, three eppendorf tubes are prepared.
3. To each eppendorf tube, 50 μL of YidC-proteoliposomes (10 μg/μL) in buffer N are added.
4. Reconstitution is analysed by adding 0.2 mg/mL of trypsin to sample 2 and 3 and incubating for 90 min at 4°C.
5. For a control, 4% (v/v) of Triton X-100 is added to sample 3.
6. After 90 min of digestion the reaction is stopped by adding 1 mg/mL of trypsin inhibitor and the reaction is incubated at 4°C for 15 min.
7. The experiment is conducted as described in **Section 3.3**.
8. The samples are loaded on a SDS-PAGE gel (*see* step 6 of **Section 3.2.1**).

3.9. Generation of a Membrane Potential Across (Proteo)Liposomes

1. The liposomes and YidC-proteoliposomes, respectively, are prepared as previously described (see **Section 3.7**, steps 1–8) in buffer N.
2. The liposomes and YidC-proteoliposomes, respectively, are centrifuged (120,000 *g*, 10 min, 4°C).
3. The liposomes and proteoliposomes are carefully resuspended in buffer K.
4. To generate a membrane potential the antibiotic valinomycin is used. Valinomycin is a specific potassium carrier and induces an influx of K^+ ions into the vesicles resulting in a positive inside potential (*see* **Section 3.10.7** or **3.11.8**).
5. The membrane potential is detected using the fluorescent dye oxonol VI as described in **Section 3.10**.

3.10. The Measurement of a Membrane Potential

1. The liposomes are prepared in the sodium containing buffer N (*see* **Section 3.7**, steps 3–7).
2. The liposomes are then centrifuged for 10 min at 120,000 *g*, 4°C.
3. The supernatant is discarded and the pellet resuspended in the potassium containing buffer K.
4. Add 200 μg liposomes resuspended in buffer K and 0.35 μL oxonol VI to 980 μL of buffer K and mix thoroughly.
5. The sample is transferred into the cuvette and stirred with 100 rpm.
6. The fluorescence is measured at 25°C for 1,200 s (*see* **Note 12**).
7. To generate a membrane potential, 35 μL valinomycin solution is added to the sample 25 s after the start of measuring the fluorescence.

8. The cuvette has to be washed twice after the measurement with 2-propanol and H_2O.

3.11. Membrane Insertion of M13 Procoat H5

1. The multilamellar vesicles are diluted 1:1 (v/v) in 2x concentrated buffer N (**Fig. 3.4**).

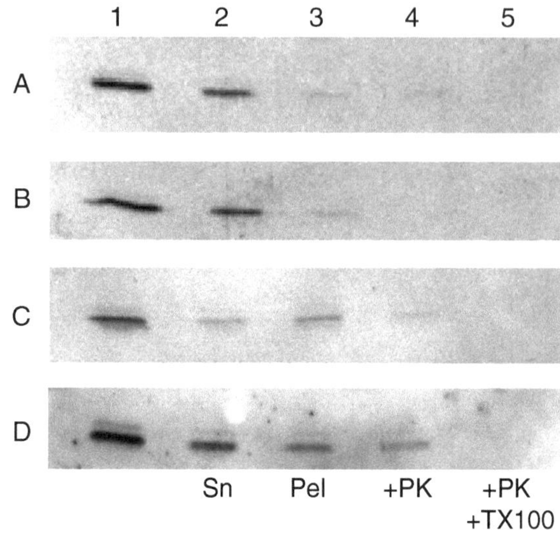

Fig. 3.4. Translocation of M13 procoat H5 (*lanes 1*) into liposomes (**A**, **B**) and YidC-proteoliposomes (**C**, **D**), with (**B**, **D**) or without (**A**, **C**) a membrane potential, respectively. The supernatant (*lane 2*), vesicle pellet (*lane 3*) and the proteinase K-treated vesicles (*lane 4*) are shown. For control, the vesicles opened by detergent were digested by proteinase K (*lane 5*).

2. The multilamellar vesicles are passed through an extruder to generate unilamellar vesicles (*see* **Section 3.7**).
3. To generate the YidC-proteoliposomes, purified YidC protein is added to the lipid solution and is then extruded.
4. The liposomes and YidC-proteoliposomes, respectively, are centrifuged for 10 min at 120,000 *g* at 4°C.
5. The supernatant is discarded and the pellet resuspended in buffer N or to generate a membrane potential in buffer K.
6. For the membrane insertion assay four eppendorf tubes are prepared.
7. To each eppendorf tube containing 210 μL of either buffer N or K depending on the desired membrane potential, 20 μL of liposomes or proteoliposomes (10 μg/μL), respectively, is added at 37°C under gentle shaking.

8. In case of a membrane potential (the sample in buffer K), 0.25 μL of valinomycin is added to the four tubes.
9. Purified M13 procoat protein is diluted 1:1 in denaturation buffer and is placed into four new eppendorf tubes at 37°C.
10. The warm liposomes or proteoliposomes suspension, respectively, are slowly added to the four tubes containing the purified M13 procoat protein and slowly shaken for 30 min at 37°C.
11. After the incubation step, the samples are cooled down on ice to stop the reaction.
12. To distinguish between the bound and the translocated protein, 0.3 mg/mL of proteinase K is added to samples 3 and 4, and incubated for 30 min at 4°C.
13. 4% (v/v) of Triton X-100 is added to sample 4.
14. Sample 2 is centrifuged for 10 min at 120,000 *g* at 4°C.
15. The supernatant of sample 2 is retained in a new tube, whereas the pellet is resuspended in 230 μL of buffer N or buffer K, respectively to the conditions that had been used.
16. The experiment is continued as described in **Section 3.3**.
17. The samples are loaded on a Tricine-SDS-PAGE (*see* step 8 of **Section 3.2.2**).

3.12. Labelling Proteins at a Single Cysteine Residue with Fluorescent Dyes

1. 10 μg of the protein is dissolved in 1 mL labelling buffer in a 1 mL glass tube. A mini stir bar is added (2 mm in diameter) to mix the sample gently. The sample is covered with a layer of nitrogen gas to avoid the oxidation of reduced disulfide bonds.
2. Disulfide bonds are reduced by adding 400 μM TCEP under nitrogen atmosphere. The sample is stirred gently for 15 min at room temperature.
3. The fluorescent dye (*see* step 1 or 2 of **Section 2.12**) is dissolved in DMSO (one crystal to 10 μL DMSO). 1 μL of the solution is diluted into 1 mL 100% MeOH (*see* **Note 13**).
4. The concentration is determined by measuring the absorption from 450 nm to 600 nm in a spectrophotometer against MeOH.

   The concentration is calculated as follows:

$$A_\lambda = \varepsilon_\lambda \, c \, x \tag{1}$$

A_λ = absorption at the absorption maximum of the dye

ε_λ = extinction coefficient ($M^{-1}cm^{-1}$) ($\varepsilon_{\lambda \text{BodipyFL}}$ = 79,000, $\varepsilon_{\lambda \text{Atto520}}$ = 110,000)

c = concentration (Mol/L)

x = length of the cuvette (cm)

5. 50 mM of the fluorescent dye is added (final concentration 500 μM) and stirred for 1 h at 37°C.
6. To remove the unreacted dye that is not bound to the protein, 100 μL Dowex slurry is added to a 1 mL sample of freshly labelled protein for 2 h on the rotating wheel or a magnetic stirrer at room temperature. The free dye is retained in the Dowex beads.
7. To remove the Dowex the sample is centrifuged at low speed (1,000 rpm, 1 min) and the supernatant is transferred into a new glass tube.
8. The labelled proteins are stored at 4°C.
9. To detect the fluorescently labelled proteins, 1 μL of the sample is loaded on a 22% SDS PAGE gel containing urea (*see* **Section 3.2.3**). To get less diffuse bands, the gel is run at 4°C (*see* **Note 14**).
10. The proteins are then visualized with the Typhoon confocal scanner (Laser: 488 nm, emission filter: BP 520 40 CY2, ECL+, Blue FAM, PMT: 600520 nm). See **Fig. 3.5A**.

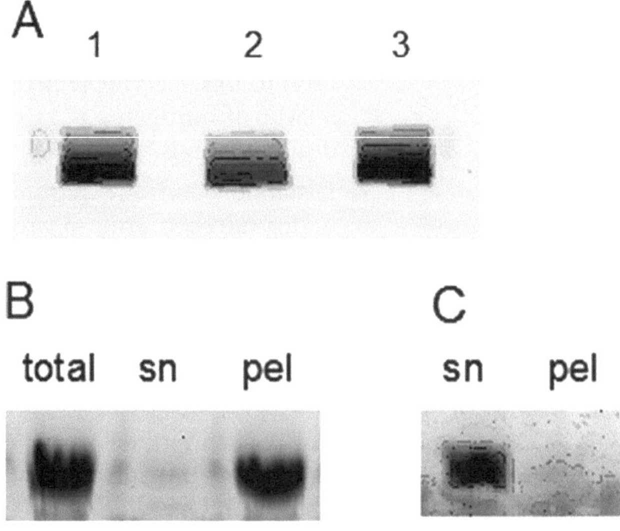

Fig. 3.5. The fluorescently labelled Pf3 coat mutant proteins visualized by the Typhoon scanner. Atto520-maleimide (**A**, **B**) and BodipyFL-maleimide (**C**) were used to label the coat proteins. NC-Pf3 coat protein (**A**, *lane 1*), 3L-NC-Pf3 coat protein (**A**, *lane 2*) and 3L-CC-Pf3 coat protein (**A**, *lane 3*) are shown. (**B**) The Atto-labelled 3L-NC-Pf3 coat protein binds to the liposomes, whereas the Bodipy-labelled 3L-NC-Pf3 does not (**C**). sn: supernatant; pel: pellet.

3.13. Fluorescent Protein Binding to the Membrane

1. Liposomes are prepared as described in **Section 3.7** (**Fig. 3.5B**).
2. The fluorescently labelled protein is added to 200 μg of liposomes in a total volume of 230 μL buffer N and incubated for 30 min at 37°C with mild shaking.
3. The samples are chilled and the liposomes are sedimented at 120,000 *g* at 4°C for 10 min.
4. The supernatant is retained in a new tube, whereas the pellet is resuspended in 230 μL buffer N.
5. The experiment is continued as described in **Section 3.3**.
6. The samples are loaded on a 22% SDS PAGE gel containing urea (*see* step 5 of **Section 3.2.3**). The gel is run at 4°C to get clear bands.
7. The proteins in the supernatant or in the liposome fraction are then visualised with the Typhoon confocal scanner (Laser: 488 nm, emission filter: 520 nm).

4. Notes

1. Purification of the proteins needs to be performed in the cold to avoid proteolysis.
2. The Pf3 coat protein has to be stored in a glass tube because the hydrophobic protein sticks to an eppendorf tube.
3. APS should be added to the gel solutions right before pouring the gel to initiate polymerization.
4. Polymerization of the SDS-Urea-PAGE separating gel is recommended overnight, whereas the stacking gel is polymerized for only 15 min. If polymerization is longer the comb is difficult to be removed.
5. Developing of the silver staining reaction should not last more than 45 min.
6. If the background of a silver stained gel is yellow during developing, you should immediately stop the staining reaction to minimize the background.
7. It is very important to keep the *E. coli* lipids in the dark and under nitrogen atmosphere to avoid oxidation.
8. Chloroform and methanol should be of ultra pure quality for lipid preparation.
9. *E. coli* lipids should not be stored for more than 1 year.

10. The iodine used to stain the lipids should be removed before ninhydrin staining. To remove the iodine the TLC plates are dried using a hair dryer.
11. The resuspension of the liposomes has to be done very carefully to prevent breaking of the vesicles.
12. When the membrane potential is measured using a fluorometer, the room light should be turned off before the measurement chamber containing the sample is opened.
13. For the fluorescence studies it is important that every step is done in the dark and every tube is protected by aluminium foil to avoid bleaching of the dye.
14. Use loading dye without bromophenol blue (BPB) for running a gel with fluorescently labelled samples because the BPB also emits light and is detected as a band.

References

1. Wolfe, P.B., Rice, M. and Wickner, W. (1985) Effects of two sec genes on protein assembly into the plasma membrane of Escherichia coli *J Biol Chem* **260**, 1836–1841.
2. Rohrer, J. and Kuhn, A. (1990) The function of a leader peptide in translocating charged amino acyl residues across a membrane *Science* **250**, 1418–1421.
3. Van Der Laan, M., Bechtluft, P., Kol, S., Nouwen, N. and Driessen, A.J. (2004) F1F0 ATP synthase subunit c is a substrate of the novel YidC pathway for membrane protein biogenesis *J Cell Biol* **165**, 213–222.
4. Facey, S.J., Neugebauer, S.A., Krauss, S. and Kuhn, A. (2007) The mechanosensitive channel protein MscL is targeted by the SRP to the novel YidC membrane insertion pathway of Escherichia coli *J Mol Biol* **365**, 995–1004.
5. Voss, N.R., Gerstein, M., Steitz, T.A. and Moore, P.B. (2006) The geometry of the ribosomal polypeptide exit tunnel *J Mol Biol* **360**, 893–906.
6. Kuhn, A. and Wickner, W. (1985) Conserved residues of the leader peptide are essential for cleavage by leader peptidase *J Biol Chem* **260**, 15914–15918.
7. Kiefer, D. and Kuhn, A. (1999) Hydrophobic forces drive spontaneous membrane insertion of the bacteriophage Pf3 coat protein without topological control *EMBO J* **18**, 6299–6306.
8. Serek, J., Bauer-Manz, G., Struhalla, G., Van Den Berg, L., Kiefer, D., Dalbey, R. et al. (2004) Escherichia coli YidC is a membrane insertase for Sec-independent proteins *EMBO J* **23**, 294–301.
9. Samuelson, J.C., Chen, M., Jiang, F., Moller, I., Wiedmann, M., Kuhn, A. et al. (2000) YidC mediates membrane protein insertion in bacteria *Nature* **406**, 637–641.
10. Studier, F.W., Rosenberg, A.H., Dunn, J.J. and Dubendorff, J.W. (1990) Use of T7 RNA polymerase to direct expression of cloned genes *Methods Enzymol* **185**, 60–89.

Chapter 4

Membrane Protein Insertion in *E. coli*

Jijun Yuan, Ross E. Dalbey, and Andreas Kuhn

Abstract

Integral membrane proteins typically span the lipid bilayer with hydrophobic α helices. These proteins can span the membrane once or multiple times with hydrophilic domains facing both sides of the membrane. In *Escherichia coli*, the insertion of proteins into the membrane is catalyzed by the Sec translocase and the YidC membrane insertase. YidC can function on its own to insert proteins or together with the Sec translocase to facilitate membrane protein insertion. In this chapter, we will describe the construction of a YidC depletion strain that can be used to examine whether YidC is required for membrane protein insertion. We will also present assays for determining whether a region of a membrane protein is inserted across the membrane.

Key words: YidC, protease mapping, AMS gel shift assay, membrane protein insertion.

1. Introduction

In *Escherichia coli*, the majority of proteins are inserted into the cytoplasmic membrane by the action of the SecYEG protein-conducting channel (for review see (1)). In addition to SecYEG, the Sec translocase consists of the membrane-embedded SecD-FYajC and YidC components, and the peripheral protein SecA. SecA is required for membrane insertion of substrates with large periplasmic domains. SecDFYajC functions to make membrane protein insertion more efficient. YidC is critical for insertion of the membrane protein CyoA and subunit a of the F_1F_OATPase but is not needed for insertion of many Sec substrates.

The Sec-independent proteins require the YidC insertase for membrane insertion (for review see (2)). This pathway is evolutionarily conserved with family members in bacteria, mitochondria, and chloroplasts, all catalyzing membrane protein

A. Economou (ed.), *Protein Secretion*, Methods in Molecular Biology 619,
DOI 10.1007/978-1-60327-412-8_4, © Springer Science+Business Media, LLC 2010

insertion (3). In *E. coli*, YidC recognizes and binds to the membrane-embedded transmembrane region of the substrates during insertion. YidC is sufficient for membrane insertion since proteoliposomes containing only YidC were shown to be capable of facilitating the insertion of substrates into the vesicle's membrane (4).

During the last two decades, several approaches have been developed to determine whether a protein component such as YidC is required in the cell for membrane protein insertion. We will describe the use of an arabinose-regulated YidC strain to determine whether YidC is essential for cell growth and is needed for membrane protein assembly. In addition, we will describe several assays including signal peptide processing, protease accessibility, and a protein gel shift method to determine whether the membrane protein inserts across the membrane.

2. Materials

2.1. Construction of YidC Depletion Strain

2.1.1. Construction of JS71 Strain

1. Restriction enzymes
2. PCR cycle: 5 min at 95°C (1 cycle); 1 min at 95°C, 1 min at 60°C, 3 min at 70°C (30 cycle); 15 min at 70°C (1 cycle).
3. A Techne PROGENE thermocycler was used for all PCR reactions.
4. PCRII vector was purchased from Invitrogen.
5. pCD13PKS25 and helper plasmid pPICK were provided by Gregory Phillips.

2.1.2. Construction of JS7131 Strain

1. Chloramphenicol is dissolved in absolute ethanol.
2. pMAK705 was provided by Sidney Kushner.

2.2. In Vivo Expression Levels of YidC in JS7131

1. LB growth media (*see* step 1 of **Section 2.4.**) is supplemented with 0.2% arabinose (final concentration) or 0.2% glucose (final concentration). Stock solutions of arabinose and glucose are prepared at the concentration of 20% (w/v) and filtered by using a steril 0.2 μm filter. The stock is 1:100 back diluted when preparing the growth media.
2. Transfer buffer (10X): 1L of 10X transfer buffer is prepared with 385 mM glycine, 48 mM Tris and 0.37% (w/v) SDS. The buffer is stored at room temperature. For preparing 1X transfer buffer, 100 mL 10X transfer buffer and 200 mL CH_3OH are added to 700 mL H_2O.

3. TBS buffer (10X): 1.5 M NaCl, 100 mM Tris-HCl, 26.8 mM KCl, pH 7.6. Store at room temperature.
4. TBS-T buffer: 1 L of TBS-T buffer is prepared by adding 100 mL 10X TBS buffer and 50 μL Tween-20 to 900 mL H_2O.
5. Blocking buffer: 5–10% non-fat milk in TBS-T buffer.
6. SuperSignal West Pico Chemiluminescent Substrate (CL) (from Pierce).

2.3. Assay for Complementation of JS7131

1. LB agar plates are prepared by adding 15 g of agar to 1L of LB media before autoclaving. Arabinose (0.2% final concentration) or glucose (0.2% final concentration) is added to the LB growth media at 60°C just before pouring the plates.
2. JS7131 is the YidC depletion strain (7).

2.4. Signal Peptide Processing: Membrane Insertion Assay I

1. LB growth media: Prepare by dissolving 25 g LB (EMD Biosciences, Inc) in 1 L ddH_2O and autoclaving immediately.
2. M9 minimal media: Dissolve 11.3 g M9 salts (Sigma) in 1 L ddH_2O, add 50 μg/mL of 19 amino acids (without methionine), 0.5% (w/v) fructose, 1 μg/mL thiamine, and 1 mM $MgSO_4$. Filter immediately by using a steril 0.2 μm filter.
3. Isopropyl β-D-1-thiogalactopyranoside (IPTG) is prepared at concentration of 100 mM as stock.
4. Trans-[^{35}S]-label is a mixture of 85% [^{35}S]-methionine and 15% [^{35}S]-cysteine with 1,000 Ci/mmol.
5. 20% (v/v) trichloroacetic acid (TCA) and acetone are kept on ice before use.
6. Tris-SDS buffer: 10 mM Tris-HCl, pH 8.0, 2% (w/v) sodium dodecyl sulfate (SDS).
7. 2X SDS PAGE loading buffer: 50 mM Tris-HCl, pH 6.8, 2% (w/v) SDS, 0.1% bromophenol blue, 20% (v/v) glycerol. The loading buffer is stored at room temperature. Before being used, 10% (v/v) β-mercaptoethanol is added to the loading buffer.

2.5. Protease-Mapping Study: Insertion Assay II

1. Proteinase K (Invitrogen) stock is stored at 4°C at a concentration of 20 mg/mL.
2. EDTA is prepared at a concentration of 100 mM, pH 8.0; lysozyme (Sigma) is prepared for the spheroplast buffer. Lysozyme (0.5 mg/mL) is 1:100 back diluted in the spheroplast buffer.
3. Spheroplast buffer: 33 mM Tris-HCl, pH 8.2, 40% (w/v) sucrose, 1 mM EDTA, 5 μg/mL lysozyme. Spheroplast buffer is kept at 4°C.

2.6. AMS Gel-Shift: Insertion Assay III

1. Prepare a 200 mM solution of AMS (acetoamido-4′ maleimidylstilbene-2, 2-disulfonic acid disodium salt) from Molecular Probes in DMSO. The solution must be freshly prepared.
2. Prepare a 20 mM solution of DTT (dithiothreitol) from Sigma in M9 minimal medium.

2.7. Immunoprecipitation of Proteins

1. IP buffer: 10 mM Tris-HCl, pH 8.0, 5 mM EDTA, 150 mM NaCl, 2.5% (v/v) Triton X-100. 5X IP buffer can be prepared as stock solution and stored at room temperature.
2. Staph A (Calbiochem).
3. Antibodies: GroEL (Sigma), YidC antiserum, C-terminal YidC antiserum, OmpA and leader peptidase antiserum (our laboratory collection).

2.8. SDS-PAGE (Sodium Dodecyl Sulfate–Polyacrylamide Gel Electrophoresis)

1. 15% SDS-PAGE Separating gel: prepare 3.5 mL separating gel mix by adding 770 μL H_2O, 910 μL of 1.5 M Tris-HCl, pH 8.8, 1.75 mL of 30% acrylamide (Bio Rad), 35 μL of 10% (w/v) SDS, 35 μL of 10% (w/v) ammonium persulfate (APS), 1.5 μL N, N, N′, N′-tetramethylethylenediamine (TEMED). APS initiates the polymerization.
2. SDS-PAGE Stacking gel: prepare 2.6 mL of stacking gel mix by adding 1.55 mL H_2O, 650 μL of 0.5 M Tris-HCl, pH 6.8, 350 μL of 30% acrylamide (Bio Rad), 25 μL of 10% (w/v) SDS, 25 μL of 10% (w/v) APS, 2.5 μL TEMED.
3. SDS-PAGE running buffer: 25 mM Tris, 192 mM glycine, 0.1% (w/v) SDS. 10X SDS-PAGE running buffer stock can be prepared and stored at room temperature.
4. Fixing buffer: 50% (v/v) CH_3OH and 10% (v/v) acetic acid. The buffer is stored at room temperature.
5. Gel dryer: Bio-Rad Model 583 Gel Dryer.
6. Filter paper: Fisher Pure Cellulose Chromatography Paper (0.35 mm thick).
7. Phosphorimager screen (Molecular Dynamics).

3. Methods

Integral membrane proteins span the membrane once or multiple times with at least one hydrophilic domain exported to the non-cytoplasmic side of the membrane. To test whether the proteins insert by the Sec/YidC pathway or the YidC pathway, SecE,

SecDF, and YidC depletion strains are available (5–7). These strains have the Sec or YidC component under the control of the *araBAD* promoter and have the endogenous *sec/yidC* genes knocked out. The strains are arabinose-dependent for growth. **Figure 4.1** and **Section 3.1** describe the construction of the YidC depletion strain JS7131. When the depletion strains are grown in LB media supplemented with glucose the expression of the *araBAD*-promoter controlled components are tightly repressed and the proteins are depleted during further growth. When using these arabinose-dependent strains, it is important to show that it grows on LB plates with arabinose but not with glucose. Also, it should be verified by Western blot that the arabinose-dependent

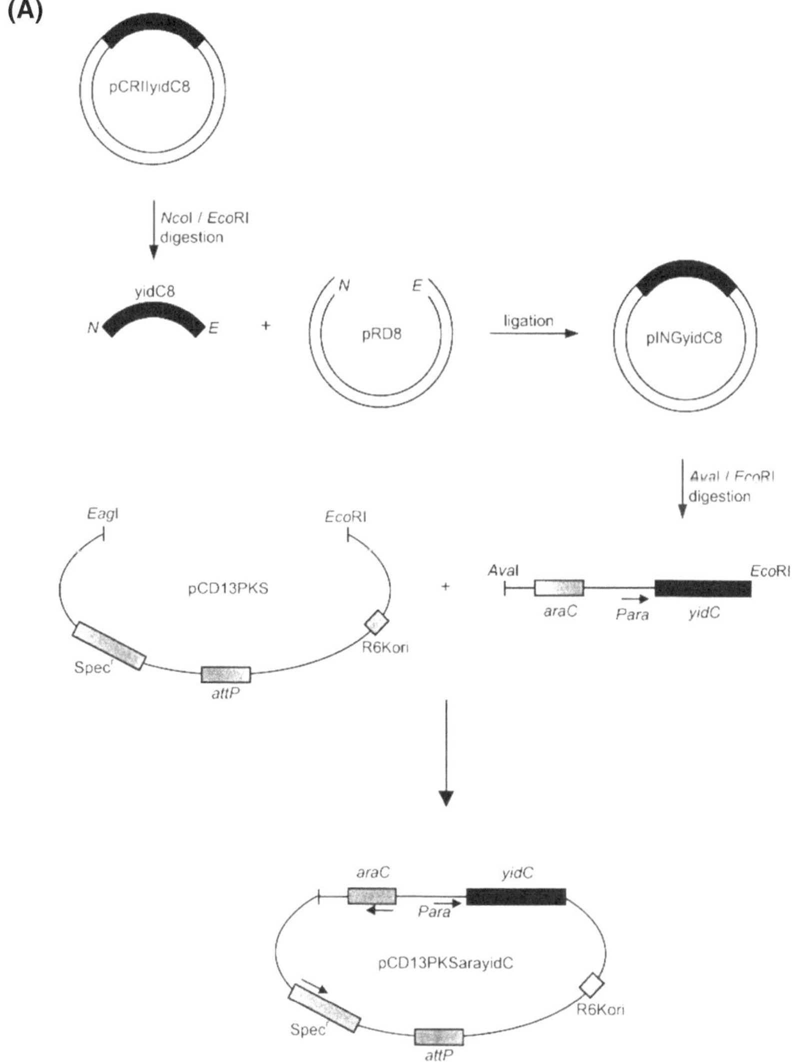

Fig. 4.1. (**A**) Construction of vector pCD13PKSara-yidC. (**B**) Construction of YidC-depletion strain JS7131.

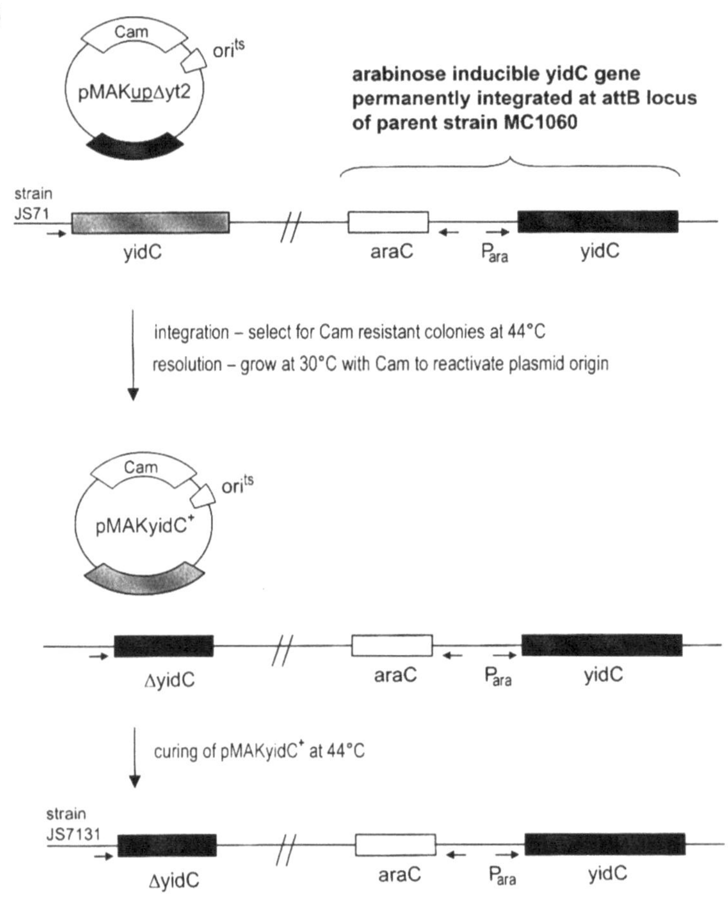

Fig. 4.1. (continued)

component is depleted under glucose conditions (**Fig. 4.2**). In addition to the Sec or YidC depletion strains, there are also some temperature and cold-senstive strains available (for review see (8)).

In *E. coli*, several assays are commonly used to monitor the membrane protein insertion. The most widely used is the protease accessibility method where protease digestion of a periplasmic loop is employed (9). This method requires first converting *E. coli* to spheroplasts. It allows the externally added proteinase K

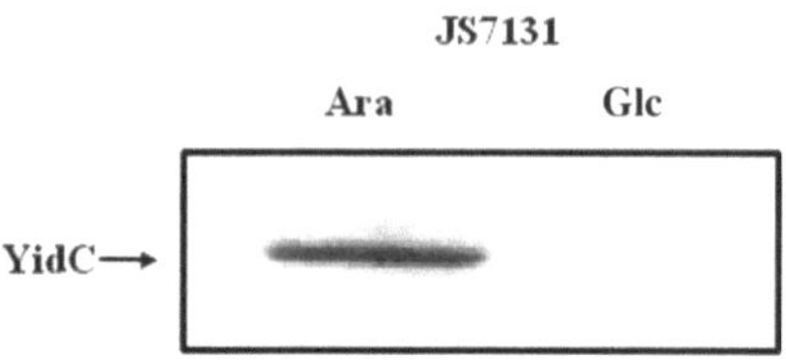

Fig. 4.2. YidC depletion. Strain JS7131 grown in LB with 0.2% arabinose (*left lane*) or 0.2% glucose (*right lane*) was analyzed for YidC by Western blotting.

to gain access to the outer surface of the inner membrane. A positive control is OmpA (an outer membrane protein marker) which is digested only in spheroplasts and not in intact cells. Digestion gives an indication of the efficiency of spheroplasts formation. GroEL (a cytoplasmic protein) is used as a negative control for lysis. We show in **Fig. 4.3** using the JS7131 YidC depletion strain that subunit a (with a Lep P2 domain at the C-teminus) is inserted into the membrane when YidC is present (ara), but not when YidC is depleted (glc). Another powerful method, which can be used if the membrane protein is synthesized with a cleavable signal peptide, is to examine signal peptide processing (10). Signal peptide processing occurs only when the signal peptide cleavage region of the preprotein is inserted across the membrane. The M13 procoat protein and CyoA, which are made in a precursor form, require YidC for membrane protein insertion and signal peptide processing (7, 10, 11). A third method to examine the translocation of a periplasmic loop across the membrane is to add

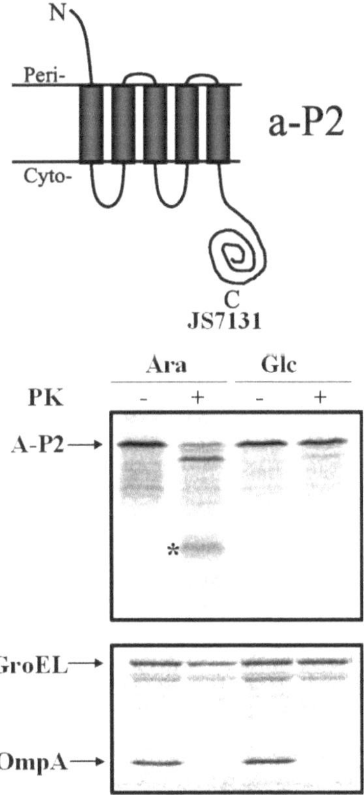

Fig. 4.3. F_1F_0ATPase subunit a-P2 protease mapping study. JS7131 containing a plasmid encoding the F_1F_0ATPase subunit a-P2 was grown in minimal M9 medium containing 0.2% arabinose (*lanes 1, 2*) or 0.2% glucose (*lanes 3, 4*) and labeled with [^{35}S]-methionine. The proteinase-treated samples (*lanes 2, 4*) show that YidC is required for the insertion of subunit a-P2. For controls, GroEL and OmpA were analyzed (*lower panel*).

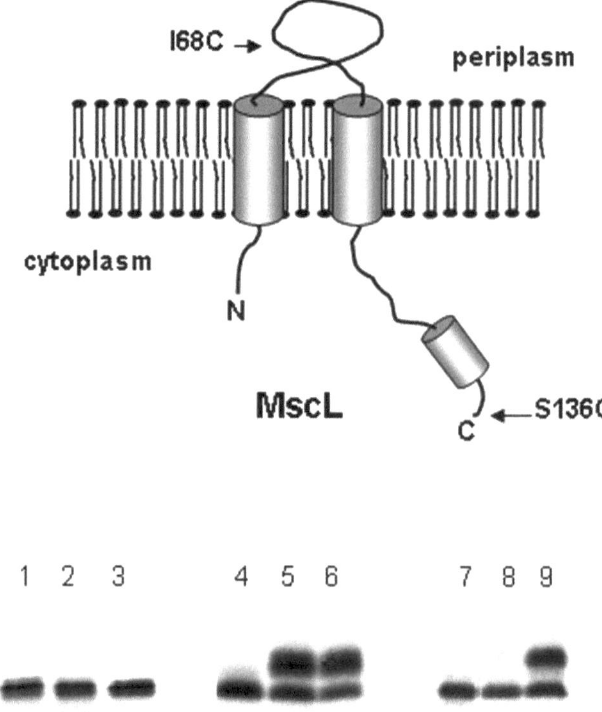

Fig. 4.4. MscL mapping study by AMS modification. JS7131 containing a plasmid encoding the MscL protein was grown in minimal M9 medium and labeled with [^{35}S]-methionine. Wild-type (*lanes 1–3*) and MscL mutants with a cysteine at position 68 (*lanes 4–6*) and 136 (*lanes 7–9*) were analyzed. The untreated samples (1,4,7), AMS treated samples (2,5,8) and the AMS-treated samples after disrupting the cells are shown.

a maleimide, which is membrane impermeable, and test whether it can modify a single cysteine located in the translocated loop. The modification of the cysteine with the maleimide results in an increase of the molecular weight leading to a shift of the protein band on the gel. **Figure 4.4** shows that AMS modification was used to monitor the translocation of the MscL periplasmic loop containing a cysteine at position 68. In contrast, the cytoplasmic cysteine at position 136 was only modified when the cell membrane was disrupted by sonication. The cysteine at position 68 was not modified in the absence of YidC, Ffh, and FtsY (12).

3.1. Construction of YidC Depletion Strain

3.1.1. Construction of JS71 Strain

1. The *yidC* gene was amplified from *E. coli* strain MC1060, by the polymerase chain reaction (PCR) with *Nco*I/ *Eco*RI site at the ends.
2. The 1755 base pair fragment was subcloned into a PCRII vector by TA cloning.
3. After removing the *Nco*I site within the *yidC* gene by site-directed mutagenesis, the *yidC* gene was subcloned into

pRD8 vector using *Nco*I and *Eco*RI (**Fig. 4.1A**). This new construct was named pINGyidC8 and the *yidC* gene was under the control of *araBAD* operator/promoter.

4. The *ara-yidC* 4-kb fragment of pINGyidC8 was subcloned into pCD13PKS25(Spec^r) in an orientation opposite to that of all other transcription within the vector.
5. The resulting clone pCD13PKSara-yidC (**Fig. 4.1A**) was permanently integrated into the *attB* site of strain MC1060 (*Δ(codB-lac)3, galK16, galE15,⁻, relA1, rpsL150, spoT1, hsdR2, ara*$^+$) with the helper plasmid pPICK.
6. This created strain JS71 (MC1060, *attB::R6Kori, ParaBAD⁻yidC*$^+$, (Spec^r)).

3.1.2. Construction of the YidC-Depletion JS7131 Strain

1. YidC was cloned into a replication temperature-sensitive plasmid, pMAK705 (Cam^r).
2. The base pairs 745-1566 of *yidC* gene were deleted by an in frame oligo-directed deletion. The new *YidC* knock out vector was called pMAKupΔyt2 (**Fig. 4.1B**).
3. pMAKupΔyt2 was transformed into strain JS71 and chromosomal integrants were isolated by plating on LB agar at 44°C in the presence of 20 μg/mL chloramphenicol.
4. Individual integrants were grown for several generations at 30°C in LB media with arabinose and chloramphenicol to reactivate the plasmid origin. Upon allelic exchange at the *yidC* locus, the knockout vector would carry *yidC*$^+$and the strain of interest would carry *ΔyidC*. This strain is arabinose-dependent for growth at the non-permissive temperature (**Fig. 4.1B**).
5. This strain was selected by streaking at 44°C on 0.2% arabinose-LB plates versus 0.2% glucose-LB plates.
6. The final step of the strain construction was to eliminate pMAKyidC$^+$ by a curing process combining cycloserine enrichment with growth at 44°C.
7. This creates plasmid-free JS7131 strain (JS71, Δ*yidC*) (**Fig. 4.1B**). This strain is a YidC depletion strain, which is dependent on arabinose for growth.

3.2. In Vivo Expression Levels of YidC in JS7131

1. Overnight JS7131 culture is washed with LB media and 1:50 back diluted into 1 mL LB media supplemented with 0.2% glucose or 0.2% arabinose (*see* **Note 1**).
2. JS7131 cells are grown to the mid-log phase in LB media for 3 h (*see* **Note 2**).
3. The cells are then spun down and resuspended in 100 μL 2X SDS-PAGE loading buffer.

4. The samples are placed in heating block at 95°C for 5 min (*see* **Note 3**).
5. Samples are analyzed on a 15% SDS-PAGE (*see* **Section 3.8**) The gel is run at 200 V for 1.5 h.
6. After electrophoresis, the gel is placed into transfer buffer for 5 min.
7. A transfer system is set up where the gel is put into a cassette next to a nitrocellulose membrane which is the same size. There is also one sheet of foam and two sheets of Whatman 3MM paper on each side (*see* **Note 4**). The cassette is placed into the tank such that the nitrocellulose membrane is between the gel and the anode. The membrane transfer is accomplished at 30 mA for 2 h.
8. The nitrocellulose membrane is removed and placed in 50 mL blocking buffer for 1 h at room temperature or overnight in cold room (*see* **Note 5**). The nitrocellulose membrane is washed with TBS-T buffer 3 times (25 min, 5 min, 5 min).
9. The nitrocellulose membrane is then incubated with 10 mL of TBS-T buffer with 1:10,000 diluted anti-YidC antiserum at room temperature for 1 h. The nitrocellulose membrane is washed 3 times in TBS-T buffer, as in step 8.
10. The nitrocellulose membrane is incubated in 10 mL of TBS-T buffer with 1:10,000 diluted secondary antibody at room temperature for 1 h. The nitrocellulose membrane is washed 3 times (as described in step 8).
11. 1 mL of mixed SuperSignal West Pico Chemiluminescent Substrate (CL) (from Pierce) is added to the surface of nitrocellulose membrane and incubated for 1 min. The CL reagent on the membrane is then removed by Kim Wipes and the nitrocellulose membrane is covered by plastic wrap.
12. The membrane is placed in a cassette and exposed to X-ray film for a suitably short period of time.

3.3. Assay for Complementation of JS7131

The activity of YidC mutants can be tested by investigating whether they restore growth to the YidC-depletion JS7131 strain grown under glucose condition.

1. To test the functionality of a YidC mutant, a 1 mL culture of JS7131 strain bearing the plasmid encoding the mutant YidC is grown for overnight in LB media supplemented with 0.2% arabinose.
2. The overnight culture is washed with LB media (1 mL) and 1:50 back diluted into new LB media supplemented with 0.2% glucose.

3. Cells are grown for 1 h in glucose-supplemented media to deplete the chromosomal YidC and streaked on LB agar plates containing 0.2% arabinose or 0.2% glucose.
4. LB agar plates are incubated at 37°C overnight.

3.4. Signal Peptidase Processing: Insertion Assay I

1. The overnight culture grown in LB containing 0.2% arabinose is pelleted and is washed with LB media (1 mL) and diluted 1:50 into LB media (1 mL).
2. Cells are grown to the mid-log phase for 2.5 h, spun down and resuspended in M9 minimal media (1 mL).
3. Cell culture (600 μL) is transferred to disposable plastic tube and grown for an additional 30 min.
4. Expression of the plasmid-encoded membrane protein is induced for 5 min by adding 6 μL of 100 mM IPTG (1 mM final concentration).
5. Cell culture is treated with trans-[^{35}S]-label (6 μL) for 1 min to label the newly synthesized proteins. Radiolabeling is stopped by placing the culture tube on ice (*see* **Note 6**).
6. Add an equal volume of ice-cold 20% (v/v) TCA and incubate on ice for 1 h. The sample is then centrifuged at 15,000 *g* for 10 min. The supernatant is removed with a syringe and transferred to the radioactive waste (*see* **Note 7**). The pellet is washed with ice-cold acetone (0.6 mL). The sample is vortexed, centrifuged at 15,000 *g* for 5 min, and the supernatant is transferred to the radioactive waste.
7. To dry the sample it is left on the bench for 15 min or in heating block at 95°C for 5 min (*see* **Note 8**).
8. The dried pellet is dissolved in 100 μL of Tris-SDS buffer.
9. For further analysis the sample can be processed for immunoprecipitation (*see* **Section 3.7**) and SDS-PAGE analysis (*see* **Section 3.8**).

3.5. Protease-Mapping Study: Insertion Assay II

1. Cell cultures are prepared (*see* steps 1 and 2 of **Section 3.4.1**) and the plasmid-encoded proteins expressed (*see* step 4 of **Section 3.4**) and radiolabeled (*see* step 5 of **Section 3.4**) as described in **Section 3.4**.
2. The [^{35}S]-labeled culture is carefully transferred to an eppendorf tube to pellet the cells for preparation of spheroplasts (*see* **Note 9**). The cells are resuspended in 0.60 mL of ice-cold spheroplast buffer and incubated on ice for 15 min (*see* **Note 10**).
3. The spheroplasts are aliquoted into two tubes with or without 0.4 mg/mL proteinase K and incubated on ice for 1 h.

4. The samples are precipitated with TCA, acetone washed, and resuspended in Tris-SDS buffer (*see* steps 6–8 of **Section 3.4**)
5. The samples are then analyzed by immunoprecipitation (*see* **Section 3.7**) and subjected to 15% SDS-PAGE and phosphorimaging (*see* **Section 3.8**).

3.6. AMS Gel-Shift: Insertion Assay III

1. Cell cultures are prepared (*see* steps 1 and 2 of **Section 3.4**) and the plasmid-encoded proteins expressed (*see* step 4 of **Section 3.4**) and [^{35}S]-labeled (*see* step 5 of **Section 3.4**) as described in **Section 3.4**.
2. The [^{35}S]-labeled material is portioned (2 × 250 μL) and transferred into two eppendorf tubes on ice.
3. To one tube add 3.2 μL of 200 mM AMS (final concentration 2.5 mM) and carefully mix. The other sample is acid-precipitated as described (*see* step 6 of **Section 3.4**).
4. Incubate on ice for 20 min and stop the AMS reaction by the addition of 250 μL of 20 mM DTT in M9 medium.
5. Incubate sample for 10 min on ice, acid-precipitate, and treat both samples as described (*see* steps 6–8 of **Section 3.4**).
6. The samples are then analyzed by immunoprecipitation (*see* **Section 3.7**) and subjected to SDS-PAGE (*see* **Note 11**) and phosphorimaging (*see* **Section 3.8**).

3.7. Immunoprecipitation of Proteins

1. The Tris-SDS solubilized samples (100 μL) prepared in **Sections 3.4**, **3.5**, and **3.6** for the membrane protein insertion assays are incubated with IP buffer (900 μL).
2. Staph A (30 μL) is added to the samples to presorb any radioactive proteins that bind to Staph A non-specifically (*see* **Note 12**). The incubation is for 15 min on ice. The samples are centrifuged for 30 s at 15,000 *g* to pellet Staph A and the supernatant is transferred to a new tube.
3. The supernatant can be aliquoted into two or more tubes (*see* **Note 13**). One tube is incubated on ice for 1 h with antibody against the protein being tested for membrane insertion. The other tube is incubated with antibody against GroEL (a cytoplasmic marker) and outer membrane protein A (an outer membrane marker).
4. Staph A (30 μL) is added to each tube and incubated on ice for another hour. The sample is centrifuged at 15,000 *g* for 30 s to pellet the Staph A antibody protein complex and the supernatant is removed. 1 mL of IP buffer is added and the pellet is resuspended by vortexing.
5. Repeat the wash and the centrifugation in step 4 twice.

6. After removing the IP buffer, 50 μL of 2X SDS loading buffer is added to the pellet and the sample is placed on heating block for 5 min. The sample is centrifuged at 15,000 *g* for 1 min to pellet the Staph A, prior to loading on a SDS-PAGE gel. Typically, 15 μL of sample is loaded onto the gel.

3.8. SDS-PAGE Analysis

1. This protocol is for the Biorad PAGE system
2. Clean the glass plates with 70% ethanol and allow the plates to air dry for 10 min.
3. Prepare 3.5 mL SDS-PAGE separating gel and mix (*see* **Section 2.8.1**). Pour 3.2 mL of the gel mix and carefully overlay with isopropanol. The gel should polymerize within 30 min.
4. The isopropanol is removed after the polymerization.
5. Prepare 2.6 mL SDS-PAGE stacking gel and mix (*see* **Section 2.8.2**). Pour 2.3 mL of the gel mix to the top and quickly insert the comb. The stacking gel will polymerize within 30 min.
6. Once the stacking gel has polymerized, remove the comb and rinse the wells with SDS-PAGE running buffer.
7. Complete the assembly of the PAGE unit and fill with SDS-PAGE running buffer.
8. Load the sample (15 μL) on the gel. The SDS-PAGE is run at 200 V for the appropriate time.
9. After running the gel, it is prepared for phosphorimaging. The gel is soaked in fixing buffer for 10 min. The gel is transferred to a piece of filter paper and dried at 70°C in the gel dryer for 1 h. The dried gel is transferred to a film exposure cassette, covered by a phosphorimaging screen (Molecular Dynamics) and exposed for at least 3 h.
10. The phosphorimaging screen is scanned by a phosphorimager (e.g., Typhoon Imager, GE Healthcare).

4. Notes

1. When the overnight culture is back diluted into LB media, the new media should be pre-warmed before back dilution. Cold LB will affect the growth of the cells.
2. The YidC depletion strain, JS7131, can have a growth defect in LB media containing 0.2% glucose. Therefore, it is critical for the immunoblotting experiment that the amount of cells are adjusted by checking $OD_{600}$of the cultures containing arabinose and glucose.

3. It is important to heat the sample at 95°C for 5 min to solubilize the protein before loading onto the SDS gel.
4. For setting up a Western blot it is essential to check the orientation of the nitrocellulose membrane during the membrane transfer. Wrong orientation will cause the protein to be transferred into the buffer instead of onto the nitrocellulose membrane.
5. Overnight incubation in blocking buffer in the cold room helps to eliminate the background in a Western blot experiment.
6. Be aware that the half life for [^{35}S] is 87.4 days and 1 month old trans-[^{35}S] will loose roughly 30% of the signal.
7. In the sample preparation TCA needs to be carefully removed by washing with ice-cold acetone. Residual TCA in the sample can cause the pH to change in the gel loading buffer and this can affect the gel electrophoresis. Traces of acetone remaining after the aspiration step can be removed by leaving the sample at room temperature for 15 min.
8. The pellet after TCA precipitation and acetone washing steps may be difficult to dissolve in the Tris-SDS buffer. To dissolve the sample, it is best to leave the sample in the Tris-SDS at room temperature for overnight.
9. When transferring spheroplasts to a new tube, the top of the pipetman tip is cut to minimize damaging the spheroplasts. This manipulation does affect the pipetted volume but only minimally.
10. When preparing the spheroplasts, never vortex the sample after adding lysozyme. Spheroplasts are very fragile and easy to break by shaking. Freshly prepared lysozyme is important for making a good preparation of spheroplasts.
11. To resolve the AMS-modified proteins from the non-modified proteins by PAGE a high-resolution gel is required that is 40 cm long.
12. In the immunoprecipitation procedure, Staph A is first added before the addition of antibody in order to minimize the radioactive proteins that bind non-specifically to Staph A. This step also removes any radioactive debris that was not solubilized in the Tris-SDS solution.
13. During proteinase K mapping, GroEL (cytoplasmic control) and OmpA (outer membrane protein A) are typically used as spheroplasts control. The immunoprecipitation of GroEL and OmpA can both be performed at the same time in the same tube.

Acknowledgments

We thank Dr. Sandra Facey for providing Figure 4.4. This work was supported by National Institute of Health Grant (GM63862-05) to R. E. D. and the DFG grant Ku749/6-1 to A. K.

References

1. Xie, K. and Dalbey, R.E. (2008) Inserting proteins into the bacterial cytoplasmic membrane using the Sec and YidC translocases *Nat Rev Microbiol* **6**, 234–244.
2. Kiefer, D. and Kuhn, A. (2007) YidC as an essential and multifunctional component in membrane protein assembly *Int Rev Cytol* **259**, 113–138.
3. Luirink, J., Samuelsson, T. and de Gier, J.W. (2001) YidC/Oxa1p/Alb3: evolutionarily conserved mediators of membrane protein assembly *FEBS Lett* **501**, 1–5.
4. Serek, J., Bauer-Manz, G., Struhalla, G., Van Den Berg, L., Kiefer, D., Dalbey, R. et al. (2004) Escherichia coli YidC is a membrane insertase for Sec-independent proteins *EMBO J* **23**, 294–301.
5. Traxler, B. and Murphy, C. (1996) Insertion of the polytopic membrane protein MalF is dependent on the bacterial secretion machinery *J Biol Chem* **271**, 12394–12400.
6. Pogliano, J.A. and Beckwith, J. (1994) SecD and SecF facilitate protein export in Escherichia coli *EMBO J* **13**, 554–561.
7. Samuelson, J.C., Chen, M., Jiang, F., Moller, I., Wiedmann, M., Kuhn, A. et al. (2000) YidC mediates membrane protein insertion in bacteria *Nature* **406**, 637–641.
8. Dalbey, R., E., Chen, M., and Wiedmann, M. (2002) Methods in Protein Targeting, Translocation and Transport *In Protein Targeting, Translocation and Transport* (Dalbey, R. E. and Von Heijne, G., ed.), AP Press, London, UK, pp. 5–34.
9. Dalbey, R.E. and Wickner, W. (1986) The role of the polar, carboxyl-terminal domain of Escherichia coli leader peptidase in its translocation across the plasma membrane *J Biol Chem* **261**, 13844–13849.
10. Celebi, N., Yi, L., Facey, S.J., Kuhn, A. and Dalbey, R.E. (2006) Membrane biogenesis of subunit II of cytochrome bo oxidase: contrasting requirements for insertion of N-terminal and C-terminal domains *J Mol Biol* **357**, 1428–1436.
11. van Bloois, E., Haan, G.J., de Gier, J.W., Oudega, B. and Luirink, J. (2006) Distinct requirements for translocation of the N-tail and C-tail of the Escherichia coli inner membrane protein CyoA *J Biol Chem* **281**, 10002–10009.
12. Facey, S.J., Neugebauer, S.A., Krauss, S. and Kuhn, A. (2007) The mechanosensitive channel protein MscL is targeted by the SRP to the novel YidC membrane insertion pathway of Escherichia coli *J Mol Biol* **365**, 995–1004.

Chapter 5

Study of Polytopic Membrane Protein Topological Organization as a Function of Membrane Lipid Composition

Mikhail Bogdanov, Philip N. Heacock, and William Dowhan

Abstract

A protocol is described using lipid mutants and thiol-specific chemical reagents to study lipid-dependent and host-specific membrane protein topogenesis by the substituted-cysteine accessibility method as applied to transmembrane domains (SCAM^{TM}). SCAM^{TM} is adapted to follow changes in membrane protein topology as a function of changes in membrane lipid composition. The strategy described can be adapted to any membrane system.

Key words: Membrane protein, topology, lipid-dependent topogenesis, phospholipids, lactose permease, SCAM^{TM}.

1. Introduction

Membrane proteins represent at least 30% of the all currently sequenced genomes and represent more than half the drug targets pursued by pharmaceutical companies (1). Effective drug design is dependent on understanding membrane protein structure and the rules that govern the folding and assembly of native and mutant membrane proteins. A fundamental aspect of the structure of polytopic membrane proteins is membrane protein topo logy. Membrane protein topology describes the way a polypeptide chain is arranged in the membrane, i.e. the number of transmembrane domains (TMs) and their orientation in the membrane. Final protein topology is determined by topogenic signals in the nascent polypeptide chain that are recognized and decoded not

A. Economou (ed.), *Protein Secretion*, Methods in Molecular Biology 619,
DOI 10.1007/978-1-60327-412-8_5,

only by the protein insertion machinery (2) but by the lipid profile (3–9).

Although the methodology for obtaining high-resolution structures for membrane proteins is improving, the need to determine low-resolution organizational information on membrane proteins in a native membrane continues. Given the enormous number of sequences that are available in genome-sequencing projects, it is not realistic to assume that the structures of all the encoded proteins will be generated by crystallographic approaches, especially for membrane proteins. Moreover purification, crystallization, and structure determination of membrane proteins still remain a challenge. Crystal structures are static and may be distorted due to purification and crystallization constraints. Information on interactions with other proteins and the lipid environment are also lost during purification. During protein purification, molecular interactions with lipids are replaced by non-native detergent interactions. Heterologous expression in a host strain with a different lipid composition than the native host can also result in loss of proper lipid–protein interactions, which can affect topological organization and function. Therefore, non-crystallographic approaches have been developed to determine lower resolution topological arrangement of membrane spanning segments in whole proteins as a function of membrane lipid composition (3, 6, 8).

Dynamic aspects of protein structure as a function of the physiological state of the cell is best probed in whole cells or membranes. *Escherichia coli* strains genetically altered in their lipid composition (10) (**Fig. 5.1**) and thiol-specific chemical reagents have been developed to study lipid-dependent and host-specific membrane protein topogenesis by the substituted-cysteine accessibility method as applied to TMs (SCAM^{TM}) (3, 5–8). In this approach cysteine residues replace individual amino acids that reside in the putative extracellular or intracellular loops connecting TMs of a membrane protein expressed in a strain in which lipid composition can be changed either before (3, 8) or after (3, 5, 9) membrane protein synthesis and assembly. Combining of these techniques provides a system in which to study the role of lipid–protein interactions in determining the structure,

Fig. 5.1. (continued) © 2009 The American Society for Biochemistry and Molecular Biology. *Lane 6* was reproduced from (31) © 2004 The American Society for Microbiology. (**b**) Phospholipid composition as a function of *pssA* gene induction. Strain AT2033 was grown first in the absence of aTc (time 0) followed by growth in the presence of aTc for 3 h. When grown without aTc, AT2033 has a greatly reduced level of PE and contains elevated levels of PG and CL. Growth in the presence of aTc results a progressive increase in PE to normal levels (5). Lipid composition is shown as mole% of the total major phospholipid species. Figure was reproduced from (5) © Bogdanov et al., 2008 J. Cell Biol. doi:10.1083/jcb.200803097 originally published as Fig. S1.

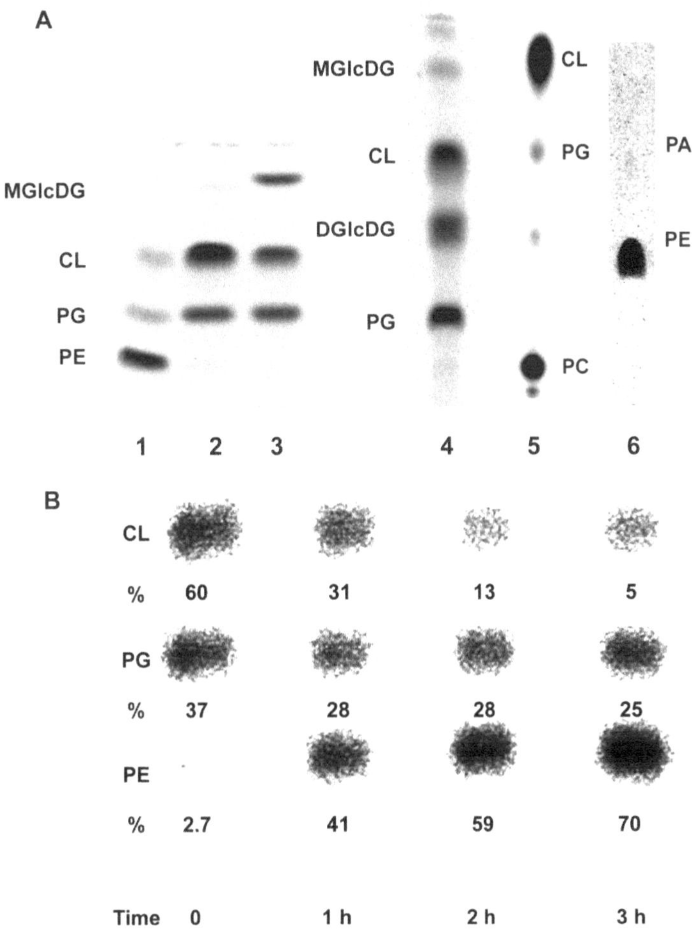

Fig. 5.1. Lipid profiles of *E. coli* mutants with altered lipid compositions. (**a**) *Lane 1*: AL95 (*pss93*::*kanR lacY*::*Tn9*)/pDD72 GM (*pssA*$^{+}$ *gm*^{R}) has wild-type phospholipid composition (80 mole % PE and 20 mole % PG plus CL) due to complementation by a plasmid (pDD72) copy of the null allele of the *pssA* gene that encodes the committed step to PE biosynthesis (3). *Lane 2*: AL95 (*pss93*::*kanR lacY*::*Tn9*) is PE-lacking due to the a null allele of the *pssA* gene and contains mainly CL and PG (3). *Lane 3*: Introduction of plasmid pTMG3 (*ALmgs amp*^{R}) into strain AL95 results in 35 mole% monoglucosyldiacylglycerol (MGlcDAG) due to the expression of the *Acholeplasma laidlawii* MGlcDAG synthase gene. The remaining lipids are primarily PG (35 mole%) and CL (25 mole%) (8). *Lane 4*: Introduction into AD931 (*lac Y*328^{am} *pss93::kan*^{R} *recAsrl*::Tn10) of the genes from *A. laidlawii* that synthesize MGlcDAG and diglucosyldiacylglycerol (DGlcDAG) results in about 30–40 mole% DGlcDAG with less than 1 mole% MGlcDAG (18). *Lane 5*: Introduction of the *pcs* gene (placed under OP_{araB} regulatory control) from *Legionella pneumophila* (30) that confers the ability to synthesize PC results in about 70 mole% PC with the remainder being PG (2.5 mole %) plus CL (26 mole %) and other minor lipids (P. N. Heacock and W. Dowhan, unpublished). *Lane 6*: UE54 (*pgsA*::FRT-*kan*-FRT *lpp-2* $\triangle$*ara714 rcsF*::mini-*Tn10cam*) carries a null allele of the *pgsA* gene encoding the committed step to PG and CL biosynthesis making it devoid of PG and CL and containing about 90 mole% PE, 4.0 mole% phosphatidic acid (PA) and 3.2% CDP-diacylglycerol. Figure and compositional results are taken from (31). *Lanes 1–3* were reproduced from (8) © 2006 The American Society for Biochemistry and Molecular Biology. *Lane 4* was reproduced from (18)

assembly, and function of membrane proteins. By combining SCAM™ with mutants of *E. coli* in which membrane phospholipid composition can be systematically controlled, the role of phospholipids as determinants of membrane protein topological organization was established (3, 8, 9). In addition, the ability to change lipid composition post-assembly of a membrane protein demonstrated the potential for polytopic membrane proteins to change their topological organization after insertion and assembly in the membrane (3, 5, 9).

The following protocol describes (1) preparation of derivatives of the target protein, (2) description and use of hosts with varied lipid composition, and (3) use of SCAM™ to map the topological organization of the target protein.

2. Materials

2.1. Transformation Protocol

1. 10X RbCl/$CaCl_2$ Transformation Salts for the preparation of transformation competent bacterial cells (MP Biomedicals, USA).
2. Centrifuge 5417R (Eppendorf).

2.2. Growth of E. coli Strains

1. Luria–Bertani (LB) medium.
2. 2.5 M $MgCl_2$.
3. Ampicillin (amp) (100 mg/ml).
4. 500 mM isopropyl-ß-D-thiogalactoside (IPTG).
5. Anhydrotetracycline (aTc) (Spectrum) (10 mg/ml).

2.3. SCAM™

1. Buffer A: 100 mM HEPES/KOH buffer, 250 mM sucrose, 25 mM $MgCl_2$, 0.1 mM KCl, adjusted to pH 7 or 10.5.
2. 10 mM 3-(*N*-maleimidylpropionyl) biocytin (MPB) (Invitrogen-Molecular Probes) freshly dissolved in dimethyl sulfoxide (DMSO).
3. 2 M ß-mercaptoethanol.
4. Digital Sonifier (Branson, USA).
5. Beckman Coulter TLA-100 ultracentrifuge equipped with TLA-55 rotor.
6. Microfuge Polyallomer Tubes (natural tint, capacity 1.5 ml) (Beckman Coulter).
7. Centrifuge 5417R (Eppendorf).

2.4. Membrane Protein Solubilization Buffer

1. Solubilization buffer: 50 mM Tris-HCl buffer (pH 8.1), 2% SDS, 1 mM EDTA.
2. Fisher Vortex Genie 2 equipped with microtube foam rack for multiple polyallomer tubes (Fisher Scientific).

2.5. SDS-PAGE

1. 2X SDS gel loading (sample) buffer: 10 mM Tris-HCl (pH 6.8), 5.6% SDS, 200 mM dithiothreitol, 10% glycerol, 0.01% bromophenol blue.
2. READY GEL: Tris-HCl precast gel for SDS Polyacrylamide Electrophoresis, 12.5% polyacrylamide (BIO-RAD Laboratories).

2.6. Immunoprecipitation (IP) Buffer

1. 50 mM Tris-HCl (pH 8.1), 0.15 M NaCl, 2% Lubrol-PX (Nacalai Tesque, Japan), 0.4% SDS, 1 mM EDTA (*see* **Note 1**).

2.7. Immunoblot Analysis (Western Blotting)

1. 0.45 μm Protran BA 85 Nitrocellulose transfer membranes (Whatman-Schlicher and Schuell)
2. Blocking buffer: 5% bovine serum albumin (BSA) (Fraction V, heat-shock treated) (Fisher Scientific) in Tris Buffered Saline (TBS) (10 mM Tris-HCl, pH 7.4, 0.9 % NaCl).
3. ImmunoPure^{R} Avidin Horseradish Peroxidase (Avidin-HRP): Avidin linked to horse radish peroxidase (reconstituted to concentration of 2 mg/ml according manufacture suggestion) (Thermo Scientific).
4. SuperSignal West Pico chemiluminescent substrates for detection of HRP (Thermo Scientific).
5. Labconco Semi-Dry blotting system (W.E.P, Company Seattle, Washington, USA).
6. Anode buffer No.1: 0.3 M Tris (pH 10.4) in 10% methanol.
7. Anode buffer No. 2: 25 mM Tris (pH 10.4) in 10% methanol.
8. Cathode buffer: 25 mM Tris, 40 mM glycine (pH 9.4) in 20% methanol.
9. Wash Buffer: 10 mM Tris-HCl (pH 7.4), 0.9 % NaCl containing, 0.05% Nonidet P40 (Igepal TM CA-630) (USB Corporation, Cleveland, OH, USA).
10. Chromatography Paper: 3 mm Cr (Whatman).

2.8. Alkali Treatment Solution

1. 0.2 N NaOH.
2. Beckman Coulter TLA-100 ultracentrifuge equipped with TLA-55 rotor.

2.9. Image Acquisition and Data Processing

1. Fluor-S Max^{TM} MultiImager (Bio-Rad) equipped with a CCD camera and a Nikon 50 mm 1:1.4 AD (F 1.4) lens at the ultrasensitive chemiluminescence setting which cools the camera to –33°C.

3. Methods

3.1. Lipid Mutants as Biological Reagents

The ability to regulate membrane lipid composition under steady state conditions (10) (**Fig. 5.1a**) coupled with determination of membrane protein orientation with respect to the plane of the membrane bilayer is a powerful approach to establish membrane protein topology or observe changes in topology as a function of membrane lipid environment (3, 8, 9) and the amino acid sequence of membrane proteins (5). The utilization of strains in which lipid composition is under control of a tightly inducible promoter (**Fig. 5.1b**) reveals surprising topological dynamics of a polypeptide after stable membrane insertion (3, 5, 9).

3.2. SCAM™

This approach is based on introduction of cysteine residues one at a time into putative extracellular or intracellular loops of a cysteine-less membrane protein of interest followed by chemical modification with a membrane impermeable thiol-specific probe either before or after compromising cell membrane integrity to determine cysteine membrane sidedness. Accessibility in whole cells establishes extracellular location while accessibility only after cell disruption establishes intracellular location (*see* **Notes 6**, **7**, and **9**). The accessibility of extramembrane domains flanking a TM then establishes orientation of the TM with respect to the plane of the membrane bilayer (6).

3.2.1. Construction of Plasmids Expressing Single Cysteine Derivatives

Cysteine is a relatively hydrophobic, small amino acid, and its introduction at most positions in a membrane protein is likely to be tolerated. Furthermore, cysteine has little preference for a particular secondary structure. As an example an Amp^R plasmid encoding a derivative of LacY (lactose permease of *E. coli*) in which endogenous cysteines are replaced by serine or alanine (cysteine-less LacY) is constructed using the site-directed mutagenesis Quickchange XL kit from Stratagene (11). Using the same method single cysteine replacements of amino acids in putative extramembrane domains are constructed (5, 8, 12). All amino acid substitutions are verified by DNA sequencing. Functional analysis and expression level by Western blotting of each derivative should be carried out if possible. Ideally gene expression should be under control of an inducible promoter such as OP_{tac} to minimize continuous expression of potentially disruptive gene products especially in an altered lipid environment.

The process of choosing suitable residues for replacement by cysteine is often empirically determined, and the rationale for deciding which residues to alter is aided by the following considerations. Secondary structure predicted by computer-aided hydropathy analysis (reviewed (13) and thus far 60–70%

reliable) is an initial starting point for the likelihood that a particular residue is in an extramembrane domain. The native cysteine residues are usually changed into alanine or serine residues which are small, commonly found in membrane proteins and appear to be tolerated at most positions thus rendering an active protein. Replacement of charged residues is generally not advised because these have a high probability of being topogenic signals or may be involved in long-range interactions. Consideration should be given to whether the replacement will be well-tolerated based on structural and functional information about the protein. Therefore, a prerequisite for each cysteine replacement is retention of function that provides assurance of retention of near native structure. If the protein contains stretches of residues of intermediate hydrophobicity that cannot unambiguously be identified as membrane spanning, substitutions should be made for approximately every 10 residues. The cysteine-less protein serves as the starting template for introducing single cysteine residues at desired positions as well as a negative labeling control to assure that residues such as lysine and histidine are not labeled by the reagents. Alternatively, templates containing natural cysteines can be utilized in this assay if they do not react with the thiol-specific reagents due to steric hindrance or membrane residency.

To obtain a minimal topological map a single cysteine replacement in each of the putative extramembrane loops should be expressed from a plasmid and analyzed in a host with wild-type lipid composition. In practice, several cysteine replacements or complete cysteine scanning across extramembrane loops and into TM segments is required for a more precise mapping of topology. The host strain for plasmid expression should be deleted of the target protein gene if it contains native cysteines and is expressed at levels high enough to be detected in the assay. Since the target protein is expressed from a multicopy plasmid, it is often possible to analyze a protein in its normal host without deletion of the native protein.

3.2.2. Transformation of Phosphatidylethanolamine-Deficient E. coli Cells with Plasmids

Single cysteine replacements are expressed in the appropriate lipid host. The transformability of an engineered *E. coli* strain (AL95) lacking its major non-bilayer-prone and zwitterionic lipid phosphatidylethanolamine (PE, 75 mole% of total phospholipid) is compromised by a requirement for divalent cations (10–50 mM Mg^{+2}) for viability, cell integrity, proper cell division, membrane impermeability, and osmotic stress response (14–17). Most of the above phenotypes are corrected by introduction of the "foreign" non-bilayer-prone neutral glycolipids α-monoglucosyldiacylglycerol (MGlcDAG) or bilayer-prone diglucosyldiacylglycerol (GlcGlcDAG, but increased passive membrane permeability of the inner and outer membranes remains (17, 18). Therefore, variations of conventional transformation protocols are required to make different lipid hosts competent for plasmid DNA uptake. Electroporation, which requires suspension of cells in very low ionic strength media, cannot be used. In most cases cells can be made competent for transformation using a $RbCl/CaCl_2$-containing solution, which is suitable for strains with normal as well as altered lipid compositions.

1. Dilute 0.1 ml of a fresh overnight culture of PE-lacking host cells grown in LB medium with 50 mM $MgCl_2$ into 5 ml of 37°C LB media supplemented with 50 mM $MgCl_2$ to support growth in the absence of PE (15 ml tube). Cells lacking PE must be kept in the presence of at least 20 mM $MgCl_2$ or $RbCl/CaCl_2$ during all manipulations (*see* **Note 2**). $MgCl_2$ can be eliminated in the above procedure for cells containing PE and more standard methods of transformation can be used. Cells containing plasmid pDD72 (encodes wild type gene required for PE synthesis) or its derivatives must be grown at 30°C because the plasmid is temperature sensitive for replication.
2. Grow cells with vigorous aeration for 3–4 h until cells reach mid-log phase (OD_{600}= 0.6).
3. Place 1 ml of culture into a sterile microfuge tube and incubate at 4°C for 30 min. Centrifuge in pre-chilled (+4°C) bench centrifuge 5417R (Eppendorf) at 14,000 rpm (20,800 *g*) for 1 min to pellet cells and gently resuspend in 0.5 ml of ice-cold 1X $RbCl/CaCl_2$ transformation salts.
5. After 30 min on ice, centrifuge to pellet cells and gently resuspend in 0.1 ml of ice-cold 1X $RbCl/CaCl_2$ transformation salts. Keep cell suspension on ice.
6. Cells are now competent. Add plasmid DNA (100–200 ng) and incubate 30 min on ice.
7. Heat shock for 2 min at 42°C and place on ice for 2 min. Add 1 ml LB containing 50 mM $MgCl_2$ and grow at 37°C for at least 2 h.
8. Pellet the cells and resuspend in 50 μl LB containing 50 mM $MgCl_2$ and plate on LB agar plates containing 50 mM $MgCl_2$ and 100 μg/ml of ampicillin. Plates should be fresh or no more than a few days old for best results. Grow it for at least 2 days at 37°C. Individual colonies are streaked for single colonies on LB plates containing ampicillin and either with or without $MgCl_2$. Single colonies are screened to verify $MgCl_2$ dependence to eliminate potential contaiminants.

3.2.3. Cell Growth and Regulated Expression of LacY and PssA Genes

1. Cells (*see* **Fig. 5.1a** for strain description) carrying plasmids expressing single cysteine replacements in cysteine-less LacY under OP_{tac} control are grown for at least two generations in LB medium supplemented with 50 mM $MgCl_2$ to

support growth in the absence of PE (*see* **Note 2**), ampicillin (100 μg/ml) to maintain LacY plasmids, and IPTG (1 mM) to induce LacY expression.

2. To independently regulate expression of LacY and PE, strain AT2033 ($P_{LtetO\text{-}1}$-*pssA*$^+$ *pss93*::*kan*^{R} *lacY*::*Tn9 recA srl*::*Tn10*) is used. In this strain the level of aTc in the growth medium regulates chromosomal *pssA* (encodes phosphatidylserine synthase for PE synthesis) expression and PE content of the cell. Expression of a plasmid copy of OP_{tac}-*lacY* is regulated by IPTG. In order to determine the effect of a change in membrane lipid composition on the topological organization of LacY, cells are grown first in the presence of IPTG without aTc to allow synthesis and membrane insertion of LacY in the absence of PE. Then cells are switched to growth without IPTG in presence of aTc in order to permit biosynthesis of new PE in the absence of newly synthesized LacY.
   (i) Cells of strain AT2033 containing different plasmids expressing derivatives of LacY are first grown overnight at 37°C in LB medium supplemented with ampicillin (100 μg/ml, required for plasmid maintenance) and 50 mM $MgCl_2$ to support growth at residual levels (*ca.* 2–3 mole%) of PE.
   (ii) Then cells are diluted to OD_{600} of *ca.* 0.05 into 200 ml of medium supplemented with ampicillin (100 μg/ml) and 50 mM $MgCl_2$, and the expression of the LacY derivatives are induced by growth in the presence of 1 mM IPTG for at least two generations (OD_{600} reading 0.20).
   (iii) One-half of the cell culture (100 ml) is pelleted by centrifugation (3,000 *g*) and stored on ice for LacY for SCAM ^{TM} analysis.
   (iv) Remaining cells (100 ml) are pelleted by centrifugation and washed twice by centrifugation with pre-warmed sterile LB medium supplemented with 50 mM $MgCl_2$ to remove IPTG and then re-suspended in 100 ml of pre-warmed medium supplemented with ampicillin (100 μg/ml) and 50 mM $MgCl_2$. Expression of the *pssA* gene is induced by addition of aTc (1 μg/ml).
   (v). After 3 h of growth in the presence of inducer, cells are harvested and subjected to SCAM^{TM} analysis.

3.2.4. General Protocol for SCAM^{TM}

The chemical nature of the reactive portion of a labeling reagent should be highly reactive with and selective for thiol groups and form a stable non-exchangeable or non-hydrolysable derivative.

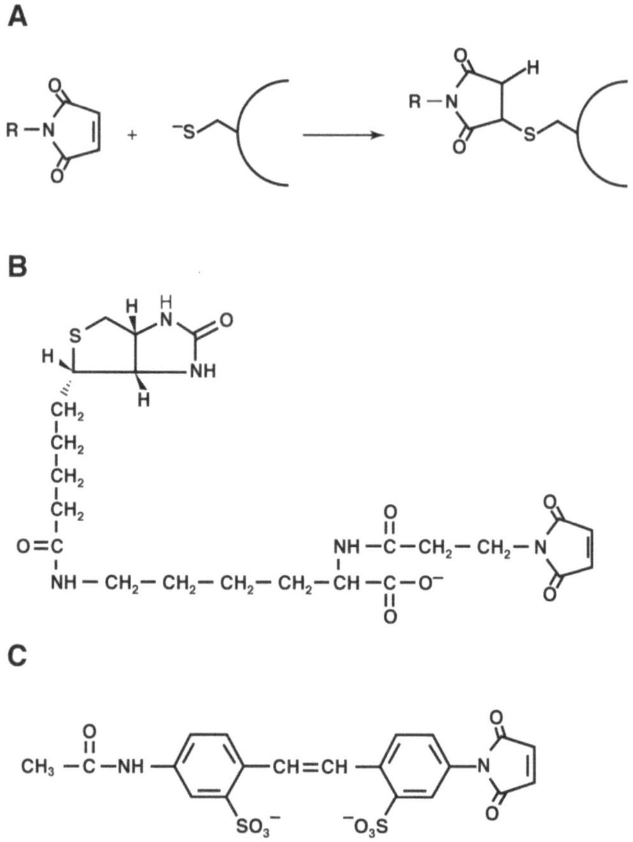

Fig. 5.2. Structure of thiol-modifying reagents and their reaction with a thiol. (**a**) Reaction of a maleimide with the thiolate of a protein-bound cysteine to form a covalent adduct. (**b**) Structure of MPB. (**c**) Structure of AMS. Figure reproduced from (6) © 2008 Elsevier B.V.

Maleimide-based thiol reagents, which are available in a wide variety of forms, are particularly suited for SCAM^{TM} (6). Maleimide reacts with the ionized form of a thiol group (**Fig. 5.2a**), and this reaction requires a water molecule as a proton acceptor. Maleimides are virtually unreactive until they encounter an available thiol group. For most water-exposed cysteine residues in proteins, pKa of the thiol of cysteine lies in the range of 8–9, and formation of cysteinyl thiolate anions is optimum in aqueous rather in a non-polar environment where the pKa of the thiol of cysteine is around 14. Therefore, the reaction rate of different sulfhydryls is controlled primarily by their water exposure making the residues that reside in regions of TM helices unfavorable for the generation of thiolate anions. Thus, the labeling characteristics of intramembrane (unreactive) and extramembrane (reactive) cysteines should be consistent with their localization either in a non-polar or polar environment, respectively (6). In most cases the unreactive cysteine residues are located within the membrane hydrophobic core or in a sterically hindered environment as described below.

SCAM[TM] using thiol-specific membrane–impermeable MPB (**Fig. 5.2b**) has been extensively used to probe the topological organization of LacY (3, 5, 6, 8), high affinity phenylalanine permease (PheP) (**9**) and γ-aminobutyric acid permease (GabP) (19) as a function of membrane lipid composition. In this assay sonication of cells is used to disrupt cell membranes, making both periplasmic and cytoplasmic cysteines accessible to MPB, whereas cysteines located within a TM domain are still protected from labeling (6). Derivatization of cysteines in whole cells indicates periplasmic exposure while derivatization only during sonication indicates cytoplasmic exposure, since desintegration allows thiol-specific reagent access to both sides of the membrane. It is important to note that the extent of biotinylation should be the same before and after sonication for an extracellular cysteine. If sonication results in an increase in biotinylation this may indicate a mixed topology.

Although single cysteine replacements could affect protein structure and expression, conclusions are based on comparison of the extent of labeling in whole cells and disrupted cells for the same protein in two different lipid environments. This approach simplifies interpretation of data obtained with a series of protein derivatives that may express at different levels since conclusions about topology are based on relative reactivity of cysteines in the same sample before and after cell disruption. Since sample pairs are analyzed on the same Western blot, no signal intensity normalization is required and the results are not dependent on the expression level of the derivative. The level of expression of any given derivative will affect the absolute intensity of labeling but not the ratio of the labeling between the sample pairs.

1. Harvest 100 ml mid-log phase ($OD_{600} \sim 0.4–0.6$) cells (*see* **Note 2**) expressing a single cysteine derivative of LacY or protein of interest by centrifugation and suspend cell pellets in 1.5 ml of buffer A (adjusted to pH 7, 9 or 10.5 as indicated). Divide the sample into two equal aliquots (0.75 ml) in Microfuge polyallomer tubes (Beckman). To increase the reactivity of diagnostic cysteine residues (particularly those that might be in sterically hindered extramembrane domains), the reaction with MPB is carried out at pH 9 or 10.5 (*see* **Note 8**).

2. Treat one set of samples with MPB at a final concentration of 100 μM (7.5 μl of 10 mM stock solution) (*see* **Notes 3** and **4**) for 5 min at room temperature to label cysteines exposed to the periplasmic side of the inner membrane. Quench the reaction by the addition of ß-mercaptoethanol to 20 mM

(7.5 μl of 2 M stock solution). After labeling, cells are sonicated for 1 min using an amplitude of 15%.

3. To simultaneously label cysteines exposed to both sides of cell membrane, subject the remaining sample to sonication for 1 min in the presence of MPB at a final concentration of 100 μM. Incubate for 4 min at room temperature and quench the reaction by the addition of ß-mercaptoethanol to 20 mM.
4. All sonicated samples are centrifuged at 4°C at 38,000 rpm (65,000 *g*) (TLA-55 Beckman Coulter rotor) for 10 min followed by resuspension in 100 μl of Buffer A containing 20 mM ß-mercaptoethanol by vortexing for 2 h at room temperature using a Fisher Vortex Genie 2. The membranes are solubilizated by detergent, and the target protein is isolated by immunoprecipitation or affinity purification as described below.
5. Following modification with MPB and isolation by immunoprecipitation, the target protein is resolved by SDS PAGE, transferred to a solid support, and detected by Western blotting using avidin linked to horse radish peroxidase. SuperSignal West Pico chemiluminescent substrate (Thermo Scientific) is used to visualize biotinylated proteins. Biotinylation of exposed cysteine residues of whole cells (periplasmic exposure) or only during sonication (cytoplasmic exposure) is detected using a Fluor-S^{TM} MultiImager (Bio-Rad).

3.2.5. Sample Solubilization

After labeling, isolated membranes are solubilized with the appropriate detergent or detergent mixture such as SDS alone, Triton X-100 alone, SDS and Triton-X-100, CHAPS, octylglucoside, deoxycholate, cholate and Tween 20, ß-D-dodecylmaltoside or nonidet P-40 and sodium deoxycholate (6). Conditions must be empirically determined that yield a non-aggregated soluble target protein throughout the remainder of the procedure. Many membrane proteins aggregate if boiled in SDS so that solubilization should be done between 37°C and 55°C. LacY forms irreversible polydisperse aggregates if solubilized by Triton X-100 alone. Therefore, a membrane-embedded LacY sample (100 μl) is solubilized by addition of an equal volume of solubilization buffer followed by vigorous vortexing for 15 min at room temperature, incubation at 37°C for 15 min, and an additional 15 min vortexing at room temperature. Prior to immunoprecipitation, the sample is diluted with buffer containing non-ionic detergent to neutralize the denaturing properties of SDS (as described below) and cleared by centrifugation in a pre-chilled (+4°C) bench centrifuge 5417R (Eppendorf) at 14,000 rpm (20, 800 *g*) for 10 min.

3.2.6. Isolation of Derivatized Target Protein

The thiol reagents react with cysteine residues present in all other proteins in the membrane. Immunoprecipitation of the membrane protein of interest or a rapid purification step is necessary to eliminate other labeled proteins. A biotin-maleimide labeled protein can be recovered from cell lysates directly with streptavidin-agarose beads (20) and can then be detected by Western blotting using a target-specific antibody. For immunoprecipitation of labeled protein from solubilized samples, polyclonal and monoclonal antibodies have been widely utilized. Antigen-antibody complexes can be isolated using precipitation with Pansorbin (*Staphylococcus aureus* cells) (3), protein A agarose (21) or protein A or G Sepharose beads (22, 23). If antibodies specific to the protein under study are not available, then epitope tags such as His_6 can be incorporated at the C-terminus of the target protein for either immunoprecipitation or isolation by Ni^{2+} chelated affinity resin (3, 24, 25). Use of affinity methods with His-tagged proteins and small-scale batch purification procedures is becoming the method of choice since the labeled protein can be directly extracted from the resin with SDS-containing buffers followed by SDS-PAGE (26). Of course protein function or topology should not be compromised by the presence of the tag. To allow appropriate antibody interaction with solubilized protein, SDS should be diluted with appropriate non-ionic detergent. In case of LacY, samples (200 μl from above) are diluted with 300 μl of cold Immunoprecipitation buffer followed by empirically established immunoprecipitation protocols (3).

Following modification and isolation, the target protein is resolved by SDS-PAGE, transferred to a solid support, and detected by Western blotting or one of the following techniques (6). Thiol reagents are available that contain a biotin group, a fluorescent group, or a radiolabel allowing detection of labeled proteins by avidin linked to horse radish peroxidase (avidin-HRP) and indirect chemilumiscence detection, fluorescence, or autoradiography, respectively. Signals can be quantified using available imaging systems and software.

3.2.7. SDS-PAGE

The final immunoprecipitates are solubilized in 30 μl of SDS sample buffer (10 mM Tris–HCl (pH 6.8), 5.6% SDS, 200 mM DTT, 10% glycerol, 0.01% bromophenol blue) by vigorous vortexing for 15 min at room temperature, incubation at 37°C for 15 min, and an additional 15 min vortexing at room temperature using a Fisher Vortex Genie 2. Samples are subjected to SDS-PAGE using standard protocols and then transferred from the gel to a solid support by Western blotting.

3.2.8. Western Blot Analysis

Protein samples are transferred from the SDS polyacrylamide gel to Protran BA 85 nitrocellulose membranes by electroblotting using a semi-dry electroblotting system.

1. After electrophoresis, place the SDS-PAGE gel in the cathode buffer to equilibrate for 10 min before blotting.
2. Cut out a piece of Protran BA 85 nitrocellulose transfer membrane and Whatman filter papers sized to fit the SDS polyacrylamide gel. See **Fig. 5.3**, which shows the procedure for building the blotting sandwich.

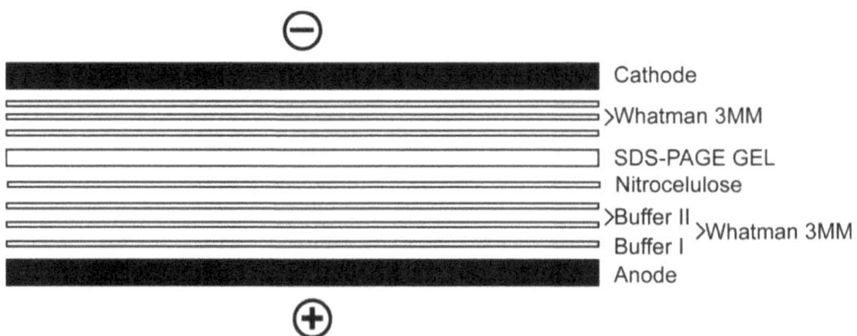

Fig. 5.3. Schematic illustration of the Western blotting sandwich used with a semi-dry electroblotting system.

3. Soak the bottom layer of chromatography paper (Whatman 3 mm Cr) in Anode buffer No.1 and allow excess buffer to drain from the paper. Place a drop of this buffer in middle of positive electrode surface of the blotting apparatus, then lay down the soaked filter paper. Center and smooth out by rolling with a glass tube to establish good contact and eliminate air bubbles.
4. Place two layers of chromatography paper wetted with Anode buffer No. 2 on top of paper soaked in Anode buffer No.1. Roll a glass tube over the chromatography paper to smooth and remove trapped air bubbles.
5. Place a Protran BA 85 nitrocellulose transfer membrane wetted with distilled water on top of the three layers of chromatography paper and roll with a glass tube.
6. Now center the acrylamide gel onto the transfer membrane making sure that air bubbles are not trapped in between the transfer membrane and acrylamide gel. Use a glass tube dipped in Cathode buffer to gently roll out any trapped air.
7. Place three layers of chromatography paper soaked in Cathode buffer on top of the acrylamide gel. Chromatography papers, gels, and transfer membrane must be in the same dimensions to achieve a uniform transfer.
8. Place the lid (negative electrode) over the trans-unit sandwich making sure that the electrode rests on the sandwich.
9. Attach the power cables on Labconco Semi-Dry blotting system to a constant voltage power supply capable of supplying sufficient current for transfer. Adjust the power

supply to provide initial current density of 2.5 mA/cm2 of gel area. The voltage reading should be approximately 10 volts with new electrodes.

10. Transfer time is somewhat dependent on the size of the proteins and percentage of gel used. In most cases complete transfer is achieved in 90 min.

3.2.9. Blocking Procedure

The nitrocellulose transfer membrane is then blocked overnight with blocking buffer. Note that 0.05% Nonidet P-40 is omitted during blocking prior to staining of biotinylated proteins with avidin-HRP.

3.2.10. Staining with Avidin-HRP

MPB is a thiol reagent containing a biotin group, which makes possible indirect chemilumiscence detection of labeled proteins after treatment with avidin linked to horseradish peroxidase (avidin-HRP).

1. The nitrocellulose transfer membranes are washed once with TBS buffer containing 0.3% BSA for 10 min.
2. Avidin–HRP is added at a final dilution of 1: 5000–10,000 of 2 mg/ml stock solution in TBS buffer containing 0.3% BSA and incubated for at least 1 h.
3. The sheets are washed two times with TBS buffer containing 0.3% BSA for 15 min each followed by another two washes with TBS/Nonidet P40 buffer and only once with TBS buffer.
4. To visualize biotinylated proteins, the sheets are incubated for 3 min with SuperSignal West Pico chemiluminescent substrates (Thermo Scientific) mixed immediately prior to use at a ratio of 1:1, and biotinylated proteins are visualized using a Fluor-S Max^{TM} MultiImager (Bio-Rad).

3.2.11. Image Acquisition and Processing

Western blots are imaged using a Fluor-S Max^{TM} MultiImager (Bio-Rad). Bio-Rad software Quantity One^{TM} versions 4.6.5.094 and 4.4.1 are used to collect and store the images as TIFF files, which can be imported later into Adobe Illustrator to construct figures. Images are expanded or reduced so that the horizontal strip containing the target protein images is sized appropriately and masked to show only the target protein results, which are then aligned with images from other gels and labeled. The only valid comparison in intensity is between whole cell and sonicated sets (images treated identically) run on the same gel.

3.2.12. Data Analysis and Interpretation

The criteria used for determining the location of an introduced cysteine are as follows. Labeling of a cysteine residue with a membrane-impermeable sulfhydryl reagent before disruption of whole cells is indicative of a periplasmic cysteine residue provided accessibility to a cytoplasmically localized control protein (*see* **Note 4**) and a cysteine-less derivative of the target protein are negative (*see* **Note 8**). Absence of labeling in whole cells but labeling during cell disruption indicates a cytoplasmic location for the cysteine-containing domain. No labeling with a sulfhydryl reagent before or during cell disruption implies localization to a hydrophobic membrane environment or unfavorable local orientation/positioning of introduced thiol groups which may prevent access by the reagent or result in an increase of the pKa of the thiol group as discussed below (*see* **Note 6**).

3.2.13. Topological Assignment of Problematic Regions

No definitive conclusion can be made for the location of a domain based on lack of reaction of the cysteine (*see* **Notes 5** and **6**). An unreactive cysteine can be due to its location within a TM or proximal environmental effects, which affect the thiol pKa or sterically restrict access (6). Cysteine scanning in the neighborhood of the unreactive cysteine may be required to differentiate between a TM or an unreactive cysteine within a larger exposed domain. Unfavorable orientation of a thiol group due to local secondary structure or properties of neighboring amino acids may restrict or prevent access by large thiol reagents. Cysteine residues closer to the membrane interface generally react slower than those near the center of extramembrane domains. Since derivatization is favored by formation of the cysteinyl thiolate anion (**Fig. 2a**), increasing the reaction buffer pH would favor alkylation of an extramembrane cysteine as well as disrupt local restrictive secondary structure while truly membrane-imbedded cysteines would not be expected to react. Moreover diagnostic cysteines can be hidden within membrane domains that partially penetrate the membrane, which have been termed membrane-dipping, shallow-penetrating, re-entry, mini-, or U-shaped loops. Such unusual topological arrangements are present in aquaporins, potassium chloride channels and protein conducting channels (27). Under strongly alkaline conditions (pH >11), soluble and peripheral membrane proteins (i.e., those that do not contain any domains that span the membrane bilayer) are released in soluble forms. Integral membrane proteins remain embedded in the lipid bilayer and can be isolated by centrifugation. Extramembrane domains that are sterically hindered or exhibit elevated pKa's can be derivatized with increasing pH up to 10.5 without compromising membrane integrity. However, appropriate controls such as inaccessibility of known cytoplasmically exposed domains and lack of label of a cysteine-less target protein should be done (*see* **Note 8**). The same may apply for TMs that form aqueous channels across or part way across the membrane. Cysteines in mini-loop domains are exposed by NaOH with loss of membrane integrity, but a true TM remains unreactive

To expose hindered cysteines the pH of the reaction mix is raised either stepwise to pH 10.5 in a series of reactions (*see* **Note 8**). Cell integrity is maintained at pH 10.5 as evidenced by labeling of cytosolic domains only during sonication. Cysteine-less protein and protein with an intramembrane cysteine are not labeled at pH 10.5 (5). To detect cysteines in a mini-loop domain, cells are treated with NaOH prior to reaction with a thiol reagent.

1. Cell aliquots in buffer A are mixed with an equal volume of cold 0.2 N NaOH, incubated for 5 min on ice, and separated into a pellet and supernatant fraction by centrifugation at 40,000 rpm (70,000 *g*) (TLA-55 Beckman Coulter rotor) for 10 min.
2. The pellets are washed three times by suspension in Buffer A using sonication for 1 min at an amplitude of 15% followed by centrifugation as above. The final pellet is resuspended in Buffer A and subjected to SCAM^{TM} at pH 7.5.

3.2.14. Indirect and Direct Labeling and Identification of Mixed Topologies

Central to the method is the use of detectable thiol-specific reagents to differentiate intracellular from extracellular domains of membrane protein which in some cases may adopt mixed or dual TM topologies within the same membrane (28) (*see* **Note 9**). Confirmation of labeling of external water-exposed cysteines by MPB can be achieved by first blocking putative external cysteines in intact *E. coli* cells with a thiol-specific reagent that is transparent in the detection phase of the procedure. Such a pre-blocking step also allows selective labeling of luminal (exposed to cytoplasm) cysteines after cell permeabilization or disruption and therefore detection of mixed topologies co-existing within the same cell membrane (3, 6). A set of impermeable blocking reagents that effectively react with thiols exposed to solvent but are transparent in the detection phase of the procedure is available for SCAM^{TM} (6). One such reagent is 4-acetamido-4′-maleimidylstilbene-2,2′-disulfonic acid (AMS, **Fig. 5.3c**) which is membrane-impermeable due to its size, two charged sulfonate groups, and high solubility in water.

The degree of mixed topology can be assessed with a two-step protocol as follows. Intact cells are treated either with or without AMS followed by labeling with MPB of intact cells and cells during disruption. Excess AMS is removed prior to any subsequent treatments by centrifuging the reaction mixture through a small column of gel filtration resin to avoid lysis of fragile cell preparations (6). In the case of mixed topology, biotinylation of the target cysteine by MPB will occur with whole cells not pre-treated with AMS and to a greater extent after cell disruption. AMS treatment will prevent any biotinylation with whole cells and reduce the amount of biotinylation observed in disrupted cells. The protective effect of AMS is almost complete and has been used to quantify the degree of mixed topology (29).

1. To confirm labeling of external (periplasmic) water-exposed cysteines by MPB and detect possible mixed topology of LacY or protein of interest cell pellets derived from 200 ml of mid-log phase (OD_{600} ~ 0.4–0.6) cells expressing a single cysteine derivative are resuspended in 3.5 ml of buffer A and divided into four equal aliquots (0.75 ml) in Microfuge polyallomer tubes (Beckman).
2. Two cell aliquots are incubated for 30 min at 25°C with AMS at a final concentration of 5 mM (37.5 μl of 100 mM aqueous stock solution) to block periplasmic water-accessible cysteine residues from the outside of cells. Then AMS is removed either by centrifugation through a small gel filtration column or by two cycles of centrifugation and resuspension in 0.75 ml of buffer A.
3. Two cell aliquots, one pre-treated with AMS from step 2 and one not treated with AMS, are biotinylated by adding MPB at a final concentration of 100 μM (7.5 μl of 10 mM stock solution) followed by incubation for 5 min at 25°C and quenching of the reaction by the addition of ß-mercaptoethanol to 20 mM. Cells are then disrupted by sonication.
4. The two remaining samples, one pre-treated with AMS from step 2 and one not treated with AMS, are biotinylated by adding MPB at a final concentration of 100 μM (7.5 μl of 10 mM stock solution) during sonication for 1 min followed by incubation for another 4 min at room temperature before quenching the reaction by the addition of ß-mercaptoethanol to 20 mM.
5. All samples are processed for detection of biotinylation as described above.

4. Notes

1. Pre-chilled (+4°C) 50 mM Tris-HCl (pH 8.1) should be used to prepare IP buffer with SDS/non-ionic detergent ratio 1/5. Lubrol-PX can be successfully substituted by Thesit® (Fluka Chemical Co.) at exactly the same concentration.
2. Cells lacking PE must not be exposed to solutions with $MgCl_2$ concentrations less than 10 mM. However, when comparing cells with different lipid compositions to wild-type cells, the $MgCl_2$ concentration should be the same for all samples and in the range of 10–50 mM with the higher concentration being optimal for growth of most PE-lacking strains.

3. Final concentration of DMSO used to dissolve MPB should never exceed 0.5%.
4. SCAM^{TM} is based on the controlled membrane permeability of sulfhydryl reagents. The results of SCAM^{TM} are valid only if the modifying reagent is thiol-specific and membrane impermeable, cells are intact and cell disruption does not expose sterically hindered or water inaccessible cysteine residues (3, 5, 6). Various reagents including MPB will cross the membrane in a concentration, time, and temperature dependent manner and permeability varies with genetic background of the host cells. Therefore, conditions must be empirically determined to minimize derivatization of intracellular cysteines. Cell lysis during a labeling experiment or during manipulation of cells will also result in labeling of intracellular cysteines. Pre-blocking external cysteines with AMS before labeling intact cells with MPB provides a means of estimating the degree of labeling due to low permeability under different experimental conditions.

   The membrane permeability of a thiol-specific labeling reagent can be tested and labeling conditions (concentration, time, and temperature) established by quantification of the degree of labeling of an abundant cytoplasmic protein that is rich in surface exposed cysteine residues. In *E. coli* ß-galactosidase is ideal for this purpose due to its content of cysteine, mobility in a region on SDS-PAGE devoid of other major proteins, availability of mutants lacking the enzyme as a control, and availability of antibody against ß-galactosidase (Sigma ImmunoChemicals). Other cytosolic bacterial markers such a glutathione or elongation factor Tu have been used to access membrane permeability. Significant differences in the permeability of different host strains emphasize the need to screen host strains for reagent permeability prior to initiating experiments (6–8).

   Whole and disrupted cells are treated with various concentrations of reagent from 10 μM to 1 mM, at temperatures from 0°C to 25°C, and for various lengths of time from 5 min to 1 h. In most cases low concentration of MPB (100 μM) and relatively short incubation period (5 min) at room temperature favor biotinylation of extramembrane thiol groups. In this case a labeling experiment with both intact cells and disrupted cells is carried out except that soluble proteins rather than the membrane fraction are retained for immunoprecipitation with antibody against ß-galactosidase followed Western blotting analysis.
5. A major problem is the variability between samples due to mechanical loss during the work up or due to differences in expression level of individual replacements. Since

conclusions are only made by comparing sample pairs (labeling without or during cell disruption), differences in expression or labeling efficiency between individual replacements are not important. Corrections for mechanical loss can be made by quantitative Western blot analysis using a target specific antibody. However, if no labeling occurs with the thiol reagent, it is important to verify that the protein was expressed and is present on the Western blot. To verify the presence of the target protein or to ensure the presence of equal amounts of target protein in each paired sample, blots can be stripped using Restore™ Western Blot Stripping Buffer (Thermo Scientific) and reprobed with appropriated antibody.

6. Caution must be used in assigning an intramembrane location to a cysteine residue because it is unreactive to hydrophilic thiol reagent in both intact and disrupted cells. Lack of or low levels of labeling may result from any of the following reasons: (1) steric hindrance due to local secondary structure; (2) internalization into the compact fold of the protein; (3) lack of ionization of the thiol group due to a hydrophobic environment; (4) local environment with the same charge as the thiol reagent; (5) increased pKa of the thiol due to the high negative charge density of neighboring residues or anionic lipids; (6) facing other helices. Periplasmic extramembrane domains tend to be shorter (sometimes only three amino acids in length) than cytoplasmic domains. Therefore, there may be little or no protrusion of these loops into the extracellular space, thus preventing reaction of the cysteine residues in these locations with relatively bulky reagents. Alkylating reagents appear to react better with cysteines toward the middle of extended hydrophilic loops than near the TM interfacial domain. Cysteine scanning across a domain is an effective means for identifying useful replacement sites and differentiating between local effects and true TMs. Scanning can be coupled with alkaline treatment as discussed above. Finally, reagents of different size and charge can be tested (6).
7. Conclusions based on full reactivity of diagnostic residues should be made with caution, since cysteine residues facing a hydrophilic pore or near a substrate-binding site maybe within a TM segment but chemically reactive due to water channels or pockets. Therefore an extramembrane domain and an aqueous pore are not easily distinguished by SCAM™. Substitution at positions crucial to overall protein structure and stability cannot be used, but such substitutions often result in low levels of the protein or loss of activity and are informative.

8. Since the formation of cysteinyl thiolate anions is favored by increasing the solution pH (optimum pH 8.0–8.5), increasing the pH during labeling should favor the reaction. However, maleimides are known to react with primary amines at pH values above 7.5. Therefore, attempts to increase efficiency of labeling by raising the pH of the assay should be thoroughly controlled in order to ensure that the modification is confined to cysteine. An effective control to rule out non-thiol modifications is to use a cysteine-less target protein.
9. TM topology studies assume that all copies of the target protein have the same orientation. The labeling patterns are the result of end-point titrations and assume a relatively fixed conformation for extramembrane loops, ignoring the presence of regions with heightened mobility and flexibility or the possibility of topological inversions on the time scale of the labeling. Topological models derived from accessibility patterns depict a static TM topology whereas the actual structure in a membrane is more likely to be dynamic. Through the application of SCAM^{TM} some examples exist of dynamic changes in topological organization induced post-assembly of membrane proteins as well as some proteins that appear to exist with multiple topological organizations (6). Even with optimal assays and reagents, membrane proteins that assume multiple conformations either within the same or between different membranes may yield confusing and conflicting results. Therefore, the static nature of methods that measure topological organization may have missed more dynamic properties of membrane proteins. In particular low yield of modification due to slow rate of reaction may be due to dynamic movement of a domain into and out of an accessible region. There are examples of cryptic intramembrane regions that become exposed to the aqueous phase and extramembrane domains that translocate to the opposite side of the membrane which limit topology studies in intact cells where the protein is turning over or metabolic conditions influence organization (6).

Acknowledgements

This work was supported by NIH grant GM20478 and funds from the John S. Dunn Foundation awarded to W. D.

References

1. Drews, J. (2006) What's in a number? *Nat. Rev. Drug Discov.* **5,** 975.
2. Goder, V., Junne, T. and Spiess, M. (2004) Sec61p contributes to signal sequence orientation according to the positive-inside rule. *Mol. Biol. Cell* **15,** 1470–1478.
3. Bogdanov, M., Heacock, P.N. and Dowhan, W. (2002) A polytopic membrane protein displays a reversible topology dependent on membrane lipid composition. *EMBO J.* **21,** 2107–2116.
4. Bogdanov, M., Mileykovskaya, E. and Dowhan, W. (2008) Lipids in the Assembly of Membrane Proteins and Organization of Protein Supercomplexes: Implications for Lipid-linked Disorders. *Subcell. Biochem.* **49,** 197–239.
5. Bogdanov, M., Xie, J., Heacock, P. and Dowhan, W. (2008) To flip or not to flip: lipid-protein charge interactions are a determinant of final membrane protein topology. *J. Cell Biol.* **182,** 925–935.
6. Bogdanov, M., Zhang, W., Xie, J. and Dowhan, W. (2005) Transmembrane protein topology mapping by the substituted cysteine accessibility method (SCAM^{TM}): application to lipid-specific membrane protein topogenesis. *Methods* **36,** 148–171.
7. Wang, X., Bogdanov, M. and Dowhan, W. (2002) Topology of polytopic membrane protein subdomains is dictated by membrane phospholipid composition. *EMBO J.* **21,** 5673–5681.
8. Xie, J., Bogdanov, M., Heacock, P. and Dowhan, W. (2006) Phosphatidylethanolamine and monoglucosyldiacylglycerol are interchangeable in supporting topogenesis and function of the polytopic membrane protein lactose permease. *J. Biol. Chem.* **281,** 19172–19178.
9. Zhang, W., Bogdanov, M., Pi, J., Pittard, A.J. and Dowhan, W. (2003) Reversible topological organization within a polytopic membrane protein is governed by a change in membrane phospholipid composition. *J. Biol. Chem.* **278,** 50128–50135.
10. Dowhan, W. (2009) Molecular Genetic Approaches to Defining Lipid Function. *J. Lipid Res.* **50,** S305–S310.
11. van Iwaarden, P.R., Pastore, J.C., Konings, W.N. and Kaback, H.R. (1991) Construction of a functional lactose permease devoid of cysteine residues. *Biochemistry* **30,** 9595–9600.
12. Frillingos, S., Sahin-Toth, M., Wu, J. and Kaback, H.R. (1998) Cys-scanning mutagenesis: a novel approach to structure function relationships in polytopic membrane proteins. *Faseb J.* **12,** 1281–1299.
13. Elofsson, A. and von Heijne, G. (2007) Membrane Protein Structure: Prediction vs Reality. *Annu. Rev. Biochem.* **76,** 125–140.
14. DeChavigny, A., Heacock, P.N. and Dowhan, W. (1991) Sequence and inactivation of the *pss* gene of *Escherichia coli.* Phosphatidylethanolamine may not be essential for cell viability. *J. Biol. Chem.* **266,** 5323–5332.
15. Mileykovskaya, E. and Dowhan, W. (2005) Role of membrane lipids in bacterial division site selection. *Curr. Opin. Microbiol.* **8,** 135–142.
16. Rietveld, A.G., Chupin, V.V., Koorengevel, M.C., Wienk, H.L., Dowhan, W. and de Kruijff, B. (1994) Regulation of lipid polymorphism is essential for the viability of phosphatidylethanolamine-deficient Escherichia coli cells. *J. Biol. Chem.* **269,** 28670–28675.
17. Wikström, M., Xie, J., Bogdanov, M., Mileykovskaya, E., Heacock, P., Wieslander, Å. and Dowhan, W. (2004) Monoglucosyldiacylglycerol, a foreign lipid, can substitute for phosphatidylethanolamine in essential membrane-associated functions in *Escherichia coli. J. Biol. Chem.* **279,** 10484–10493.
18. Wikström, M., Kelly, A., Georgiev, A., Eriksson, H., Rosen-Klement, M., Bogdanov, M., Dowhan, W. and Wieslander, Å. (2009) Lipid-engineered *Escherichia coli* membranes reveal critical lipid head-group size for protein function. *J. Biol. Chem.* **284,** 954–965.
19. Zhang, W., Campbell, H.A., King, S.C. and Dowhan, W. (2005) Phospholipids as determinants of membrane protein topology. Phosphatidylethanolamine is required for the proper topological organization of the gamma-aminobutyric acid permease (GabP) of *Escherichia coli. J. Biol. Chem.* **280,** 26032–26038.
20. Sato, Y., Zhang, Y.W., A., A.-T. and Rudnick, G. (2004) Analysis of transmembrane domain 2 of rat serotonin transporter by cysteine scanning mutagenesis. *J. Biol. Chem.* **279,** 22926–22933.
21. Wada, T., Long, J.C., Zhang, D. and Vik, S.B. (1999) A novel labeling approach supports the five-transmembrane model of subunit a of the Escherichia coli ATP synthase. *J. Biol. Chem.* **274,** 17353–17357.

22. Cao, W. and Matherly, L.H. (2003) Characterization of a cysteine-less human reduced folate carrier: localization of a substrate-binding domain by cysteine-scanning mutagenesis and cysteine accessibility methods. *Biochem. J.* **374,** 27–36.
23. Zhu, Q., Lee, D.W. and Casey, J.R. (2003) Novel topology in C-terminal region of the human plasma membrane anion exchanger, AE1. *J. Biol. Chem.* **278,** 3112–3120.
24. Fujihira, E., Tamura, N. and Yamaguchi, A. (2002) Membrane topology of a multidrug efflux transporter, AcrB, in Escherichia coli. *J. Biochem.* **131,** 145–151.
25. Long, J.C., DeLeon-Rangel, J. and Vik, S.B. (2002) Characterization of the first cytoplasmic loop of subunit a of the *Escherichia coli* ATP synthase by surface labeling, cross-linking, and mutagenesis. *J. Biol. Chem.* **277,** 27288–27293.
26. Valiyaveetil, F.I. and Fillingame, R.H. (1998) Transmembrane topography of subunit a in the Escherichia coli F1F0 ATP synthase. *J. Biol. Chem.* **273,** 16241–16247.
27. Lasso, G., Antoniw, J.F. and Mullins, G.L. (2006) A combinatorial pattern discovery approach for the prediction of membrane dipping (re-entrant) loops. *Bioinformatics* **22,** 290–297.
28. Gafvelin, G. and von Heijne, G. (1994) Topological "frustration" in multispanning *E. coli* inner membrane proteins. *Cell* **77,** 401–412.
29. Kimura, T., Ohnuma, M., Sawai, T. and Yamaguchi, A. (1997) Membrane topology of the transposon 10-encoded metal-tetracycline/H+ antiporter as studied by site-directed chemical labeling. *J. Biol. Chem.* **272,** 580–585.
30. Martinez-Morales, F., Schobert, M., Lopez-Lara, I.M. and Geiger, O. (2003) Pathways for phosphatidylcholine biosynthesis in bacteria. *Microbiology* **149,** 3461–3471.
31. Shiba, Y., Yokoyama, Y., Aono, Y., Kiuchi, T., Kusaka, J., Matsumoto, K. and Hara, H. (2004) Activation of the Rcs signal transduction system is responsible for the thermosensitive growth defect of an *Escherichia coli* mutant lacking phosphatidylglycerol and cardiolipin. *J. Bacteriol.* **186,** 6526–6535.

Chapter 6

In Vivo Analysis of Protein Translocation to the *Escherichia coli* Periplasm

Dominique Belin

Abstract

General protein export requires the cooperation of two elements, the Sec translocase and a signal sequence. The interactions of both wild type and mutant components can be studied in vivo using a number of genetic systems. Signal sequence mutations that prevent export have been characterized ("down mutations"). Suppressors of these signal sequence mutations, known as *prl* mutations, have been isolated in most *sec* genes. More recently, inactive N-terminal regions of cytoplasmic proteins were converted into active signal sequences ("up mutations").

Alkaline phosphatase (PhoA), an enzyme only active after export to the periplasm, provides the best and most versatile quantitative reporter for protein translocation studies. Cleavable signal sequences can be used to monitor protein export in a time frame of 15–120s. Chimeric proteins expressed from an inducible promoter can be used to measure kinetics of enzyme accumulation in a time frame of 10–100 min. Finally, the export activity of PhoA-chimeras can be visualized in a semi-quantitative way by staining colonies growing on Petri dishes with a chromogenic substrate, in the time frame of 10–40 h.

Key words: Signal sequence, alkaline phosphatase, *E. coli*, translocase, *prl* mutations, gene fusions, leader peptide, leader peptidase.

1. Introduction

Protein secreted into the periplasm, or inserted into the bacterial inner membrane, are usually synthesized with a targeting element, the signal sequence, that interacts with SecYE(G), a complex of three integral membrane proteins that constitutes the preprotein translocase core (1). A number of assays can be used to study in vivo the consequence of interactions between a signal sequence and the translocase.

A. Economou (ed.), *Protein Secretion*, Methods in Molecular Biology 619,
DOI 10.1007/978-1-60327-412-8_6,

The first assay of protein export was pioneered by Blobel and Dobberstein (2) and is based on cleavage of N-terminal signal sequences by leader peptidase (*see* **Section 2.2**). Short biosynthetic pulse-labeling, immunoprecipitation, and SDS-PAGE detect the precursor and mature forms of a secreted protein. Essentially any periplasmic or outer membrane protein can be used in this assay, although most studies have been restricted to the few *Escherichia coli* proteins for which antibodies are commercially available, such as PhoA and MalE. Thus, chimeric proteins containing a cleavable signal sequence fused to the mature portion of PhoA are easily detected with anti-PhoA antibodies. One limitation of this approach is that uncleaved signal sequences, such as those of many inner proteins and a few secreted proteins, cannot be analyzed by this method. Furthermore, there is an intrinsic technical limitation with this assay since relative amounts of precursor lower than 5% of the amount of mature protein cannot be accurately determined. Finally, signal sequence cleavage is not strictly linked to the translocation process, as evidenced for instance by the fact that Lep, the main leader peptidase, does not associate with the translocase. Pulse-labeling experiments are routinely performed in export studies, but they offer an additional tool that becomes critical when one wants to compare different chimeric proteins. Indeed, the absolute values of an enzymatic assay cannot be directly compared unless the proteins are synthesized at the same rate (*see* **Fig. 6.2**). The factors that govern translation efficiency are poorly understood and translations rates must be experimentally determined (3).

The second type of assays is based on the properties of PhoA, the alkaline phosphatase of *E. coli* (4, 5). PhoA is a non-specific phospho-monoesterase that is normally active only in the periplasm. This compartment-dependent enzymatic activity is based on the fact that PhoA contains four Cys residues that must be oxidized for the enzyme to be active. Since disulfide bond formation normally occurs only in the periplasm, the fraction of PhoA that remains in the cytoplasm is entirely inactive. This offers an exceptional dynamic range were the enzymatic activity of a given signal sequence can be accurately measured over a greater than 1,000-fold range. Furthermore, the mature portion of PhoA is an independent folding unit and can be fused at either end of protein sequences with little or no effects on enzymatic activity. When a chimeric protein containing a signal sequence fused to the mature portion of PhoA is expressed from an inducible promoter, such as the P_{BAD} promoter, maximal rates of transcription are reached within minutes, but the enzyme level only reaches steady state after about 1 h (6). This provides a second in vivo assay for export that allows the visualization of signal sequence-specific or translocase mutation-specific effects (7) (*see* **Fig. 6.1**).

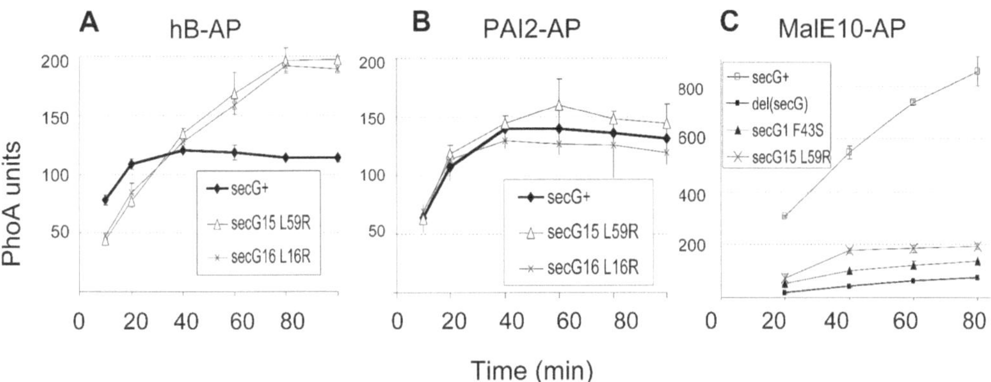

Fig. 6.1. Mutations in signal sequences or in *sec* genes affect the export kinetics of PhoA chimera. (**A** and **B**) An internal deletion in the N-region of the PAI-2 (SERPINB2) signal sequence (hB-AP) is exported more slowly in strains carrying the indicating missense suppressor mutations in *secG*. Since hB-AP expression is lethal in wild type cells, export is rapid at initial times and arrested after 20 min. This effect is specific for hB-AP since the chimera, since the export of PAI2-AP is not affected by these *secG* alleles (adapted from (20)). Other *secG* alleles that selectively suppress the toxicity of PAI2-AP also slow down its export (7). (**C**) The *malE10* mutation (L10P) weakly affects export in wild type cells (21). This signal sequence was fused to the mature portion of PhoA and export kinetics determined in strains carrying the indicated *secG* alleles. The two missense mutations affect export of this protein but they are clearly different from a null *secG* allele.

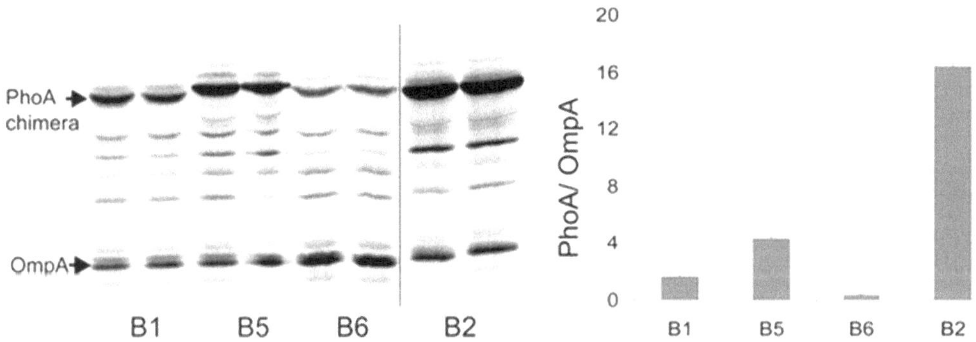

Fig. 6.2. The effect of different N-terminal regions on the synthetic rates of PhoA chimera. The N-terminal regions of several chicken proteins members of the ovalbumin-related serine protease inhibitors (SERPINs; (22)) were fused to the mature portion of PhoA. Duplicate cultures were of pulse labeled and lysates were immuno-precipitated with anti-PhoA and anti-OmpA antibodies and the eluates were analyzed by SDS-PAGE. All chimeras contain the same number of Met residues. The mean ratios of PhoA/ OmpA are shown in the chart.

Finally, cells exporting a PhoA chimera can convert XP, a chromogenic substrate of PhoA in a blue precipitate whose intensity is related to the amount of periplasmic PhoA. While this colony-based assay is very sensitive and can detect subtle differences (*see* **Fig. 6.3**), it is not quantitative and heavily affected by several factors including colony size, colony density, and moisture of the plate. With the use of strains containing appropriate signal sequence mutations, this method provides a genetic screen to isolate *prl* mutations that allow translocase mutants to compensate

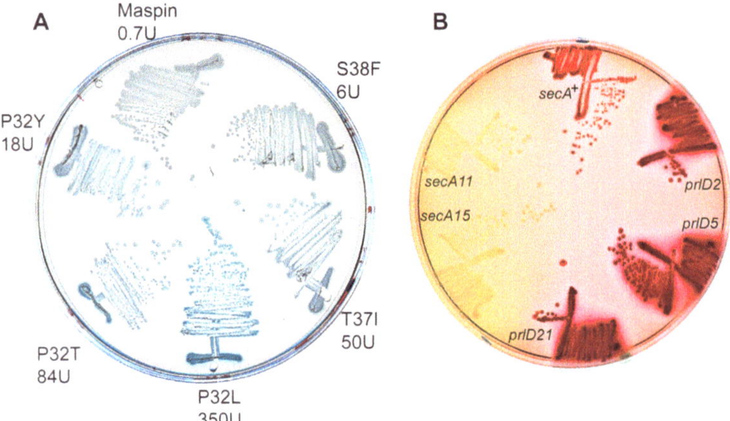

Fig. 6.3. Identification of mutant signal sequences and mutant *sec* genes on indicator plates. (**A**) The N-terminal region of murine maspin (SERPINB5) is devoid of signal sequence activity, but can be converted into an efficient signal sequence (3). The export activities of the indicated mutants are visualized on arabinose-XP indicator plates. The activity of each PhoA chimera is indicated. (**B**) The strains contain the *malE14* signal sequence mutation (21), which only allow a weak fermentation on Mac Conkey maltose indicator plates, resulting in the formation of *pink* colonies. *secA* mutation that improve export (*prlD* alleles (23)) form *dark red* colonies, while mutations that selectively decrease the export activity of mutant signal sequences (24) form *white* colonies.

for signal sequence mutations. Qualitative determination of Prl and Sec phenotypes conferred by mutations in *sec* genes can be also scored on color indicator plates (*see* **Fig. 6.3**, panel B).

2. Materials

Most materials and techniques for molecular biology and bacteriology have been described (8, 9).

2.1. PhoA assay

1. LB: 10 g/l tryptone, 5 g/l yeast extract, 10 g/l NaCl, adjusted to pH 7.4 with 5 M NaOH, and sterilized (20 min at 121°C).
2. Iodoacetamide: 0.5 M stock solution. Store at −20°C, protect from light with aluminum foil to avoid radical cleavage of the I–C bond.
3. MOPS 3X: 200 mM 3-(N-morpholino)propanesulfonic acid, 250 mM NaCl, 48 mM NH_4Cl. Adjust the pH to 7.2 with 1 M KOH. Store at room temperature and protect from light.
4. 1 M $MgCl_2$: filter or autoclave. Store at room temperature.

5. MOPS-buffer: dilute MOPS 3X 3-fold with sterile water. Add $MgCl_2$ to 10 mM. Keep on ice and store at 4°C.
6. TZ: 1 M Tris-HCl pH 8.1, 1 mM $ZnCl_2$. Store at room temperature.
7. PNPP: stock solution is 40 mg/ml *p*-nitro-phenyl-phosphate (Fluka, 771768; the powder is stored desiccated at 4°C) in 1 M Tris-HCl pH8.0. Store at –20°C.
8. AP-STOP: mix 1 volume of 0.5 M EDTA, pH 8.0 and 4 volumes of 2.5 M K_2HPO_4. Store at room temperature.

2.2. Pulse Labeling and Immunoprecipitation

1. M63 5X: 15 g/l KH_2PO_4, 35 g/l K_2HPO_4, 10 g/l $(NH_4)_2SO_4$ and 2.5 ml/l of 1 mg/ml $FeSO_4$; if needed, adjust the pH to 7.0 with KOH.
2. Hydrophobic L-amino acids (50X): dissolve 250 mg each of Ile, Phe, Trp, and Tyr in 100 ml of sterile water. Dissolve by heating in a boiling water bath. Adjust the pH to 8.0 with Na_2CO_3. Sterilize by filtration while the solution is still hot. The solution can be stored at 4°C or at −20°C (*see* **Note 1**).
3. Hydrophilic L-amino acids (100X): dissolve 250 mg each of the other 14 amino acids (no Met and Cys) in 50 ml of water. Dissolve by heating in a boiling water bath. Sterilize by filtration. The solution can be stored at 4°C or at −20°C.
4. Vitamin B1: 2 mg/ml in water. Store at 4°C and protect from light.
5. M63+: M63 1X, hydrophobic amino acids 1X (50 μg/ml each), hydrophilic amino acids 1X (50 μg/ml each), 0.2% glycerol, 4 μg/ml vitamin B1, 1 mM $MgCl_2$. Store at 4°C.
6. [^{35}S]-methionine: either pure methionine or mixtures of methionine and cysteine can be used, the latter being much less expensive. We use [^{35}S]Met-label (IS103) from Hartmann Analytic (http://www.hartmann-analytic.de) which contains 70–80% methionine at a specific activity of >1000 Ci/mmol. Aliquot and store at −80°C. The label can be used for at least 4 months after the reference date provided that amounts are corrected for radioactive decay.
7. Unlabeled Met: 5 mg/ml L-methionine in M63 1X. Prepare freshly.
8. SDS-buffer: 10 mM Tris-HCl pH 8.1, 1 mM EDTA and 1% SDS. Store at room temperature.
9. KI-buffer: 50 mM Tris-HCl pH 8.1, 150 mM NaCl and 2% Triton-X100. Store at room temperature.
10. 1 M DTT: store at −20°C in aliquots.

11. Iodoacetamide: 1 M stock solution. Store at −20°C, protect from light with aluminum foil to avoid radical cleavage of the I–C bond.
12. HS-buffer: 50 mM Tris-HCl pH 8.1, 1 M NaCl, 1 mM EDTA and 1%Triton-X100. Store at room temperature.
13. Anti-PhoA antibodies are from Millipore (AB1204) or from Biodesign/ Meridian Life Science (K59134R). Anti-OmpA antibodies are currently not available commercially but can be obtained from the author. Anti-MalE antiserum is from New England Biolabs (E8030S). IgG aggregate upon repeated cycles of freezing and thawing. To minimize aggregation, IgG should be stored in aliquots, rapidly frozen in liquid nitrogen or ethanol/CO_2, and rapidly thawed in a 27°C water bath. Alternatively, IgG or antisera can be diluted with glycerol (50% concentration final) and stored at −20°C.
14. IgGsorb: Fixed *Staphylococcus aureus* (IGSL 10, The Enzyme Center, Malden, Mass). Reconstitute the suspension in 10 ml of water and incubate several hours on ice. Centrifuge 10 min at 5,000 *g* and 4°C. Resuspend the pellet in 20 ml of KI-buffer and centrifuge again. Resuspend the pellet with 9 ml of KI-buffer (10% suspension). Store in aliquots at –20°C. Fixed cells can also be purchased from Calbiochem (Pansorbin) (*see* **Note 2**).
15. Sample buffer: 50 mM Tris-HCl pH 6.8, 2% w/v SDS, 10% v/v glycerol, 0.01% w/v bromophenol blue. For analysis of proteins under reducing conditions, include 5% β-mecraptoethanol.

2.3. Detection of Improved Export with Mutations in Signal Sequences or in the Translocase

Plasmids and strains are available from the author upon request.

1. pBADslAP: the signalless-PhoA construct contains the KpnI-XbaI fragment from pSWFII (10) cloned into the cognate sites of pBAD24 (6). The KpnI site encodes a Val-Pro linker. The plasmid has the origin of pBR322 (~50 copies per cell), confers resistance to ampicillin, and encodes the positive regulator AraC (3).
2. DHB3: an *ara*, Δ*phoA* MC100-derived strain (11).
3. Arabinose: make a 20% w/v solution, and sterilize by filtration. Store at room temperature.
4. XP: 5-bromo-4-chloro-3 indolyl phosphate (XP) was purchased from Biosynth (http://www.biosynth.com/). The powder is stored at −20°C and protected from light. A 20 mg/ml stock solution is prepared in dimethylformamide and stored at −20°C; protect from light. The stock solution is stable for several months.

5. pKY248: a SecY-encoding plasmid (12). The plasmid has the origin of p15A (~25 copies per cell) and confers resistance to chloramphenicol (*see* **Note 3**).
6. psecG+: a SecG-encoding plasmid (13). The plasmid has the origin of p15A (~25 copies per cell) and confers resistance to chloramphenicol (*see* **Note 3**).
7. pDB707: a SecE-encoding plasmid. The pGB2-derived plasmid (14) has the origin of pSC101 (~5 copies per cell) and confers resistance to spectinomycin (*see* **Note 3**).
8. pBE2: a SecA-encoding plasmid (15). The plasmid has the origin of p15A (~25 copies per cell) and confers resistance to chloramphenicol (*see* **Note 3**).
9. DB636: a strain derived from MC4100, *leu*+, Δara_{714}, *malE18*, *secG*::Kn. The strain grows at 37°C on M63 minimal glucose plates but not on minimal maltose plates because of the combined effects of the *malE* and *secG* mutations (13). The strain forms white colonies on Mac Conkey maltose plates.
10. Minimal plates: dissolve 14 g/l of bacto-agar in 800 ml of water and autoclave. When the melted agar temperature is at ~45–55°C, add 200 ml of M63 5X, 1 ml of 1 M $MgCl_2$, 2 ml of vitamin B1 2 mg/ml, the appropriate sugar (0.2% final concentration), and the desired antibiotics.
11. Maltose: make a 20% w/v solution, and sterilize by filtration. Store at room temperature.
12. Mac Conkey plates: Mac conkey agar is obtained from Difco (http://www.bd.com/ds/technicalCenter/inserts/difcoBblManual.asp). Medium containing 50 g/l is autoclaved. Sugars are added to a final concentration of 1% (*see* **Note 4**).
13. DB502: a strain derived from MC4100, *leu*+, Δara_{714}, *malE14*. The strain forms pink colonies on Mac Conkey maltose plates.

3. Methods

3.1. PhoA Assay

1. Inoculate LB supplemented with the appropriate antibiotic(s) with isolated colonies of the strain to be tested, and grow overnight cultures (*see* **Note 5**).
2. Dilute the cultures 1/100 in fresh medium and grow with aeration at 37°C to an $A_{600\text{ nm}}$ of 0.2–0.4 (*see* **Note 6**).
3. Induce the expression of the reporter gene for 1 h (steady state assay) or various times (kinetics of export). For kinetic

studies, appropriate time points are usually taken between 10 and 90 min after induction (*see* **Fig. 6.1**).

4. During cell growth, prepare two (or more) assay tubes per culture, which contain 0.9 ml de TZ, 25 μl 0.1% SDS and 25 μl of $CHCl_3$. Keep the tubes at room temperature (*see* **Note 7**).
5. Arrest export by collecting 1–1.5 ml on ice in an Eppendorf tube containing 4–6 μl de iodoacetamide (~2 mM final) (*see* **Note 8**).
6. Centrifuge 1 min at 12,000 *g* and 4°C. Discard the supernatant and resuspend the cells in approximately 1 ml of MOPS-buffer. Repeat the washing once more and resuspend thoroughly the cells in 1 ml of MOPS- buffer. Keep the cells on ice (*see* **Note 9**).
7. Read the cell density at 600 nm either directly or after dilution in MOPS-buffer. ($A_{600\ nm}$ in equation).Cell density can be read shortly after the beginning of the incubation.
8. Add the appropriate volume of cells to the assay tubes (*see* **Note 10**), vortex add 100 μl de PNPP 4 mg/ ml (dilution 1:10 of the stock) in TZ vortex, and incubate in a water bath at 28°C until the tube turns slightly yellow (*see* **Note 11**).
9. Stop the reaction with 120 μl of AP-STOP solution, mix thoroughly, and record the incubation time ($t_{[min]}$ in equation).
10. Stop the second tube before it becomes deep yellow and record the incubation time (*see* **Note 12**).
11. Just before reading, centrifuge the cells for 5 min at 12,000 *g* and 4°C (*see* **Note 13**).
12. Remove carefully 0.8 ml from the upper portion and read the *p*-nitrophenol absorbance at 420 nm ($A_{420\ nm}$ in equation).
13. The PhoA units are calculated as:
PhoA units = $1000 \times A_{420\ nm} / (t_{[min]} \times V_{[ml]} \times A_{600\ nm})$

3.2. Pulse Labeling and Immunoprecipitation

1. Inoculate LB supplemented with the appropriate antibiotic(s) with isolated colonies of the strain to be tested, and grow 6–8 h late log phase cultures.
2. Dilute cultures 1/50 in M63+ minimal medium supplemented with the appropriate antibiotic(s) and grow with aeration at 37°C overnight.
3. Dilute the cultures 1/40 in fresh medium and grow at 37°C to an $A_{600\ nm}$ of 0.2 to 0.4 (*see* **Note 14**).
4. Induce the expression of the reporter gene for 10 min. Label 1 ml of cells for 15 s to 1 min with 20–50 μCi/ml

of ^{35}S-methionine. Short pulses ensure that the amount of precursor is maximal.

5. Stop incorporation with 100 μl/ml of unlabeled Met, and collect the cultures on ice (*see* **Note 15**).
6. Centrifuge the cells for 2 min at 12,000 *g* and 4°C; discard the supernatant in radio-active waste.
7. Resuspend the cells in 50 μl of SDS-buffer by vortexing.
8. Lyse the cells by incubating for 2 min in a 90°C water bath; let the tubes cool at room temperature for 5 min.
9. Add 800 μl of KI-buffer mix thoroughly and incubate on ice for at least 10 min.
10. Centrifuge for 10 min at 12,000 *g* and 4°C and transfer the supernatant to a fresh tube. Lysates may be frozen and stored at −80°C.
11. To discriminate between the oxidized (exported) and reduced (cytoplasmic) forms of PhoA chimera, lysates may be treated with 30 mM DTT or untreated, and incubated for 30 min at 37°C. After addition of 0.1 M iodoacetamide, to block all free SH groups, the samples are incubated for 30 min on ice. In this case, the immune complexes must be eluted without β-mercaptoethanol (*optional*).
12. Incubate 100 to 300 μl of lysate with the appropriate antibodies for 1 h at room temperature or overnight on ice.
13. Bind the immune complexes to 30 μl IgGsorb for 30 min on ice with occasional vortexing.
14. Collect the bound complexes by centrifugation for 20 s at room temperature; discard the supernatant in radioactive waste.
15. Thoroughly vortex to resuspend the pellet in 1 ml of HS-buffer and collect the bound complexes by centrifugation for 20s at room temperature (*see* **Note 16**).
16. Repeat the washing procedure once.
17. Thoroughly resuspend the pellet in 1 ml of 10 mM Tris-HCl pH 8.1 and collect the bound complexes by centrifugation for 20s.
18. Resuspend the pellet in 40 μl sample buffer with or without β-mercaptoethanol and elute at 90°C for 2 min (*see* **Note 17**).
19. Centrifuge the cells for 5 min at 12,000 *g* and room temperature and load the eluate on an SDS-PAGE (*see* **Fig. 6.2**).

3.3. Detection of Improved Export with Mutations in Signal Sequences or in the Translocase

3.3.1. Screening of Signal Sequence Mutations

1. Clone the DNA fragment encoding a putative signal sequence or the peptide of interest in pBADslAP so that the N-terminal peptide is in frame with the mature portion of PhoA.
2. The DNA fragment can be mutagenized by PCR (16), with hydroxylamine (17). The plasmid can also be propagated in a mutator strain (9).
3. After transformation in DHB3 cells, plate the cells on LB plates supplemented with 0.2% arabinose and 40 mg/l XP. Incubate at 37°C for 14–16 h. If needed, allow further color development at room temperature.
4. Quantify export as described in **Section 3.1**. Verify that altered export does not result from changes in translation efficiency, as described in **Section3.2** (*see* **Fig. 6.2**) (*see* **Note 18**).

3.3.2. Isolation of *prl* Mutations that Improve the Export Activity of Defective Signal Sequences

1. Although *prl* mutations can be isolated in chromosomal genes, they are dominant and it is easier to work with individual *sec* genes cloned on a plasmid. Plasmids encoding SecY, SecE, SecG, or SecA are described in **Section 2.3**.
2. The *sec* gene of interest, or a fragment thereof, can be mutagenized as described above.
3. Transform strain DB636 carrying pBADslAP with the mutagenized plasmid (*see* **Note 19**). After expression of the antibiotic resistance during a 1 h growth in LB, the cells are centrifuged and washed twice in M63 1X.
4. Plate the cells on minimal maltose plates (up to 10^7/plate) supplemented with the appropriate antibiotic(s), and grow at 37°C for 24–72 h (*see* **Notes 20** and **21**). Purify the colonies on the same plates.
5. Isolate the plasmid encoding the *sec* gene; verify that it confers the expected Prl phenotype by retransforming DB636 cells.
6. Verify the Prl phenotype by streaking on Mac Conkey maltose plates (*see* **Note 22**).
7. Quantify the effect by assaying PhoA activity as described in **Section 3.1** (*see* **Note 23**).
8. Improved signal sequence cleavage can be assessed in strain DB502 using anti-MalE antibodies, as described in **Section 3.2**.

4. Notes

1. The hydrophobic aminoacids tend to precipitate upon storage with no consequence on the experiments. Mix thoroughly or heat before use.
2. Fixed *S. aureus* cells can be replaced by Protein-A Sepharose. While this usually gives less background, the low cost, ease, and rapidity of collecting fixed cells make them more attractive.
3. pKY248, psecG$^+$, pDB707, and pBE2 are all compatible with pBADslAP.
4. Mac Conkey agar contains lactose, but all strains used here are *lac*. Color development and stability is less reliable with Mac Conkey agar base.
5. To ensure a reliable assay, three independent cultures are routinely assayed. Cultures can also be grown in minimal medium (*see* **Section 3.2**), but values obtained in rich and minimal media often vary, particularly with plasmid-borne constructs (3).
6. Cells should be in mid- to late-log phase of growth (less than 1 $A_{600\ nm}$).
7. SDS and $CHCl_3$ are not strictly required since active PhoA is periplasmic; this probably represents a leftover from the β-galactosidase assay. However, they ensure sterility of the assay tubes, particularly during long incubations.
8. The addition of iodoacetamide blocks free cysteines and ensures that cytoplasmic PhoA does not become activated during the assay (18). Some authors also treat the cultures with 1 mM NaF to inhibit acid phosphatase(s), but this is not necessary with most *E. coli* strains.
9. The washing procedure is particularly critical with cells grown in phosphate containing media such as the M63 minimal medium.
10. For cultures with more than 100 U of PhoA, use 10 μl of cells, otherwise use 50–100 μl.
11. The optical density should not exceed 1$A_{420\ nm}$. Usually, two times points are taken for each culture and enzyme activity should be within 10% for the two time points. Incubations for less than 30 min are not reliable. Incubations for 10–15 h are required for samples with less than 5 PhoA units.

12. The samples can be stored in the dark and at 4°C for several hours before reading.
13. This procedure alleviates the need to measure the absorbance at 550 nm, which should otherwise be subtracted in the equation: the term $A_{420\ \mathrm{nm}}$ is then replaced by $(A_{420\ \mathrm{nm}} - (1.75 \times A_{550\ \mathrm{nm}}))$.
14. The medium may contain up to 2% LB with little effect on incorporation for cells that grow poorly in minimal medium.
15. When cells are pulse-labeled for 15s, the best way to arrest processing is to add 0.7 ml of labeled cells to a pre-chilled tube containing 0.6 ml of M63+ and 0.1 ml of unlabeled Met. This ensures a very rapid drop of the culture temperature to about 18°C, a temperature at which translation initiation is essentially prevented while elongation proceeds. Pre-chilling is best achieved in an ice–water mixture. With longer labelings, 20–60s, addition of unlabeled Met to the culture and transfer to an ice–water mixture is adequate. If the samples are to be chased to follow the conversion of the precursor into the mature form, incubate the cultures at 37°C for 1 min or longer.
16. Failure to completely resuspend the pellets results in high backgrounds.
17. β-mercaptoethanol may be omitted from the sample buffer either to allow distinction between the oxidized and reduced forms of the protein or to improve the resolution of proteins that co-migrate with the IgG heavy chains. Membrane proteins precipitate upon boiling in the presence of SDS. This must be eluted by a 15 min incubation at 37°C.
18. While translation efficiency is most affected by alterations near the AUG start codon, mutations, deletions, or insertions as far as 200 nt away have been found to affect expression levels.
19. Expression from p15A derived plasmid is usually sufficiently high. Another SecE-encoding plasmid is also available (pJS51 (19)). However it contains a pBR322 origin and cannot be stably maintained with pBADslAP.
20. Some *prl* mutations exhibit a synthetic lethal phenotype with the *secG* deletion (Boulfekhar et al., unpublished data). These mutations can be isolated in derivatives of DB636. In addition, introduction of a *secG*$^{+}$ gene in DB636 restores growth on minimal maltose at 37°C. In both cases, the selection can be performed at 23°C, at temperature at which the *malE18* mutation prevents maltose utilization in otherwise wild type cells.

21. Although DB636 does not form visible colonies on minimal maltose plates, residual growth often allow for the appearance of prl suppressor mutations during prolonged incubation. These may occur on the plasmid, but also on the chromosome.
22. Weak prl mutations may not give rise to detectably pink colonies. They may be more easily detected in strain DB502 (*see* **Fig. 6.3**).
23. If the selection is done in the absence of pBADslAP, the reporter plasmid can be introduced later.

Acknowledgments

The author thanks Luzma Guzman, Alan Derman, and the other members of the Beckwith lab for sharing their enthusiasm and expertise on alkaline phosphatase. Sandrine Bost identified the power of the kinetic assay to study mutant phenotypes. The experiments in the lab were mostly performed with the invaluable help of Filo Silva. This work was supported by grants of the FNS and by the Canton de Genève.

References

1. Van den Berg, B., Clemons, W.M., Jr., Collinson, I., Modis, Y., Hartmann, E., Harrison, S.C., and Rapoport, T.A. (2004) X-ray structure of a protein-conducting channel. *Nature* **427**, 36–44.
2. Blobel, G., and Dobberstein, B. (1975) Transfer of proteins across membranes. I. Presence of proteolytically processed and unprocessed nascent immunoglobulin light chains on membrane-bound ribosomes of murine myeloma. *J. Cell Biol.* **67**, 835–851.
3. Belin, D., Guzman, L.M., Bost, S., Konakova, M., Silva, F., and Beckwith, J. (2004) Functional activity of eukaryotic signal sequences in Escherichia coli: the ovalbumin family of serine protease inhibitors. *J. Mol. Biol.* **335**, 437–453.
4. Manoil, C. (1991) Analysis of membrane protein topology using alkaline phosphatase and beta-galactosidase gene fusions. *Methods Cell Biol.* **34**, 61–75.
5. Manoil, C., and Beckwith, J. (1986) A genetic approach to analyzing membrane protein topology. *Science* **233**, 1403–1408.
6. Guzman, L.M., Belin, D., Carson, M.J., and Beckwith, J. (1995) Tight regulation, modulation, and high-level expression by vectors containing the arabinose PBAD promoter. *J. Bacteriol.* **177**, 4121–4130.
7. Bost, S., and Belin, D. (1995) A new genetic selection identifies essential residues in SecG, a component of the *Escherichia coli* protein export machinery. *EMBO J.* **14**, 4412–4421.
8. Maniatis, T., Fritsch, E.F., and Sambrook, J. 1982. *Molecular cloning: a laboratory manual.* Cold Spring Harbor, N.Y.: Cold Spring Harbor Laboratory.
9. Miller, J.H. 1992. *A short course in bacterial genetics.* Cold Spring Harbor, NY: Cold Spring Harbor Laboratory Press.
10. Ehrmann, M., Boyd, D., and Beckwith, J. (1990) Genetic analysis of membrane protein topology by a sandwich gene fusion approach. *Proc. Natl. Acad. Sci. USA* **87**, 7574–7578.
11. Boyd, D., Manoil, C., and Beckwith, J. (1987) Determinants of membrane protein topology. *Proc. Natl. Acad. Sci. USA* **84**, 8525–8529.
12. Taura, T., Baba, T., Akiyama, Y., and Ito, K. (1993) Determinants of the quantity of the

stable SecY complex in the Escherichia coli cell. *J.Bacteriol.* **175**, 7771–7775.

13. Bost, S., and Belin, D. (1997) *prl*mutations in the *Escherichia coli sec G* gene. *J.Biol.Chem.* **272**, 4087–4093.
14. Churchward, G., Belin, D., and Nagamine, Y. (1984) A pSC101-derived plasmid which shows no sequence homology to other commonly used cloning vectors. *Gene* **31**, 165–171.
15. Kim, Y.J., Rajapandi, T., and Oliver, D. (1994) SecA protein is exposed to the periplasmic surface of the E. coli inner membrane in its active state. *Cell* **78**, 845–853.
16. Spee, J.H., de-Vos, W.M., and Kuipers, O.P. (1993) Efficient random mutagenesis method with adjustable mutation frequency by use of PCR and dITP. *Nucleic.Acids Res.* **21**, 777–778.
17. Humphreys, G.O., Willshaw, G.A., Smith, H.R., and Anderson, E.S. (1976) Mutagenesis of plasmid DNA with hydroxylamine: isolation of mutants of multi-copy plasmids. *Mol. Gen. Genet.* **145**, 101–108.
18. Derman, A.I., and Beckwith, J. (1995) *Escherichia coli* alkaline phosphatase localized to the cytoplasm slowly acquires enzymatic activity in cells whose growth has been suspended: a caution for gene fusion studies. *J.Bacteriol.* **177**, 3764–3770.
19. Pohlschroder, M., Murphy, C., and Beckwith, J. (1996) In vivo analyses of interactions between SecE and SecY, core components of the Escherichia coli protein translocation machinery. *J.Biol.Chem.* **271**, 19908–19914.
20. Bost, S., Silva, F., Rudaz, C., and Belin, D. (2000) Both transmembrane domains of SecG contribute to signal sequence recognition by the *Escherichia coli* protein export machinery. *Mol. Microbiol.* **38**, 575–587.
21. Bedouelle, H., Bassford, P.J.J., Fowler, A.V., Zabin, I., Beckwith, J., and Hofnung, M. (1980) Mutations which alter the function of the signal sequence of the maltose binding protein of Escherichia coli. *Nature* **285**, 78–81.
22. Benarafa, C., and Remold-O'Donnell, E. (2005) The ovalbumin serpins revisited: perspective from the chicken genome of clade B serpin evolution in vertebrates. *Proc Natl Acad Sci USA* **102**, 11367–11372.
23. Huie, J.L., and Silhavy, T.J. (1995) Suppression of signal sequence defects and azide resistance in *Escherichia coli* commonly result from the same mutations in secA. *J. Bacteriol.* **177**, 3518–3526.
24. Khatib, K., and Belin, D. (2002) A novel class of *secA* alleles that exert a signal-sequence-dependent effect on protein export in *Escherichia coli. Genetics* **162**, 1031–1043.

Chapter 7

Sorting of Bacterial Lipoproteins to the Outer Membrane by the Lol System

Shin-ichiro Narita and Hajime Tokuda

Abstract

Bacterial lipoproteins comprise a subset of membrane proteins with a lipid-modified cysteine residue at their amino termini through which they are anchored to the membrane. In Gram-negative bacteria, lipoproteins are localized on either the inner or the outer membrane. The Lol system is responsible for the transport of lipoproteins to the outer membrane.

The Lol system comprises an inner-membrane ABC transporter LolCDE complex, a periplasmic carrier protein, LolA, and an outer membrane receptor protein, LolB. Lipoproteins are synthesized as precursors in the cytosol and then translocated across the inner membrane by the Sec translocon to the outer leaflet of the inner membrane, where lipoprotein precursors are processed to mature lipoproteins. The LolCDE complex then mediates the release of outer membrane-specific lipoproteins from the inner membrane while the inner membrane-specific lipoproteins possessing Asp at position 2 are not released by LolCDE because it functions as a LolCDE avoidance signal, causing the retention of these lipoproteins in the inner membrane. A water-soluble lipoprotein–LolA complex is formed as a result of the release reaction mediated by LolCDE. This complex traverses the hydrophilic periplasm to reach the outer membrane, where LolB accepts a lipoprotein from LolA and then catalyzes its incorporation into the inner leaflet of the outer membrane.

Key words: Lipoprotein, outer membrane, Lol system, ABC transporter, spheroplasts, reconstitution, *Escherichia coli.*

1. Introduction

Gram-negative bacteria have many species of lipoproteins on the outer membrane. Most outer membrane lipoproteins are transported to the outer membrane by the Lol system (1), while some are directly transported to the outer surface of the outer membrane through specific mechanisms (2). The transport of lipoproteins by the Lol system can be monitored both in vivo

A. Economou (ed.), *Protein Secretion*, Methods in Molecular Biology 619,
DOI 10.1007/978-1-60327-412-8_7,

and in vitro. Inactivation of the Lol system results in mislocalization of outer membrane lipoproteins to incorrect compartments, depending on the species of impaired Lol proteins. Depletion of LolA (3)or LolCDE (4)results in the accumulation of outer membrane lipoproteins in the inner membrane (3, 4), while depletion of LolB results in the accumulation of outer membrane lipoproteins in the periplasm and then in the inner membrane (5). Thus, the Lol system is responsible for proper localization of outer membrane lipoproteins and essential for *Escherichia coli* growth. Transient mislocalization of lipoproteins in the inner membrane through inhibition of the Lol protein function is detected by pulse labeling of newly synthesized lipoproteins, followed by fractionation of the membranes into inner and outer ones by sucrose gradient centrifugation. Lipoprotein transport can be dissected in vitro. The release of lipoproteins from the inner membrane requires both the LolCDE complex and LolA, and is examined in spheroplasts. Because spheroplasts lack periplasmic materials including LolA, newly synthesized lipoproteins accumulate on the periplasmic surface of the inner membrane while non-lipidated periplasmic proteins and outer membrane proteins are secreted into the spheroplast supernatant. When purified LolA is added to the spheroplasts, lipoproteins are released into the medium, enabling examination of the activities of LolCDE and LolA. The lipoprotein-sorting signal is also examined using this assay system. The lipoprotein releasing activity can be reconstituted in proteoliposomes from purified LolCDE and phospholipids, and assayed in the presence of LolA. Incorporation of lipoproteins into the outer membrane is examined in vitro with the lipoprotein–LolA complex, which is prepared through the in vitro release reaction mentioned above. This step requires the LolB function.

2. Materials

2.1. Membrane Localization of Lipoproteins in E. coli

2.1.1. Pulse-Labeling of Lipoproteins and Separation of Membranes

1. M63 minimal media: For 5 mL, mix 50 μL of 200 mM $MgSO_4$, 50 μL 2 mg/ml thiamine, 50 μL 4 mg/ml each of thymine and uracil, 250 μL 0.8 mg/ml each of all amino acids except methionine and cysteine, 100 μL 20% glucose or maltose, and 4.5 mL M63 salt. Each solution is sterilized in an autoclave prior to mixing. For 900 mL of M63 salt, add 2 g KH_2PO_4, 7 g K_2HPO_4, 1 g $(NH_4)_2SO_4$, 0.5 g Na_3-citrate-$2H_2O$, and 5 g NaCl.
2. Tran^{35}S-label, a mixture of 70% [^{35}S]methionine and 20% [^{35}S]cysteine (1000 Ci/mol, MP Biomedicals, cat. No. 0151006.4).

3. 25–55% (w/w) sucrose linear gradient: Prepare 25 and 55% sucrose solutions in 20 mM Tris-HCl, pH 7.5, containing 1 mM EDTA. Overlay 1 mL 55% sucrose solution in a 11 × 34 mm ultracentrifuge tube (Seton Scientific, Sunnyvale, CA) with 1 mL 25% sucrose solution. A linear gradient is made with a Gradient Master (Biocomp Instruments, Fredericton, Canada). Alternately, a linear gradient can be made manually with a two-chambered linear gradient maker (Hoefer, San Francisco, CA).
4. Bath type sonicator such as a Bransonic cleaner (Branson Ultrasonics, Danbury, CT), whose chamber is filled with ice-water.

2.1.2. Immunoprecipitation

1. IP buffer: 50 mM Tris-HCl, pH 7.5, 2% Triton X-100, 150 mM NaCl, 1 mM EDTA. Store at 4°C. Add 1 mM *p*-amidinophenylmethylsulfonylfluoride immediately prior to use.

2.1.3. SDS-Polyacrylamide Gel Electrophoresis (SDS-PAGE)

1. Stock solutions: 37.5% acrylamide and 0.8% bis-acrylamide. Store at 4°C.
2. Running buffer: 100 mM Na-phosphate, pH 7.2, 6% SDS.
3. HI sample buffer (5X): 50 mM Na-phosphate, pH 7.2, 5% SDS, 50% glycerol, 0.025% bromophenol blue, 5% (v/v) 2-mercaptoethanol. Prepare without 2-mercaptoethanol and store at −80°C. Add 2-mercaptoethanol before use.

2.2. Release of Lipoproteins from Sphcroplasts

1. Tris-sucrose solution: 10 mM Tris-HCl, pH 7.5, 750 mM sucrose. Freshly prepare and chill at 4°C.
2. Labeling mix: Dissolve 430 mg sucrose in M63 salt and fill up to 5 mL. Add 50 μL 10 mg/mL DNase I and 66 μCi Tran^{35}S-label immediately prior to use. Labeling mix should be prewarmed to 30°C.
3. Chase mix: 4% each of methionine and cysteine dissolved in labeling mix without Tran^{35}S-label.
4. Peristaltic pump (Atto, Tokyo, Japan).

2.3. Reconstitution of Lipoprotein-Releasing Apparatus

1. *E. coli* lipid: Dissolve 5 g of *E. coli* total lipid extract (Avanti Polar Lipids, Alabaster, AL) in a 15-mL mixture of chloroform and methanol (9: 1). Add 500 mL acetone containing 2 mM 2-mercaptoethanol to precipitate polar lipids. Dissolve the precipitate in 400 mL diethyl ether containing 2 mM 2-mercaptoethanol. After removal of diethyl ether in an evaporator, dissolve lipids in 2 mM 2-mercaptoethanol to give a final concentration of about 50 mg/mL. Dispense aliquots into glass tubes, fill the tubes with N_2 gas, and store at −80°C.

2. LolA: His-tagged LolA is purified from *E. coli* TT016 harboring pAM201 as described (6) and stored at −80°C in 20 mM Tris-HCl, pH 8.0.
3. LolCDE complex: The LolCDE complex containing His-tagged LolD is purified from *E. coli* JM83 cells harboring pKM301 and pKM402 as described (7), and stored at -80°C in 50 mM Tris-HCl, pH 7.5, containing 0.01% *n*-dodecyl-β-D-maltopyranoside, 10% glycerol and 2 mM ATP.
4. Pal: Outer membrane lipoprotein Pal is purified from *E. coli* JM83 cells harboring pTAN21 (1) as described (8). Purified Pal is dissolved in 10 mM Tris-HCl, pH 7.5, containing 2% sucrose monocaprate.
5. Buffer A: 50 mM Tris-HCl, pH 7.5, 5 mM $MgSO_4$, 100 mM NaCl.

2.4. Incorporation of Lipoproteins into the Outer Membrane

1. Outer membrane: Crude membrane fractions are prepared as described (9) and the outer membrane fraction is separated by sucrose density gradient as described (10). Outer membranes devoid of LolB are prepared from *E. coli* SM704 cells grown in the absence of IPTG. The *lolB* gene in this strain is under the control of the *lac* promoter/operator (11).

3. Methods

3.1. Membrane Localization of Lipoproteins in E. coli

To examine the localization of lipoproteins in vivo, total membranes are separated into inner and outer ones by sucrose density gradient centrifugation (for example, *see* **Fig. 7.1**). Cellular proteins are pulse-labeled with [^{35}S]methionine in vivo. The labeled cells are converted into spheroplasts and disrupted in a microcentrifuge tube in a bath-type sonicator. The inner and outer membranes are fractionated into distinct fractions, which are confirmed by Western blotting with antibodies against inner and outer membrane proteins (10). Radiolabeled lipoproteins in each fraction are immunoprecipitated and analyzed by SDS-PAGE, followed by autoradiography. For analysis of the major outer membrane lipoprotein Lpp, comprising 58 amino acids, a particular SDS-PAGE system called HI gel (12) should be used. This system was developed for the analysis of small molecular proteins such as Lpp

3.1.1. Pulse-Labeling of Lipoproteins and Separation of Membranes

1. *E. coli* cells are cultivated in a test tube containing 5 mL M63 minimal medium until the culture OD at 660 nm reaches 0.5 to 1.0. If cells carry a chromosomal *lol* gene under a controllable promoter, for example the *lac*

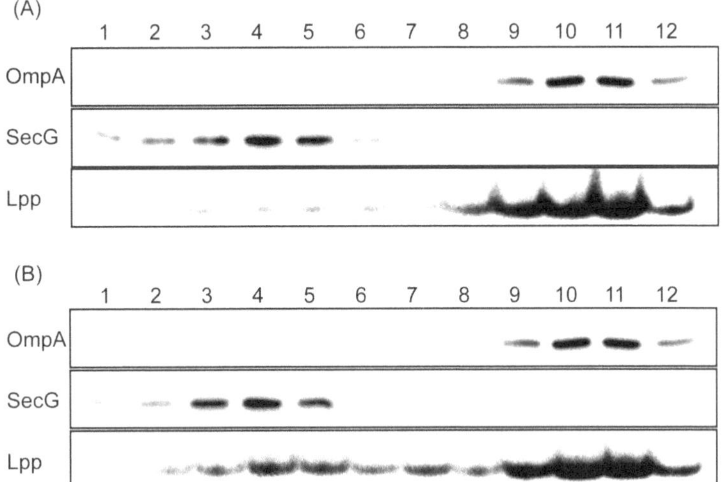

Fig. 7.1. Mislocalization of Lpp in the inner membrane (14). *E. coli* DLP79-22 cells were incubated without (**A**) or with (**B**) 300 μg/mL LolA-inhibitor CCT-00431 at 37̊C for 5 min and labeled with Tran ^{35}S-label. A total membrane fraction was prepared from the cells and subjected to sucrose density gradient centrifugation. Lpp in each fraction was immunoprecipitated with anti-Lpp antiserum, and analyzed by SDS-PAGE and autoradiography. OmpA and SecG were detected by immunoblotting with the respective antisera, as markers for the outer and inner membranes, respectively. Partial inhibition of the LolA function by CCT-00431 caused mislocalization of a part of Lpp in the inner membrane.

promoter, the Lol protein can be depleted by growing the cells in the absence of IPTG until the cells cease to grow.

2. If the gene encoding a lipoprotein of interest is cloned on a plasmid under the controllable promoter, induce the expression of the lipoprotein for 5 min.
3. Transfer the cell culture to a 14 mL polypropylene tube (BD Biosciences, San Jose, CA), to which 50 μCi Tran^{35}S-label has been added, and incubate at an appropriate temperature for 1 min.
4. Labeling can be chased for an appropriate period by the addition of a mixture of 12 mM each of methionine and cysteine.
5. Labeling is terminated by the addition of 5 mL crushed ice, followed by shaking until the ice dissolves.
6. Cells are harvested in microcentrifuge tubes, washed with 1 mL 20 mM Tris-HCl, pH 7.5, and then resuspended in 250 μL of 20 mM Tris-HCl, pH 7.5, containing 750 mM sucrose and 30 μg/mL DNase I. Cells are transferred to a 14-mL polypropylene tube. If a centrifuge for 14-mL tubes is available, it is convenient to harvest cells in such tubes.
7. Cells are converted to spheroplasts by the addition of 12.5 μL 2 mg/mL lysozyme and 500 μL 1.5 mM EDTA and then kept on ice for 10 min.

8. After the addition of 15 μL 100 mM $MgCl_2$, spheroplasts are disrupted in a bath type sonicator filled with ice water until the suspension becomes clear.
9. Cleared spheroplast suspensions are centrifuged at 10,000 *g* for 1 min to remove unbroken cells, and then at 100,000 *g* for 30 min in a microcentrifuge (Beckman Optima TLX ultracentrifuge) with a TLA-100.3 rotor to obtain total membranes as pellets.
10. Total membrane fractions are suspended in 100 μL 20 mM Tris-HCl, pH 7.5, containing 1 mM EDTA by pipetting up and down. Brief sonication helps to yield a homogeneous suspension (*see* **Note 1**).
11. Prepare a 2 mL 25–55% (w/w) linear sucrose gradient and chill it at 4°C.
12. Overlay total membranes on the sucrose gradient and centrifuge it overnight at 45,000 *g* in a Beckman TLS-55 swing rotor in an Optima TLX ultracentrifuge.
13. The sucrose gradient is fractionated from the top to the bottom by carefully removing aliquots with a micro pipetter.

3.1.2. Immunoprecipitation

1. To immunoprecipitate lipoproteins from the fractionated samples, rabbit antisera raised against specified lipoproteins and IgG sorb (fixed *Staphylococcus aureus* cells, Enzyme Center, Lawrence, MA) are used. Alternatively, Protein A-agarose, in which protein A is attached to agarose beads, can be used. Lyophilized IgG sorb is dissolved in distilled water. This solution may be kept at 4°C for up to about 2 months.
2. Membrane proteins in each fraction (200 μL) of the sucrose gradient are precipitated by the addition of a 1/10 volume of 100% trichloroacetic acid (TCA). After 10-min incubation on ice, samples are centrifuged at 16,000 *g* for 10 min, and the supernatants are discarded. The precipitates are washed with acetone, dried, and then dissolved in 50 μL 50 mM Tris-HCl, pH 7.5, containing 1% SDS and 1 mM EDTA. One third of the dissolved samples is analyzed by SDS-PAGE as described by Laemmli (13), followed by Western blotting with anti-SecG and -OmpA antisera to determine the positions of the inner and outer membranes.
3. To the rest of the dissolved samples, add 1 mL IP buffer and keep on ice for 10 min.
4. Materials precipitated on the addition of IP buffer are removed by centrifugation in a microcentrifuge at 16,000 *g* for 5 min. The supernatants are transferred to new microcentrifuge tubes and treated with 1.5 μL antisera raised against

a specific lipoprotein by repeated inversion of the tubes 10 times, followed by incubation at 4°C overnight.

5. Add 50 μL IgG sorb solution to each tube and mix with a Vortex. Keep the mixture at 4°C for 20 min.
6. The IgG sorb is precipitated by centrifugation at 4,000 *g* for 1 min and the supernatant is discarded.
7. Add 500 μL IP buffer to the precipitate, and mix with the Vortex until the IgG sorb is suspended homogenously. The IgG sorb is again precipitated by centrifugation at 16,000 *g* for 1 min, and the supernatant is discarded.
8. Add 500 μL 0.05% SDS, mix briefly with the Vortex, and centrifuge at 16,000 *g* for 1 min. Discard the supernatant. This step is important for removal of Triton X-100 present in the IP buffer, because it disturbs SDS-PAGE carried out with the HI gel system.
9. Add 15 μL HI sample buffer, and then heat at 98°C while mixing the sample at 1,400 rpm with a Thermomixer (Eppendorf). After cooling to room temperature, centrifuge the samples in microcentrifuge tubes at 16,000 *g* for 5 min. Analyze the supernatants by SDS-PAGE.

3.1.3. SDS-PAGE

1. The protocol for the HI gel system is as follows. For one mini-gel, mix 3 mL acrylamide/bis-acrylamide solution, 900 μL 1 M Na-phosphate, pH 7.2, 450 μl 10% SDS, 4.65 mL water, 13.5 mg ammonium persulfate, and 9 μL *N,N,N,N′*-tetramethyl-ethylenediamine (TEMED). Pour in the mixture, leaving a space of about 2 mm for a stacking gel, and overlay with ethanol. The gel should polymerize within about 1 h.
2. Discard the ethanol and rinse the top of the gel with water.
3. Prepare the stacking gel by mixing 200 μL acrylamide/bis-acrylamide solution, 20 μL 1 M Na-phosphate, pH 7.2, 100 μL 10% SDS, 1.68 mL water, 4 mg ammonium persulfate, and 2 μL TEMED. Pour in the stacking gel solution and insert the comb. The gel should polymerize within about 30 min.
4. Carefully remove the comb and clean the wells with distilled water. Using an aspirating device, remove remaining water in the wells.
5. Pour the running buffer into the lower chamber of the electrophoresis unit. Load 10 μL of a sample into each well. Carefully overlay the wells with the running buffer using a pipette. Pour the running buffer into the upper chamber.
6. Run 50 mA through the stacking gel and 100 mA through the separating gel. Stop the electrophoresis before

the dye fronts run off the gel. Cover the gel with a Whatman 3MM paper and dry it with a vacuum heated slab gel dryer. Radiolabeled lipoproteins can be visualized by autoradiography.

3.2. Release of Lipoproteins from Spheroplasts

The first step of Lol-mediated reactions is the detachment of outer membrane-specific lipoproteins from the inner membrane, and involves LolCDE and LolA. This step can be examined in spheroplasts, from which outer membrane-specific lipoproteins are released by the addition of LolA. After conversion of *E. coli* cells to spheroplasts, newly synthesized lipoproteins are labeled with ^{35}S-methionine. Lipoproteins released into the spheroplast supernatant are immunoprecipitated, and detected by SDS-PAGE and autoradiography. Confirm that the release of non-lipidated proteins such as maltose binding protein or OmpA is LolA-independent, while that of outer membrane-specific lipoproteins requires LolA. This proves that cells are properly converted to spheroplasts, which do not burst during experiments.

3.2.1. Preparation of Spheroplasts

1. *E. coli* MC4100 cells are cultivated in a test tube containing 5 mL of M63 minimal medium until the culture OD at 660 nm reaches 0.8 (*see* **Note 2**).
2. Cells are harvested in microcentrifuge tubes at 16,000 *g* for 1 min and resuspended in 500 μL Tris-sucrose solution. The cell suspensions are then transferred to 50-mL conical flasks.
3. 25 μL 2 mg/mL lysozyme is added and the flask is kept on ice for 2 min.
4. 1 mL of 0.9 mM EDTA, pH 7.5, is gradually added through a peristaltic pump while gently rotating the flask on ice. *E. coli* cells are converted to spheroplasts at this step (*see* **Note 3**). Incubate spheroplasts on ice for an additional 5 min.

3.2.2. Detection of Pulse-Labeled Lipoproteins that Are Released from Spheroplasts by LolA

1. 300 μL spheroplast suspension is transferred to a 14 mL polypropylene tube and incubated on ice for 3 min after addition of 1 μL 1 mg/mL LolA (*see* **Note 4**).
2. The tube is then transferred to a 30°C water bath and incubated for 1 h.
3. 750 μL labeling mix is added to the tube, which is then incubated at 30°C for 2 min.
4. 75 μL chase mix is added and the tube is incubated at 30°C for 2 min.
5. The reaction is terminated by placing the tube into an ice–water bath for 1 min. Each spheroplast suspension is transferred to a microcentrifuge tube and centrifuged at

16,000 *g* for 2 min. The supernatant fraction is transferred to a new microcentrifuge tube.

6. To precipitate proteins in the supernatant fraction, 120 μL 100% (w/v) TCA is added. To the spheroplasts recovered as pellets, 100 μL 10% (w/v) TCA is added and then spheroplasts are disrupted in a bath type sonicator. Precipitates are dissolved as described in **Section 3.1.2**. Both fractions are then subjected to immunoprecipitation with anti-OmpA and -lipoprotein antisera, followed by SDS-PAGE and autoradiography.

3.3. Reconstitution of Lipoprotein-Releasing Apparatus

The LolA-dependent release of lipoproteins can be reconstituted in proteoliposomes from LolCDE, lipoproteins and phospholipids (**Fig. 7.2A**). The efficiency of lipoprotein release from

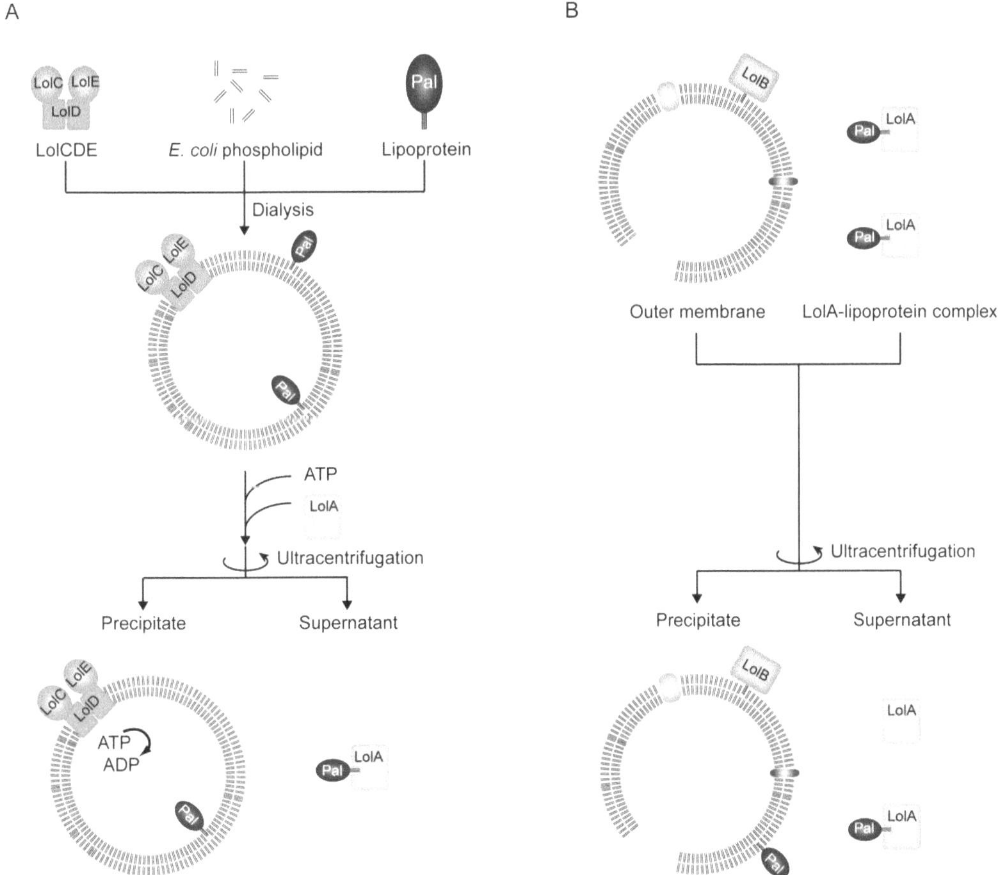

Fig. 7.2. Schematic representation of lipoprotein transport. (**A**) The lipoprotein-releasing apparatus is reconstituted in proteoliposomes from *E. coli* phospholipids, Pal and LolCDE by means of the sucrose monocaprate dilution method. LolA-dependent release is then examined by measuring the amount of Pal in the supernatant. (**B**) The LolA–Pal complex is incubated with the outer membrane at 37̊C for 1 h. Incorporated lipoproteins are detected in the precipitate after ultracentrifugation.

proteoliposomes is not high, presumably because the orientation of the reconstituted LolCDE and lipoproteins is random; about 10% of reconstituted lipoproteins is released from proteoliposomes. However, the reconstitution system is essential for examining the release reaction as to the requirement for nucleotides, the effect of the phospholipid composition, and the amounts and structures of lipoproteins. Among *E. coli* lipoproteins so far tested, Pal is the most suitable substrate, although the reason for this is not completely known.

3.3.1. Reconstitution

1. 800 μg *E. coli* phospholipids, 1 μg purified LolCDE, and 2 μg purified Pal are mixed in 100 μL buffer A containing 1.2% sucrose monocaprate, followed by incubation on ice for 10 min (*see* **Note 5**).
2. The reconstitution mixture is diluted with 900 μL buffer A and then dialyzed against 2 L buffer A at 4°C overnight.
3. The dialyzed sample is centrifuged at 150,000 *g* for 2 h in a TLA-100.3 rotor in an Optima TLX ultracentrifuge.
4. The precipitates are suspended in 100 μL buffer A by pipetting up and down. The suspensions are frozen in a dry ice–ethanol bath and then thawed at room temperature, followed by brief sonication in a bath type sonicator.

3.3.2. Assay for Lipoprotein-Releasing Activity of Reconstituted LolCDE

1. 250 μL buffer A containing 4 μg LolA and 2 mM ATP is added to 50 μL proteoliposome suspension in a 1.5-mL ultracentrifuge tube (*see* **Note 6**).
2. The lipoprotein release reaction is started by incubation at 30°C and terminated by transferring the reaction mixture on ice for 10 min.
3. Centrifuge the reaction mixture at 150,000 *g* for 2 h in a TLA-100.3 rotor in an Optima TLX ultracentrifuge. Proteins in the supernatant and pellets are precipitated by treatment with TCA (*see* **Note 7**).
4. One-third of the supernatants and 1/50 of the precipitates are analyzed by SDS-PAGE according to Laemmli (13), and proteins are transferred to polyvinylidene difluoride membranes, followed by immunodetection with anti-Pal antiserum.

3.4. Incorporation of Lipoproteins into the Outer Membrane

Incorporation of lipoproteins into the outer membrane is dependent on the function of LolB. Therefore, lipoproteins accumulate in the periplasm as a complex with LolA when the LolB function is depleted. Lipoproteins are transferred from LolA to the outer membrane via LolB (**Fig. 7.2B**). To examine this reaction in vitro, the LolA–lipoprotein complex is prepared from the spheroplast supernatant obtained after the LolA-dependent release assay described in **Section 3.2**.

1. Initiate the release of ^{35}S-labeled lipoproteins from spheroplasts by the addition of LolA as described above. After removal of spheroplasts by centrifugation at 16,000 *g* for 2 min, the supernatant is further centrifuged at 100,000 *g* for 30 min to remove insoluble materials (*see* **Note 8**).
2. 1 mL of the supernatant is mixed with 0.2 mg outer membranes prepared from wild-type cells or *lolB*$^{-}$ cells.
3. The mixture is incubated at 30°C for 1 h.
4. The reaction is terminated by chilling the mixture on ice. After centrifugation at 100,000 *g* for 30 min, the supernatants and pellets are treated with TCA.
5. [^{35}S]-labeled lipoproteins in each fraction are subjected to immunoprecipitation, SDS-PAGE, and autoradiography.

4. Notes

1. Thorough homogenization of membranes prior to loading on the sucrose gradient is important for successful fractionation of the inner and outer membranes. Use of a glass-Teflon homogenizer is recommended if proteins are not radiolabeled.
2. If a lipoprotein is cloned into a plasmid under the control of an inducible promoter, cultivate the cells in the presence of an inducer for 5 min before the cells are harvested. The inducer should be also added to the labeling mix.
3. Use of a peristaltic pump prevents a rapid increase in EDTA concentration, which may induce the bursts of spheroplasts. Some strains are highly sensitive to an osmolarity change. In such a case, consider addition of sucrose to the EDTA solution.
4. Spheroplasts are fragile. When the spheroplast suspensions are transferred, use a pipette with a wide-open tip.
5. So far as examined, sucrose monocaprate is the best detergent for reconstituting LolCDE into proteoliposomes by the detergent dilution method. Octylglucoside destabilizes the LolCDE complex and cannot be used.
6. Proteoliposomes reconstituted by this method are considerably permeable to ATP under the conditions used.
7. To avoid contamination of the supernatant fraction by the precipitate, take a 4/5 volume from the top and discard the rest of the supernatant.

8. Various outer membrane lipoproteins labeled with ^{35}S are released from spheroplasts as a complex with LolA. The homogeneous LolA–Pal complex can be formed in vitro as described in **Section 3.3.2**. In this case, Pal-depleted outer membranes should be used since reconstituted Pal is not labeled and thus not detectable on Western blotting.

Acknowledgements

We wish to thank Rika Ishihara for the help in the preparation of this chapter. The work described here was supported by grants to H. T. from the Ministry of Education, Science, Sports, and Culture of Japan.

References

1. Yokota, N., Kuroda, T., Matsuyama, S., and Tokuda, H. (1999) Characterization of the LolA-LolB system as the general lipoprotein localization mechanism of *Escherichia coli*. *J. Biol. Chem.* **274**, 30995–30999.
2. Pugsley, A.P., d'Enfert, C., Reyss, I., and Kornacker, M.G. (1990) Genetics of extracellular protein secretion by Gram-negative bacteria. *Annu. Rev. Genet.* **24**, 67–90.
3. Tajima, T., Yokota, N., Matsuyama, S., and Tokuda, H. (1998) Genetic analyses of the in vivo function of LolA, a periplasmic chaperone involved in the outer membrane localization of *Escherichia coli* lipoproteins. *FEBS Lett.* **439**, 51–54.
4. Narita, S., Tanaka, K., Matsuyama, S., and Tokuda, H. (2002) Disruption of *lolCDE*, encoding an ATP-binding cassette transporter, is lethal for *Escherichia coli* and prevents release of lipoproteins from the inner membrane. *J. Bacteriol.* **184**, 1417–1422.
5. Tanaka, K., Matsuyama, S., and Tokuda, H. (2001) Deletion of *lolB*, encoding an outer membrane lipoprotein, is lethal for *Escherichia coli* and causes accumulation of lipoprotein localization intermediates in the periplasm. *J. Bacteriol.* **183**, 6538–6542.
6. Miyamoto, A., Matsuyama, S., and Tokuda, H. (2001) Mutant of LolA, a lipoprotein-specific molecular chaperone of *Escherichia coli*, defective in the transfer of lipoproteins to LolB. *Biochem. Biophys. Res Commun.* 287, 1125–1128.
7. Ito, Y., Kanamaru, K., Taniguchi, N., Miyamoto, S., and Tokuda, H. (2006) A novel ligand-bound ABC transporter, LolCDE, provides insights into the molecular mechanisms underlying membrane detachment of bacterial lipoproteins. *Mol. Microbiol.* **62**, 1064–1075.
8. Mizuno, T. (1979) A novel peptidoglycan-associated lipoprotein found in the cell envelope of *Pseudomonas aeruginosa* and *Escherichia coli*. *J. Biochem. (Tokyo)* **86**, 991–1000.
9. Kaback, H. R. (1971) Bacterial membranes. *Methods Enzymol.* **22**, 99–120.
10. Matsuyama, S., Tajima, T., and Tokuda, H. (1995) A novel periplasmic carrier protein involved in the sorting and transport of *Escherichia coli* lipoproteins destined for the outer membrane. *EMBO J.* **14**, 3365–3372.
11. Matsuyama, S., Yokota, N., and Tokuda, H. (1997) A novel outer membrane lipoprotein, LolB (HemM), involved in the LolA (p20)-dependent localization of lipoproteins to the outer membrane of *Escherichia coli*. *EMBO J.* **16**, 6947–6955.
12. Hussain, M., Ichihara, S. and Mizushima, S. (1980) Accumulation of glyceride-

containing precursor of the outer membrane lipoprotein in the cytoplasmic membrane of *Escherichia coli* treated with globomycin. *J. Biol. Chem.* **255**, 3707–3712.

13. Laemmli, U. K. (1970) Cleavage of structural proteins during the assembly of the head of bacteriophage T4. *Nature.* **227**, 680–685.

14. Ito, H., Ura, A., Oyamada, Y., Yoshida, H., Yamagishi, J., Narita, S., Matsuyama, S., and Tokuda, H. (2007) A new screening method to identify inhibitors of the Lol (localization of lipoproteins) system, a novel antibacterial target. *Microbiol. Immunol.* **51**, 263–270.

Chapter 8

Purification and Functional Reconstitution of the Bacterial Protein Translocation Pore, the SecYEG Complex

Ilja Kusters, Geert van den Bogaart, Janny de Wit, Viktor Krasnikov, Bert Poolman, and Arnold Driessen

Abstract

In bacteria, proteins are secreted across the cytoplasmic membrane by a protein complex termed translocase. The ability to study the activity of the translocase in vitro using purified proteins has been instrumental for our understanding of the mechanisms underlying this process. Here, we describe the protocols for the purification and reconstitution of the SecYEG complex in an active state into liposomes. In addition, fluorescence based in vitro assays are described that allow monitoring translocation activity discontinuously and in real time.

Key words: Protein secretion, reconstitution, SecYEG, SecA, in vitro translocation, fluorescence, quenching.

1. Introduction

Protein translocation across the cytoplasmic membrane of bacteria is mediated by a protein complex termed translocase (for review *see* 1). Translocase consists of the membrane embedded protein conducting channel SecYEG (2), the associated soluble motor protein SecA (3, 4), and a chaperone, SecB. Secretory proteins synthesized at the ribosome are bound as nascent chains by SecB which prevents their folding and aggregation. SecB targets these so-called precursor proteins to the SecYEG bound SecA (4, 5). Subsequent protein translocation is driven by SecA motor through repeated cycles of ATP binding and hydrolysis whereby the precursor protein is threaded through the SecYEG pore (6).

A. Economou (ed.), *Protein Secretion*, Methods in Molecular Biology 619,
DOI 10.1007/978-1-60327-412-8_8, © Springer Science+Business Media, LLC 2010

Major advances in our understanding of this process have been achieved by studying the function of the components of the translocase by in vitro methods. In this chapter we describe methods to express, purify, and functionally reconstitute the translocase into proteoliposomes and assays to monitor in vitro translocation activity discontinuously and in real time.

2. Materials

2.1. Isolation of Inner Membrane Vesicles (IMVs)

1. LB broth supplemented with 0.1 mg/mL ampicillin
2. *E. coli* SF100 transformed with pET84 (for over expression of $SecY_{G295C}$ EG with N-terminal His-Tag on SecY, ApR)
3. Isopropyl-β-d-thiogalactopyranoside (IPTG), 1 M
4. PMSF 100 mM in 96% Ethanol, 1 M Dithiothreitol (DTT), 100 mg/mL DNase and RNase each.
5. Cell disrupter (French press)
6. Tris-sucrose: 50 mM Tris-HCl pH8, 20% (w/v) sucrose
7. Tris-sucrose for sucrose gradient: Tris-sucrose with 55, 51, 45 and 36 % (w/v) sucrose
8. Hepes-KOH pH 7, 50 mM

2.2. Purification and Fluorescent Labeling of SecYEG

1. Buffer S: 50 mM Hepes-KOH pH 7, 20% (v/v) glycerol, 100 mM KCl, 2% (w/v) n-Dodecyl-β-maltoside (DDM)
2. Buffer W: Buffer S with 0.1% DDM and 10 mM Imidazol
3. Buffer E: Buffer W with 0.3 M Imidazol
4. HIS-Select Nickel Affinity Gel (Sigma-Aldrich)
5. Fluorescein-maleimide (Invitrogen, Molecular Probes), 100 mM in dimethylformamide
6. Bio-Spin Chromatography column, empty (Bio-Rad)

2.3. Reconstitution of SecYEG into E. coli Total Lipid Liposomes

1. SecYEG-buffer: 50 mM Tris pH 8, 50 mM KCl, 10% glycerol, 1 mM DTT
2. *E. coli* total lipids (Avanti Polar lipids, Inc.) 4 mg/mL in SecYEG-buffer
3. Bio-Beads SM-2 adsorbents (Bio-Rad), washed and equilibrated (*see* step 5 of **Section 3.3**)
4. Bio-Spin Chromatography column, empty (Bio-Rad)
5. Bath sonicator or membrane extruder

2.4. Purification of SecA

1. LB broth supplemented with 0.1 mg/mL ampicillin
2. *E. coli* DH5α transformed with plasmid pMKL18 (unpublished, gift of R. Freudl, SecA gene cloned in pUC19 vector, expression of SecA, ApR) (7).
3. Cell disrupter (French press)
4. SecA buffer: 50 mM Tris-HCl pH 7.6, 10% glycerol, 1 mM DTT and SecA buffer supplemented with 1 M NaCl
5. FPLC system (ÄKTA explorer^{TM}, GE Healthcare or equivalent)
6. HiTrap Q HP columns (5 mL)
7. Centriprep® YM-50 centrifugal filter unit (Millipore)
8. Superdex 200 XK26/60 column (GE Healthcare)
9. SDS-PAGE

2.5. Purification of ProOmpA from Inclusion Bodies

1. LB broth supplemented with 0.1 mg/mL ampicillin
2. *E. coli* DH5α transformed with pET503 (over expression of proOmpA C290S, ApR)
3. Isopropyl-β-d-thiogalactopyranoside (IPTG), 1M
4. Tris-HCl pH 7, 50 mM
5. Sonicator MSE Soniprep 150 (Sanyo Biomedical Europe) or other cell disruptor.
6. ProOmpA buffer: 8 M urea, 50 mM Tris-HCl pH 7.0

2.6. Fluorescent Labeling of ProOmpA

1. Tri(2-carboxyethyl)phosphine (TCEP, Invitrogen) 100 mM in 100 mM Tris-HCl, pH 7.0 (*see* **Note 1**)
2. Fluorescein-5-maleimide (Invitrogen) 40 mM in dimethylformamide (DMF), (*see* **Note 2**)
3. ProOmpA in 8 M urea 1 mg/mL in 50 mM Tris-HCl pH 7
4. Dithiothreitol (DTT), 1 M, cold Acetone (−20°C) and 20% (w/v) trichloroacetic acid
5. ProOmpA buffer: 8 M urea, 50 mM Tris-HCl pH 7

2.7. In Vitro Translocation of Fluorescently Labeled ProOmpA (Discontinuously)

1. 10-fold Translocation buffer: 500 mM HEPES-KOH pH 7.4, 300 mM KCl, 5 mg/mL BSA, 100 mM dithiothreitol (DTT), 50 mM $MgCl_2$
2. *E. coli* inner membrane vesicles (IMVs) or proteoliposomes with reconstituted SecYEG
3. Purified *E. coli* SecA, SecB (*see* **Note 3**) and fluorescently labeled proOmpA
4. Energy mix (50 mM creatine phosphate, 0.1 mg/mL creatine kinase), 100 mM ATP in 100 mM Tris-HCl pH 7.5,

and 1 mg/mL proteinase K, 20% (w/v) trichloroacetic acid (TCA), cold Acetone (−20°C)

5. Materials for running a 12% SDS-PAGE gel and 2x SDS-PAGE loading buffer
6. Roche Lumi-imager F1 (Roche Diagnostics) or equivalent imager

2.8. In Vitro Translocation of Fluorescently Labeled ProOmpA (Real Time)

1. Materials from 1 to 4 from **Section 2.7**.
2. Spectrofluorimeter, e.g. Aminco Bowman Series 2 (SLM Instruments).

3. Methods

All components of the Sec translocase can be purified and reconstituted to yield an active in vitro translocation system. The SecYE complex reconstituted into liposomes together with SecA represents the minimal translocase that is sufficient for protein translocation (2, 8). SecG co-purifies with SecE with SecY and enhances the translocation efficiency (2, 8). Several methods for the reconstitution of membrane proteins into proteoliposomes have been described. Here, we describe a mild method for the reconstitution of SecYEG that leads to a high number of protein-containing proteoliposomes with a homogeneous SecYEG distribution and little, if any, protein aggregates. We made use of a dual-color laser-scanning confocal microscope and dual-color fluorescent-burst analysis (DCFBA, for review and description of the method *see* 9) to determine the distribution of reconstituted SecYEG among proteoliposomes supplemented with the fluorescent lipid analog DiD. The fluorescent signal distribution of individual proteoliposomes follows a log normal distribution indicating a random reconstitution and the absence of large lipid-free SecYEG aggregates (**Fig. 8.1**). In addition, 80% of SecY co-migrates with DiD containing liposomes. The remaining 20% may be assigned to small proteoliposomes with non-detectable DiD content or proteoliposomes with high protein/lipid ratio that migrate with some distance from the center of the confocal volume. The presence of small SecYEG aggregates cannot be excluded. With the method described here, 30–40% of the liposomes contain SecYEG (*see* **Note 4**) as determined with DCFBA (number of DiD containing liposomes co-migrating with SecY, data not shown).

Protein translocation into proteoliposomes or IMVs can be assayed in vitro in two ways: i.e., by (i) protection of translocated precursor proteins from an externally added protease and

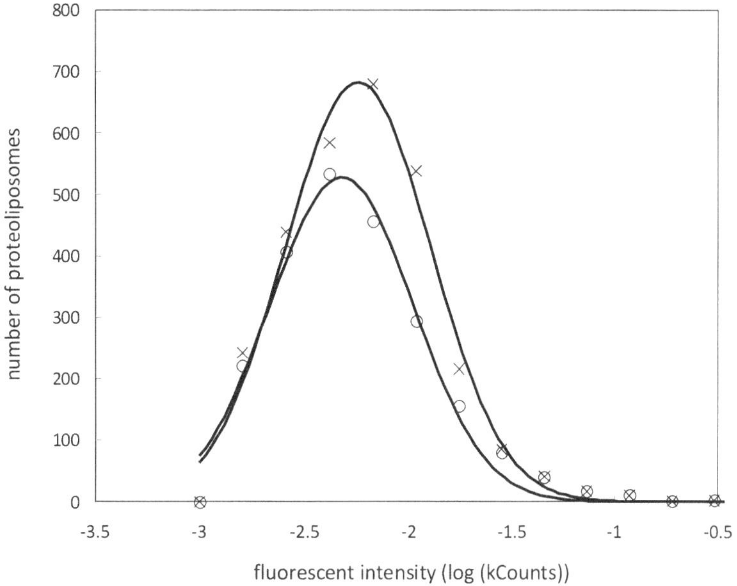

Fig. 8.1. Distribution of reconstituted SecYEG in liposomes and in vitro translocation of proOmpA. Distribution of fluorescein labeled and reconstituted SecYEG in liposomes of *E. coli* total lipids supplemented with the hydrophobic fluorophore DiD. Single proteoliposomes were detected by a dual-color laser-scanning confocal microscope. The fluorescent intensities of the SecY signals alone (×) and the SecY signals that overlapped with the DiD signals (corresponding to the SecY containing liposomes, o) are shown. Both data sets are fitted to a log normal distribution (*solid* and *dashed lines*). The overlap of both curves shows that 80% of SecY co-migrates with DiD containing liposomes.

(ii) fluorescent quenching of fluorophores attached to the precursor protein once it is inside the vesicles. While in the first method translocated (protease protected) precursor proteins are visualized by SDS-PAGE and in gel fluorescence (**Fig. 8.3**) or Western blotting, the second method allows real time observation of fluorescently labeled precursor proteins. Here, fluorescein derivates attached to proOmpA are quenched when translocated together with the precursor protein into the IMVs (**Fig. 8.4**) or proteoliposomes resulting in a decreasing total fluorescence. This decrease is observed only when the system is energized with ATP and when SecA is present (10).

3.1. Isolation of Inner Membrane Vesicles (IMVs)

1. One liter of LB supplemented with 0.1 mg/mL ampicillin is inoculated with a starting $OD_{600\ nm}$ of 0.05 of an overnight culture of *E. coli* SF100 transformed with pET84 and grown at 37°C.
2. Expression of SecYEG is induced at an $OD_{600\ nm}$ of approximately 0.6 by addition of 0.5 mM IPTG. The cells are grown for 2 h longer and collected by centrifugation.

3. The pellet is resuspended in 10 mL Tris-sucrose, frozen in liquid nitrogen and stored at −20°C (*see* **Note 5**).
4. To prevent proteolysis all further steps are carried out at 4°C.
5. The suspension is defrosted in ice water and supplemented with 1 mM PMSF, 2 mM DTT, 1 mg/mL DNase and RNase each (*see* **Note 6**).
6. Cells are lysed by two passages through a cell disrupter (French press) at 8000 psi. After the first passage the PMSF concentration is raised to 2 mM.
7. Unbroken cells are removed by 15 min centrifugation at 5000 *g* and membranes are collected from the supernatant by ultracentrifugation at 125,000 *g* for 90 min.
8. The pellet is resuspended to a total volume of 2.4 mL Tris-sucrose and 800 μL is loaded on each of a sucrose gradient consistent of 0.9 mL 55%, 2 mL 51%, 1.4 mL 45%, and 0.9 mL 36% Tris-sucrose (*see* **Note 7**). The gradients are centrifuged for 30 min at 250,000 *g* at 4°C.
9. The upper brownish band at around 45% sucrose contains the IMVs and is removed from the gradient and diluted 5-fold with 50 mM Tris-HCl pH8. The IMVs are recollected by ultracentrifugation (250,000 *g*, 30 min 4°C).
10. The IMVs are resuspended in 0.5 mL 50 mM Hepes-KOH pH7, frozen in liquid nitrogen and stored at −80°C.

3.2. Purification and Fluorescent Labeling of SecYEG with Fluorescein-Maleimide

1. All steps are carried out at 4°C.
2. Two milligram IMVs (**Fig. 8.2**, lane 2) are diluted to 1 mL with Buffer S followed by 30 min incubation under gentle mixing. Non-solubilized material is removed by 30 min centrifugation (14,000 *g*).
3. If no fluorescent labeling is required, proceed with step 6.
4. The supernatant is supplemented with 2 mM TCEP and incubated for 30 min.
5. Three times 5 μL fluorescein-maleimide is added, each addition followed by 30 min incubation under gentle mixing. The labeling reaction is continued for 2 h (*see* **Note 2**).
6. The NiNTA beads (150 μL) are washed according to the specifications of the manufacturer and equilibrated with Buffer W. Subsequently, the beads are added to the labeling reaction or solubilized membranes. The suspension is incubated for 1 h under gentle mixing.
7. An empty BioSpin column is used to separate the beads from the solution and to perform washing and elution steps. The NiNTA beads are washed five times with 1 mL Buffer W

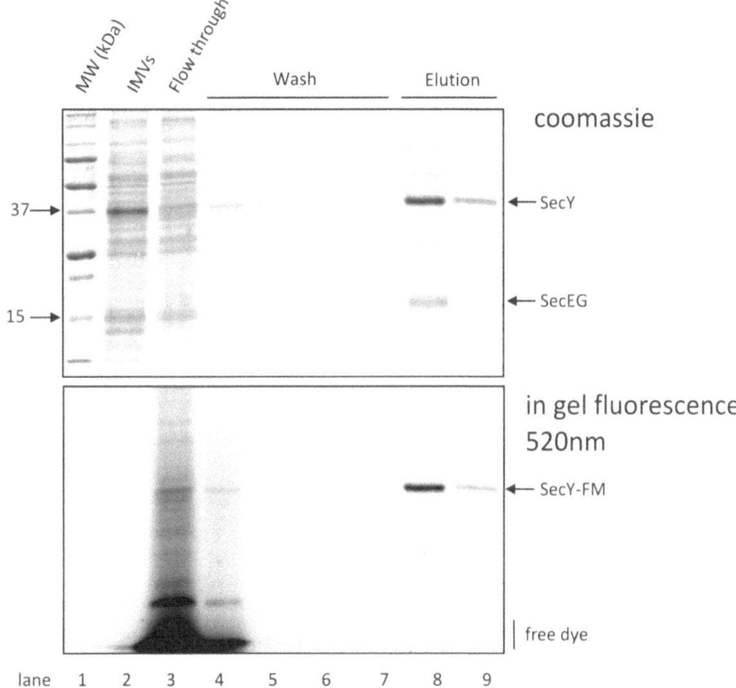

Fig. 8.2. Labeling and purification of SecYEG as described in **Section 3.2**. Coomassie stained gel and in-gel fluorescence (520 nm). IMVs: inner membrane vesicles containing overexpressed SecYEG, Fth: flow through. SecY migrates around 37 kDa, SecE and SecG as double band around 15 kDa. Specific labeling of SecY with fluorescein (SecY-FM) is displayed by in-gel fluorescence at 520 nm.

and SecYEG is eluted with 300 μL Buffer E (*see* **Note 8** and **Fig. 8.2**, lanes 4–7 and 8–9).

3.3. Reconstitution of SecYEG into E. coli *Total Lipid Liposomes*

1. All steps are carried out at 4°C or on ice.
2. In order to form small unilamellar vesicles, 1 mg liposomes is sonicated for 15 min in a bath sonicator or extruded through a polycarbonate filter with a pore size of 200 nm.
3. Solubilization of the liposomes is achieved by incubation with 0.2 % dodecylmaltoside (DDM) for 15 min on ice.
4. Depending on the protein concentration, up to 100 μL of purified SecYEG (for protein concentrations of <0.2 mg/mL) is added to the solubilized liposomes and incubated under gentle mixing for 30 min.
5. Two times 100 mg Bio-Beads are washed twice in methanol, twice in ethanol, and thrice in demineralized water which is evaporated using a vacuum centrifuge or 80°C oven after the last washing step. The Bio-Beads are equilibrated with 100 μL SecYEG-buffer prior use.

6. The suspension from step 4 is added to 100 mg Bio-Beads and incubated for 5 h under gentle mixing.
7. A short centrifugation step (30 s, 8000 *g*) separates the Bio-Beads from the solution which is transferred to a fresh tube containing 100 mg equilibrated Bio-Beads. The reconstitution reaction is incubated over night under gentle mixing.
8. An empty BioSpin chromatography column is used to separate the beads from the solution and the Bio-Beads are washed with 2 mL SecYEG-buffer. The flow-through and wash fractions are pooled and the proteoliposomes are collected by ultracentrifugation (250,000 *g* for 30 min at 4°C).
9. The proteoliposomes are resuspended in 100 μL SecYEG-buffer, frozen in liquid nitrogen and stored at −80°C.

3.4. Purification of WT-SecA

1. Five milliliters of an overnight culture of *E. coli* DH5α transformed with pMKL18 is inoculated in 300 mL LB-Amp supplemented with 0.5% glucose for 8 h at 37°C.
2. 20 mL of the over day culture is added to 1 L of LB-Amp and the cells are grown over night at 37°C.
3. The cells are collected by centrifugation and resuspended in 20 mL Buffer A. DNase, RNase (each 1 mg / mL), 1 mM PMSF, and 2 mM DTT are added (*see* **Note 6**).
4. Cell-lysis is achieved by two passes through a French press at 8000 psi. Membranes and unbroken cells are removed by ultracentrifugation at 125,000 *g* for 1 h at 4°C.
5. The supernatant is supplemented with 150 mM NaCl and diluted with Buffer A to a protein concentration of 5 mg/mL.
6. Two HiTrap Q HP columns (5 mL) are combined, assembled in a FPLC system, and equilibrated with 50 mL SecA buffer supplemented with 150 mM NaCl.
7. The cell free extract is loaded on the columns at a flow rate of 1 mL/min.
8. The columns are washed with 100 mL SecA buffer containing 180 mM NaCl at a flow rate of 2 mL/min and elution is achieved with 100 mL of a linear gradient from 180 to 400 mM NaCl in SecA buffer.
9. Using ultrafiltration (Centriprep® YM-50) the SecA containing fractions are concentrated to 5 mL.
10. A Superdex 200 XK26/60 column is equilibrated with 360 mL SecA buffer at a flow rate of 0.5 mL/min (overnight) and the concentrated SecA fractions are loaded and eluted with SecA buffer at the same flow speed.

11. Purity of SecA can be analyzed on a 10% SDS-PAGE where it migrates at a mass of around 100 kDa.

3.5. Purification of ProOmpA from Inclusion Bodies

1. One liter of LB supplemented with 0.1 mg/mL ampicillin is inoculated with 25 mL of an overnight culture of *E. coli* DH5α transformed with pET503.
2. At an $OD_{600\ nm}$ of approximately 0.6 proOmpA expression is induced by addition of 1 mM IPTG and the cells are grown for 2 h longer.
3. The cells are collected by centrifugation, washed once in 100 mL 50 mM Tris, pH7, and resuspended in 5 mL of the latter buffer.
4. Cells are lysed by sonication (20 cycles of 30 s sonication and 30s pause) and the inclusion bodies are separated from the lysate by centrifugation (1500 *g* for 7 min at 4°C).
5. The proOmpA pellet is resuspended in 10 mL proOmpA buffer, frozen in liquid nitrogen, and stored at −80°C (*see* **Note 9**).

3.6. Fluorescent Labeling of ProOmpA

1. Steps 2–4 are carried out at room temperature.
2. Urea dissolved proOmpA is reduced by addition of 2 mM TCEP and incubation for 30 min.
3. A 100-fold molar excess of fluorescein-5-maleimide is added and the labeling reaction is incubated for 2 h in the dark (*see* **Note 2**).
4. In order to end the labeling reaction, 10 mM DTT is added and incubated for 30 min.
5. ProOmpA is precipitated by addition of two volumes 20% (w/v) TCA and incubation for 30 min on ice. The precipitate is collected by centrifugation (16,000 *g*, 30 min, 4°C).
6. The pellet is washed with 1 mL ice cold acetone and recollected by centrifugation (16,000 *g* for 15 min at 4°C) (*see* **Note 10**).
7. ProOmpA-fluorescein is resuspended in 50 μL proOmpA buffer, frozen in liquid nitrogen, and stored at −80°C (*see* **Note 11**).

3.7. In Vitro Translocation of Fluorescently Labeled ProOmpA (Discontinuously)

1. The translocation mixture consists of 5 μL 10-fold translocation buffer, 3.2 μL energy mix, 50 μg/mL IMVs or 10 μL proteoliposomes, 80 μg/mL SecB, 50 μg/mL SecA, and 20 μg/mL proOmpA-fluorescein and is adjusted to a final volume of 49 μL with demineralized water. The mixture is incubated for 3 min at 37°C prior to the start of the translocation reaction by addition of 2 mM ATP.

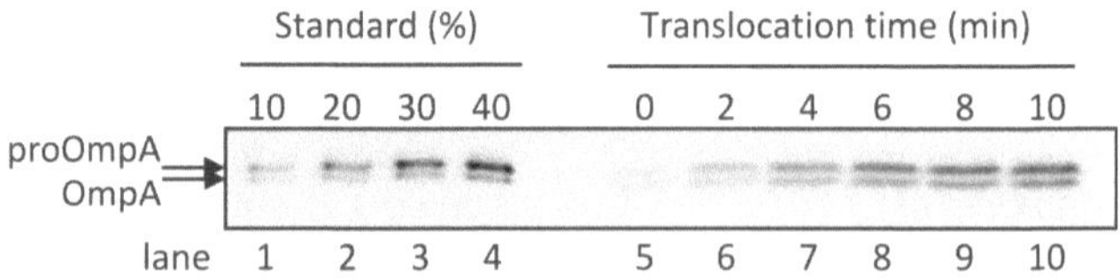

Fig. 8.3. In vitro translocation of fluorescein-labeled proOmpA into IMVs containing over expressed SecYEG as described in **Section 3.7**. ProOmpA is partially processed by leader peptidase yielding the mature OmpA which is the lower protein band.

2. After 1 to 20 min time intervals, the translocation is stopped by adding 40 μL of the reaction mixture to a vial containing 5 μL 1 mg/mL proteinase K followed by incubation for 15 min on ice (*see* **Fig. 8.3**, lanes 5–10).
3. Four microliters of the reaction mixture is mixed with 4 μL 2x SDS-sample buffer and serves as a standard (10%) to determine the translocation efficiency (*see* **Fig. 8.3**, lanes 1–4).
4. In order to precipitate the proteins, 100 μL 20% TCA is added and incubated on ice for 30 min. The precipitate is then collected by centrifugation (16,000 *g* for 30 min at 4°C), washed with 1 mL ice cold acetone, and recollected by centrifugation (16,000 *g* for 15 min at 4°C) (*see* **Note 10**).
5. The pellet (often invisible) is resuspended in 15 μL 2x SDS-sample buffer and boiled at 95°C for 4 min.
6. Both the 10% standard (see step 3) and the various translocation mixtures are run on a 12% SDS-PAGE and in-gel fluorescence is visualized using the Roche Lumi-imager F1. With appropriate software, the proOmpA-fluorescein bands can be quantified and related to the 10% standard to determine the translocation efficiency. An example of a discontinuous in vitro translocation assay is shown in **Fig. 8.3**.

3.8. In Vitro Translocation of Fluorescently Labeled ProOmpA (Real Time)

1. An Aminco Bowman Series 2 spectrofluorimeter is set as followed: excitation wavelength 490 nm, emission 520 nm, slitwidths 4 nm, and for measuring mode time traces of 10 min with a 1 s sampling time.
2. A translocation mixture (*see* step 1 of **Section 3.7**) of 150 μL is preincubated in a 120 μL thermo stated microcuvette at 37°C for 3 min before the reaction is started by addition of 2 mM ATP. An example of a real time in vitro translocation assay is shown in **Fig. 8.4**.
3. This method can be combined with the discontinuous assay (*see* **Section 3.7**) by taking samples at different timepoints (**Fig. 8.4**, reversed triangles) and proceeding from **Section 3.7** step 2.

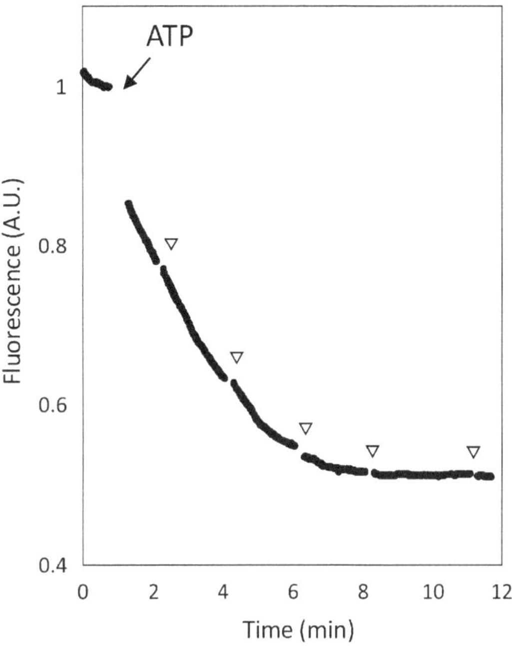

Fig. 8.4. Real time in vitro translocation of proOmpA labeled with fluorescein into IMVs containing over expressed SecYEG as described in **Section 3.8**. Translocation of proOmpA-fluorescein is induced by addition of ATP and results in quenching of the fluorescein inside the vesicles. For comparison, samples were taken at the indicated time points and treated as described in **Section 3.7** and shown in **Fig. 8.3.**

4. Notes

1. TCEP is very acidic. Adjust pH with 1 M NaOH.
2. Maleimide and fluorescein are light sensitive. Protect from light as much as possible. Wrap vials in aluminum foil.
3. SecB with C-terminal His-tag can be purified by NiNTA-chromatography. SecB is not essential for in vitro translocation but enhances the efficiency.
4. SecYEG containing proteoliposomes can be separated from empty liposomes by spinning through a sucrose gradient made of 40%, 30%, 20%, 10% (w/v) Tris-sucrose in equal volumes (*see* **Note** 7). The upper band contains the proteoliposomes.
5. Cells in Tris-sucrose can be stored at −20°C for several months.
6. PMSF is not stable in water. Proceed with the next steps as soon as possible.

7. The sucrose gradient is made by gently pipetting the different sucrose solutions on top of each other starting with 55% (w/v) sucrose.
8. Incubating the beads with elution buffer for 15 min increases the elution efficiency.
9. ProOmpA from inclusion bodies is usually sufficiently pure for the described translocation assays. In order to purify proOmpA further, it can be applied on a HiTrap™ Q HP column (GE Healthcare) equilibrated with proOmpA-buffer at pH 8. Under these conditions, proOmpA does not bind to the column and elutes with the wash fractions while other proteins stay bound to the column.
10. To remove all acetone the pellet can be incubated for up to 10 min at 37°C.
11. This procedure removes most of the free fluorescein-5-maleimide. For further cleaning repeat steps 5–7.

Acknowledgements

This work was supported by NanoNed, a national nanotechnology program coordinated by the Dutch Ministry of Economical Affairs and the Zernike Institute for Advanced Materials.

References

1. Driessen, A.J. and Nouwen, N. (2008) Protein translocation across the bacterial cytoplasmic membrane. *Annu Rev Biochem.* 77, 643–67
2. Brundage, L., Hendrick, J. P., Schiebel, E., Driessen, A. J., and Wickner, W. (1990) The purified *E. coli* integral membrane protein SecY/E is sufficient for reconstitution of SecA-dependent precursor protein translocation. *Cell* **62**, 649–657.
3. Economou, A. and Wickner, W. (1994) SecA promotes preprotein translocation by undergoing ATP-driven cycles of membrane insertion and deinsertion. *Cell* **78**, 835–843.
4. Hartl, F. U., Lecker, S., Schiebel, E., Hendrick, J. P., and Wickner, W. (1990) The binding cascade of SecB to SecA to SecY/E mediates preprotein targeting to the *E. coli* plasma membrane. *Cell* **63**, 269–279.
5. Lecker, S.H., Driessen, A. J., Wickner, W. (1990) ProOmpA contains secondary and tertiary structure prior to translocation and is shielded from aggregation by association with SecB protein. *EMBO J.* **9**(7), 2309–14
6. Schiebel, E., Driessen, A. J. M., Hartl, F. U., and Wickner, W. (1991) $\Delta\mu_{H+}$ and ATP function at different steps of the catalytic cycle of preprotein translocase. *Cell* **64**, 927–939.
7. van der Wolk, J. P., Klose, M., de Wit, J. G., den, B. T., Freudl, R., and Driessen, A. J. M. (1995) Identification of the magnesium-binding domain of the high-affinity ATP-binding site of the Bacillus subtilis and Escherichia coli SecA protein. *J. Biol. Chem.* **270**, 18975–18982.
8. van der Does, C., Manting, E. H., Kaufmann, A., Lutz, M., and Driessen, A. J. M. (1998) Interaction between SecA and

SecYEG in micellar solution and formation of the membrane-inserted state. *Biochemistry.* **37**, 201–210.

9. van den Bogaart, G., Kusters, I., Velasquez, J., Mika, J. T., Krasnikov, V., Driessen, A. J., Poolman, B. (2008) Dual-color fluorescent-burst analysis to study pore formation and protein-protein interactions. *Methods.* Epub ahead of print.
10. de Keyzer, J., van der Does, C., and Driessen, A. J. M. (2002) Kinetic analysis of the translocation of fluorescent precursor proteins into *Escherichia coli*membrane vesicles. *J. Biol. Chem.* **277**, 46059–46065.

Chapter 9

Reconstitution of the SecY Translocon in Nanodiscs

Kush Dalal and Franck Duong

Abstract

Secretory proteins are transported across the bacterial envelope using a membrane protein complex called the SecY channel or translocon. Major advances in understanding this transporter have been accomplished with methods including purification, crystallization, and reconstitution of the translocation reaction *in vitro*. We here describe the incorporation of the SecY complex into supported nanometer scale lipid bilayers called Nanodiscs. These nanoparticles mimic a membrane environment and circumvent many of the technical problems typically observed with liposomes and detergent micelles. The technology is simple, yet should lead to additional new progresses in the field of membrane protein transport.

Key words: Translocon, SecY complex, detergent, lipids, nanodiscs, nanolipoprotein particles, membrane scaffold protein.

1. Introduction

Proteins are transported out of the cytosol using the SecY translo con (1, 2). The core of the translocon is a channel formed by three membrane proteins—SecY, SecE and SecG—associated together in a relatively stable complex (3). The translocation of prepro-tein substrate through the SecY channel is driven by SecA and cycles of ATP binding and hydrolysis (4). Alternatively, in the co-translational mode of translocation, the nascent chain on the ribo-some is targeted to the membrane and pushed through the SecY complex during elongation (5).

Traditional methods for studying the SecY channel involve lipid vesicles, but the oligomeric state of the SecY complex can-not be controlled. Alternatively, it involves detergent micelles, although this may artificially stabilize or destabilize the association

A. Economou (ed.), *Protein Secretion*, Methods in Molecular Biology 619,
DOI 10.1007/978-1-60327-412-8_9,

of the complex with its cytosolic partners. A novel method termed Nanodiscs is currently developed. The technology uses a modified form of the apolipoprotein A-I, with the optimized capacity to wrap around small patch of lipid bilayers and detergent purified membrane proteins (6). Hence, it becomes possible to study the SecY complex without the need for liposomes or detergents. In this chapter, we describe a protocol for expression and purification of the SecY complex, followed by reconstitution into Nanodiscs.

2. Materials

2.1. Purification of the SecY Complex

1. Strain *Escherichia coli* BL21 transformed with the plasmid pBAD22 encoding for His_6-tagged SecE, SecY and SecG under the control of the *araBAD* promoter.
2. Luria Bertani (LB) broth (Sigma, St. Louis, MO).
3. 150 mM Phenylmethylsulfonylfluoride (PMSF) (Sigma).
4. 20% (w/v) L-arabinose (Alfa Aesar, Ward Hill, MA).
5. Lysis buffer: 50 mM Tris pH 7.9, 300 mM NaCl, 10% glycerol.
6. French Press cell (SLM Aminco) and hydraulic pump (*see* **Note 1**).
7. 10% (v/v) Triton® X-100 (Bioshop Canada, Burlington, Ontario) in MilliQ H_2O.
8. 0.3 M $NiSO_4$ in MilliQ H_2O.
9. IMAC (Immobilized metal affinity chromatography) running buffer: 50 mM Tris pH 7.9, 300 mM NaCl, 10% glycerol, 0.03% (w/v) n-dodecylbeta-D-maltoside (DDM; Anatrace, Maumee, OH).
10. IMAC running buffer supplemented with 30 mM imidazole.
11. IMAC running buffer with 50 mM NaCl (instead of 300 mM).
12. IMAC elution buffer: 50 mM Tris pH 7.9, 50 mM NaCl, 10% glycerol, 0.03% (w/v) DDM, 500 mM imidazole.
13. Ion exchange running buffer: same as item 11.
14. Ion exchange gradient buffer: 50 mM Tris pH 7.9, 600 mM NaCl, 10% glycerol, 0.03% (w/v) DDM.
15. FPLC system (ÄKTA purifier™ or equivalent apparatus [GE Healthcare Biosciences AB, Uppsala, Sweden]).

16. IMAC FPLC column (10 mL of packed Chelating Sepharose™ beads in a 10/10 Tricorn column [GE Healthcare Biosciences]).
17. HiTrap™ SP HP 5 mL cation exchange column (GE Healthcare Biosciences).
18. Amicon® Ultra centrifugal filter, MWCO 50 kDa (Millipore, Billerica, MA).
19. Sodium dodecyl sulfate (SDS)-polyacrylamide gel electrophoresis (PAGE) buffers and apparatus for running a 12% SDS-PAGE gel.

2.2. Membrane Scaffold Protein (MSP)

1. ~100 mg of lyophilized MSP powder (MSP1D1, 24.7 kDa; Ref. 6).
2. Reconstitution buffer: 20 mM Tris-HCl pH 7.9, 100 mM NaCl, 0.5 mM EDTA.

2.3. Lipids

1. Purified *E. coli* total lipid extract (Avanti polar-lipids, Alabaster, AL).
2. 100% chloroform solution (Fisher Scientific Canada, Ottawa, ON).
3. Standard nitrogen (N_2) gas supply.
4. Vacuum dessicator (Corning Life Sciences, Lowell, MA).

2.4. Reconstituting the SecY Complex into Nanodiscs

1. Concentrated SecY complex at ~3g/L in 0.03% (w/v) DDM detergent micelles (*see* **Section 3.1**, item 19).
2. Membrane scaffold protein (*see* **Section 3.2**).
3. 10% (w/v) and 1% (w/v) DDM solutions in MilliQ H_2O.
4. 1000 nmole of dried *E. coli* lipids in a 1.5 mL screwcap microtube.
5. Lipid reconstitution buffer: 50 mM Tris pH 7.9, 50 mM NaCl.
6. Branson 2510 Ultrasonic water bath or equivalent (Branson, Danbury, CT).
7. Nanodisc reconstitution buffer: *same as* Ion exchange gradient buffer (*see* **Section 2.1**, item 14).
8. Bio-beads SM-2 Adsorbent (BioRad, Hercules, CA). 10 mL of dry beads are washed successively with 50 mL of ethanol, methanol, and MilliQ H_2O. Beads are stored in 25 mL 50 mM Tris pH 7.9, 50 mM NaCl.
9. Rotating shaker apparatus.

2.5. Size Exclusion Chromatography

1. FPLC system (ÄKTA purifier™ [GE Healthcare Biosciences]).

2. Superdex-200 prep grade resin packed in a Tricorn 10/10 column (GE Healthcare Biosciences).
3. S200 buffer: 50 mM Tris pH 7.9, 100 mM NaCl, 5% glycerol.
4. Amicon® Ultra centrifugal filter, MWCO 50 kDa (Millipore).
5. Native-PAGE buffers and apparatus for running a 4–12% gradient native gel.

2.6. SDS-PAGE

1. 5X SDS-PAGE sample buffer: 250 mM Tris-HCl pH 7.0, 10% (w/v) SDS, 10% (v/v) β-mercaptoethanol (BME), 50% glycerol, 0.05% bromophenol blue. Stored at −20°C.
2. 4X SDS-PAGE separating gel buffer: 1.5 M Tris-HCl pH 8.8.
3. 4X SDS-PAGE stacking gel buffer: 0.5 M Tris-HCl pH 6.8.
4. 10% Ammonium persulfate (APS) in MilliQ H_2O.
5. 40% mixed Acrylamide/bis solution (37:5:1) (BioRad).
6. *N,N,N,N′*-tetramethyl-ethylenediamine (TEMED) (BioRad).
7. 100% Isopropanol.
8. 10X SDS-PAGE electrophoresis running buffer: 250 mM Tris base, 1.9 M glycine, 0.15 (w/v) SDS. Stored at room temperature. To be diluted to 1X with MilliQ H_2O before use.
9. PAGE staining buffer: 40% methanol, 10% acetic acid, 0.025% (w/v) Commassie Blue R250 (BioRad).
10. PAGE destaining buffer: 20% ethanol, 10% acetic acid.
11. BioRad Mini PROTEAN® 3 apparatus, or equivalent (BioRad).

2.7. Native PAGE

1. 10X Native-PAGE loading buffer: 50% glycerol, 0.005% bromophenol blue. Stored at −20°C.
2. 4X Native-PAGE gel buffer: 1.5 M Tris-HCl pH 8.8.
3. 10% APS and TEMED.
4. 40% acrylamide solution and 2% bis-acrylamide solution (BioRad).
5. 1X native PAGE electrophoresis running buffer: 25 mM Tris, 190 mM Glycine, pH 8.8. Stored at 4°C.
6. PAGE staining and destaining buffer (*see* **Section 2.6**, items 9 and 10).
7. BioRad Mini PROTEAN® 3 apparatus and Multi Casting Chamber, or equivalent (BioRad).

8. 2-well gel mixing chamber.
9. Peristaltic pump (Pump P-1, Pharmacia Biotech).

3. Methods

The essential component of the Nanodisc reconstitution is an amphipathic helix termed membrane scaffold protein (MSP). The MSP surrounds the acyl chains of the lipids in which the SecY complex is embedded (termed Nanodisc-SecY complex) (**Fig. 9.1A**). The reconstitution requires highly purified, detergent soluble SecY protein complex. The purification of the SecY complex is accomplished by IMAC chromatography (eluted SecY fractions shown in **Fig. 9.1B**). A cation ion exchange step is required to further purify the SecY complex to sufficient quality for Nanodisc reconstitution (eluted SecY fractions shown in **Fig. 9.1C**).

The Nanodisc reconstitution involves mixing lipids, membrane scaffold protein and the soluble SecY complex. Detergent is removed by adsorption onto Biobeads during overnight incubation. The progressive removal of the detergent initiates the self-assembly process, resulting into the association of the SecY complex with the MSP. The separation of the Nanodisc-SecY complex from the adducts is obtained by size exclusion chromatography (**Fig. 9.1D**). The chromatography is resolutive enough to separate aggregates and "empty" discs (not containing the SecY complex). The fractions are analyzed by SDS-PAGE (**Fig. 9.1E**) to identify those containing the Nanodisc-SecY complex (i.e., SecY, SecE/G, and MSP). Alternatively, the Nanodisc-SecY complex is detected by non-denaturing native-PAGE (**Fig. 9.1F**).

3.1. Purification of the SecY Complex

1. 3L of LB, supplemented with 0.1 g/L ampicillin, are inoculated with 30 mL of an overnight culture of *E. coli* BL21(DE3) transformed with plasmid pBAD22-hisEYG.
2. The cells are shaken at 37°C until an $OD_{600\ nm}$ of 0.6.
3. The cells are induced with 0.2% (w/v) L-arabinose for 3 h.
4. The cells are collected at 5000 *g* for 10 min at 4°C. The pellet is resuspended with ~50 mL of lysis buffer and 200 μL of 150 mM PMSF. All subsequent steps are carried out at 4°C.
5. The cells are broken by passage through a French press (3X) at 8000 psi. The unbroken cells are removed by centrifugation at 5000 *g* (10 min). The supernatant is spun again at 125,000 *g* (60 min) to isolate the crude membranes.

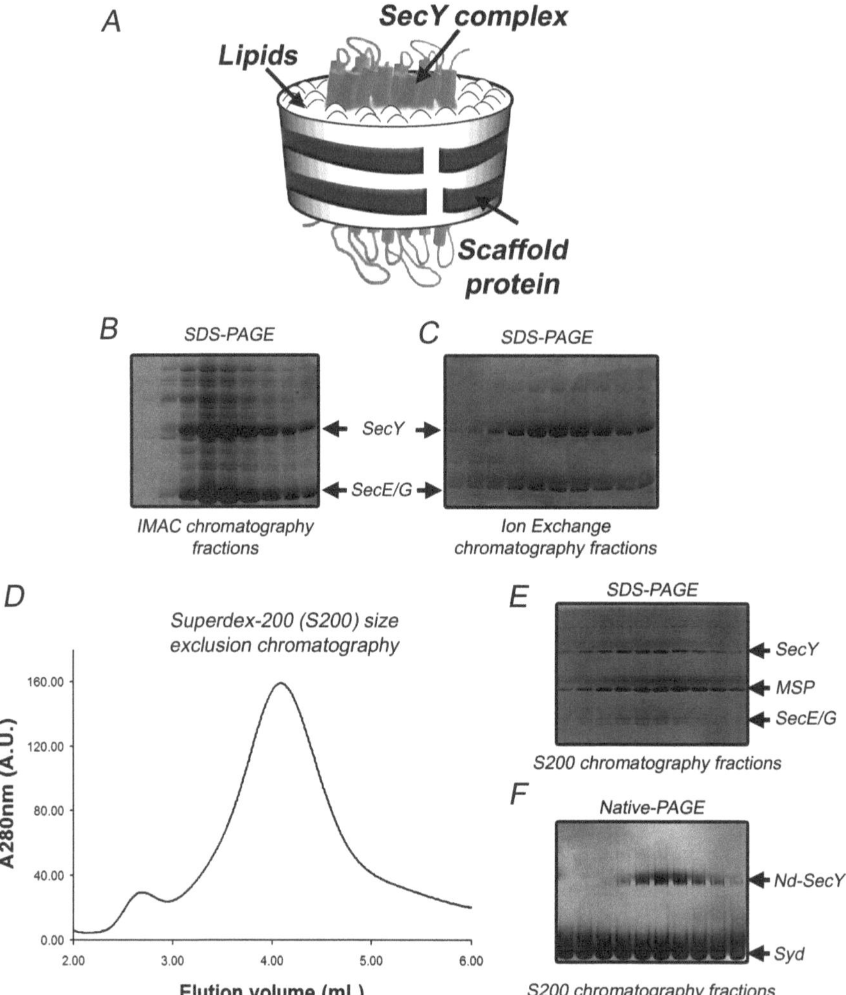

Fig. 9.1. Purification and reconstitution of the SecY complex in Nanodisc. (**A**) The acyl chains of the phospholipids are surrounded by the amphipathic helical membrane scaffold protein. (**B**) IMAC chromatography of the his-tagged SecY complex and analysis of the eluted fractions by 12% SDS-PAGE and Coomassie staining. (**C**) Additional purification of the SecY complex using cation exchange chromatography. The SecY complex was eluted from the HiTrap^{TM} HP SP cation exchange column with a linear 50–600 mM NaCl gradient (*see* **Section 3.1**). (**D**) Superdex-200 size exclusion chromatogram of the Nanodisc-SecY reconstitution (*see* **Section 3.5**). Aggregates appear in the column void volume (2.8 ml), and the Nanodisc-SecY complex is eluted at 4.1 mL. (**E**) Native-PAGE and (**F**) SDS-PAGE analysis of the Nanodisc-SecY fractions recovered from the size exclusion chromatography. The SecY-Nanodisc complex that was pipetted into each lane of the native gel was first mixed with 1 μg of purified His_6-tagged Syd protein (*see* **Note 6**).

6. The crude membrane pellet is resuspended to 10 g/L in lysis buffer, and then solubilized with 1% Triton® X-100. The solution is gently rotated overnight at 4°C.
7. The membranes are spun at 125,000 *g* for 60 min to remove unsolubilized material. The supernatant is collected and kept on ice.

8. The pumps of the FPLC ÄKTA Purifier are washed in MilliQ H_2O, and a Tricorn 10/10 IMAC column (*see* **Section 2.1**, item 16) is mounted. All steps are performed at a flow rate of 2 mL/min.
9. The resin is washed with 2 column volumes of MilliQ H_2O, then chelated with 20 mL of 0.3 M $NiSO_4$, and further washed to remove unbound metal. Protein concentrations are monitored at $A_{280\ nm}$ (absorbance at 280 nm wavelength).
10. The column is equilibrated with 2 column volumes of IMAC running buffer, followed by injection of the supernatant from step 7.
11. The column is washed with 3 column volumes of IMAC running buffer or until the $A_{280\ nm}$ value becomes stable.
12. To remove non-specifically bound protein, the column is further washed with 3 volumes of IMAC running buffer supplemented with 30 mM imidazole. An important elution of non-specifically bound proteins is observed at this point.
13. The column is finally equilibrated with 2 volumes of IMAC running buffer, with the NaCl concentration reduced to 50 mM. This is a preparative step for ion exchange chromatography (steps 16–19) since a low salt concentration is needed to ensure binding of the SecY complex onto the HiTrap^{TM} HP SP column.
14. The elution of the SecY complex is achieved with the IMAC elution buffer, and approximately 30 fractions of 1 mL are collected. About 20 μL of each fraction are mixed with 5 μL of 5X SDS-PAGE sample buffer and analyzed by 12% SDS PAGE followed by Coomassie blue staining (*see* **Section 3.6**).
15. The fractions enriched for the SecY complex (shown in **Fig. 9.1B**) are pooled and stored at –80°C until ready for cation exchange chromatography.
16. A 5 ml HiTrap^{TM} SP HP cation exchange column is mounted on the chromatography system and washed with 5 column volumes of Ion exchange buffer (containing 50 mM NaCl). All subsequent steps are carried out at 1 mL/min.
17. The pooled SecY complex IMAC fractions from step 15 are loaded onto the column, and eluted with a linear gradient of NaCl (50-600 mM) prepared in Ion exchange buffer.
18. For each 750 μL fraction collected, 20 μL are mixed with 5 μL of 5X SDS-PAGE sample buffer and analyzed by a 12% SDS-PAGE (*see* **Section 3.6**).

19. Fractions containing purified SecY complex (shown in **Fig. 9.1C**) are pooled and concentrated to approximately 3 g/L using an Amicon® Ultra centrifugal filter. The concentrated SecY complex is stored at −80°C until needed for the Nanodisc reconstitution (*see* **Section 3.4**). The purification yield is about 1.5 mg/L of culture.

3.2. Membrane Scaffold Protein (MSP)

1. The MSP are solubilized in reconstitution buffer to a stock concentration of 15 g/L and stored at −80°C.
2. A working solution of MSP is diluted to 5 g/L (with reconstitution buffer) from the stock solution and stored at −80°C until needed for the Nanodisc reconstitution (*see* **Section 3.4**).

3.3. Lipids

1. *E. coli* total phospholipids contained in the manufacturer vial are dissolved in chloroform at a concentration of 20 nmole/μL (*see* **Note 2**).
2. Aliquots of 1000 nmole of the chloroform soluble lipids (50 μL of stock) are placed in 1.5 mL screwcap microtubes.
3. The lipids are dried under a gentle stream of nitrogen. A pipetman tip is fixed to the outlet tubing to ease the application of the nitrogen stream.
4. The lipids are further dried in a vacuum dessicator overnight.
5. Lipids are stored at −20°C until needed (*see* **Section 3.4**).

3.4. Reconstituting the SecY Complex into Nanodiscs

1. A typical reconstitution consists of mixing together of SecY complex at a SecY:MSP:lipid molar ratio of 1:4:100 (*see* **Note 3**). The reaction is carried out in a final volume of 1 mL at 4°C.
2. 75 μL of 1% DDM are added to ensure initial solubilization of all components.
3. 1000 nmole of dried lipids are resuspended to 5 nmole/μL with 200 μL of lipid reconstitution buffer using alternating vortexing and sonication. The lipids are solubilized with 0.5 % DDM by pipetting 10 μL of 10% DDM into the suspension. Exactly 120 μL (600 nmole) of lipids are added to the reconstitution mixture.
4. Followed by the addition of 24 nmole of MSP (564 μg, or 113 μL of a 5 g/L solution).
5. Followed by the addition of 6 nmole of detergent soluble SecY complex (450 μg protein, or 150 μL of a 3 g/L solution).
6. The Nanodisc reconstitution buffer is finally added to bring the total volume to 1 mL.

7. This is followed by the addition of 300 μL of BioBeads.
8. The mixture is gently rotated overnight at 4°C.
9. The following day, the microtube is placed on ice for 10 min, so that the BioBeads settle at the bottom of the tube. The supernatant is next transferred to a new 1.5 mL ultracentrifuge tube.
10. The sample is then spun at 100,000 *g* for 20 min at 4°C to remove aggregates.
11. The supernatant is collected into a new 1.5 mL microtube and stored at 4°C until the size exclusion chromatography step (*see* **Note 4**).

3.5. Size Exclusion Chromatography

1. The ÄKTA purifier™ pumps are washed with MilliQ H_2O, then with S200 buffer (*see* **Section 2.5**, item 3).
2. A Tricorn 10/10 column packed with S200 prep grade resin is mounted and equilibrated with 2 column volumes of S200 buffer. The flow rate for all steps is 0.5 mL/min.
3. 300 μL maximal volume of the reconstituted Nanodisc-SecY complex (*from* **Section 3.4**, item 11) are injected onto the column (*see* **Note 5**). The protein concentration is monitored at $A_{280\ nm}$.
4. Fractions of 250 μL are collected as the Nanodisc-SecY complex is eluted. About 20 μL of each fraction are mixed with 5 μL of 5X SDS-PAGE loading buffer and analyzed by 12% SDS-PAGE (**Section 3.6** and **Fig. 9.1E**). In parallel, 20 μL of each fraction is mixed with 1 μg of Syd protein (*see* **Note 6**) and allowed to incubate at room temperature for 5 min. Subsequently, a few μl of 10X native-PAGE loading buffer (~2 μL) is added to each fraction and the samples are analyzed by 4–12% gradient native-PAGE (**Section 3.7** and **Fig. 9.1F**).
5. The profile of the size exclusion chromatogram indicates that aggregates are eluting in the void volume of the column (2.8 mL), whereas the Nanodisc-SecY complex is eluting in the latter fractions (**Fig. 9.1D**).
6. The remaining crude Nanodisc-SecY complex (from step 3) can be purified by repeating steps 2–4. From beginning to end, the efficiency of the reconstitution is approaching 25%.
7. The fractions containing the Nanodisc-SecY complex are pooled and concentrated to approximately 1.5 g/L using an Amicon® Ultra centrifugal device. The prep is stored at 4°C and is stable for a few days (*see* **Note 7**).

3.6. SDS-PAGE

1. The reader is referred to the instruction booklet provided online (http://www.bio-rad.com/LifeScience/pdf/Bulletin_4006193A.pdf) concerning the mounting and running of the SDS-PAGE Mini PROTEAN® 3 system (BioRad).

3.7. Native PAGE

1. The following steps describe how to run a native-PAGE using the BioRad Mini PROTEAN® 3 system (BioRad). The volumes indicated are for the preparation of six minigels using the PROTEAN® 3 Multi Casting Chamber.
2. A 12% native-PAGE gel is prepared by mixing 14.6 mL of 40% acrylamide, 8 mL of 2% bis-acrylamide, 12.5 mL of 4X native-PAGE gel buffer, 10 mL of 100% glycerol and 4.9 mL of MilliQ H_2O (50 mL final volume).
3. A 4% native-PAGE gel is prepared by mixing 4.9 mL 40% acrylamide, 2.7 mL 2% bis-acrylamide, 12.5 mL of 4X native-PAGE gel buffer and 29.9 mL of MilliQ H_2O (50 mL final volume).
4. 27.5 mL of the 12% and 4% native gel mixes (from steps 2 and 3, respectively) are added to the gel mixing chamber. Tubing is attached to the gel mixing chamber and to the peristaltic pump, which is attached in serial to the mulitcaster chamber.
5. Polymerization of the 12% solution is initiated with 58 μL of 10% APS and 5.8 μL of TEMED. Similarly, 145 μL of 10% APS and 14.5 μL of TEMED are added to the 4% solution.
6. The separating valve on the gel mixing device is set to the position "mix" and the pump is started at a medium flow rate. Both chambers are stirred with magnetic bars for adequate mixing and to prevent the premature polymerization of acrylamide. Combs for 12 or 15 lanes are placed on the top of the gels after complete filling of the gel caster. The polymerization should occur within 40 min.
7. Each gel is mounted on the BioRad apparatus and sufficient amount of native-PAGE running buffer is added to the inner and outer chambers of the apparatus. The electrophoresis is run at 20 mA at constant voltage (500 V max) until the dye from the sample buffer reaches the bottom of the gel.
8. The gel is removed and incubated in 15 mL of PAGE staining buffer for 30 min. The gel is rinsed with water and destained with 15 mL of PAGE destaining buffer until the protein bands become clearly visible.

4. Notes

1. The large cell of the French press holds a maximum volume of approximately 35 mL. It should be cooled down to 4°C before use, and the piston lubricated with 100% glycerol.
2. Screwcap microtubes possess o-ring lids that will prevent the evaporation of chloroform.
3. Different SecYEG:MSP:lipid ratios are possible. Reconstitution without lipids is also feasible. The purification and the reconstitution will work for certain SecY mutants, but only if the mutation does not destabilize the heterotrimeric SecYEG associations. Different lipid mixtures can be used such as cardiolipin, phophatidylcholine, and phosphatidylgycerol.
4. Nanodiscs will partially precipitate after a few days at 4°C. The size exclusion chromatography following the reconstitution should be performed as soon as possible.
5. Warming up the Nanodiscs to 42°C for 5 min before loading onto the gel filtration column reduces aggregation.
6. The Nanodisc-SecY complex migrates as a smeary band on native-PAGE. To obtain sharper bands (as shown in **Fig. 9.1C**), the protein Syd is added to the sample. Syd is a small 23 kDa SecY-binding protein (7). Syd forms complex and alters the isoelectric point of the Nanodisc-SecY particles, resulting in a better migration on the native gel (8).
7. Concentrating the Nanodisc-SecY complex beyond 1.5 mg/mL may lead to increased aggregation.

Acknowledgments

The authors thank Dr. Stephen Sligar for providing samples of membrane scaffold protein. KD was supported by the Pacific Century Graduate Scholarship from the BC provincial government. The laboratory is funded by the Canadian Institutes for Health Research (CIHR) and the Natural Sciences and Engineering Research Council of Canada (NSERC).

References

1. Rapoport TA (2007) Protein translocation across the eukaryotic endoplasmic reticulum and bacterial plasma membranes. *Nature* **450**, 663–669.
2. Eichler, J., and Duong, F. (2004) Break on through to the other side – the sec translocon. *Trends Biochem Sci* **29**, 221–223.

3. Brundage, L., Hendrick, J. P., Schiebel, E., Driessen, A. J., and Wickner, W. (1990) The purified *E. coli* integral membrane protein SecY/E is sufficient for reconstitution of SecA-dependent precursor protein translocation. *Cell* **62**, 649–657.
4. Economou, A. and Wickner, W. (1994) SecA promotes preprotein translocation by undergoing ATP-driven cycles of membrane insertion and deinsertion. *Cell* **78**, 835–843.
5. Halic, M. and Beckmann, R. (2005) The signal recognition particle and its interactions during protein targeting. *Curr Opin Struct Biol* **15**, 116–125.
6. Denisov, I.G., Grinkova, Y.V., Lazarides, A.A., and Sligar, S.G. (2004) Directed self-assembly of monodisperse phospholipid bilayer Nanodiscs with controlled size. *J Am Chem Soc* **126**, 3477–3487.
7. Shimoike, T., Taura, T., Kihara, A., Yoshihisa, T., Akiyama, Y., Cannon, K., and Ito, K. (1995) Product of a new gene, syd, functionally interacts with SecY when overproduced in *Escherichia coli*. *J Biol Chem* **270**, 5519–5526.
8. Alami, M., Dalal, K., Lelj-Garolla, B., Sligar, S.G., and Duong, F. (2007) Nanodiscs unravel the interaction between the SecYEG channel and its cytosolic partner SecA. *EMBO J* **26**, 1995–2004.

Chapter 10

In Vitro Assays to Analyze Translocation of the Model Secretory Preprotein Alkaline Phosphatase

Giorgos Gouridis, Spyridoula Karamanou, Marina Koukaki, and Anastassios Economou

Abstract

Almost one-third of the proteins synthesized in the cytosol of cells ends up in membranes or outside the cell. Secretory polypeptides are synthesized as precursor proteins that carry N-terminal signal sequences. Secretion is catalyzed by the "translocase" that comprises a channel-clamp protein and an ATPase motor. Translocase activities have been fully reconstituted in vitro. This provided powerful tools to examine the role of each component in the reaction. Here we describe protocols for the purification of the secretory preprotein alkaline phosphatase and a series of in vitro assays developed in order to examine the binding of alkaline phosphatase to the translocase, its ability to stimulate ATP hydrolysis, and finally its transfer across the membrane. The assays are applicable to any similar study of secretory preproteins.

Key words: *E. coli* translocase, alkaline phosphatase, preprotein translocasion, mature domain targeting.

1. Introduction

Bacterial polypeptide translocase comprises a membrane embedded SecYEG preprotein conducting channel and a peripheral ATPase, SecA that catalyzes chemomechanical energy conversion for the process (1). The integral membrane SecYEG heterotrimer is capable of translocating proteins not only through but also allows membrane proteins to escape laterally, into the lipid bilayer (1). SecA, the "molecular" motor of the translocase contains a helicase motor domain that performs ATP hydrolysis, and two specificity domains to which the hydrolysis energy must be transferred and somehow converted to mechanical work (1). In a

A. Economou (ed.), *Protein Secretion*, Methods in Molecular Biology 619,
DOI 10.1007/978-1-60327-412-8_10,

recent landmark study the x-ray crystal structure of the translocase has been solved (2).

Secretory preproteins after being released from the ribosome (post-translational translocation) must be targeted to the translocase (1). Some nascent chains of secretory preproteins are bound by SecB that prevents their folding and aggregation and targets them to the translocase via their membrane receptor SecA (3, 4). Using purified components in in vitro reconstituted reactions it has been possible to dissect the energetics and mechanism of the translocase at work.

2. Materials

2.1. Genetic Constructs Used for Expression of ProPhoA and PhoA

1. pIMBB882: The gene encoding wild type proPhoA (Protein Data Bank; ID: P00634) was isolated by PCR from the chromosome of *E. coli* JM109 (DE3) and inserted to the *Nde*I-*Xho*I sites of vector pET22b. pET 22b(+) is a high copy number vector, confers resistance to ampicillin, and has a pBR322 origin (5). The vector carries an N-terminal *pelB* signal sequence for potential periplasmic localization, plus optional C-terminal His_6 tag sequence (*see* **Note 1**).
2. pIMBB953: The gene encoding the mature domain of PhoA (phoA Δ2-22) (6–8) was isolated using the same strategy as in the case of pIMBB882.

2.2. Expression of ProPhoA and PhoA

1. LB medium: Dissolve 10 g tryptone, 5 g of yeast extract, 10 g NaCl in 800 mL distilled water, adjust to pH 7.4 with 10 N NaOH and bring to 1 L with distilled water. Autoclave (20 min at 121°C) and store at room temperature.
2. Ampicillin: Dissolve in sterile and distilled water, filter and store at −20°C at a concentration of 100 mg/mL.
3. IPTG (Isopropyl-β-d-thiogalactopyranoside): Dissolve in sterile and distilled water, filter and store at −20°C at a concentration of 1 M.
4. *E. coli* competent cells BL/21.19 (9).
5. Centrifuge: Beckman Coulter Avanti J-26 XP (or analogous).

2.3. Purification of ProPhoA and PhoA from Inclusion Bodies

1. Tris (Tris-(hydroxymethyl)-amino methane): Dissolve in distilled water, adjust to pH 8.0 with 10 N NaOH; store at 4°C at a concentration of 1 M.
2. NaCl: Dissolve in distilled water, store at room temperature at a concentration of 5 M.

3. KCl: Dissolve in distilled water, store at room temperature at a concentration of 4 M.
4. Imidazol: Dissolve in distilled water, store at 4°C at a concentration of 1 M.
5. PMSF (phenylmethanesulphonylfluoride or phenylmethylsulphonyl fluoride): (*see* **Section 2.1**).
6. Glycerol: 87% (v/v) or 100% (v/v), use directly to make the buffers.
7. Urea: use powder directly to make the buffers.
8. EDTA (ethylenediaminetetraacetic acid): Dissolve in distilled water, adjust to pH:8.0 with 10 N NaOH; store at −20°C at a concentration of 0.5 M.
9. DNase (deoxyribonuclease): Dissolve in sterile distilled water, filter and stored at −20°C at a concentration of 50 mg/mL.
10. Buffer A: 50 mM Tris-HCl, pH 8.0, 0.5 M NaCl, 5% glycerol; Buffer B: 50 mM Tris-HCl, pH 8.0, 0.5 M NaCl, 8 M urea, 5% glycerol; Buffer C: 50 mM Tris-HCl, pH 8.0, 0.5 M NaCl, 6 M urea, 5% glycerol; Buffer D: 50 mM Tris-HCl, pH 8.0, 50 mM NaCl, 6 M urea, 5% glycerol; Buffer E: 50 mM Tris-HCl, pH 8.0, 50 mM NaCl, 6 M urea, 100 mM imidazol 5% glycerol; Buffer F: 50 mM Tris-HCl, pH 8.0, 50 mM KCl, 6 M urea, 1 mM EDTA, 5% glycerol.
11. Ni^{2+}-Nitrilo Triacetic superflow resin (Qiagen; http://www1.qiagen.com/literature/).
12. Ultrasonic cell disrupter: Soniprep 150 (or analogous; *see* **Note 2**).
13. Dialysis tubing: Medicell International LTD (12 kDa–14 kDa cut off; or analogous).
14. Concentrators: Amicone Millipore (10 kDa–12 kDa cut off; or analogous).
15. Dounce homogenizer.

2.4. Removal of Urea from ProPhoA and PhoA

1. Buffer G: 50 mM Tris-HCl, pH 8.0, 50 mM KCl, 4 M urea, 10% glycerol, 1 mM EDTA; Buffer H: 50 mM Tris-HCl, pH 8.0, 50 mM KCl, 2 M urea, 10% glycerol, 1 mM EDTA; Buffer I: 50 mM Tris-HCl, pH 8.0, 50 mM KCl, 1 M urea, 10% glycerol, 1 mM EDTA; Buffer J: 50 mM Tris-HCl, pH 8.0, 50 mM KCl, 0.5 M urea, 10% glycerol, 1 mM EDTA; Buffer K: 50 mM Tris-HCl, pH 8.0, 50 mM KCl, 0 M urea, 10% glycerol, 1 mM EDTA.

2.5. Phosphatase Activity of ProPhoA and PhoA to Monitor Their Folded State

1. Reaction buffer: 1 M Tris-HCl, pH 8.0, 5 mM $MgCl_2$, 5 mM $ZnCl_2$.
2. pNPP (p-nitrophenyl phosphate): dissolve in 1 M Tris-HCl, pH 8.0, protect with aluminum foil and store at −20°C at a concentration of 1 M.
3. AP-stop: mix 1 volume of 0.5 M EDTA pH, 8.0 and 4 volumes of 2.5 M K_2HPO_4. Store at room temperature.

2.6. Dermination of Dissociation Constants (K_D) of ProPhoA and PhoA with the Translocase

1. Purified SecA (*see* **Section 3.4**); IMVs (*see* **Section 3.1**); proPhoA and PhoA (*see* **Section 3.2**).
2. TNT transcription coupled to translation system (Promega Quick synthesis).
3. [^{35}S]-methionine: pure methionine, 10 mCi/ml (Perkin Elmer).
4. BSA: dissolve in 5 mM Tris-HCl, pH 8.0; store at −20°C at a concentration of 100 mg/mL.
5. 10X reaction buffer B: 500 mM Tris-HCl, pH:8.0, 500 mM KCl, 50 mM $MgCl_2$.
6. 4X reaction mix (reaction buffer B / BSA): mix 0.8 mL of BSA 100 mg/mL; 8 mL of 10X reaction buffer B and distilled water up to 20 mL.
7. BSA / sucrose cushion: mix 1.37 g of sucrose with 5 mL of the 4X reaction mix and dissolve in distilled water up to 20 mL by constant stirring at room temperature.
8. Nitrocellulose: Whatman PROTRAN (or analogous).
8. Beckman TLX120 ultracentrifuge (or analogous).
9. Vacuum manifold: Bio-Dot® Microfiltration Apparatus (Bio-rad).
10. Prism (Graph Pad).
11. Storm 840 phosphorimager (GE Healthcare) or analogous.

2.7. In Vitro ATPase Assay

1. BSA: (*see* **Section 2.6**).
2. ATP: dissolve in distilled and sterile water, adjust pH to 8.0 with 10 N NaOH, store aliquots at −20°C at a concentration of 0.1 M.
3. Purified SecA (see Driessen 3.1); IMVs (see Driessen 3.4); proPhoA and PhoA (*see* **Section 3.2**).
4. Citric acid: dissolve powder in distilled water, store at room temperature at a concentration of 37% (w/v).
5. 10X reaction buffer B (*see* **Section 2.6**).
6. Malachite Green reagent (MGR).

Prepare by adding 250 μL of 20% Triton X-100 to 50 mL of malachite green stock solution (MGSS). Store at 4°C up to 1 week.

7. Malachite green stock solution (MGSS): make 1 L stock solution and store at 4°C for at least 2–3 months as described below:
   a. Prepare Solution 1 and 2.
   b. Mix solutions 1 and 2 and make up to 1 L with distilled water.
   c. Allow to clarify on ice for at least 4 h (constant stirring).
   d. Filter through Whatman 3MM blotting paper.
   e. Store in plastic bottles at 4°C.
   Solution 1: Dissolve 340 mg of Malachite Green (Sigma) in 75 mL of distilled water.
   Solution 2: Dissolve 10.5 g of Ammonium Molybdate tetrahydrate (Sigma) in 250 mL 4 N HCl.

2.8. *In Vitro* Translocation Reactions

1. Proteinase K: dissolve in 20 mM Tris-HCl, pH:7.4, 1 mM $CaCl_2$, 50% glycerol; store at −80°C at a concentration of 20 mg/mL.
2. TCA (Trichloroacetic acid): dissolve powder in distilled water, store at 4°C at a concentration of 100% (w/v).
3. Acetone 100% (v/v).
4. Triton X-100: make a 20% (v/v) stock solution using distilled water. Store at room temperature and protect with aluminum foil.
5. α-PhoA antibody. Specific rabbit antibodies were raised against purified proPhoA. proPhoA was purified and concentrated (*see* **Section 3.2**) to 20 mg/mL, and 5 mg were used for antibody production.
6. Laemmli sample buffer 6X: mix 3.5 mL of 1 M Tris-HCl, pH:6.8, 3 mL of glycerol 100%, 1 g of SDS, and 6 mg of bromophenol blue. Add water up to 8 mL and stir constantly at room temperature. Divide into aliquots of 0.8 mL and store at −20°C. Prior to use add 200 μL of β-mercaptoethanol to each aliquot.

3. Methods

3.1. Expression of ProPhoA and PhoA

1. Transform plasmids (described in **Section 2.1**) in BL 21.19 *E.coli* cells that grow at 30°C (10).
2. Inoculate LB supplemented with ampicillin (100 μg/μL) with isolated colonies, and grow overnight cultures.

3. Dilute the cultures 1:100 in fresh medium and grow with aeration at 30°C until an $OD_{600\ nm} \approx 0.6–0.8$.
4. Induce the expression of the genes for 3–4 h with 1 mM of IPTG.
5. Harvest bacterial cells by centrifugation at 10,000 *g*, and weigh the cell pellet.
(*see* **Note 3**).

3.2. Purification of ProPhoA and PhoA from Inclusion Bodies

Following the expression conditions (*see* **Section 3.1**), 70–90% of the expressed protein is present in inclusion bodies (determined by SDS-PAGE; *see* step 9). These inclusion bodies are then isolated, and the proteins of interest are purified using metal affinity chromatography (11).

1. Resuspend the bacterial cell pellet in buffer A (*see* **Section 2.3**) by stirring at 4°C (*see* **Note 4**).
2. Add PMSF (2.5 mM final concentration) and DNase (50 μg/mL final concentration).
3. Disrupt resuspended bacterial cells (use an amplitude of 25–30 microns, cool the sample every 3–4 min in a dry ice/ethanol bath to keep the temperature below 8°C).
4. Centrifuge at 50,000 *g* for 30 min at 4°C in order to pellet the inclusion bodies.
5. Keep supernatant and resolubilize the pellet in buffer B (*see* **Section 2.3**; *see* **Note 5**).
6. Add to the resuspended pellet PMSF (2.5 mM final concentration).
7. Centrifuge at 50,000 *g* for 30 min at 22°C.
8. Keep the urea-solubilized pellet and resuspend the new pellet in an equal volume of buffer B (*see* **Section 2.3**).
9. Load the supernatant, the urea-solubilized pellet, and the pellet in a 12% SDS-PAGE gel (12) (>70% of the protein of interest should be found in the urea-solubilized pellet).
10. Estimate the amount of protein of interest in the urea-solubilized pellet and pack the appropriate column with the Ni^{2+}-NTA superflow resin (*see* **Note 6**).
11. Pre-equilibrate the column with 12 column volumes of buffer C (*see* **Section 2.3**).
12. Dilute the urea-solubilized pellet with buffer A (*see* **Section 2.3**) to a final concentration of 6 M urea, add 5 mM β-mercaptoethanol, and load to the column immediately (*see* **Note** 7).
13. Wash the resin with 10 column volumes of buffer C (*see* **Section 2.3**).

14. Wash the resin with 15 column volumes of buffer D (*see* **Section 2.3**).
15. Elute protein with 10 column volumes of buffer E (*see* **Section 2.3**) and run a 12% SDS-PAGE gel to monitor the degree of purification (a representative purification of proPhoA is seen in **Fig. 10.1**).
16. Incubate the protein with 30 mM EDTA for at least 3 h.
17. Perform dialysis in buffer F (*see* **Section 2.3**; *see* **Note 8**).
18. Concentrate the protein (if needed) with Amicon Millipore concentrator.

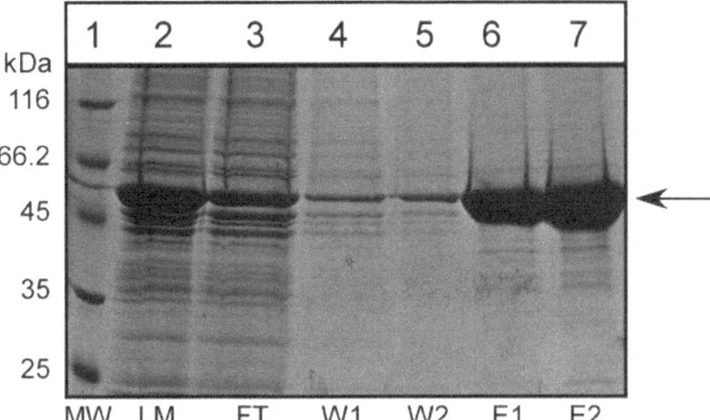

Fig. 10.1. Coomassie-stained SDS-PAGE gel from purification of alkaline proPhoA (performed as described in **Section 3.2**). His_6-tagged alkaline phosphatase (52 kDa; *arrow*) was purified from the urea-solubilized pellet (**Section 3.2**) derived from 3 L of *E. coli* culture (**Section 3.1**) on 13 mL of Ni^{2+}-NTA Superflow resin. Total yield was 100 mg. *Lane 1*: molecular weight marker (M.W.); *lane 2*: solubilized urea pellet from proPhoA producing cells (Loading material, LM); *lane 3*: non-retained material (Flow-through, FT); *lane 4*: Wash 1 (W1); *lane 5*: Wash 2 (W2); *lanes 6, 7*: Eluates (E1, E2).

3.3. Removal of Urea from ProPhoA and PhoA

Alkaline proPhoA and PhoA are purified and kept in 50 mM Tris-HCl, pH 8.0, 50 mM KCl, 6 M urea, 10% glycerol, 1 mM EDTA. These two can not only be used in the assays directly out of 6 M urea, but they also remain soluble if urea is completely removed (*see* **Note 9**). The procedure used to remove the urea is performed using the buffers described in **Section 2.4**: Dialyze an appropriate amount of protein in buffer G (for 2 h. Proceed by dialyzing the sample sequentially against buffers H, I, J, K (*see* **Section 2.4**). Each dialysis step should last for 2 h at least. *All the buffers should be cooled (4° C)* and the procedure must be accomplished in a cold room.

3.4. Phosphatase Activity of ProPhoA to Monitor Its Folded State

1. Use proPhoA and PhoA in buffer E (*see* **Section 2.4**) after removal of urea (*see* **Section 3.3**).
2. Dilute proteins from buffer F (*see* **Section 2.3**) into 1 mL of the reaction buffer (*see* **Section 2.5**).

3. Add pNPP (15 mM) and incubate at 37°C for the appropriate time ($time_{[min]}$ in equation) until a strong yellow color is observed.
4. Stop the reaction with AP-stop and read the absorbance of p-nitrophenol at 420 nm (A_{420} in equation).
5. Determine the units of alkaline phosphatase (*see* **Note 10**) using the formula:

$$\text{PhoA}_{\text{(activity)}} = \frac{1000 \times A_{420}}{\text{time}_{[\text{min}]}}$$

(Results from this experiment are presented in **Fig. 10.2**).

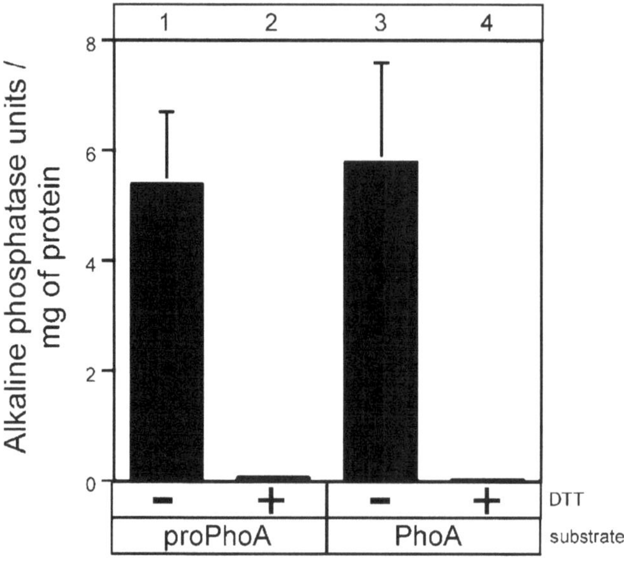

Fig. 10.2. Phosphatase activity of proPhoA and PhoA (performed as described in **Section 3.3**]. 50 μg of proPhoA or PhoA protein (diluted in buffer K; **Section 2.4**) was either treated with 1 mM DTT for 3 h at 4°C or left at the same temperature without any treatment. Treated and untreated samples were assayed for their ability to hydrolyze the pNPP substrate as described, and units of alkaline phosphatase per mg of protein were determined (*see* **Note 14**). Phosphatase activity does not depend on the presence or absence of the signal peptide (compare *lanes 1, 3*), but on the ability of the protein to form disulfide bonds (compare *lanes 2, 4*). The state of the protein that is able to perform hydrolysis is defined as "native" whereas the one that is not is defined as "non-native" (19).

3.5. Determination of Dissociation Constants (K_D) of ProPhoA and PhoA with the Translocase

1. Synthesize radio-labeled [^{35}S]-prophoA and [^{35}S]-phoA using the vectors described in **Section 2.1** and the TNT transcription coupled to translation system (use Promega Quick protocol; *see* **Note 11**).
2. Perform serial dilutions of proPhoA and PhoA in buffer F (2.3) in order to add into the reactions the appropriate amount of protein from each dilution (*see* **Note 12**).

3. Make the SecA/SecY complex by mixing IMVs containing over-expressed SecYEG with purified SecA. After incubation of the complex on ice for 10 min, add the proper amount of it in the reactions (*see* **Note 13**).
4. Prepare 20 eppendorf tubes; tube no.1 will contain only radio-labeled protein and tubes 2–20 will contain the same concentration of radio-labeled protein together with increasing concentrations of chemical amounts of protein, e.g. 0.5, 1, 10,20. . ..100, 200. . .. . .1000 nM.
4. Perform the reactions according to **Table 10.1**.

Table 10.1
Reaction setup for K_D determination (*see* Section 3.5)

Reaction mix (4X)	5
SecA/IMVs complex	9.3
Purified protein (each dilution)	0.7
Radio-labeled protein dilution	5
Total	20 μL

5. Incubate the reactions on ice for 20 min.
6. Isolate the membrane bound material (*see* **Note 14**).
7. Immobilize the membrane bound material using the vacuum manifold on nitrocellulose (*see* **Note 15**).
8. Analyze results and calculate K_D (13) (*see* **Note 16**). ($K_D$for proPhoA and PhoA are presented in **Table 10.2**).

Table 10.2
K_D values for "native" and "non-native" proPhoA and PhoA

K_D (μM)

ProPhoA	ProPhoA	PhoA	PhoA
Non-native	Native	Non-native	Native
0.228 ± 0.02	–	0.609 ± 0.08	–

Binding of [^{35}S]-proPhoA and [^{35}S]-PhoA to the translocase (performed as described in **Section 3.5**). [^{35}S]-proPhoA, [^{35}S]-PhoA "native," and "non-native" form (obtained as in **Fig. 10.2**) were tested for their ability to be targeted to the translocase. As observed from **Table 10.2** proPhoA and PhoA bind to the translocase with similar affinities in their "non-native" conformation, while they both fail to be targeted in their "native" conformation. Targeting seems to be almost unaffected from the presence or absence of the signal peptide and severely affected from the conformation of proPhoA.

3.6. In Vitro ATPase Assay

ATPase activity of SecA in solution (Basal activity), when bound to membranes (Membrane activity), and when it is translocating substrates (Translocation activity) can be determined by the following procedure (14, 15).

1. Prepare stock solution according to **Table 10.3**.

Table 10.3
Reaction setup for ATPase activity determination (*see* Section 3.6)

Stock solution	Basal	Membrane	Translocation
10X BB	5	5	5
BSA (10 mg/mL)	2	2	2
SecA (2 mg/mL)	1	1	1
ATP (0.1 M)	0.5	0.5	0.5
dH_2O	37.5	37.5	37.5
Aliqout/tube	46	46	46
IMVs (6 M urea-treated; ≈2.5 mg/mL)	–	2	2
proPhoA (≈8 mg/mL)	–	–	2
1X BB	2	–	–
Buffer F (2.1.3)	2	2	–
Total	50 μL	50 μL	50 μL

2. Aliquot stock solution, vortex briefly to mix all contents.
3. Aliquot IMVs, sonicate, and vortex briefly every minute.
4. Aliquot proPhoA substrates.
5. Incubate at 37°C for 20 min (or other appropriate time).
6. Terminate reactions by transferring the tubes on ice.
7. Add to every reaction, 800 μL of malachite green molybdate reagent to terminate reactions (*see* **Note 17**).
8. Five minutes after termination, add 100 μL of 37% (w/v) citric acid per tube to avoid further ATP hydrolysis and to optimize color development.
9. Keep tubes at room temperature for 40 min.
10. Measure absorbance at $OD_{660\ nm}$ (*see* **Note 17**). (Results obtained are presented in **Fig. 10.3**).

3.7. In Vitro Translocation Reaction

1. Follow the same procedure (described in **Section 3.6**) for translocation ATPase activity but double the reactions volume for each condition (100 μL instead of 50 μL; *see* **Note 18**).

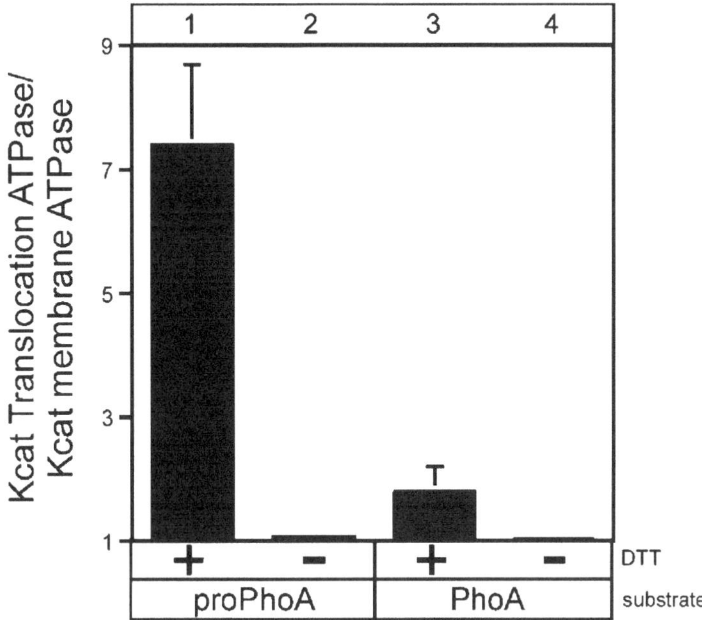

Fig. 10.3. In vitro ATPase assay for proPhoA and PhoA (performed as described in **Section 3.6**). proPhoA and PhoA, "native" and "non-native" forms were tested for their ability to induce stimulation of membrane ATPase activity of SecA (translocation stimulation). To estimate translocation stimulation, Membrane and Translocation ATPase activities were determined (as described in **Section 3.6**). Kcat values were expressed as pmol Pi released per pmol of SecA protomer in 1 min. "Non-native" proPhoA proficiently stimulates translocase ATPase activity, whereas "non-native" PhoA slightly stimulates translocase ATPase, despite the fact that it is bound to the translocase (**Fig. 10.3**). As expected "native" proPhoA and PhoA forms fail completely to stimulate translocase ATPase activity since they are not recognized by the translocase.

2. After incubation transfer the reactions on ice (as described in **Section 3.6**).
3. To control for bona fide translocation into the lumen of the IMVs, add to one reaction 5 μL Triton X-100 20% (v/v).
4. Add proteinase K (5 μL of the stock) and incubate on ice for 20 min.
5. Add 30 μL of TCA stock and incubate on ice for 30 min.
6. Centrifuge the samples at 23,000 *g* for 35 min and remove carefully the supernatant.
7. Wash the pellet with 1 mL of 100% cold acetone, incubate at −20°C for 15 min and remove the acetone after centrifugation at 23,000 *g* for 15 min. Repeat the procedure.
8. Resuspend the acetone washed pellets in 25 μL Laemmli sample buffer (2X) and analyze the samples in a SDS-PAGE gel (13%).

9. Visualize the proteolysis protected material by Western blotting (16) using α-PhoA antibody.
   (Results obtained are presented in **Fig. 10.4**).

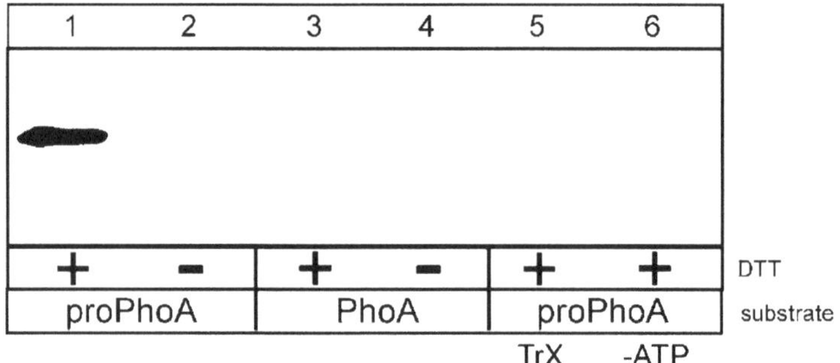

Fig. 10.4. In vitro translocation of proPhoA and PhoA (performed as described in **Section 3.7**). proPhoA and PhoA, "native" and "non-native" forms were tested for their ability to be translocated into the lumen of inverted membrane vesicles. Reactions (*lanes 1–4*) were performed (as described in **Section 3.7**). Two identical reactions to that of *lane 1* were performed, the one of them treated with Triton X-100 20% (v/v) prior addition of the protease (*lane 5*) while in the other addition of ATP was omitted (see **Note 14**). Only "non-native" proPhoA can be translocated into the lumen of IMVs as it is the only substrate capable to stimulate translocase ATPase activity.

4. Notes

1. Using restriction enzymes *Nde* I and *Xho* I to clone the PCR fragment in the vector, the *pelB* signal sequence for periplasmic localization was removed and the His_6 tag fused C-terminally to the proteins of interest.
2. For volumes up to 10 mL use an exponential microprobe; for volumes up to 50 mL use solid probe 9.5 mm in diameter; for volumes up to 200 mL use a solid probe 19 mm in diameter.
3. Use freshly transformed cells. The yield of proPhoA and PhoA is about 50–80 mg of expressed protein per liter of culture.
4. Resuspend in 10 mL of buffer A per gram of cell paste.
5. For efficient resolubilization use an equal volume of buffer B (*see* **Section 2.3**) with that of the supernatant (don't reduce the volume of buffer B more than 50% of the volume of the supernatant). Use also a Dounce homogenizer. Try first to resolubilize the pellet with the loose pestle and then with the tight pestle. After this procedure stir the suspension for at least 45 min, and during this time pass the suspension from the Dounce homogenizer at least twice.

6. The binding capacity of the Ni^{2+}-NTA resin for proPhoA and PhoA is about 5–7 mg of protein/ mL of resin.
7. Flow rate for optimal binding is approximately 1 mL/min.
8. When performing dialysis for buffer exchange, a 100X volume of dialysis buffer over that of the volume of the sample is used.
9. For protein concentrations from 30 μM to 3 mM, proPhoA and derivatives remain soluble after removal of urea only below 12°C. For concentration below 30 μM proPhoA and derivatives remain soluble even at 37°C.
10. One unit of PhoA hydrolyzes 1.0 μmole of p-nitrophenyl phosphate/min at 37°C as determined spectroscopically in the reaction buffer (*see* **Section 2.5**).
11. After the synthesis a buffer exchange has to be made using G-50 resin equilibrated with buffer F (**Section 2.3**). Also add glycerol to a final concentration of 30%. Taking into account the average yield of the TNT synthesis, ≈ an 8-fold dilution has to be made in 1X reaction buffer B (**Section 2.6**) just before the addition of the diluted radio-labeled protein to the reactions. Adjust the dilution of the radio-labeled protein according to the signal that you will get after the detection.
12. A defined volume of protein (0.7 μL) is added from each serial dilution to the final reaction (volume 20 μL) in order to achieve the proper protein concentration. The protein gets diluted 29 times so as to obtain a final urea concentration of 0.2 M. This means that each dilution should have a protein concentration that is 29 folds over the desired concentration. The concentrations of protein used in the reactions in order to determine the K_D depends on the values of the K_D.If, for example, the K_D of proPhoA for the translocase ($K_D \approx 0.2–0.3$ μM) is to be calculated, a concentration of 1 μM has to be reached in order to have saturation of the binding sites of the membrane receptors. This will allow a good estimation of the K_D value. The most concentrated dilution in that case should be 1 μM × 29 = 29 μM and all the others should be prepared by serial dilutions of that one. For a $K_D \approx 0.2–0.3$ μM a range of concentrations from 0.05 μM to 1 μM must be covered.
13. IMVs containing 2 mg/mL total membrane protein are diluted 25 times in 1x reaction buffer B (2.6), and 5 μL of these are mixed with 4.3 μL of SecA (0.200 μg/μL). Make a master mix of the complex and after incubation add 9.3 μL of that to each reaction.

14. Preblock centrifuge tubes with 200 μL of 50 mg/mL BSA and incubate them for at least 30 min at 4°C. Remove BSA and overlay reactions on equal volume of BSA/sucrose cushion. Centrifuge on Beckman TLX120 ultracentrifuge (320,000 *g*; 30 min; 4°C). Transfer the supernatant in a new tube and bring it to 300 μL by adding reaction buffer B and keep on ice. Resuspend the pellet in 150 μL of reaction buffer B (bath-sonicate; 3 times for 20 ffs each time), transfer into a new tube. Add again 150 μL reaction buffer B and repeat sonication procedure. Transfer into the same tube with the previous 150 μL.
15. Spot the resuspended pellets on a nitrocellulose membrane using a vacuum manifold (Bio-rad). Instructions on how to apply the apparatus can be found on the web site: http://www.biorad.com/cmc_upload/Literature/13324/M1706545B.pdf. Quantitate radioactivity by phosphorimaging (use Image quant application; Molecular Dynamics) using Storm 840 phosphorImager. (detailed instruction on how to use the Storm 840 phosphorImager and perform quantitation of the results can be found on the web site: http://virgil.ruc.dk/kurser/Gene Technology/dokumenter/PhosphorImaging.pdf).
16. Analyze results by non-linear regression fitting for one binding site using Prism (Graph Pad). The program will estimate the K_D and will determine statistical parameters (http://www.graphpad.com/prism/Prism.htm) (17, 18).
17. For each ATPase activity (Basal, Membrane, Translocation) also perform an identical control reaction without adding SecA, and subtract the value of the control reaction from all other values. Adjust the incubation time (step 5) and/or the amount of the reaction added to malachite green molybdate reagent (step 7) according to the $OD_{660\ nm}$ measurement. Values below 1.700 are in the linear range while at higher values the color development is saturated.
18. The procedure is identical with that of the translocation ATPase assay, but instead of measuring the released phosphate, we visualize by Western blotting the protease protected material that has been translocated into the lumen of the IMVs. In that case it is important to include two controls: a reaction in which we add Triton X-100 20% (v/v) prior addition of the protease in order to be sure that we visualize only material that has been entirely translocated to the lumen of IMVs, and a reaction in which we do not add ATP since the translocation reaction is ATP dependent.

Acknowledgments

We are grateful to D. Boyd (Beckwith Lab) and S. Schulman (Rapoport Lab) for gifts of strains, plasmids, and protocols. The research leading to these results has received funding from the European Community's Sixth Framework Programme (*FP6/2002–2007*) under grant agreement no. LSHC-CT-2006-037834/Streptomics (to A.E.) and the Greek General Secretariat of Research and the European Regional Development Fund (01AKMON46 and PENED03ED623; to A.E.). G.G. is an Onassis Foundation predoctoral fellow.

References

1. Papanikou, E., Karamanou, S., and Economou, A. (2007) Bacterial protein secretion through the translocase nanomachine. *Nat Rev Microbiol* **5,** 839–851.
2. Zimmer, J., Nam, Y., and Rapoport, T. A. (2008) Structure of a complex of the ATPase SecA and the protein-translocation channel. *Nature* **455,** 936–943.
3. Hartl, F. U., Lecker, S., Schiebel, E., Hendrick, J. P., and Wickner, W. (1990) The binding cascade of SecB to SecA to SecY/E mediates preprotein targeting to the E. coli plasma membrane. *Cell* **63,** 269–279.
4. Lecker, S. H., Driessen, A. J., and Wickner, W. (1990) ProOmpA contains secondary and tertiary structure prior to translocation and is shielded from aggregation by association with SecB protein. *EMBO J* **9,** 2309–2314.
5. Hoffman, B. J., Broadwater, J. A., Johnson, P., Harper, J., Fox, B. G., and Kenealy, W. R. (1995) Lactose fed-batch overexpression of recombinant metalloproteins in Escherichia coli BL21 (DE3): process control yielding high levels of metal-incorporated, soluble protein. *Protein Expr Purif* **6,** 646–654.
6. Li, W., Schulman, S., Boyd, D., Erlandson, K., Beckwith, J., and Rapoport, T. A. (2007) The plug domain of the SecY protein stabilizes the closed state of the translocation channel and maintains a membrane seal. *Mol Cell* **26,** 511–521.
7. Derman, A. I., Puziss, J. W., Bassford, P. J., Jr., and Beckwith, J. (1993) A signal sequence is not required for protein export in prlA mutants of Escherichia coli. *Embo J* **12,** 879–888.
8. Prinz, W. A., Spiess, C., Ehrmann, M., Schierle, C., and Beckwith, J. (1996) Targeting of signal sequenceless proteins for export in Escherichia coli with altered protein translocase. *EMBO J* **15,** 5209–5217.
9. Mitchell, C., and Oliver, D. (1993) Two distinct ATP-binding domains are needed to promote protein export by Escherichia coli SecA ATPase. *Mol Microbiol* **10,** 483–497.
10. Froger, A., and Hall, J. E. (2007) Transformation of Plasmid DNA into E. coli Using the Heat Shock Method. *J Vis Exp* 253.
11. Janknecht, R., de Martynoff, G., Lou, J., Hipskind, R. A., Nordheim, A., and Stunnenberg, H. G. (1991) Rapid and efficient purification of native histidine-tagged protein expressed by recombinant vaccinia virus. *Proc Natl Acad Sci USA* **88,** 8972–8976.
12. Shapiro, A. L., Vinuela, E., and Maizel, J. V., Jr. (1967) Molecular weight estimation of polypeptide chains by electrophoresis in SDS-polyacrylamide gels. *Biochem Biophys Res Commun* **28,** 815–820.
13. Klotz, I. M. (1982) Numbers of receptor sites from Scatchard graphs: facts and fantasies. *Science* **217,** 1247–1249.
14. Karamanou, S., Gouridis, G., Papanikou, E., Sianidis, G., Gelis, I., Keramisanou, D., Vrontou, E., Kalodimos, C. G., and Economou, A. (2007) Preprotein-controlled catalysis in the helicase motor of SecA. *Embo J* **26,** 2904–2914.
15. Lill, R., Cunningham, K., Brundage, L. A., Ito, K., Oliver, D., and Wickner, W. (1989) SecA protein hydrolyzes ATP and is an essential component of the protein transloca-

tion ATPase of Escherichia coli. *EMBO J* **8,** 961–966.

16. Burnette, W. N. (1981) "Western blotting": electrophoretic transfer of proteins from sodium dodecyl sulfate–polyacrylamide gels to unmodified nitrocellulose and radiographic detection with antibody and radioiodinated protein A. *Anal Biochem* **112,** 195–203.
17. Batke, J., and Gaal, J. (1986) Displacement analysis of binding inhomogeneities in crude extracts of receptors. *J Biochem Biophys Methods* **12,** 203–212.
18. Swillens, S., Waelbroeck, M., and Champeil, P. (1995) Does a radiolabelled ligand bind to a homogeneous population of non-interacting receptor sites? *Trends Pharmacol Sci* **16,** 151–155.
19. Kim, E. E., and Wyckoff, H. W. (1991) Reaction mechanism of alkaline phosphatase based on crystal structures. Two-metal ion catalysis. *J Mol Biol* **218,** 449–464.

Chapter 11

Characterization of Interactions Between Proteins Using Site-Directed Spin Labeling and Electron Paramagnetic Resonance Spectroscopy

Jennine M. Crane, Angela A. Lilly, and Linda L. Randall

Abstract

Site-directed spin-labeling and the analysis of proteins by electron paramagnetic resonance spectroscopy provides a powerful tool for identifying sites of contact within protein complexes at the resolution of aminoacyl side chains. Here we describe the method as we have used it to study interactions of proteins involved in export via the Sec secretory system in *Escherichia coli*. The method is amendable to the study of most protein interactions.

Key words: Site-directed spin labeling, electron paramagnetic resonance spectroscopy, EPR, spin-labeled protein.

1. Introduction

1.1. The System Under Study

The secretion of proteins from their site of synthesis through a biological membrane involves crucial interactions among numerous proteins. Early studies of the Sec, or general secretory, system in *Escherichia coli* identified all of the proteinaceous components of that system. Over the last three decades sufficient knowledge concerning the interactions among components has accumulated so that the pathway taken by a precursor polypeptide destined for export can be described (for a review *see* Ref. 1). The passage through the membrane is provided by the translocon comprising a SecYEG core and the accessory proteins SecD, SecF, and YajC. The Sec system cannot translocate stably folded proteins. Therefore, soluble chaperones must capture precursors before

A. Economou (ed.), *Protein Secretion*, Methods in Molecular Biology 619,
DOI 10.1007/978-1-60327-412-8_11,

they fold. SecB is such a chaperone that binds promiscuously to many species of unfolded polypeptides. With a polypeptide ligand bound, SecB interacts specifically with SecA, which itself has affinity for SecY. In this way a precursor to be secreted is delivered to the translocation channel. SecA, an ATPase, is the motor of the translocon. It undergoes cycles of binding and hydrolysis of ATP thereby providing energy that is transduced into movement of the polypeptide through the channel.

Our earlier work focused on the interactions among the components of the system. We identified and characterized complexes between SecB and unfolded polypeptides, as well as between SecB and SecA. This work utilized several techniques including size-exclusion column chromatography (2), sedimentation velocity analytical centrifugation (3), and Fourier transform ion cyclotron resonance mass spectrometry (4). Static light scatter in line with size-exclusion chromatography allowed the determination of the stoichiometry of components within a complex (5). Isothermal titration calorimetry was used to determine the thermodynamic parameters (K_d and ΔH) governing the interactions (6, 7).

Having established a basic understanding of the complexes, we have shifted our focus to the determination of contact sites between binding partners at the resolution of aminoacyl residues (8–10). In this chapter we describe the technique we have used, site-directed spin-labeling and electron paramagnetic resonance spectroscopy (EPR). We have introduced spin probes into both SecB and SecA for these studies. The method is described in general terms so that it can be applied to any protein of interest.

1.2. Principles of EPR

We begin with a brief discussion of the theory of EPR spectroscopy (for a review *see* Ref. 11). Molecules absorb energy when incident electromagnetic radiation has an energy equal to the difference in energy between two states. The energy absorbed causes a transition from the lower energy state to the higher state. In EPR spectroscopy the energy differences studied are due to the interaction of unpaired electrons with an applied magnetic field. When the magnetic moment of the unpaired electron aligns parallel with the magnetic field it is in the lowest energy state ($-1/2$ spin state) and when it is antiparallel to the applied field it is in the highest energy state ($+1/2$ spin state). The difference in the energy states is proportional to the strength of the applied field; thus, both spin states have the same energy in the absence of an applied magnetic field and their energy levels diverge linearly as the field is increased (**Fig. 11.1**). Resonant absorption occurs if $\Delta E = h\nu$, where ΔE is the energy difference, h is Planck's constant, and ν is the frequency of the incident radiation.

An unpaired electron is sensitive to its local environment and experiences the local magnetic fields produced by the magnetic moments of nuclei in close proximity. The field from a nucleus

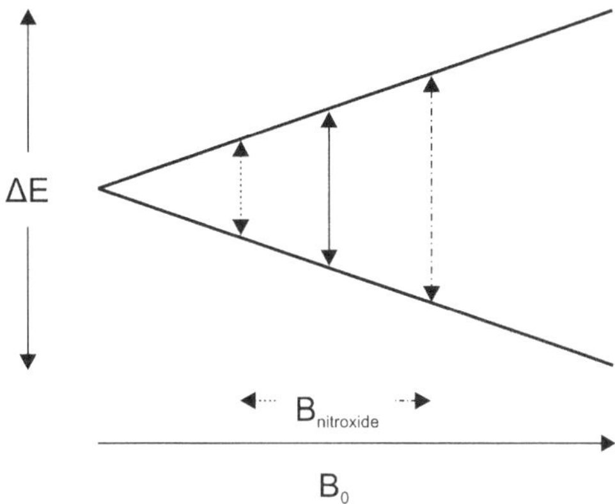

Fig. 11.1. Dependence of the energy difference between spin states as a function of the applied magnetic field.

(**Fig. 11.1**, $B_{nitroxide}$) will either add to or subtract from the applied magnetic field (**Fig. 11.1**, B_0) depending on the orientation of the nuclear dipole as illustrated by the arrows in **Fig. 11.1**. This interaction between the unpaired electron and a nitrogen nucleus (in the case of a nitroxide spin label) gives rise to splitting such that instead of a single line the absorption spectrum contains three lines.

During an experiment the applied magnetic field is slowly swept through a range of field values that include the field for resonance of the electron. For studies employing nitroxide probes we use an X-band microwave bridge with a frequency of 9.75 GHz and sweep a magnetic field of 100 gauss centered at a field for resonance of 3356 gauss. The resonance signal is amplified by encoding it in such a way that it becomes distinguishable from background noise. Modulation coils placed on both sides of the magnet generate a small magnetic field that imposes oscillation, commonly two or three gauss (chosen by the operator), on the applied field. The signal, resulting from absorption of energy at resonance, oscillates at the same frequency as the modulating field. This oscillating signal is selectively amplified whereas all other changes in microwave intensity are ignored as background noise. The modulation of the field is constant through the sweep and the instrument records a signal that is proportional to the change in amplitude of the oscillating intensity during a single modulation cycle. Thus, the signal is recorded as a first-derivative of the absorption (**Fig. 11.2**).

The energy of microwave radiation is too weak to break chemical bonds; therefore, the technique is nondestructive to biological molecules and one can study interactions among proteins in their native state. In addition, there is no size limitation so one can examine large protein complexes. However, most proteins do not contain unpaired electrons; thus, a paramagnetic probe, the spin label, must be introduced.

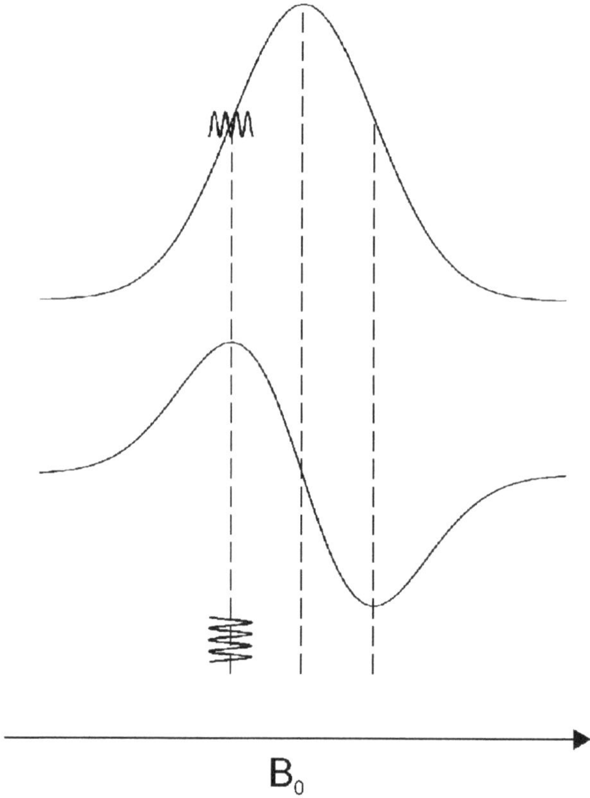

Fig. 11.2. An absorption curve (*upper*) and its derivative (*lower*), as seen in EPR.

1.3. The Approach

The approach involves construction of a collection of variants of the protein of interest each containing a single cysteine side chain so that a spin label can be introduced at a specific site using sulfhydryl chemistry. The reagent used in this study (1-oxy-2,2,5,5 tetramethyl pyrroline-3-methyl)-methanethiosulfonate and the nitroxide side chain it generates are shown in **Fig. 11.3**. This is the reagent of choice for several reasons: (1) it derivatizes the protein through a disulfide bond making it highly specific for cysteine thus eliminating the possibility of labeling other aminoacyl residues by less specific reactions, (2) the chemistry is rapid and for the proteins we have labeled to date (>100 variants of SecB and SecA) the available cysteines are

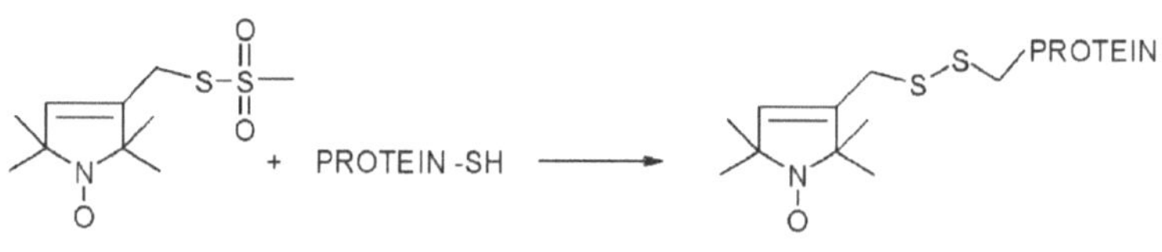

+ PROTEIN-SH → ... PROTEIN

Fig. 11.3. Reaction of the methanethiosulfonate spin label to give the nitroxide-derivatized cysteine side chain.

quantitatively labeled, (3) the nitroxide side chain generated has been extensively studied by W. L. Hubbell and his colleagues (12–15) and a large body of knowledge is available that aids in interpretation of the data.

The approach is amenable to any protein, including membrane proteins (15), provided that the protein can be engineered so that the only cysteine accessible for reaction with the nitroxide reagent is that introduced at the site of interest. If the protein contains native cysteine residues, ideally they should be substituted by another amino acid. However, in one protein we study, SecB, two of the four native cysteines, C76 and C113, could not be replaced without disrupting the structure. Fortunately, neither of these cysteines showed reactivity with the reagent and so our base protein in those studies had only two of the four native cysteines removed. All four of the native cysteines in SecA were successfully changed to serine without perturbing the protein.

In our studies we have mapped the surface of contact between proteins by using changes in line shape of a spin-labeled protein that occur when the protein forms a complex with a binding partner. The shape of an EPR spectrum contains information about the mobility of the nitroxide on a nanosecond time scale (11, 16). The motion of the nitroxide has its origin in rotation around bonds within the nitroxide side chain as well as in local backbone fluctuations. As the nitroxide goes from highly mobile to constrained or immobile, two features of the spectrum, readily seen by visual inspection, change. First, the central line width broadens (**Fig. 11.4**, ΔH_{pp}) which is also seen as a decrease in intensity of

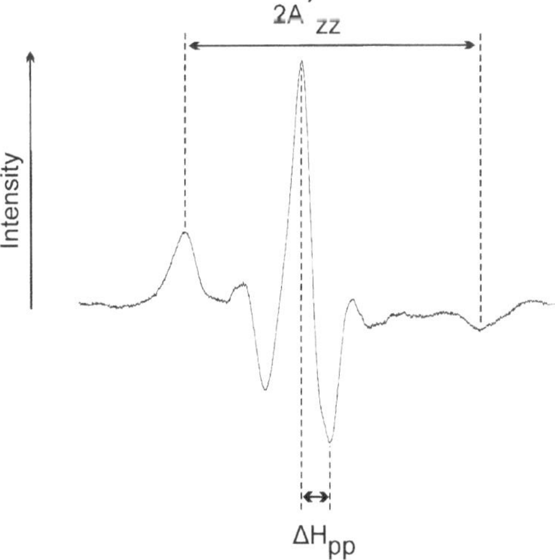

Fig. 11.4. Parameters reflecting mobility. The peak-to-peak width of the central resonance line (ΔH_{pp}) is measured as indicated and is equivalent to the peak width at half-height of an absorbance spectrum. The spectral breadth ($2A'_{zz}$) is the distance between the outermost hyperfine extrema. The spectrum used to illustrate these parameters is that of residue L126 of SecB.

the central line since all spectra are normalized. Second, the overall spectral breadth increases; that is, the total intensity is spread over a wider range of the magnetic field. For spectra reflecting slow motional states the hyperfine extrema are well-resolved and the separation of the extrema can be used to measure the spectral breadth (**Fig. 11.4**, A'_{zz}).

The solid traces in **Fig. 11.5** show nitroxides at several different positions in the chaperone SecB to illustrate the diversity of line shapes that will be observed depending on the location of the substituted amino acid within the protein structure. The line shape of the spectrum displayed in **Fig. 11.5A** is characteristic of a highly constrained residue that makes tertiary contacts with other structural elements within the protein, **Fig. 11.5B** illustrates a position exposed on the surface of a helix, and **Fig. 11.5C** shows a very mobile residue in a region of little tertiary structure. If within a complex the nitroxide probe makes contact with residues on a binding partner, either another protein or a small ligand, the mobility of the nitroxide will be constrained. When SecB is in complex with an unfolded polypeptide even the most immobile

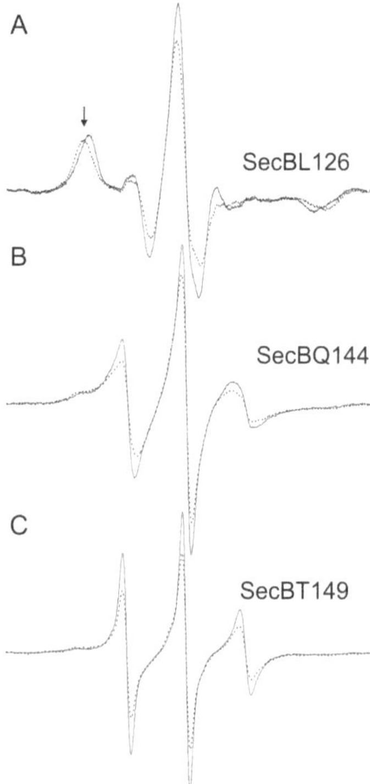

Fig. 11.5. Spectra of constrained residues with the spin label on the SecB residue indicated. (**A**) SecBL126, alone (*solid trace*) and in complex with unfolded polypeptide ligand (*dotted trace*); (**B**) SecBQ144, alone (*solid trace*) and in complex with unfolded polypeptide ligand (*dotted trace*); and (**C**) SecBT149, alone (*solid trace*) and in complex with SecA (*dotted trace*).

of these residues (L126, SecB) shows changes that are indicative of further constraint (**Fig. 11.5A**, compare the solid trace with the dotted trace). The central line amplitude drops and intensity moves out which is most clearly seen at the low field side (indicated by arrow). Constraints are also seen on residues Q144 and T149 when complexes with unfolded polypeptide or with SecA are formed.

In addition to constraints of mobility of a nitroxide, increased mobility will be observed if within a complex changes in conformation occur that result in breaking interactions of a side chain with neighboring structural elements. **Figure 11.6** shows two positions on the surface of helices that become more mobile when the spin-labeled protein, SecA, binds to the membrane-bound translocon, SecYEG. Mobilization results in movement of intensity from the extrema toward the center line. The amplitude of the center line is also increased.

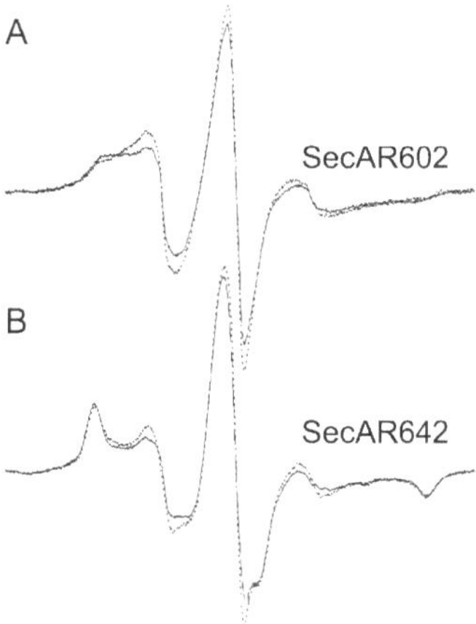

Fig. 11.6. Spectra of mobilized residues. *Solid traces* are spin-labeled SecA alone; *dotted traces* are spin-labeled SecA in complex with lipids. All spectra were gathered at 27°C. (**A**) Spin-labeled on residue R602 on SecA; (**B**) spin-labeled on residue R642 of SecA.

1.4. Considerations

Identification of the interface of contact between two proteins within a complex requires that enough residues be surveyed to cover a considerable surface of the protein under investigation. Interpretation depends on the emergence of patterns. Constraint of a single residue is likely to indicate a contact, but it is possible that contact at a distance from the nitroxide results in a conformational change that in turn constrains a residue that lies outside of the binding surface. In our study of SecB in complex with unfolded polypeptide ligands we examined 49% of the entire surface.

Observation of constraints allows one to define sites of contact and observation of mobilization such as we observed in SecA in complex with SecYEG and SecB (10) are likely to reflect conformational changes that release side chains from contacts that exist in the protein in the absence of a binding partner. It is important to find sites which show no change to define regions on the surface that are not involved in binding. In this respect when no change is observed in line shape it is crucial to be certain that substitution by the nitroxide has not inactivated the protein of interest. Each spin-labeled variant should be assayed for formation of a complex. In our studies all complexes of interest between soluble proteins could be demonstrated by size-exclusion column chromatography. SecA binds the membrane-bound translocon SecYEG and since column chromatography cannot be used to assess binding to vesicles we used a biological assay, i.e., the stimulation of SecA ATPase activity by binding to vesicles to make certain all SecA species were active (10). If simple assays of activity or complex formation are not available one should minimally check that the protein remains folded. This can usually be done by comparing the position of elution during size-exclusion chromatography of the protein before and after substitution with the nitroxide.

In order for line shapes to be interpreted in terms of local backbone fluctuations and internal motion of the nitroxide side chains, the molecular tumbling of the protein must be slow relative to the timescale of EPR spectroscopy. If the radius of hydration of the protein is known, the timescale of tumbling (τ_c) can be calculated as follows:

$$\tau_c = \frac{1}{6\,D_r};\ D_r = \frac{kT}{8\pi\eta a^3},$$

where D_r is the rotational diffusion coefficient, k is Boltzmann's constant, T is temperature, η is the viscosity of the solvent, and a is the radius of hydration of the protein. Proteins of molar mass 40 kDa or greater will tumble sufficiently slowly. Proteins of lower mass can be studied by addition of sucrose to the solution to 30%. This will increase the viscosity 2.75-fold relative to water and thereby slow tumbling accordingly.

2. Materials

2.1. Mutagenesis to Introduce a Single Accessible Cysteine

In this section we provide a detailed description of the mutagenesis procedure that we have used to create our single-cysteine variants. We do not describe purification of the proteins since the procedure of choice will depend on the species of protein under study.

1. Double-stranded DNA (dsDNA) template: a plasmid containing the gene for the protein of interest isolated from

a *dam*$^+$ *Escherichia coli* strain (*dam* encodes DNA adenine methylase, which methylates DNA duplexes on adenine in GATC sequences) suspended in either 10 mM Tris (HCl) pH 8.0 to 8.5 or H_2O at a concentration of 10 ng/μL.

2. *PfuUltra II* Fusion HS DNA polymerase and 10 X PCR reaction buffer (Stratagene).
3. *Dpn*1 endonuclease 10 units/μL (Fermentas).
4. dNTP mixture containing 10 mM of each of the deoxynucleotides, dATP, dCTP, dGTP, and dTTP (*see* **Note 1**)
5. Thermal cycler for PCR.
6. Competent cells for transformation.
7. Growth media and antibiotics appropriate for the bacterial strain used.
8. Thin-walled PCR tubes.
9. Sterile pipette tips and sterile 1.5 mL microcentrifuge tubes.
10. Sterile 15 mL Falcon tubes.
11. Desk-top microcentrifuge.
12. Plasmid preparation kit (Qiagen).
13. DNA sequencing primer to sequence region of mutagenesis.
14. Reducing agent: Dithiothreitol (DTT) or tris-(2-carboxyethyl) phosphine hydrochloride (TCEP, Molecular Probes).

2.2. Labeling of Cysteine Variants with Nitroxide Reagent

1. The nitroxide spin label reagent: (1-oxy-2,2,5,5 tetramethyl pyrroline-3-methyl)-methanethiosulfonate (Toronto Research Chemicals, Inc.).
2. Acetonitrile, HPLC grade.
3. Lyophilizer or centrifugal vacuum evaporator (Labconco).
4. Nap 10 column (Amersham Biosciences).
5. Buffer A: 10 mM Hepes (HAc), 300 mM KOAc, pH 7.0.
6. Centrifugal concentrator: Nanosep 30 (Pall Life Science) or Centricon 30 (Millipore).
7. Buffer B: 10 mM Hepes (HAc), 300 mM KOAc, 5 mM $Mg(Ac)_2$, pH 7.0.

2.3. EPR Measurements

1. Spectrometer: Bruker EMX X-band spectrometer with a high sensitivity resonator. To work at temperatures other than room temperature it needs to be equipped with a variable temperature accessory.
2. Liquid nitrogen for temperature regulation.

3. Capillaries from Fiber Optic Center, Inc. (*see* **Note 2**). Synthetic silica capillaries: 0.6 mm I.D. × 0.84 mm O.D., Supracil Cat. No. CV 6084S. These are used for gathering data. Glass capillaries: 0.6 mm I.D. × 0.84 mm O.D., Cat. No. CV 6084. These are used during isolation of the spin-labeled protein. Synthetic silica capillaries: 1 mm I.D. × 1.2 mm O.D., Cat. No. CV 1012S. These are used for making an adaptor to hold the sample capillary in the resonator.
4. Torch, natural gas, oxygen to seal capillaries at one end.
5. Critoseal, Cat. No. 8889-215003, from Oxford Labware.
6. Optional for making the capillary holder: Teflon heat shrink tubing (PTFE/FEP tube 0.06 inch ID, Part No. SMDT-060-24, Small Parts, Inc.) and a heat gun to shrink the tubing to fit tightly around the 1 mm capillary.
7. Eppendorf gel loader tips, 20 μL, and Gilson 20 μL micropippetor to fill the capillaries.
8. Benchtop low speed centrifuge which holds tubes at least 11–13 cm long so that the capillaries can be centrifuged to force solution to the bottom.
9. Software appropriate for data analysis: Labview programs written by Christian Altenbach, Jules Stein Eye Institute, Department of Chemistry and Biochemistry, UCLA. An alternative is the WinEPR software supplied by Bruker. Software for generating figures comparing spectra: Origin (OriginLab).

3. Methods

3.1. Mutagenesis to Introduce a Single Accessible Cysteine

1. Design two complementary mutagenic primers, one priming the upper strand and the other priming the lower strand of the dsDNA template (30 to 40 bases), with the desired mutation near the middle and a stretch of 15 to 18 unmodified bases on each side that are a perfect match to the template. Both of the primers must contain the desired mutation.
2. Have primers commercially synthesized and suspend at 5 μM in endonuclease free water.
3. Prepare the reaction mixture: 5 μL of dsDNA template at a concentration of 10 ng/μL; 5 μL of each of the 5 μM primers; 5 μL 10 X PCR reaction buffer; 1 μL of 10 mM dNTP mixture; endonulease-free H_2O to a final volume of 49 μL. Add 1 μL of *PfuUltra II* Fusion HS DNA

polymerase (2.5 units/μL) last and mix gently. If you do not have a heated lid on your thermocycler, overlay the reaction with a drop of mineral oil.

4. Run the reactions using the following parameters:
   a. 95°C for 2 minutes to denature;
   b. Use a three step cycle: (1) 95°C 30 seconds, (2) 55°C 30 seconds, and (3) 68°C 15 seconds per kilobase of template plasmid.
      If a single base change was used to introduce the cysteine run 14 cycles. If two or three bases were changed, run 16 cycles.
   c. 68°C for 3 minutes; 4°C hold.
5. Place the sample on ice briefly after removing it from the thermal cycler. The sample must be below 37°C before proceeding to the next step.
6. Add 1 μL of the *Dpn* I endonuclease (10 units/μL) to the sample, gently mix and incubate the sample for 1 hour at 37°C. The reaction is carried out in the PCR reaction buffer so no changes to the buffer condition are needed. The buffer supplied with the DpnI enzyme when purchased is not used.
7. All subsequent steps are to be carried out under sterile conditions.
8. Thaw competent cells on ice. Add 50 to 100 μL of competent cells to a 15 mL round-bottom Falcon tube. Add 1 to 5 μL of the *Dpn* I-treated DNA. Incubate on ice for 30 minutes to transform the cells.
9. Heat pulse the transformation mixture for 2 minutes at 42°C, then place it on ice for 2 minutes.
10. Add 0.5 mL of growth medium appropriate for the strain used, WITHOUT ANTIBIOTICS, to the Falcon tube containing the heat-treated competent cells and incubate with shaking for 1 hour at 37°C.
11. Transfer cells to a sterile 1.5 mL microcentrifuge tube and centrifuge for 1 minute at maximum speed in a desktop microcentrifuge. Remove all but approximately 50 μL of supernatant. Suspend the cells in the remaining 50 μL of supernatant and plate the entire volume onto an agar plate with the growth medium and ANTIBIOTICS appropriate for the strain used.
12. Incubate the plate at 37°C overnight or until distinct colonies are visible.
13. Select individual, isolated colonies from the overnight plate and grow separate cultures of 4 mL from each colony.

14. Isolate plasmid DNA from each culture following the instructions included with the plasmid preparation kit used.
15. Determine the absorbance of each plasmid preparation at wavelength 260 nm. Calculate the DNA concentration using a conversion factor of absorbance of 1 is equivalent to 50 μg/mL of dsDNA.
16. Have each plasmid DNA preparation sequenced commercially following the guidelines provided by the supplier of the service for the amount of sample and primers required.
17. After your DNA sequence has been verified as correct, purify the protein by the method of choice for the species under study. During purification keep the solutions reduced at all times using 2 mM DTT or 2 mM TCEP (*see* **Note 3**).

3.2. Labeling of the Cysteine Variants with the Nitroxide Reagent

3.2.1. Preparation of the Nitroxide Reagent

1. Purchase the reagent in vials of 10 mg and suspend the entire amount in 380 μL of acetonitrile to give a concentration of 100 mM nitroxide reagent.
2. Dispense the solution in portions of 20 μL into 0.65 mL Eppendorf tubes and take to dryness by lyophilization or using a centrifugal vacuum evaporator (Labconco).
3. Store the reagent in the dark at −80°C.
4. As needed add 20 μL of acetonitrile to a tube to give a solution of 100 mM nitroxide reagent. The solution can be used for several experiments and frozen between usages.

3.2.2. Removal of Reducing Agent and Exchange into Labeling Buffer

If possible one should start with 10 mg of protein so there is sufficient protein labeled not only for EPR studies but also for assays of activity (*see* **Section 1.4.**). If the protein is only available in limited quantities, smaller amounts can be labeled. We have never labeled less than 3 mg, but besides needing a good recovery so that the final sample will contain a high concentration of spin there should be no difficulty.

1. Equilibrate a Nap10 column with 15 mL of Buffer A.
2. Apply the protein to be spin labeled to the top of the column in a maximum volume of 1.0 mL. If the protein is contained in less than 1 mL do not dilute it (*see* **Note 4**), rather apply the sample and then follow with Buffer A to give a total volume of 1.0 mL. Allow this volume to flow through the column and discard it.
3. Elute the protein by addition of 1.2 mL of Buffer A. If the sample was applied in 1 mL, it will be completely recovered in 1.5 mL. However, you should not collect more than 1.2 mL so that all reducing agent is removed.

3.2.3. Substitution of Cysteine with the Nitroxide Reagent

1. If necessary, concentrate the sample using a Centricon 30 so that it is in 1 mL or less to facilitate removal of the free spin in the subsequent step no. 4 (see below).
2. Determine the concentration of the protein to be labeled by reading the absorbance at wavelength 280 nm. If the sequence of the protein is known, an extinction coefficient (ε) can be calculated using *ProtParam* (http://us.expasy.org). The concentration should be expressed in terms of molarity (i.e., not mg/mL) so that one can calculate the amount of reagent to add. Add the reagent in a 3-fold molar excess (*see* **Note 5**) to the accessible cysteine in each polypeptide chain. Care must be taken to keep the concentration of acetonitrile below 2% to avoid denaturation of the protein.
3. Incubate the protein and reagent together on ice in the dark for 2–3 hours.
4. Remove the free reagent by passage of the spin-labeled protein over a Nap10 column, equilibrated in Buffer B (*see* **Note 6**).
5. As before, see step 2 of **Section 3.2.2**, apply the protein to the top of the column. If the volume of the protein solution is less than 1 mL then Buffer B is added after the protein solution enters the column so that the total volume applied is 1 mL. Discard the initial 1 mL of volume that comes through the column.
6. Apply 1.2 mL Buffer B to the top of the column and collect the eluent in 4–5 drop fractions (approximately 0.15 mL).
7. Analyze each fraction by EPR (*see* **Section 3.3**). Determine which fractions to pool in order to maximize recovery of the labeled protein without including any free spin (*see* **Note 7**). The free spin elutes in the later fractions.
8. Determine the concentration of the spin-labeled protein by reading the absorbance at 280 nm and using the extinction coefficient as described in step 2 (see above).
9. Concentrate the spin-labeled protein using a Nanosep 30 centrifugal concentrator, if necessary. Generally, a sample containing 60 μM spin gives an excellent signal (*see* **Note 8**).

3.3. Acquiring Data and Subsequent Analysis

1. Prepare capillaries for data collection. All capillaries are sealed at one end by a brief heating with an oxygen gas torch (*see* **Note 9**). Make an adaptor to hold the 0.6 mm I.D. capillary in the resonator by sealing one end of a 1 mm I.D. capillary. This capillary is too small to be held in the smallest collet that is supplied with the Bruker resonator. It must be

further modified to increase the diameter. This can be simply done using parafilm or tape. To make a more robust adaptor one can insert the top of the capillary into a short stretch of Teflon heat shrink tubing and then use a heat gun to shrink it to fit tightly around the capillary. This capillary will now be held tightly in a small collet supplied with the resonator.

2. Fill the sealed capillary using a gel loader tip which will fit into the top of the 0.6 mm capillary. Slowly dispense 5–6 μL of the solution while withdrawing the tip. This leaves the solution at the top of the capillary. It is forced to the bottom by a brief spin in a bench top centrifuge using a test tube that is slightly longer than the capillary (10 cm) as an adaptor. If the sample is available in sufficient quantities, it is simpler to use unsealed capillaries. Immerse the tip of an open capillary into a solution allowing it to fill by capillary action. Insert the filled capillary into the Critoseal container to introduce a plug at the end.
3. Compare a series of samples containing the spin-labeled protein that are prepared in the same buffer conditions. For example, in experiments with unfolded polypeptide ligands and SecB we dilute the ligand from denaturant; thus, denaturant must be added to the same final concentration to the sample with SecB alone (*see* **Note 10**).
4. Scan over a field of 100 gauss (1 gauss = 10^{-4} Tesla) centered at 3356 gauss using an incident power of 20 mW (attenuation 10 dB).
5. We routinely gather 15 scans; the ratio of the signal intensity to the noise increases as the square of the number of scans. Whereas four scans give a 2-fold improvement relative to two scans and 16 scans give a 2-fold improvement over four scans, it is not practical to improve the signal by gathering more than 15 scans since a 2-fold increase in the signal to noise would require 225 scans.
6. Choose a field modulation appropriate for the spectrum. Set the frequency of modulation to 100 kHz and set the field of modulation to one gauss. Acquire an initial spectrum (one scan will do) and measure the center line width (ΔH_{pp}). Increase the field of modulation to improve the signal-to-noise ratio, but do not exceed the width of the center line. We usually modulate at two gauss. Scans must be gathered at the same modulation in order to be compared. However, it is not necessary to gather the same number of scans for comparison since the data are normalized during analyses.
7. Set a baseline and normalize each spectrum using WinEPR or the Labview programs.

8. Compile the data into a scientific plotting and analysis software program such as Origin. Compare the spectral line shape of a given spin-labeled variant with the line shapes of the protein in complex with the binding partners to be tested. Also compare spectra to determine changes in line shape that result from addition of components to the solution, such as denaturant.
9. Visual inspection is used to determine whether changes in line shape represent a constraint or a mobilization of the side chain as described in **Section 1.3**.

4. Notes

1. If the solution of deoxynucleotides is prepared by the investigator and not obtained from a kit, the pH must be adjusted to between pH 7.5 and pH 8.0 to neutralize the acid of the phosphate groups.
2. Synthetic silica capillaries are used for collection of data because they have no background EPR signal. For routine surveys of fractions eluted from the Nap10 column (*see* step 6 of **Section 3.2.3**) quartz capillaries can be used. The background signal is often barely detectable. The price difference is considerable ($3.32 per silica capillary versus $0.26 for quartz).
3. During protein purification all solutions must be kept reduced to prevent oxidation of the cysteines. Since oxida tion to sulfenic, sulfinic, and sulfonic acid is irreversible, it is crucial to keep solutions reduced from the start. We dis rupt our cells using a French Press. Sonication introduces air and additionally causes local heating near the probe.
4. The smaller the volume of the sample applied, the sharper the peak of elution is for the protein. Therefore, most of the protein can be recovered in 1.2 mL, well-separated from the included volume that contains the low molecular weight reducing agents.
5. This approach is applicable to proteins that contain cysteine in addition to the target cysteine as long as the target is more accessible so that it can be preferentially labeled. When other cysteines are in the protein species we use a 1.3-fold molar excess to the target cysteine and incubate on ice in the dark for 1 hour.
6. Nitroxide free in solution gives three very sharp lines. The sharp signal overlaps the line shape of the spin-labeled protein making it difficult to see changes in the line shape.

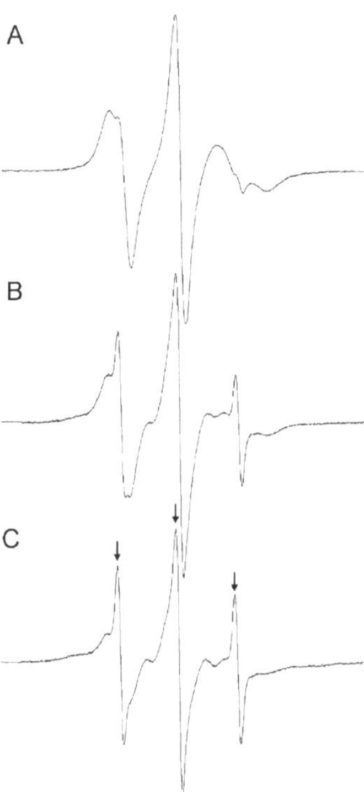

Fig. 11.7. Effect of free spin on line shape. SecB was spin-labeled on residue Q33. A sample containing 160 μM spin was loaded into a capillary and the open end sealed with Critoseal. The capillary was stored at 7°C and scanned periodically. The *arrows* show the positions of the three lines that represent the free spin. (**A**) Scanned within a week of labeling, (**B**) after 1 month at 7°C, and (**C**) after 4 months at 7°C.

The Labview software from Christian Altenbach provides a means to subtract free spin up to approximately 5%. **Figure 11.7** shows the appearance of free spin in a sample stored at 7°C for 4 months. Since the free spin has very sharp lines and a nitroxide on a protein has a much broader spectrum, a small percentage of free spin dominates the line shape.

7. When removing the free label we work in the cold room to suppress the release of the nitroxide. We find that substitution at some positions is susceptible to release of the nitroxide.
8. When the 0.6 mm diameter capillaries are inserted into the resonator only a volume of 5 to 6 μL at the bottom of the tube is in the path of the microwave radiation. Thus loading more volume does not increase the signal. The spectrometer records the total spin in the element of volume irradiated. It does not matter if it is uniformly dispersed.
9. Caution must be taken so that the sealed end does not have a ball on it that prevents insertion into the 1 mm capillary

serving as a holder. The tube should be heated as briefly as possible. After cooling, capillaries can be checked under a low power dissection microscope to be sure they are sealed. One should also test each sealed capillary by inserting it into a 1 mm capillary to be certain that it fits before using them with sample. If necessary, samples can be removed from the capillary by inserting the open end into a centrifuge tube and applying a brief spin, or the capillary can be broken off and a Hamilton syringe used to withdraw the sample.

10. SecB binds polypeptides in an unfolded state with no affinity for stably folded proteins. Therefore, the unfolded ligand must be diluted from denaturant directly into a solution containing SecB to allow formation of a complex. For this reason in our studies of SecB complexed with unfolded polypeptides, there is a low level of denaturant present. The denaturant is not necessary for any other reason.

References

1. Papanikou, E., Karamanou, S., and Economou, A. (2007) Bacterial protein secretion through the translocase nanomachine *Nat. Rev. Microbiol.* **5,** 839–851.
2. Randall, L. L., Topping, T. B., and Hardy, S. J. S. (1990) No specific recognition of leader peptide by SecB, a chaperone involved in protein export *Science* **248,** 860–863.
3. Randall, L. L., Crane, J. M., Liu, G., and Hardy, S. J. S. (2004) Sites of interaction between SecA and the chaperone SecB, two proteins involved in export *Protein Sci.* **13,** 1124–1133.
4. Bruce, J. E., Smith, V. F., Liu, C., Randall, L. L., and Smith, R. D. (1998) The observation of chaperone-ligand noncovalent complexes with electrospray ionization mass spectrometry *Protein Sci.* **7,** 1180–1185.
5. Randall, L. L., Crane, J. M., Lilly, A. A., Liu, G., Mao, C., Patel, C. N., and Hardy, S. J. (2005) Asymmetric binding between SecA and SecB two symmetric proteins: implications for function in export *J. Mol. Biol.* **348,** 479–489.
6. Randall, L. L., Hardy, S. J. S., Topping, T. B., Smith, V. F., Bruce, J. E., and Smith, R. D. (1998) The interaction between the chaperone SecB and its ligands: evidence for multiple subsites for binding *Protein Sci.* **7,** 2384–2390.
7. Patel, C. N., Smith, V. F., and Randall, L. L. (2006) Characterization of three areas of interactions stabilizing complexes between SecA and SecB, two proteins involved in protein export *Protein Sci.* **15,** 1379–1386.
8. Crane, J. M., Mao, C., Lilly, A. A., Smith, V. F., Suo, Y., Hubbell, W. L., and Randall, L. L. (2005) Mapping of the docking of SecA onto the chaperone SecB by site-directed spin labeling: insight into the mechanism of ligand transfer during protein export *J. Mol. Biol.* **353,** 295–307.
9. Crane, J. M., Suo, Y., Lilly, A. A., Mao, C., Hubbell, W. L., and Randall, L. L. (2006) Sites of interaction of a precursor polypeptide on the export chaperone SecB mapped by site-directed spin labeling *J. Mol. Biol.* **363,** 63–74.
10. Cooper, D. B., Smith, V. F., Crane, J. M., Roth, H. C., Lilly, A. A., and Randall, L. L. (2008) SecA, the motor of the secretion machine, binds diverse partners on one interactive surface *J. Mol. Biol.* **382,** 74–87.
11. Fajer, P. G. (2000) *in* "Encyclopedia of Analytical Chemistry" (Meyers, R. A., Ed.), pp. 5725-61, John Wiley & Sons Ltd., London.
12. McHaourab, H. S., Lietzow, M. A., Hideg, K., and Hubbell, W. L. (1996) Motion of spin-labeled side chains in T4 lysozyme. Correlation with protein structure and dynamics *Biochemistry* **35,** 7692–7704.

13. Hubbell, W. L., Gross, A., Langen, R., and Lietzow, M. A. (1998) Recent advances in site-directed spin labeling of proteins *Curr. Opin. Struct. Biol.* **8,** 649–656.
14. Columbus, L., Kalai, T., Jeko, J., Hideg, K., and Hubbell, W. L. (2001) Molecular motion of spin labeled side chains in alpha-helices: analysis by variation of side chain structure *Biochemistry* **40,** 3828–3846.
15. Fanucci, G. E., and Cafiso, D. S. (2006) Recent advances and applications of site-directed spin labeling *Curr Opin. Struct. Biol.* **16,** 644–653.
16. Schneider, D., and Freed, J. (1989) Spin Labeling: Theory and Application, Biological Magnetic Resonance (Berliner, L., and Reuben, J., Eds.), Vol. 8, Plenum, New York.

Chapter 12

Analysis of Tat Targeting Function and Twin-Arginine Signal Peptide Activity in *Escherichia coli*

Tracy Palmer, Ben C. Berks, and Frank Sargent

Abstract

The Tat system is a protein export system dedicated to the transport of folded proteins across the prokaryotic cytoplasmic membrane and the thylakoid membrane of plant chloroplasts. Proteins are targeted for export by the Tat system via N-terminal signal peptides harbouring an S-R-R-x-F-L-K 'twin-arginine' motif. In this chapter qualitative and quantitative assays for native Tat substrates in the model organism *Escherichia coli* are described. Genetic screening methods designed to allow the rapid positive selection of Tat signal peptide activity and the first positive selection for mutations that inactivate the Tat pathway are also presented. Finally isothermal titration calorimetry (ITC) methods for measuring the affinity of twin-arginine signal peptide–chaperone interactions are discussed.

Key words: Tat system, twin-arginine signal peptide, TMAO reductase, hydrogenase, chaperone, protein–protein interaction.

1. Introduction

The Tat (*t*win-*a*rginine protein *t*ransport) system exports folded proteins across the energy-coupling membranes of prokaryotes and plant organelles. Protein substrates are targeted to the Tat machinery by means of cleavable N-terminal signal peptides that contain a consensus S-R-R-x-F-L-K 'twin-arginine' motif, where the twin-arginines are almost invariant and are essential to initiate export by the Tat system (1, 2). In prokaryotes the best characterised Tat system is that of the model bacterium *Escherichia coli*. Detailed analysis has indicated that *E. coli* has some 27 native Tat substrates. About two-thirds of these bind catalytically

A. Economou (ed.), *Protein Secretion*, Methods in Molecular Biology 619,
DOI 10.1007/978-1-60327-412-8_12, © Springer Science+Business Media, LLC 2010

essential redox cofactors and several Tat substrates are components of anaerobic respiratory chains that are vital for the respiratory flexibility of the organism (3, 4). Well-studied complex substrates of the Tat system include the soluble periplasmic protein TMAO reductase, and the membrane-bound (periplasmically oriented) enzymes DMSO reductase, formate dehydrogenases –N and –O and the uptake hydrogenases −1 and −2 (5, 6). For each of these enzymes, suitable spectrophotometric methods exist for assaying their activity using artificial redox dyes. Coupled with robust fractionation approaches this allows the effects of inactivating *tat* (or other biosynthetic) mutations on their subcellular localisation and activity to be accurately determined.

Several of the cofactor-containing *E. coli* Tat substrates are transported as heterodimers, for example the large and small subunits of the uptake hydrogenases. In these cases there is a twin-arginine signal peptide present on only one of the subunits and the protein pairs assemble together in the cytoplasm as a complex before they are targeted to the Tat system (7). Clearly the complex processes of cofactor assimilation into Tat substrates along with partner protein binding must be carefully orchestrated to ensure they are completed prior to the transport process. To coordinate these events, most cofactor-containing Tat substrates have dedicated chaperones which bind with exquisite specificity to the twin-arginine signal peptides of their cognate proteins (8). Chaperone binding affects recognition/interaction of the signal peptide with the Tat machinery, allowing the biosynthetic processes to be completed, and the chaperone subsequently dissociates to allow the now fully assembled precursor to target to the export system. A reliable *in vitro* approach to examine signal peptide-chaperone interactions utilises isothermal titration calorimetry. This quantitative technique allows the accurate determination of binding stoichiometries, dissociation constants and also gives thermodynamic information about the interaction in the shape of Gibbs free energy values and enthalpy changes.

Other substrates of the *E. coli* Tat system are rather simpler because they either do not bind a catalytic cofactor (e.g. SufI or the cell wall amidases AmiA and AmiC), or they bind their cofactor once they reach the periplasm (e.g. copper acquisition by CueO family proteins) (2). These substrates are more useful as reporters for Tat transport in kinetic experiments, since the export process is not complicated by the processes of cofactor biosynthesis and insertion, or the interaction with dedicated signal peptide binding chaperones. The Tat-dependent amidase proteins also make useful reporters for genetic screens. Failure to export these two proteins to the periplasm results in an inability of *E. coli* to cleave the peptidoglycan septum during cell division, leading to a cell envelope defect and a marked increase in sensitivity to sodium dodecylsulphate (SDS). This defect can be phenocopied in a *tat*$^{+}$

background by combined deletion of the *amiA* and *amiC* genes, and can be rescued by production in *trans* of one or other of these genes (9). We have recently exploited this observation to develop a genetic screen for Tat signal peptide activity using the AmiA mature sequence as the reporter protein and SDS resistance in solid media as the selection for signal peptide activity (10).

Finally, heterologous reporter proteins that are compatible with export by the Tat system, for example green fluorescent protein, or maltose binding protein, are increasingly useful tools for exploring Tat function (11–13). We have described a robust reporter system that provides a facile positive selection for *E. coli* Tat system inactivation (14). This system uses the normally cytoplasmic enzyme, chloramphenicol acetyltransferase (CAT) as a reporter since when fused to a standard twin-arginine signal peptide it is efficiently exported to the periplasm. This renders cells sensitive to the antibiotic chloramphenicol because its detoxification by CAT requires acetyl coenzyme A as a co-substrate, which is not present in the periplasmic compartment. Mutations that interfere with function of the Tat system, for example inactivating point mutations in Tat components, can therefore be easily selected since, in contrast to tat^{+} strains, they will support growth of cells harbouring the Tat signal peptide-CAT fusion on media containing chloramphenicol (14).

2. Materials

2.1. Preparation of SDS Plates

1. Miller LB agar powder (Invitrogen).
2. 20% SDS solution in distilled water, autoclaved.
3. Sterile distilled water.

2.2. Preparation of Glycerol-TMAO Minimal Plates

1. M9 salts, 5X stock solution (after (15)) per litre: 64 g $Na_2HPO_4.7\ H_2O$, 15 g KH_2PO_4, 2.5 g NaCl, 5 g NH_4Cl. Sterilise by autoclaving.
2. MoSe solution: 1 mM K_2SeO_3, 1 mM $(NH_4)_6MO_7$, filter sterilise and store at 4°C.
3. 1% thiamine solution (filter sterilise and store at 4°C).
4. 50% glycerol solution. Sterilise by autoclaving.
5. 20% trimethylamine-*N*-oxide (Sigma) solution, filter sterilised.
6. 1 M $MgSO_4$ solution, autoclaved.
7. 1 M $CaCl_2$ solution, autoclaved.

8. Bactoagar (Invitrogen) made up to 2x concentration in distilled water and autoclaved.
9. Sterile distilled water.

2.3. Preparation of CR-Medium (after (16))

1. CR base medium: per litre, 5.29 g K_2HPO_4 (anhydrous), 8.24 g KH_2PO_4 (anhydrous), 5 g peptone, 2 g $(NH_4)_2SO_4$.
2. CR trace elements: per litre, 0.48 g $FeCl_3.6H_2O$, 0.33 g $MnCl_2.4H_2O$, 0.36 g $CaCl_2.2H_2O$, 2.0 g $ZnCl_2$, 0.2 g H_3BO_3, 0.1 g $CoSO_4.7H_2O$. Store at 4°C, shake well before use.
3. MoSe solution: 1 mM K_2SeO_3, 1 mM $(NH_4)_6$ MO_7, filter sterilise and store at 4°C.
4. 1% thiamine solution, filter sterilise and store at 4°C.
5. 1 M $MgSO_4$ solution, autoclaved.
6. 20% solution of casamino acids (Difco), filter sterilise and store at 4°C.
7. 50% glycerol solution. Sterilise by autoclaving.
8. 20% trimethylamine-*N*-oxide (Sigma) solution, filter sterilised OR 20% dimethylsulphoxide solution, filter sterilised OR 20% KNO_3 solution, autoclaved OR 16% fumarate solution, filter sterilised.

2.4. Fractionation of E. coli *Cells*

1. Cell washing buffer: 50 mM Tris-HCl, pH 7.6.
2. Sphaeroplast buffer: 50 mM Tris-HCl, pH 7.6, 0.5 M sucrose.
3. Membrane washing buffer: 50 mM Tris-HCl, pH 7.6, 250 mM NaCl.
3. 1 M EDTA, pH 7.6.
4. Lysozyme powder (Sigma).
5. Deoxyribonuclease I powder (Sigma).
6. 20% Triton X-100 dissolved in 100 mM HEPES pH 7.5.

2.5. Assay for TMAO Reductase Activity

1. 100 mM K-phosphate buffer, pH 7.0.
2. 200 mM benzyl viologen (Sigma) in 100 mM K-phosphate buffer, pH 7.0. Keep on ice.
3. 1% sodium dithionite solution in aqueous 1 mM NaOH (*see* **Note 1**). Keep on ice.
4. 20% solution of trimethylamine-*N*-oxide (TMAO) in water.
5. Cylinder of nitrogen gas for sparging of buffer.
6. 1.8 ml lidded cuvette (Hellma, England; **Fig. 12.2**) with 1 mm diameter hole drilled through the lid (*see* **Note 2**).
7. Glass balls, 1.5–2.5 mm diameter (BDH; **Fig. 12.2**).
8. 25 μl and 50 μl volume Hamilton syringes.

2.6. Assay for Respiratory Formate Dehydrogenase (Fdh-N/Fdh-O) Activity

1. 100 mM K-phosphate buffer, pH 7.0.
2. 3 mg/ml phenazine methosulphate solution 100 mM K-phosphate buffer, pH 7.0. This is light sensitive to keep in a foil-covered tube. Keep on ice.
3. 2.4 mM 2,6-dichlorophenolindophenol (DCPIP) solution in 100 mM K-phosphate buffer, pH 7.0. Keep on ice.
4. 1.2 M sodium formate in 100 mM K-phosphate buffer, pH 7.0.
5. Cylinder of nitrogen gas for sparging of buffer.
6. 1.8 ml lidded cuvette (Hellma, England; **Fig. 12.2**) with 1 mm hole drilled through the lid (*see* **Note 2**).
7. Glass balls, 1.5–2.5 mm diameter (BDH; **Fig. 12.2**).
8. 25 μl and 50 μl Hamilton syringes.

2.7. Assay for Uptake Hydrogenase Activity

1. 100 mM K-phosphate buffer, pH 7.0.
2. 250 mM benzyl viologen (Sigma) in 100 mM K-phosphate buffer, pH 7.0. Keep on ice.
3. 1% sodium dithionite solution in aqueous 1 mM NaOH (*see* **Note 1**). Keep on ice.
4. Cylinder of hydrogen gas for hydrogen saturation of buffer.
5. 1.8 ml lidded cuvette (Hellma, England; **Fig. 12.2**) with 1 mm hole drilled through the lid (*see* **Note 2**).
6. Glass balls, 1.5–2.5 mm diameter (BDH; **Fig. 12.2**).
7. 25 μl and 50 μl Hamilton syringes.

2.8. Pulse-Chase Analysis of Tat Substrate Export in E. coli

1. M9 salts, 5X stock solution (after (15)) per litre: 64 g $Na_2HPO_4.7H_2O$, 15 g KH_2PO_4, 2.5 g NaCl, 5 g NH_4Cl. Sterilise by autoclaving.
2. 1% thiamine solution (filter sterilise and store at 4°C).
3. 1 M $MgSO_4$ solution, autoclaved.
4. 1 M $CaCl_2$ solution, autoclaved.
5. 20% glucose solution in water, filter sterilised.
6. Sterile distilled water.
7. Amino acid mixture – contains 1% of each of the 20 standard amino acids, with the exception of cysteine and methionine (*see* **Note 3**).
8. 20 mg/ml solution of rifampicin in methanol. Store at −20°C in a foil-covered tube.
9. [^{35}S]-labelled cysteine and methionine mix (Easytag express, Perkin Elmer).

10. 25 mg/ml methionine solution in water, filter sterilised, store at 4°C.
11. Optional: Cell fractionation buffer: 30 mM Tris-HCl, pH 8.0, 20 % (w/v) sucrose, 1 mM Na_2EDTA.
12. Optional: ice-cold 5 mM $MgSO_4$.

2.9. Exploiting AmiA as a Reporter for Tat-Targeting Signal Peptides

1. Plasmid pUniAmiA.
2. *E. coli* strain MCDSSAC (As MC4100, *amiA*Δ2-33, *amiC*Δ2-32 (*see* (9)).
3. LB SDS plates prepared as described in **Section 3.1**.

2.10. Using Chloramphenicol Acetyltransferase as a Positive Reporter for Loss of Tat Function in E. coli

1. Plasmid pUNICAT-NapA.
2. Isogenic *tat*$^+$ and *tat*$^-$ *E. coli* strains.
3. LB plates supplemented with 100 μg/ml ampicillin.
4. LB plates supplemented with 100 μg/ml ampicillin and a range of chloramphenicol concentrations.

2.11. Measuring Signal Peptide-Chaperone Interactions Using Isothermal Titration Calorimetry

1. Calorimeter (e.g. Microcal VP).
2. Buffer solution (*see* **Note 4**).
3. Pure protein or peptide preparation (*see* **Note 5**).
4. Dialysis membrane (*see* **Note 5**).
5. Optional. Range of buffers with different heats of ionization (e.g. sodium or potassium phosphate, PIPES, MOPS, ACES, Tris-HCl (*see* **Note 4**).

3. Methods

3.1. Preparation of SDS Plates

The failure to export the related cell wall amidases AmiA and AmiC to the periplasm results in *tat*-deficient strains of *E. coli* K12 being sensitive to growth in the presence of the detergent SDS. Therefore a rapid, qualitative assay for loss of Tat function is to assess the ability of strains to grow on solid media containing 2% SDS (9).

1. Add LB agar powder to distilled water to a concentration of 2X and autoclave.
2. When cooled to approximately 55°C add antibiotic supplements as required, 20% SDS to give the required final concentration (*see* **Note 6**) and pre-warmed sterile distilled water to adjust the final concentration of LB agar to 1X.
3. Swirl gently to mix, and pour. Allow plates to harden and streak or plate appropriate strains (*see* **Note7**). Care should

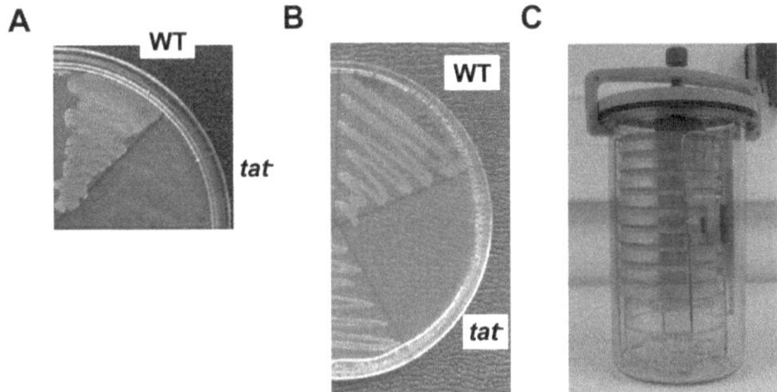

Fig. 12.1. Phenotypic analysis of *tat* mutant strains. (**A**) Strains MC4100 (36) and BØD (as MC4100, Δ*tatB*; (37)) were struck out from overnight cultures onto LB medium containing 2% SDS and incubated overnight at 37°C. (**B**) Strains MC4100 (38) and B1LKO (as MC4100, Δ*tatC*; (39)) were streaked out from overnight cultures onto glycerol-TMAO minimal plates, stacked in an anerobic jar and incubated for 3 days at 37°C. (**C**) Typical anaerobic jar used for anaerobic growth of strains.

be taken when streaking strains as the agar containing SDS tears more easily. An example of *tat*$^+$ and *tat*$^-$ strains cultured overnight on SDS-containing plates is shown in **Fig. 12.1A**.

3.2. Preparation of Glycerol-TMAO Minimal Plates

Growth of *E. coli* strains anaerobically with TMAO as sole electron acceptor is supported by two Tat-dependent molybdoenzymes – TMAO reductase and DMSO reductase (which, despite its name can readily reduce TMAO). Therefore *tat* mutants are unable to grow anaerobically on minimal medium in the presence of glycerol (a non-fermentable carbon source) and TMAO, providing a simple, qualitative assessment of Tat functionality.

1. To 500 ml of molten (55°C) 2 × bactoagar add 200 ml of pre-warmed 5 X M9 salts, 266 ml of pre-warmed sterile distilled water, 2 ml 1 M $MgSO_4$ 0.1 ml 1 M $CaCl_2$ 1 ml 1% thiamine (*see* **Note 8**) 1 ml MoSe solution (*see* **Note 8**), 10 ml of pre-warmed 50% glycerol (final concentration 0.5%), 20 ml of pre-warmed 20% TMAO (final concentration 0.4%, *see* **Note 9**) and antibiotic supplements as required.
2. Swirl gently to mix, and pour. Allow plates to harden and streak or plate appropriate strains (*see* **Note** 7). Stack plates into a gas tight jar (Becton Dickinson/BBL GasPak System; see **Fig. 12.1C**).
3. Dampen an anaerobic indicator strip and attach it to the plate rack on the inside of the jar, ensuring that the blue indicator part is not in contact with the walls of the jar.
4. Add 10 ml water to a hydrogen and carbon dioxide gas generating kit sachet (Becton Dickinson) and place in the jar. Immediately screw on the lid (finger tight only) and incubate

at the appropriate temperature for 3–4 days. An example of *tat*$^{+}$ and *tat*$^{-}$ strains cultured for 3 days on glycerol-TMAO minimal medium is shown in **Fig. 12.1B**. A typical gas-tight anaerobic jar is shown in **Fig. 12.1C**.

3.3. Preparation of CR-Medium

This semi-defined growth medium is particularly suitable for supporting *E. coli* growth by anaerobic respiration. Strains grow more quickly in CR than in fully defined minimal media, and to a higher final biomass. LB medium supplemented with glycerol and an appropriate terminal electron acceptor can be used in place of CR; however the tryptone present in LB leads to catabolite repression and may lower the specific activity of the respiratory enzyme of interest.

1. Adjust pH of CR base medium to 6.4 (if necessary) add 1 ml of CR trace elements solution and autoclave.
2. To 1 l of autoclaved CR base medium/trace elements add 2 ml 1 M $MgSO_4$, 2 ml 20% casamino acids, 1 ml 1% thiamine (*see* **Note 8**), 1 ml MoSe solution, 10 ml 50% glycerol and 20 ml of either 20% TMAO solution, DMSO solution or KNO_3 solution or 25 ml of 16% fumarate solution (*see* **Note 10**).
3. For growing anaerobic cultures bottles or flasks should be filled to the top and lightly capped to prevent exposure to oxygen. It is not necessary to degas media or grow under an oxygen-free atmosphere as any oxygen in the growth media will be rapidly used up after inoculation.

3.4. Fractionation of E. coli Cells

This method of cell fractionation is based on lysozyme/EDTA treatment in a sucrose buffer (17, 18). The EDTA removes the Mg^{2+} and Ca^{2+}ions from the outer membrane allowing the lysozyme to enter the periplasm. The lysozyme selectively digests the peptidoglycan layer enabling the efficient release of the periplasm. The resulting sphaeroplasts are maintained in an intact state by a high osmolarity sucrose buffer. Transfer of the sphaeroplasts to a buffer lacking sucrose results in lysis by osmotic shock. Lysed sphaeroplasts are subsequently separated into cytoplasm and membranes, and proteins in membrane fraction are extracted with Triton X-100 to give a homogeneous sample for subsequent enzyme assay.

1. Grow strains anaerobically at the appropriate temperature in CR medium as required (*see* **Notes 10 and 11**). All of the following steps should be carried out at 4°C unless stated otherwise.
2. Harvest cells at 6000 *g* and wash once with 50 mM Tris-HCl, pH 7.6.

3. Weigh the cell pellet (*see* **Note 12**) and then resuspend the cells in 10 ml sphaeroplast buffer per gram wet weight of cell pellet. To this, add a final concentration of 5 mM EDTA, pH 7.6 and 0.6 mg lysozyme per ml of resuspended cell volume.
4. Incubate the suspension without shaking at 37°C for 30 minutes to allow the disruption of the cell wall and peptidoglycan layer.
5. Centrifuge the suspension at 17,000 *g* for 15 minutes to pellet the sphaeroplasts. *Retain the supernatant from the centrifugation as the periplasmic fraction.*
6. Resuspend the sphaeroplast pellet in 50 mM Tris-HCl, pH 7.6 at a volume of 10 ml per gram wet weight of the original cell pellet. (The low osmotic strength buffer will cause the sphaeroplasts to lyse.) Add a few flakes of Deoxyribonuclease I powder to degrade released DNA. To ensure all sphaeroplasts are disrupted, passage the suspension twice through a cooled French press pressure cell at 8000 psi.
7. Centrifuge the French-pressed sample at 27,000 *g* for 20 minutes to remove cell debris. Carefully remove and retain the supernatant.
8. Centrifuge the supernatant for 30 minutes at 278,000 *g* in a Beckman T-100 bench top ultra-centrifuge. After centrifugation, the upper third of the supernatant containing the cytoplasm was removed. The lower portion does contain cytoplasmic proteins but also contains fragmented membranes, very small membrane vesicles and polysomes, which is best discarded.
9. After removing all of the supernatant, resuspend the membrane pellet in 50 mM Tris-HCl, pH 7.6, 250 mM NaCl and re-centrifuge at 278,000 *g* for 30 minutes.
10. Decant the supernatant and resuspend the pellet in 800 μl of 50 mM Tris-HCl, pH 7.6. To this add 200 μl of 20% Triton X-100 in 100 mM HEPES pH 7.5 to solubilise the membranes, incubate on ice for 30 minutes and centrifuge at 278,000 *g* for 30 minutes in a Beckman T-100 bench top ultra-centrifuge. Retain the supernatant as the solubilised membrane fraction.
11. The quality of fractionation should be assessed by measuring the distribution of compartment-specific marker proteins (*see* **Note 13**).

3.5. Assay for TMAO Reductase Activity

Reduction of TMAO is measured by an anaerobic colorimetric assay. The method employed is based on the coupling of TMAO reduction with the oxidation of benzyl viologen (19). Benzyl viologen is an artificial electron donor and gives a purple colour when reduced. Thus the reaction can be monitored via the colour change of benzyl viologen at 600 nm.

1. Deoxygenate the 100 mM K-phosphate buffer, pH 7.0 by bubbling with nitrogen gas for at least 30 minutes prior to commencement of activity assay.
2. Add 2–3 glass balls to the bottom of the cuvette. Pipette into this 10 μl of 200 mM benzyl viologen solution and fill cuvette to the top with sparged 100 mM phosphate buffer, pH 7.0. Seal cuvette with the lid. An example of an anaerobic cuvette is shown in **Fig. 12.2A**.
3. Through the hole in the lid, inject biological sample to be assayed using the Hamilton syringe (see **Fig. 12.2B**). Shake to mix.

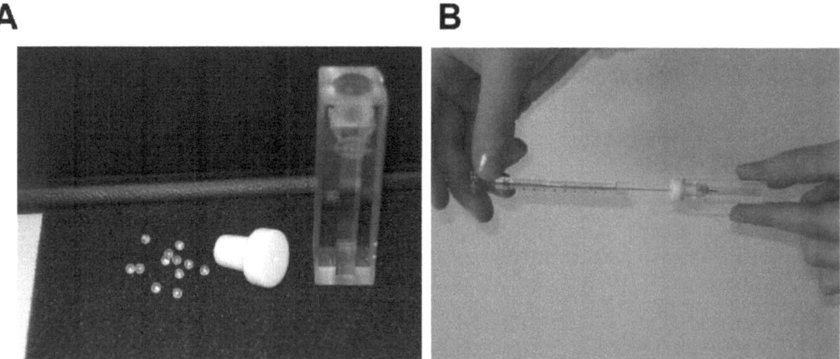

Fig. 12.2. (**A**) Typical cuvette used for anaerobic enzyme assay. Small nylon balls (*left* of the picture) are added to the cuvette to facilitate mixing of the sample. (**B**) To keep the contents of the cuvette anaerobic additions are made through a narrow bore hole in the cuvette lid.

4. Inject into this a few μl of the dithionite solution. The solution in the cuvette will turn purple as the benzyl viologen becomes reduced, but may decolorize. Add further small aliquots of dithionite until the purple colour stabilises and the solution has an optical density at 600 nm of approximately 0.8–1.0 absorbance units.
5. Continuously measure the optical density of the sample for 1–2 minutes until a flat baseline is observed. To start the reaction open the lid of the spectrophotometer (*but do not stop the measurement*), remove the cuvette, quickly add 20 μl of the 20% TMAO solution to start the reaction, mix by shaking and place the cuvette back in the spectrophotometer to continue the measurement.
6. If TMAO or DMSO reductase activity is present in the sample, the optical density at 600 nm will decrease as the enzyme oxidises the reduced benzyl viologen and the purple colour is lost.

7. Measure the TMAO-initiated optical density change for 2–3 minutes and then determine the initial rate of the reaction. The extinction coefficient for reduced benzyl viologen at 600 nm is 7400 M^{-1} cm^{-1} (*see* **Note 14**). To allow comparison of enzymatic activity in different cellular compartments the rate of reaction is calculated in μmoles TMAO reduced per minute per gram cells (weighed at the start of the fractionation procedure). For comparison of activity across similar fractions, activity can be expressed as μmoles TMAO reduced per minute per milligram protein in the sample (determined by Lowry protein assay (20)).

3.6. Assay for Respiratory Formate Dehydrogenase (Fdh-N/Fdh-O) Activity

Formate is oxidised by formate dehydrogenase-N and electrons are passed onto phenazine methosulphate (PMS). The reduced PMS then reduces dichloroindophenol (DCIP) that can be monitored in a spectrophotometric assay by a change in absorbance at 600 nm (21).

1. Deoxygenate the 100 mM K-phosphate buffer, pH 7.0 by bubbling with nitrogen gas for at least 30 minutes prior to commencement of activity assay.
2. Add 2–3 glass balls to the bottom of the cuvette – an example of an anaerobic cuvette is shown in **Fig. 12.2A**. Pipette into this 50 μl of phenazine methosulphate solution, 60 μl of DCPIP solution and fill cuvette to the top with sparged 100 mM phosphate buffer, pH 7.0. Seal cuvette with the lid and shake to mix. The solution will be blue/purple in colour.
3. Through the hole in the lid, inject biological sample to be assayed using the Hamilton syringe (*see* **Fig. 12.2B**). Shake to mix.
4. Place in spectrophotometer and monitor optical density at 600 nm until a flat baseline is achieved.
5. To start the reaction open the lid of the spectrophotometer (*but do not stop the measurement*), remove the cuvette, quickly add 25 μl of the 1.2 M formate solution, mix by shaking and place the cuvette back in the spectrophotometer to continue the measurement.
6. If formate dehydrogenase activity is present in the sample the optical density at 600 nm will decrease as the enzyme reduces the DCPIP and the sample is decolorized.
7. Measure the formate-initiated optical density change for 2–3 minutes and then determine the initial rate of the reaction. The extinction coefficient for reduction of DCPIP (a two electron reaction) is 21,000 M^{-1} cm^{-1}. Activity is expressed as either in μmoles formate oxidised per minute per gram cells (weighed at the start of the fractionation procedure) or

μmoles formate oxidised per minute per milligram protein in the sample (determined by Lowry protein assay (20)).

3.7. Assay for Uptake Hydrogenase Activity

Hydrogen oxidation (or 'uptake') involves the splitting of molecular H_2 into two protons and two electrons and this is a reaction catalysed by most hydrogenases in vitro. The resultant electrons can be passed onto viologen dyes if the enzyme allows (22). In the case of *E. coli* hydrogenases, all three isoenzymes react favourably with benzyl viologen. Thus oxidised (colourless) BV can be reduced in a hydrogen- and hydrogenase-specific manner to reduced (purple) BV and the resultant absorbance change monitored in a spectrophotometer.

1. Vacuum degas the 100 mM K-phosphate buffer for 30 minutes.
2. Bubble the degassed buffer with hydrogen gas for at least 60 minutes prior to commencement of activity assay. This should be carried out in a fume hood as hydrogen is a fire hazard.
3. Add 2–3 glass balls to the bottom of the cuvette – an example of an anaerobic cuvette is shown in **Fig. 12.2A**. Fill the cuvette with hydrogen-saturated phosphate buffer and seal with the lid.
4. Through the hole in the lid, pipette into this 100 μl of 250 mM benzyl viologen solution – see **Fig. 12.2B**. Inject into this a few μl of the dithionite solution until the sample is pale purple – the aim is to reach a stable optical density at 600 nm of 0.3–0.5.
5. Continuously measure the optical density of the sample for 1–2 minutes until a flat baseline is observed. To start the reaction open the lid of the spectrophotometer (*but do not stop the measurement*), remove the cuvette, inject biological sample to be assayed to start the reaction, mix by shaking and place the cuvette back in the spectrophotometer to continue the measurement.
6. If hydrogenase activity is present in the sample, the optical density at 600 nm increases as the enzyme reduces the benzyl viologen.
7. Measure the optical density change for 2–3 minutes and then determine the initial rate of the reaction. The extinction coefficient for reduced benzyl viologen at 600 nm is 7400 M^{-1} cm^{-1} (*see* **Note 14**). Rates are expressed as in μmoles hydrogen oxidised per minute per gram cells (weighed at the start of the fractionation procedure) or μmoles hydrogen oxidised per minute per milligram protein in the sample (determined by Lowry protein assay (20)).

3.8. Pulse-Chase Analysis of Tat Substrate Export in E. coli

For these experiments a copy of the Tat substrate of interest is encoded on a plasmid such as pT7.5 (23) with its expression placed under the control of the phage T7 promoter. The gene encoding the T7 RNA polymerase is provided in *trans* on a separate plasmid where its expression is induced by heat shock. Subsequently the expression of endogenous *E. coli* genes is inhibited by addition of rifampicin which blocks the action of the *E. coli* RNA polymerase but not the T7 enzyme. Addition of a pulse of radiolabel in the form of [^{35}S]-methionine thus permits specific labelling of the T7-controlled Tat substrate, and the radiolabel is then 'chased' out by the presence of an excess of unlabelled methionine. Prescursor and mature forms of the radiolabelled protein can be detected following SDS polyacrylamide gel electrophoresis and autoradiography. An optional step involves the fractionation of cells following the 'chase' period to confirm that the processed form of the protein resides in the periplasmic fraction. We recommend that where possible *E. coli* strain K38 is used as the host strain for these experiments as it takes up rifampicin better than most other standard laboratory strains, reducing the background labelling from native *E. coli* genes. A *tat*$^{-}$ version of this strain is also available (24).

1. Grow a 5 ml overnight culture of the host strain containing both pGP1-2 (23) and the appropriate pT7 recombinant plasmid aerobically in LB medium at 30°C with the appropriate concentrations of antibiotics.
2. Make 100 ml of M9 medium + amino acids by mixing together 20 ml of 5X M9 salts, 200 μl of 1 M $MgSO_4$ solution, 10 μl of 1 M $CaCl_2$ solution, 100 μl of 1% thiamine solution, 2 ml of 20% glucose solution, 0.01% of each of the 18 non-sulphur amino acids, and sterile distilled water to 100 ml final volume.
3. Subculture the overnight culture at a 1:80 dilution in fresh LB medium and grow at 30°C the optical density at 600 nm reaches 0.2. Harvest 1 ml of culture by centrifugation for 1 minute at top speed in a microfuge, wash the pellet once in M9 medium + amino acids.
4. Resuspend the washed cells in 5 ml of M9 medium + amino acids. Culture for a further hour at 30°C to allow cellular content of cysteine and methionine to be depleted.
5. Shift the culture rapidly from 30°C to 42°C for 15 minutes (this induces synthesis of T7 polymerase from plasmid pGP1-2, which encodes the T7 RNA polymerase from the inducible phage λ P_L promoter along with the gene encoding a constitutively expressed, temperature-sensitive λ repressor (CI857). This subsequently results in the induced expression of the Tat-dependent precursor protein

under transcriptional control of the T7 ϕ10 promoter on the second plasmid).

6. Add rifampicin (to a final concentration of 400 μg/ml, resulting in the inhibition of the *E. coli* RNA polymerase). Maintain the sample at 42°C for a further 10 minutes then shift the temperature back to 30°C for 20 minutes. All remaining steps are carried out at 30°C unless stated otherwise.
7. Add radiolabelled cysteine and methionine mix to the sample (use 0.05 mCi of radiolabel for each 5 ml culture).
8. After 5 minutes, withdraw a 0.5 ml sample and snap freeze in liquid nitrogen. To the remaining 4.5 ml immediately add 750 μg/ml unlabelled methionine. Withdraw further 0.5 ml samples being removed at defined time points into the 'chase', for example at 1, 2, 5, 10, 15, 30 minutes after addition of the unlabelled methionine. Flash-freeze each sample in liquid nitrogen as they are collected.
9. At the end of the 'chase' gently thaw the samples and pellet the cells by centrifugation. Resuspend the pellets in 75 μl SDS loading buffer and analyse the samples by autoradiography following SDS-PAGE (25). A typical pulse-chase analysis of SufI produced from plasmid pT7.5 is shown in **Fig. 12.3A**. Band intensity can be quantified using any standard quantitation package to allow the percentage of precursor and mature forms to be determined.
10. Optional step: The same method can be adapted to fractionate the cells into sphaeroplast and periplasmic samples, to examine the localisation of the precursor and mature forms of the Tat substrate. To achieve this, the cells to be fractionated are prepared as for the pulse-chase experiments described in steps 1–9 above, with the exception that only 2 ml of cell culture is radiolabelled. 'Chase' cells with unlabelled methionine for a single time-point (e.g. 10, 30, or 60 minutes) and pellet by centrifugation. Resuspend pelleted cells in 1 ml of cell fractionation buffer and incubate at 20°C for 10 minutes. Re-pellet the cells, discard the supernatant and resuspend the pellet in 133 μl of ice-cold 5 mM $MgSO_4$. After 10 minutes on ice, centrifuge the sample. Retain the supernatant from this spin as the periplasmic fraction and the pellet as the sphaeroplasts (*see* **Note 15**). An example of a fractionated sample producing radiolabelled SufI is shown in **Fig. 12.3B**.

3.9. Exploiting AmiA as a Reporter for Tat-Targeting Signal Peptides

This approach utilises the mature region of AmiA as a Tat-dependent reporter protein. AmiA is an *E. coli* Tat substrate that is incompatible with the Sec pathway. It has catalytic activity as a

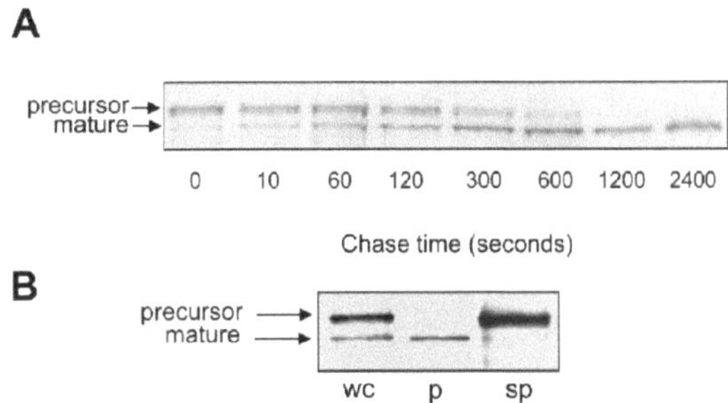

Fig. 12.3. Pulse-chase analysis of the Tat substrate protein, SufI. Strain MC4100 (38) was transformed with plasmids pGP1-2 (23) and pNR14 (24) and pulse-chase analysis was carried out as described in the text. (**A**) After a 5-min pulse with [^{35}S]methionine the cells were chased with non-radioactive methionine for a total of 40 min. Whole cell aliquots were taken at the time points indicated and reactions stopped by flash-freezing in liquid nitrogen. Proteins were separated by SDS-PAGE (12.5% (w/v)), and exposed to photographic film. (**B**) Experiment was performed as described in (**A**) with the exception that at a single time-point of 60 minutes post addition of non-radioactive methionine cells were fractionated into periplasmic (p) and sphaeroplast (sp) fractions using the cold osmotic shock method. A sample of whole cells (wc) prior to fractionation is shown for comparison.

peptidoglycan amidase and is involved in cell wall remodelling. Strain MCDSSAC is *tat*$^{+}$ but lacks periplasmic AmiA and its homologue AmiC because the twin-arginine signal peptide coding regions of these proteins have been deleted. This strain phenocopies *tat* mutant strains in terms of cell envelope leakiness and sensitivity to growth in the presence of SDS (9).

The SDS-sensitivity of strain MCDSSAC can be rescued if a copy of full length *amiA* is provided on a plasmid, but not by pUniAmiA, where the AmiA signal peptide coding sequence is absent. Cloning of DNA coding for a twin-arginine signal peptide (e.g. the TorA or NapA signal peptides, or even the signal peptide of *Haloferax volcanii* NarG) in frame with the mature region of AmiA guides export of the protein and SDS resistance of the strain, which can be directly selected on LB plates containing SDS (10). This construct can also be used for pulse-chase analysis of the AmiA construct (as described in **Section 3.8**) as there is a T7 promoter on the plasmid directly upstream of the *tat* promoter.

1. Clone DNA encoding signal peptide of interest (with its predicted signal peptidase cleavage site if appropriate) as a *Bam*HI-*Xba*I fragment into pUniAmiA (*see* **Fig. 12.4A**). Design the construct so that the *Bam*HI site is immediately adjacent to the start codon of the candidate signal peptide as this will place it at the correct distance to initiate synthesis from the *tatA* ribosome binding site which is just upstream. Ensure that the *Xba*I site is in frame with the signal peptide of interest to place the signal peptide in frame with the *amiA*

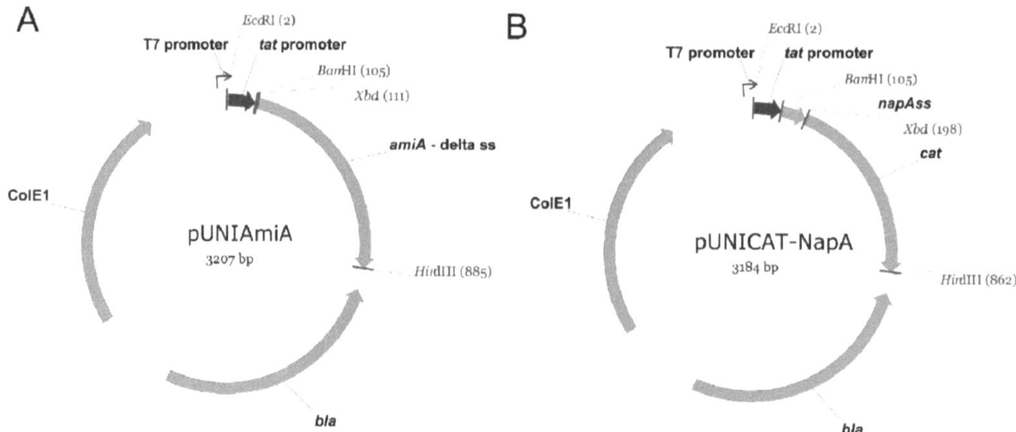

Fig. 12.4. Restriction maps of plasmids (**A**) pUniAmiA (10) and (**B**) pUNICAT-NapA (26). In each case the plasmids are based on vector pT7.5 (23) and in addition to encoding the phage T7 promoter also carry the constitutive *E. coli tatA* promoter (36). Only a few, relevant restriction sites are shown and are all unique *except* the *Eco*RI site in pUNICAT-NapA where there is a second *Eco*RI site present in the *cat* gene (not shown). *bla* encodes β-lactamase, specifying ampicillin resistance, *cat* encodes chloramphenicol acetyltransferase, amiA – delta ss is the coding region for the mature sequence of *E. coli* AmiA (i.e. lacking the signal peptide) and *napAss* is the coding region for the *E. coli* NapA signal peptide.

coding sequence. Alternatively DNA can be shotgun cloned into the *Bam*HI site as *Sau3A* fragments for random library construction.

2. Transform the construct(s) into competent MCDSSAC cells and plate onto LB plates supplemented with SDS (and ampicillin to select for the pUniAmiA-based plasmid). If colonies form after overnight incubation of the plates, this indicates Tat signal peptide activity from the encoded DNA. Alternatively transformants can be selected onto media lacking SDS, grown up in liquid media and then streaked onto SDS-containing plates as described in **Note 7**.

3.10. Using Chloramphenicol Acetyltransferase as a Positive Reporter for Loss of Tat Function in E. coli

Here the CAT enzyme fused in frame to a twin-arginine signal peptide is used as a positive reporter to select for *tat*-inactivating mutations. Fusion of the DNA encoding CAT derived from pACYC184 to the NapA signal peptide results in complete export of the enzyme to the periplasm. Cells with an active Tat system producing this fusion cannot grow in the presence of chloramphenicol whereas if the Tat system is inactive strains can grow on solid media containing up to 80 μg/ml chloramphenicol, forming the basis for the positive selection (26). We have used such an approach previously to isolate a bank of non-functional *tatA* alleles from a large *tatA* mutant library (14). In those experiments we used CAT fused to the FdnG signal peptide as a reporter, which conferred maximal resistance to 10 μg/ml chloramphenicol when the Tat system was inactive. The 'second generation' CAT construct described here confers much higher resistance when the Tat system is inactive, giving a wider selection range, thus allowing possible selection of different classes of Tat-inactivating mutation that affect Tat function to differing degrees.

1. Transform isogenic *tat*+ and *tat*− strain with pUNICAT-NapA. Select transformants on LB medium supplemented with 100 μg/ml ampicillin.
2. Grow up single colonies of transformants and plate out onto LB plates supplemented with 100 μg/ml ampicillin and a range of chloramphenicol concentrations either by serial dilution or by streaking loopfuls of culture. This step is necessary as the local composition of the water may affect the intrinsic resistance to chloramphenicol via subtle changes in outer membrane composition (27). The system is now ready to use as a screen for mutations that increase chloramphenicol resistance of the *tat*+ strain.

3.11. Measuring Signal Peptide-Chaperone Interactions Using Isothermal Titration Calorimetry

Calorimetry is a sensitive biophysical technique that is used to detect heat (enthalpy) changes during biological, chemical or physical reactions (*see* **Note 16**). In terms of protein–signal peptide interactions, this technique requires the use of an expensive specialist instrument but can provide a wealth of thermodynamic data. The calorimeter itself consists of two liquid-filled cells – a reference cell and a sample cell (**Fig. 12.5**). Standard calorimeters have a sample cell of around 1.5 ml though the latest models are much more sensitive and have sample cells of only 0.2 ml capacity. The principle of this system revolves around the gradual *titration* of ligand into a solution of a, for example, binding protein (**Fig. 12.6**). The interaction between the ligand and protein will either generate or consume a tiny amount of heat in the sample cell at which point the computer program that controls the calorimeter will match that change in the reference cell. The amount of (electrical) power used to maintain this equilibrium is directly proportional to the heat change and can be easily quantified. As the titration proceeds and the relative concentration of ligand in the sample cell increases and, usually if the dissociation constant is <0.5 mM (*see* **Note 17**), the binding sites will eventually become saturated enabling the calculation of relative dissociation (K_d)/association constants ($K_a = 1/K_d$) and an estimation of number of binding sites (*see* **Note 5**). In addition, the estimation of relative K_d allows direct calculation of the Gibbs free energy change (ΔG) for the interaction (*see* **Note 18**). The calorimeter provides the enthalpy change for the interaction (ΔH), thus combining this information with the ΔG value enables the determination of the entropy change (ΔS) during the interaction (*see* **Note 19**).

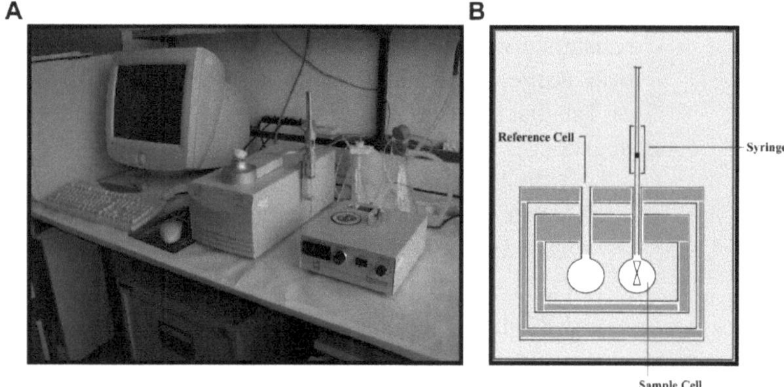

Fig. 12.5. An example of a microcalorimeter. (**A**) The MicroCal VP-ITC system located in The Bioanalytical Suite, School of Biological Sciences, University of East Anglia, Norwich NR4 7TJ, United Kingdom. (**B**) Cartoon representation of the key components of a microcalorimeter. The reference cell is usually filled with water and does not usually need to be attended to before an experiment. The sample cell must be kept clean, however, and it is recommended that it is treated with 5–10% (v/v) Decon^{TM} solution before every experiment. The syringe spins rapidly during each experiment (the end of the syringe having been adapted to double-up as a mixing device) and is programmed to inject ligand at certain time intervals. A thermoelectric detector monitors the temperature difference between the two cells and the computer modulates the power supplied to the sample cell to maintain a constant value. As ligand is titrated into the sample cell, heat is either given out or taken up from the environment. The integral of the electric power required to maintain a constant difference between the two cells, when plotted over time, is a measure of total heat resulting from the process being studied.

1. Separately prepare highly purified chaperone and highly purified twin-arginine signal peptide (usually as a fusion to maltose binding protein).
2. Dialyse both extensively against identical buffer solution.
3. Ultracentrifuge each sample at 200,000 *g* to remove aggregates.
4. Load the peptide in the sample cell at a final concentration of 10 μM.
5. Load the chaperone protein in the injection syringe at 100 μM final concentration.
6. Perform an ITC titration as per the system manufacturer's instructions. Set the mixing speed to 300 rpm in the first instance, and the temperature to 28°C or 301 K (*see* **Note 20**). Set the calorimeter to inject 1 μl at first (to expel any trapped tiny air bubbles from the syringe) followed by 30 × 8 μl injections at 250 second intervals (**Fig. 12.6**).
7. Control the experiment as carefully as possible by repeating the titration using buffer only in the syringe (and receptor protein in the sample cell) *and* by injecting ligand in the syringe into a buffer only solution.

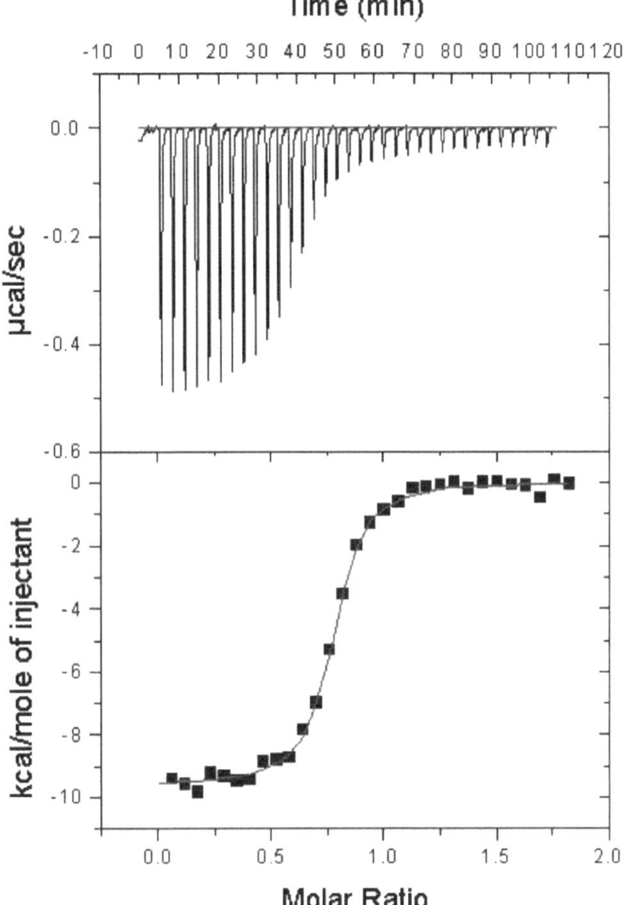

Fig. 12.6. A typical ITC experiment involving a twin-arginine signal peptide and its cognate chaperone. Here, the syringe was loaded with *E. coli* DmsD at 150 μM and the sample cell contained maltose binding protein fused at its C-terminus to the twin-arginine signal peptide of the *E. coli* DmsA protein (at 15 μM). It is common to titrate the binding protein into the ligand, especially when issues of solubility arize. In this case, it was easier to obtain a stable concentrated DmsD solution to load into the syringe. This experiment was conducted in 20 mM Tris.HCl (pH 7.6) at 28°C. The data was fitted to a one-site model using the MicroCal ITC software and the K_d was 104 nM and the n value was 0.76. The $\Delta H_{observed} = -9.62$ kcal mol^{-1}(1 kcal = 4.18 kJ, thus $\Delta H = -40.21$ kJ mol^{-1}) and was determined directly by the calorimeter. Applying the equations outlined in **Notes 18 and 19**, $\Delta G_{observed} = -40.17$ kJ mol^{-1} (-9.6 kcal mol^{-1}) and $T\Delta S_{observed} = -0.04$ kJ mol^{-1} (0.01 kcal mol^{-1}).

4. Notes

1. Sodium dithionite is exceptionally oxygen labile. Once purchased this white powder should preferentially be dispensed into aliquots of 1 g and stored in sealed bottles in a

dessicator, ideally also under vacuum or in an oxygen-free atmosphere. Sodium dithionite solution should always be prepared fresh just prior to use and stored on ice. It will slowly break down during the course of the experiment, particularly if a series of assays are performed. Therefore the experimenter should not be surprised when a larger volume of dithionite solution needs to be used to obtain the same OD_{600} measurement for later assays in an experimental series.

2. It is also possible to use non-lidded cuvettes for these assays, and to seal off the cuvette with a rubber septum of appropriate diameter. However, this is often much more fiddly to work with. For example, it is difficult to fill the cuvette completely to the top to exclude an air gap, and therefore the filled, sealed cuvette will need to be sparged with nitrogen gas for 2–3 minutes prior to adding the dithionite. For ease of experimentation we strongly recommend investing in suitable lidded cuvettes.
3. Athough a premix of the 18 non-sulphur amino acids can be readily purchased, it generally works out much cheaper to make a 1% solution of each one individually, filter sterilise and store at −20°C in 1 ml aliquots.
4. Essentially any buffer system can be used in an ITC experiment. It is also possible to use unbuffered aqueous solutions. Note that it is best to avoid reducing agents such as dithiothreitol or 2-mercaptoethanol, since these slowly decompose and affect the baseline. It is important to consider, however, that each buffer has its own characteristic heat of ionization. In the case of Tris.HCL, for example, $\Delta H_{\text{ionisation}} = 11.36$ kcal mol^{-1} (1 kcal = 4.18 kJ). This can be used to a scientist's advantage (to boost the signal during a titration), but it must be borne in mind that $\Delta H_{\text{binding}} = \Delta H_{\text{observed}} + \Delta H_{\text{ionisation}}$. For in-depth thermodynamic studies it is best to use a buffer system with a $\Delta H_{\text{ionisation}}$ close to zero (e.g. phosphate buffer $\Delta H_{ionisation} = 1.22$ kcal mol^{-1}), or to perform identical titrations in a range of buffers before plotting $\Delta H_{\text{observed}}$ against $\Delta H_{\text{ionisation}}$ and extrapolating $\Delta H_{\text{binding}}$ from that graph.
5. Proteins should be purified as carefully and completely as possible. When measuring any protein–ligand interaction, the two components in the reaction must be in identical buffers. The easiest way to do this is do dialyse the protein extensively against several changes of buffer, then to use the final dialysis buffer itself to prepare ligands for experimentation. In the case of protein–

protein interactions, both proteins should be dialysed in the same vessel. Accurate sample concentration measurement is critical to obtaining reliable ITC data. With proteins, it is best to calculate the Molar extinction coefficient at 280 nm based on the numbers of tryptophan ($\varepsilon_{280\ \text{nm}} = 5690\ \text{M}^{-1}\ \text{cm}^{-1}$), tyrosine ($\varepsilon_{280\ \text{nm}} = 1280\ \text{M}^{-1}\ \text{cm}^{-1}$), and cystine ($\varepsilon_{280\ \text{nm}} = 120\ \text{M}^{-1}\ \text{cm}^{-1}$) side-chains. Based on the ligand and receptor concentration provided, the software will integrate the titration data, fit a curve and attempt to estimate a stoichiometry of binding (n). Never 'fix' the stoichiometry value before an experiment, always leave this floating. With control experiments (such as EDTA versus magnesium) the n value will always come out around 1. With protein experiments, however, the n value can vary a lot from the theoretical, especially if a portion of ligand or receptor is denatured or inactive for some reason. Indeed, if the protein preparation is contaminated with another protein (which would interfere with accurate specific protein estimations) the resultant n value will again be misleading.

6. For strains based on MC4100, the wild type strain can tolerate SDS concentrations of up to 4%, whereas *tat*$^{-}$ strains usually fail to grow on 1% SDS. Final concentrations of SDS used in plates, however, may need to be varied depending on genotype of parental strain and probably also on water composition as low levels of divalent cations can help to increase the SDS resistance of *tat* mutant strains (28).

7. When streaking or patching strains onto selective media plates it is advisable to use a loopful of culture from cells grown in liquid media rather than taking colonies directly from solid media as this gives a more even coverage allowing for easier scoring of phenotype.

8. It is necessary to include thiamine in minimal growth media for MC4100 derivatives because the strain is a thiamine auxotroph. Likewise in the original Sambrook and Russell (15) recipe for M9 medium, MoSe solution is not included. However, since TMAO and DMSO reductases are molybdenum-containing enzymes, for anaerobic growth with TMAO we find strains grow better in the presence of this supplement. In addition selenium is essential for the activity of formate dehydrogenase.

9. TMAO can be replaced with alternative electron acceptors at 0.4% final concentration. For example, inclusion of 0.4% potassium nitrite will allow growth with nitrite to be determined – *tat* strains cannot grow with nitrite as sole electron acceptor because the exported iron sulphur protein, NrfC,

is a Tat substrate and is essential for electron transfer to the nitrite reductase (5).

10. Expression of the *torCAD* genes is maximally induced anaerobically in the presence of TMAO (29), whereas the major respiratory formate dehydrogenase, Fdh-N, is induced with anaerobic growth in the presence of nitrate (30). Although DMSO reductase catalyses the reduction of DMSO, it can also readily reduce TMAO (31). In strains derived from MG1655, for example MC4100 and derivatives, DMSO reductase activity is rather low and it is often easier to assay for benzylviologen::TMAO oxidoreductase activity because the enzyme has a higher specific activity for TMAO than DMSO. Expression of the *dmsABC* operon is not induced by either DMSO or TMAO (32); however, we strongly advise the use of 0.4% DMSO as terminal electron acceptor when growing cultures to assay for this enzyme to prevent induction of the much more highly active Tor system. Hydrogenase-1 and -2 production is repressed by nitrate (33). For assay of these enzymes, strains are generally grown with fumarate as terminal electron acceptor, although similar specific activities are observed if TMAO is used.

11. It is recommended to grow ≥ 200 ml cultures for fractionation experiments. In order to minimise errors in pipetting and weight measurements, a minimum biomass wet-weight of 0.5 g is recommended.

12. This is best achieved by noting the weight of the centrifuge pot before use and then re-weighing the tube containing the cell pellet.

13. The quality of cell fractions was traditionally determined by assessing the distribution of compartment-specific marker enzymes, using for example acid phosphatase as a periplasmic enyme marker (34) and glucose-6-phosphate dehydrogenase as a cytoplasmic marker (6). Fractionation quality can also be analysed by Western blotting using antibodies raised to proteins found in different sub-cellular locations.

14. Benzyl viologen reduces and oxidises by gain or loss of one electron; however the reduction of TMAO to TMA, or oxidation of hydrogen to water involves two electrons. Therefore the effective BV extinction coefficient for the reduction of TMAO or oxidation of hydrogen is 14,800 M^{-1} cm^{-1}.

15. This method of cell fractionation is gentler than the procedure described in **Section 3.4** as lysozyme is not included. It permeabilises the outer membrane but the cell wall remains intact. As a result of this, large proteins such as

TMAO reductase are not released efficiently because they cannot sieve through the peptidoglycan. Likewise, cell wall binding proteins are also not released. However Tat substrates such as SufI and CueO (YacK) are effectively liberated using this technique (24).

16. In theory, any type of reaction that involves heat exchange can be measured in a microcalorimeter, including steady-state enzyme kinetics. One useful control experiment to conduct before an important experiment involves the titration of 15 mM EDTA into a 1.5 mM $MgCl_2$ solution. If a classic sigmoidal binding curve is not forthcoming (note also that this is an endothermic reaction and so will give a binding-curve shaped like the number 2) then the sample cell should be cleaned overnight in a 10% solution of Decon™ at 65°C.

17. ITC is best suited to measuring interactions with K_ds in the range of 5 μM to 5 nM. The reason for this is that accuracy in ITC measurements depends on the ready determination of free versus bound ligand concentrations and this is most easily done if a sigmoidal binding curve can be fitted to the data. The best ITC experiments are designed to optimise the concentrations of binding partners such that the data points are spread evenly across a sigmoidal curve. Biophysicists have come up with a rule of thumb (the '*c*' value) to enable experimentalists to optimise sample concentrations in ITC studies: $c = [M]_{tot}/K_d$ (where $[M]_{tot}$ is the concentration of the total number of binding sites in the sample cell). Most publications (35) suggest aiming for a *c* value of 10–500 to gain convincing data by ITC. This highlights the importance of sample concentrations in ITC experiments. For measuring weaker interactions it may be impossible to get the sample concentrations high enough to attain a sensible *c* value, and for very tight interactions reducing the sample concentrations too much will challenge the sensitivity of the instrument. Note also that for protein–peptide interactions it is important to use the minimum amount of protein that will give reproducible data while maintaining, if possible, a reasonable *c* value. The concentration of ligand in the syringe is usually set at 10 × that of receptor in the sample cell. Note also that relatively weak interactions (such as those of 100 μM or above) may be better detected using 'single injection mode' ITC, where the whole ligand solution is injected into the sample cell in one smooth action. Much tighter interactions (K_d in the pM range) can be analysed using a 'competitive binding' protocol, where a weaker-binding

ligand is displaced by titration of the tighter-binding ligand.

18. $\Delta G = RT\ln K_d$, where R is the gas constant (8.3 J K^{-1} mol^{-1}), T is the absolute temperature of the experiment in Kelvin and ln is natural log.

19. $\Delta G = \Delta H - T\Delta S$, where T is the absolute temperature. Understanding whether an interaction is 'enthalpy driven' (i.e. the ΔH component contributes most to ΔG) or 'entropy driven' (i.e. the ΔS component dominates) is very useful information in deciphering what types of interaction dominate. For example, domination of entropy in an interaction can be indicative of hydrophobic interactions while domination of enthalpy usually points to hydrogen bonding and ionic interactions.

20. The absolute temperature that an experiment is conducted clearly has a direct effect on the data produced (*see* **Notes 18 and 19**). It is best practise to select a temperature that is both physiologically relevant to the system under investigation and also maintains some stability of the proteins over the time course of the experiment. If there is a need to vary the temperature, bear in mind that all controls must be repeated at all temperatures tested.

Acknowledgements

We would particularly like to thank Prof. Gary Sawers (Halle-Wittenberg) for his help with developing some of the early methods for analysis of the *E. coli* Tat system, and all the members of our laboratories past and present. Work in our laboratories is or has been funded by the BBSRC, the MRC, the Wellcome Trust, the European Union and the Royal Society.

References

1. Theg, S. M., Cline, K., Finazzi, G., and Wollman, F. A. (2005) The energetics of the chloroplast Tat protein transport pathway revisited *Trends Plant Sci* **10,** 153–154.
2. Berks, B. C., Palmer, T., and Sargent, F. (2003) The Tat protein translocation pathway and its role in microbial physiology *Adv Microb Physiol* **47,** 187–254.
3. Berks, B. C., Palmer, T., and Sargent, F. (2005) Protein targeting by the bacterial twin–arginine translocation (Tat) pathway *Curr Opin Microbiol* **8,** 174–181.
4. Tullman-Ercek, D., DeLisa, M. P., Kawarasaki, Y., Iranpour, P., Ribnicky, B., Palmer, T., and Georgiou, G. (2007) Export pathway selectivity of Escherichia coli twin arginine translocation signal peptides *J Biol Chem* **282,** 8309–8316.
5. Weiner, J. H., Bilous, P. T., Shaw, G. M., Lubitz, S. P., Frost, L., Thomas, G. H., Cole, J. A., and Turner, R. J. (1998) A novel and ubiquitous system for membrane targeting and secretion of cofactor-containing proteins *Cell* **93,** 93–101.

6. Sargent, F., Bogsch, E. G., Stanley, N. R., Wexler, M., Robinson, C., Berks, B. C., and Palmer, T. (1998) Overlapping functions of components of a bacterial Sec-independent protein export pathway *EMBO J* **17,** 3640–3650.
7. Rodrigue, A., Chanal, A., Beck, K., Muller, M., and Wu, L. F. (1999) Co-translocation of a periplasmic enzyme complex by a hitchhiker mechanism through the bacterial tat pathway *J Biol Chem* **274,** 13223–13228.
8. Jack, R. L., Buchanan, G., Dubini, A., Hatzixanthis, K., Palmer, T., and Sargent, F. (2004) Coordinating assembly and export of complex bacterial proteins *EMBO J* **23,** 3962–3972.
9. Ize, B., Stanley, N. R., Buchanan, G., and Palmer, T. (2003) Role of the Escherichia coli Tat pathway in outer membrane integrity *Mol Microbiol* **48,** 1183–1193.
10. Ize, B., Coulthurst, S. J., Buchanan, G., Hatzixanthis, K., Caldelari, I., Barclay, E. C., Richardson, D. J., Palmer, T., and Sargent, F. (2009) Remnant signal peptides on non-exported enzymes: implications for the evolution of prokaryotic respiratory chains *Microbiology* **155,** 3992–4004.
11. Thomas, J. D., Daniel, R. A., Errington, J., and Robinson, C. (2001) Export of active green fluorescent protein to the periplasm by the twin-arginine translocase (Tat) pathway in Escherichia coli *Mol Microbiol* **39,** 47–53.
12. Santini, C. L., Bernadac, A., Zhang, M., Chanal, A., Ize, B., Blanco, C., and Wu, L. F. (2001) Translocation of jellyfish green fluorescent protein via the Tat system of Escherichia coli and change of its periplasmic localization in response to osmotic up-shock *J Biol Chem* **276,** 8159–8164.
13. Blaudeck, N., Kreutzenbeck, P., Freudl, R., and Sprenger, G. A. (2003) Genetic analysis of pathway specificity during post-translational protein translocation across the Escherichia coli plasma membrane *J Bacteriol* **185,** 2811–2819.
14. Hicks, M. G., Lee, P. A., Georgiou, G., Berks, B. C., and Palmer, T. (2005) Positive selection for loss-of-function tat mutations identifies critical residues required for TatA activity *J Bacteriol* **187,** 2920–2925.
15. Sambrook, J., and Russell, D. W. (2001) Molecular Cloning: a laboratory manual, Cold Spring Harbor Laboratory Press, New York.
16. Cohen, G. N., and Rickenberg, H. V. (1956) Concentration specifique reversible des amino acides chez Escherichia coli. *Ann Inst Pasteur (Paris)* **91,** 693–720.
17. McEwan, A. G., Jackson, J. B., and Ferguson, S. J. (1984) Rationalization of properties of nitrate reductases in Rhodopseudomonas capsulata *Arch Microbiol* **137,** 344–349
18. Osborn, M. J., Gander, J. E., and Parisi, E. (1972) Mechanism of assembly of the outer membrane of Salmonella typhimurium. Site of synthesis of lipopolysaccharide *J Biol Chem* **247,** 3973–3986.
19. Silvestro, A., Pommier, J., and Giordano, G. (1988) The inducible trimethylamine-N-oxide reductase of Escherichia coli K12: biochemical and immunological studies *Biochim Biophys Acta* **954,** 1–13.
20. Lowry, O. H., Rosebrough, N. J., Farr, A. L., and Randall, R. J. (1951) Protein measurement with the Folin phenol reagent *J Biol Chem* **193,** 265–275.
21. Lester, R. L., and DeMoss, J. A. (1971) Effects of molybdate and selenite on formate and nitrate metabolism in Escherichia coli *J Bacteriol* **105,** 1006–1014.
22. Ballantine, S. P., and Boxer, D. H. (1985) Nickel-containing hydrogenase isoenzymes from anaerobically grown Escherichia coli K-12 *J Bacteriol* **163,** 454–459.
23. Tabor, S., and Richardson, C. C. (1985) A bacteriophage T7 RNA polymerase/promoter system for controlled exclusive expression of specific genes *Proc Natl Acad Sci USA* **82,** 1074–1078.
24. Stanley, N. R., Palmer, T., and Berks, B. C. (2000) The twin arginine consensus motif of Tat signal peptides is involved in Sec-independent protein targeting in Escherichia coli *J Biol Chem* **275,** 11591–11596.
25. Laemmli, U. K. (1970) Cleavage of structural proteins during the assembly of the head of bacteriophage T4 *Nature* **227,** 680–685.
26. Maillard, J., Spronk, C. A., Buchanan, G., Lyall, V., Richardson, D. J., Palmer, T., Vuister, G. W., and Sargent, F. (2007) Structural diversity in twin-arginine signal peptide-binding proteins *Proc Natl Acad Sci USA* **104,** 15641–15646.
27. Li, H., Lin, X. M., Wang, S. Y., and Peng, X. X. (2007) Identification and antibody-therapeutic targeting of chloramphenicol-resistant outer membrane proteins in Escherichia coli *J Proteome Res* **6,** 3628–3636.
28. Caldelari, I., Palmer, T., and Sargent, F. (2008) Escherichia coli tat mutant strains are able to transport maltose in the absence of an active malE gene *Arch Microbiol.*
29. Pascal, M. C., Burini, J. F., and Chippaux, M. (1984) Regulation of the trimethylamine N-oxide (TMAO) reductase in Escherichia coli:

analysis of tor::Mud1 operon fusion *Mol Gen Genet* **195,** 351–355.
30. Berg, B. L., and Stewart, V. (1990) Structural genes for nitrate-inducible formate dehydrogenase in Escherichia coli K-12 *Genetics* **125,** 691–702.
31. Bilous, P. T., Cole, S. T., Anderson, W. F., and Weiner, J. H. (1988) Nucleotide sequence of the dmsABC operon encoding the anaerobic dimethylsulphoxide reductase of Escherichia coli *Mol Microbiol* **2,** 785–795.
32. Cotter, P. A., and Gunsalus, R. P. (1989) Oxygen, nitrate, and molybdenum regulation of dmsABC gene expression in Escherichia coli *J Bacteriol* **171,** 3817–3823.
33. Richard, D. J., Sawers, G., Sargent, F., McWalter, L., and Boxer, D. H. (1999) Transcriptional regulation in response to oxygen and nitrate of the operons encoding the [NiFe] hydrogenases 1 and 2 of Escherichia coli *Microbiology* **145,** 2903–2912.
34. Atlung, T., Nielsen, A., and Hansen, F. G. (1989) Isolation, characterization, and nucleotide sequence of appY, a regulatory gene for growth-phase-dependent gene expression in Escherichia coli *J Bacteriol* **171,** 1683–1691.
35. Turnbull, W. B., and Daranas, A. H. (2003) On the value of c: can low affinity systems be studied by isothermal titration calorimetry? *J Am Chem Soc* **125,** 14859–14866.
36. Jack, R. L., Sargent, F., Berks, B. C., Sawers, G., and Palmer, T. (2001) Constitutive expression of Escherichia coli tat genes indicates an important role for the twin-arginine translocase during aerobic and anaerobic growth *J Bacteriol* **183,** 1801–1804.
37. Sargent, F., Stanley, N. R., Berks, B. C., and Palmer, T. (1999) Sec-independent protein translocation in Escherichia coli. A distinct and pivotal role for the TatB protein *J Biol Chem* **274,** 36073–36082.
38. Casadaban, M. J., and Cohen, S. N. (1979) Lactose genes fused to exogenous promoters in one step using a Mu-lac bacteriophage: in vivo probe for transcriptional control sequences *Proc Natl Acad Sci USA* **76,** 4530–4533.
39. Bogsch, E. G., Sargent, F., Stanley, N. R., Berks, B. C., Robinson, C., and Palmer, T. (1998) An essential component of a novel bacterial protein export system with homologues in plastids and mitochondria *J Biol Chem* **273,** 18003–18006.

Chapter 13

Site-Specific Cross-Linking of In Vitro Synthesized *E. coli* Preproteins for Investigating Transmembrane Translocation Pathways

Sascha Panahandeh and Matthias Müller

Abstract

A method is described for the preparation and usage of an *E. coli* cell-free translation system primed to incorporate the commercially available photoreactive analogue of phenylalanine, *p*Bpa, into newly synthesized proteins. Incorporation is achieved by means of an amber suppressor tRNA specifically charged with *p*Bpa. The method is exemplified for the site-specific photocross-linking of the signal sequence of a Tat (twin-arginine translocation) precursor protein to the Tat translocase in the cytoplasmic membrane of *E. coli*.

Key words: Site-specific cross-linking, photoprobes, *p*-benzoyl-phenylalanine, amber suppressor, twin-arginine translocation, Tat, protein export, in vitro transcription-translation system, inner membrane vesicles, *Escherichia coli*.

1. Introduction

During the past 20 years, site-specific photocross-linking has repeatedly been used to probe the molecular environment of secretory and membrane proteins during their synthesis, transport, and membrane integration. Site-specific photocross-linking of proteins involves the targeted introduction of photoreactive derivatives of amino acids into proteins. This is in contrast to methods of chemical cross-linking, in which for example amino group-specific bifunctional compounds can cross-link essentially any free amino group of a protein to its nearest neighbours.

A. Economou (ed.), *Protein Secretion*, Methods in Molecular Biology 619,
DOI 10.1007/978-1-60327-412-8_13, © Springer Science+Business Media, LLC 2010

The common principle of various strategies of photocross-linking developed over the years was the incorporation of photoreactive derivatives of lysine or phenylalanine into the protein of interest during its synthesis in cell-free translation systems. Irradiation with UV light would then result in the formation of a covalent bond between the newly synthesized secretory or membrane protein and any contacting component, be it of cytosolic or membranous origin. Initial protocols made use of chemically modifying the ε-amino group of lysine with azido-(nitro)-benzoyl or trifluoromethyldiazirino-benzoyl moieties. Modifications were performed after the lysine had been charged onto isolated tRNA. In this approach, positioning of the photoprobe was restricted to the places of naturally occurring lysine codons in the mRNA that coded for the protein of interest. A more versatile strategy allowing a wider selection of photoprobe positions was the use of amber stop codon-suppressing tRNAs that were chemically charged with trifluoromethyldiazirine-phenylalanine (Tmd-Phe). In this way, the photoprobe could be placed at any position in the polypeptide chain, whose corresponding codon had been replaced by the amber stop codon TAG.

When applied to the study of eukaryotic secretory and membrane proteins, these photocross-linking strategies revealed the 54 kDa-subunit of the signal recognition particle (SRP), the α-subunit of the Sec61 translocon, the TRAM-protein (translocating chain-associating membrane protein) of the endoplasmic reticular membrane, as well as phospholipids as interacting partners of ribosome-bound, nascent secretory and membrane proteins (1–6). These findings could be recapitulated for bacterial secretory and membrane proteins using the same photocross-linking protocols, the only exception being that membrane proteins integrating into the bacterial cytoplasmic membrane contact YidC instead of the eukaryote-specific TRAM-protein (7–10).

Site-specific photocross-linking was recently also applied to investigate what is called the twin-arginine translocation (Tat)-pathway of bacteria. The Tat-pathway (11–14) is dedicated to the export of secretory proteins harbouring an almost invariant twin-arginine sequence motif in their N-terminal signal peptides. Moreover, it has the remarkable ability to export proteins in a fully folded conformation. In many bacteria, the Tat-specific export is achieved by three functionally individual membrane proteins, termed TatA, TatB, and TatC. TatC and TatB form a complex that is involved in recognition of the Tat-signal sequences and their insertion into the membrane. TatA is believed to mediate the actual translocation event, but it is virtually unclear what kind of protein-conducting device the TatABC proteins provide.

To investigate this in more detail, we have performed photocross-linking to pick up interactions between a Tat-substrate protein and the individual subunits TatA, TatB, and

TatC (15, 16). Initially Tmd-Phe was used which, however, has the considerable disadvantage of not being commercially available and requiring numerous steps of chemical synthesis part of which give only little yield of product. A technically much simpler and more easily applicable approach of acylating the amber suppressor tRNA directly with a photoreactive analogue of phenylalanine became recently available. The Schultz lab constructed plasmids that encode the orthogonal pair of an amber suppressor tRNA, which specifically accepts the photoreactive derivative of Phe, *p*-benzoyl-L-phenylalanine (*p*Bpa), as well as its cognate *p*Bpa-specific amino acyl-tRNA synthetase (17). Here we describe how cell-free extracts can be prepared from *E. coli* strains harbouring these plasmids and therefore expressing the pair of *p*Bpa-specific amber suppressor tRNA/tRNA-synthetase. We further detail how these extracts can be used to incorporate externally added *p*Bpa into an amber mutant Tat-substrate protein and how this is employed for efficient photocross-linking to Tat subunits present in membrane vesicles of *E. coli.*

2. Materials

2.1. Preparation of an Amber Stop Codon-Suppressing S-135 Cell Extract from E. coli

1. Growth medium (S-30 medium): 9.0 g/L tryptone/peptone (pancreatic digest of casein; Carl Roth, Karlsruhe, Germany), 0.8 g/L yeast extract, 5.6 g/L NaCl, 1 mL/L 1 M NaOH. Prepare 4–6 L in 1-L batches, each contained in a 5-L Erlenmeyer flask covered with aluminium foil and autoclave. Prepare an additional 100 mL of medium in a 0.5 L Erlenmeyer flask to be used as starter culture and autoclave (*see* **Note 1**).
2. 5 mg/ml Tetracycline in 70% (v/v) ethanol p.a. stored in 1-ml aliquots at −20°C (*see* **Note 2**).
3. 20% Glucose solution, autoclaved.
4. 1 M Triethanolamine acetate (TeaOAc) adjusted to pH 7.5 with acetic acid, filtered and stored at 4°C (*see* **Note 3**).
5. 1 M Magnesium acetate ($Mg(OAc)_2$), filtered and stored at 4°C.
6. 4 M Potassium acetate (KOAc) also adjusted to pH 7.5 with acetic acid, filtered and stored at 4°C.
7. 1 M Dithiothreitol (DTT) stored in 1-mL aliquots at −20°C.
8. S-30 buffer: 10 mM TeaOAc pH 7.5, 14 mM $Mg(OAc)_2$, 60 mM KOAc, 1 mM DTT, stored at 4°C.

9. Phenylmethylsulfonyl fluoride (PMSF; Roche): freshly prepare about 1 mL of a 0.1 M solution in ethanol before use (*see* **Note 4**).
10. A mix of 18 amino acids (without methionine and cysteine) in water each at a concentration of 1 mM.
11. 1 mM Methionine.
12. 1 mM Cysteine.
13. 0.25 M ATP neutralized with 1 M KOH.
14. 0.2 M Phosphoenol pyruvate tri(cyclohexylammonium) salt.
15. 2 mg/mL Pyruvate kinase solution (Roche).
16. Supplemented S-30 (for degradation of endogenous mRNA): per mL of S-30, add 60 μL 1M TeaOAc pH 7.5, 0.6 μL 1M DTT, 1.6 μL 1 M $Mg(OAc)_2$, 6 μL 1 mM 18 amino acid mix, 6 μL 1 mM methionine, 6 μL 1 mM cysteine, 2 μL 0.25 M ATP (neutralized), 27 μL 0.2 M phosphoenol pyruvate, and 2.4 μL 2 mg/mL pyruvate kinase.
17. Dialysis tubing with a width of 25 mm and a molecular weight cut off of 14,000 Da (Visking; Carl Roth) (*see* **Note 5**). Two dialysis tubing clips.
18. For preparation of dialysis tubing: 2% $NaHCO_3$, 1 mM ethylenediamine tetraacetic acid (EDTA).

2.2. Preparation of Inverted Inner Membrane Vesicles (INV)

1. Growth Medium (INV medium): 10 g/L each of yeast extract and tryptone/peptone (pancreatic digest of casein; Carl Roth), 28.9 g/L K_2HPO_4 anhydrous, 5.6 g/L KH_2PO_4 anhydrous, 10 g/L glucose. Prepare 4 × 5-L Erlenmeyer flasks, each containing 10 g yeast extract and 10 g tryptone/peptone dissolved in 753 mL H_2O, autoclave. In addition, prepare one 0.5-L Erlenmeyer flask containing 1 g yeast extract and 1 g tryptone/peptone dissolved in 75.3 mL H_2O, autoclave.
2. 1 M K_2HPO_4, autoclave.
3. 1 M KH_2PO_4, autoclave.
4. 25% Glucose, autoclave.
5. Starter culture medium (100 mL): to 75.3 mL yeast extract and tryptone/peptone (*see* **Section 2.2**, item 1) add 4.1 mL 1 M KH_2PO_4, 16.6 mL 1 M K_2HPO_4, and 4 mL 25% glucose.
6. Complete INV medium: to 753 mL yeast extract and tryptone/peptone (*see* **Section 2.2**, item 1) add 41 mL 1 M KH_2PO_4, 166 mL 1 M K_2HPO_4, and 40 mL 25% glucose.

7. 1 M TeaOAc adjusted to pH 7.5 with acetic acid, filtered and stored at 4°C.
8. 0.2 M EDTA-KOH, pH 7.0, filtered and stored at 4°C.
9. 2.5 M Sucrose ultrapure (MP Biomedicals, Solon, OH), heat slightly for better dissolution, store at room temperature.
10. 1 M DTT stored in 1-mL aliquots at −20°C.
11. 0.1 M PMSF freshly prepared in ethanol.
12. 1 M Isopropyl-β-D-thiogalactopyranoside (IPTG).
13. Buffer A: 50 mM TeaOAc, pH 7.5, 250 mM sucrose, 1 mM EDTA-KOH, pH 7.0, and 1 mM DTT. Prepare fresh.
14. Buffer B: 0.5 M TeaOAc, pH 7.5, 10 mM EDTA-KOH, pH 7, and 10 mM DTT. Prepare fresh.
15. Sucrose solutions for sucrose gradient centrifugation, freshly prepared. 0.77 M sucrose: 10 mL buffer B, 30.8 mL 2.5 M sucrose, H_2O to 99.5 mL, 0.5 mL 0.1 M PMSF added last; 1.44 M sucrose: 10 mL buffer B, 57.6 mL 2.5 M sucrose, H_2O to 99.5 mL, 0.5 mL 0.1 M PMSF added last; 2.02 M sucrose: 10 mL buffer B, 80.8mL 2.5 M sucrose, H_2O to 99.5 mL, 0.5 mL 0.1 M PMSF added last.
16. INV buffer: 50 mM TeaOAc, pH 7.5, 250 mM sucrose, and 1 mM DTT. Cool on ice.

2.3. Site-Specific Cross-Linking of In Vitro Synthesized E. coli Precursor Proteins Using a pBpa-Specific Amber Suppressor tRNA and Its Cognate Amino Acyl-tRNA Synthetase

2.3.1. In Vitro Transcription-Translation Reaction and Site-Specific Cross-Linking

1. Template DNA: Plasmid DNA prepared by Qiagen plasmid maxi kit is suitable for in vitro synthesis (*see* **Note 6**). Prepare DNA in TE buffer (10 mM Tris HCl, pH 8.0, 1 mM EDTA) at about 1 μg/μL and store at 4°C. For site specific incorporation of the photo-reactive cross-linker *p*-benzoyl-*L*-phenylalanine (*p*Bpa), introduce at selected positions TAG stop codons into the DNA sequence encoding the protein of interest (*see* **Note 7**). This can be done for instance by using the PCR-based QuikChange Site-Directed Mutagenesis Kit system (Stratagene, Cedar Creek, TX, USA) following the manufacturer's instruction.
2. 1 M TeaOAc adjusted to pH 7.5 with acetic acid, filtered and stored at 4°C.
3. 4 M KOAc also adjusted to pH 7.5 with acetic acid, filtered and stored at 4°C.
4. 1 M $Mg(OAc)_2$, filtered and stored at 4°C.
5. 25 mM $Mg(OAc)_2$, filtered and stored at 4°C.
6. 0.1 M Spermidine trihydrochloride (Sigma, St. Louis, MO), dissolved in water and stored in single-use aliquots at −20°C (*see* **Note 8**).

7. 40% (w/v) Polyethylene glycol 6000–8000, dissolved in water and stored in 1-mL aliquots at −20°C.
8. 1 mM (each) of 18 amino acids (without methionine and cysteine), dissolved in water and stored in 1-mL aliquots at −20°C.
9. 0.2 M D TT, dissolved in water and stored in 10-μL aliquots at −20°C (*see* **Note 8**).
10. 0.2 M Phosphoenol pyruvate, dissolved in water and stored in 50-μL aliquots at −20°C (*see* **Note 8**).
11. 0.5 M Creatine phosphate, dissolved in water and stored in 10-μL aliquots at −20°C.
12. 10 mg/mL Creatine phosphokinase, dissolved in water and stored in 10-μL aliquots at −20°C.
13. Neutralized nucleotide (NTP) stock (50 mM ATP and 10 mM each of GTP, CTP, UTP): prepare by mixing equal volumes of 250 mM ATP, 50 mM each of GTP, CTP, UTP, and 1 M KOH. Make all solutions in water and store the NTP stock in 10-μL aliquots at −20°C (*see* **Note 9**).
14. EasyTag^{TM} Express [^{35}S]-Protein Labelling Mix, 407 MBq (11 mCi)/mL (Perkin Elmer, USA). This mixture contains 73% [^{35}S]-methionine and 22% [^{35}S]-cysteine; store in 50-μL aliquots at −80°C. (Radioactive material is hazardous. Avoid ingestion or contact with skin or clothing. Always wear gloves when handling. Monitor hands, equipment, and bench frequently.)
15. T7 RNA Polymerase. Commercially available preparations (e.g. from Promega, Madison, WI, USA) are fine; large quantities are also reasonably easy to prepare from overproducing *E. coli* strains (18).
16. 2 mM *p*Bpa (H-p-Bz-Phe-OH, Bachem AG, Switzerland): immediately before use, prepare a fresh solution of 1 M *p*Bpa in 1 *N* NaOH and dilute to 2 mM with water. Keep on ice before use and protect from light (*see* **Note 10**).
17. 11 mM Puromycin neutralized with 1 M KOH, stored in 20-μL aliquots at −20°C.
18. 10% Trichloroacetic acid: prepare a 100% solution and use for further dilutions.
19. UV-lamp, $\lambda = 365$ nm , 6 W (e.g. VL-6.L, Vilbert Lourmat Deutschland GmbH, Germany).

2.3.2. Sodium Dodecyl Sulfate-Polyacrylamide Gel Electrophoresis (SDS-PAGE)

1. Separating gel buffer: 2 M Tris-HCl, pH 8.8. Filter and store at 4°C.
2. Stacking gel buffer: 0.5 M Tris-HCl, pH 6.8. Filter and store at 4°C.

3. 25% (w/v) SDS. Store at room temperature.
4. 30% Acrylamide/0.8% bisacrylamide solution (Rotiphorese Gel 30, Carl Roth). Acrylamide is a neurotoxin; always wear gloves when handling acrylamide solutions and gels.
5. *N,N,N′,N′*-Tetramethylethylene diamine (TEMED).
6. Ammonium peroxodisulfate. Prepare 10% solution in water and store at 4°C. It is stable for several days.
7. Running buffer (5x): dissolve 150 g Tris base and 720 g glycine in 5 L water and store at room temperature.
8. Solution 1: 2 mL 1 M Tris base, 1 mL 0.2 M EDTA, pH 8.0, 7 mL water.
9. Solution 2: 4 mL 25% SDS, 1 mL 1 M Tris base, 3.5 mL 100% glycerol, 3.5 mL 0.1% bromophenol blue.
10. Solution 3: 1 M DTT.
11. Prepare PAGE-loading buffer: mix five parts of solution 1, four parts of solution 2, and one part of solution 3. Always prepare fresh.
12. Prestained molecular weight marker: Precision Plus Protein Standards (Bio-Rad, Hercules, CA).
13. Fixing solution: 35% ethanol, 10% acetic acid.

3. Methods

3.1. Preparation of an S-135 Cell Extract from *E. coli* for In Vitro Synthesis and Site-Specific Cross-Linking

1. Pre-cool the French press cell by placing it at 4°C.
2. Supplement 100 mL of S-30 medium with 0.4 mL autoclaved 20% glucose solution and 0.5 mL of 5 mg/mL tetracycline. Inoculate from plates or glycerol stocks with an *E. coli* strain carrying plasmid pDULE-pBpa (*see* **Note 11**). Grow cells overnight at 37°C with sufficient aeration in a rotary shaker (cover flask with aluminium foil). Use this culture to inoculate at a 1:100 ratio 4–6 L growth medium supplemented with 4 mL/L of 20% glucose and 0.5 mL/L of 5 mg/mL tetracycline (**see Note 12**) and grow the cells in a rotary shaker to late log-phase (optical density at 600 nm = 1.0–1.2U/mL).
3. Prepare 4 L of S-30 buffer and store at 4°C.
4. Chill the cell cultures quickly by placing the flasks in an ice water bath and harvest the cells at 4°C in a cooled SCL3000 rotor (Sorvall) for 10 min at 8650 g (7000 rpm). All subsequent steps should be done at 4°C or on ice. Resuspend the cell pellets in S-30 buffer (*see* **Note 13**).

Combine the cell suspensions in one or two tared centrifuge bottles and centrifuge again. Determine the wet weight of the cell pellet (approx 2 g/L medium).

5. Resuspend the cell pellet in 1 volume (1 mL/g wet cell mass) of S-30 buffer containing 0.5 mM PMSF (add PMSF from a fresh 0.1 M stock in ethanol).
6. For breakage of the cells pass the cell suspension two to three times through a French pressure 40 k cell (Spectronic Unicam, Cambridge, UK) at 8000 psi. This corresponds to a gage pressure setting of 500 when using the 1-in. piston cell at the "high ratio" selection (*see* **Note 14**).
7. After cell breakage centrifuge the suspension in a pre-cooled SS34 rotor (Sorvall) for 30 min at 30,000 g (15,500 rpm) at 4°C. Remove supernatant (S-30) carefully (*see* **Note 15**).
8. To allow degradation of endogenous mRNA in the S-30, perform a readout of polysomal mRNA. To this end prepare supplemented S-30 according to **Section 2.1**, item 16. Incubate at 37°C for 1 h. Afterwards chill the S-30 on ice.
9. Dialyze the S-30 three times against 1 L of cold S-30 buffer for 1 h each at 4°C (*see* **Note 16**).
10. Prepare S-135 from the S-30 (*see* **Note 17**) by pipetting 1-mL aliquots of S-30 into tubes of a Beckman TLA 100.2 rotor and spin at 287,600 g (90,000 rpm) for 13 min at 4°C. Remove 750 μL (*see* **Note 18**) of each supernatant, combine (= S-135), and quick-freeze in aliquots of 50- to 100-μL in liquid nitrogen (*see* **Note 19**). Store the S-135 at –80°C.

3.2. Preparation of INV from a TatABC-Overproducing *E. coli* Strain

1. Pre-cool the French press cell by placing it at 4°C.
2. Inoculate 100 mL of starter culture medium from plates or glycerol stocks with an *E. coli* strain harbouring the *tatABC* genes cloned under an inducible promoter (*see* **Note 20**). Grow cells overnight at 37°C with sufficient aeration in a rotary shaker (cover flask with aluminium foil).
3. Inoculate 4 flasks containing 1 L complete INV medium with 20 mL of starter culture each. Grow cells at 37°C. Expression of the TatABC proteins is induced at an optical density of 0.5 by adding 1 mM isopropyl-β-D-thiogalactopyranoside (IPTG) and growth is continued for 3–4 h or until an optical density at 600 nm of 1.5–1.8 is reached (*see* **Note 20**).
4. Prepare 50 mL buffer A and cool on ice.

5. Chill the cell cultures quickly by placing the flasks in an ice water bath and harvest the cells at 4°C in a cooled SCL3000 rotor (Sorvall) for 10 min at 8650 g (7000 rpm). All subsequent steps should be done at 4°C or on ice. Resuspend the cell pellets in buffer A (*see* **Note 13**). Combine the cell suspensions to one or two tared centrifuge bottles and centrifuge again. Determine the wet weight of the cell pellet (*see* **Note 21**).
6. Resuspend the cell pellet in 1 volume (1 mL/g wet cell mass) of buffer A containing 0.5 mM PMSF (add PMSF from a fresh 0.1 M stock in ethanol).
7. For breakage of the cells, pass the cell suspension two to three times through a French pressure 40 k cell at 8000 psi. This corresponds to a gage pressure setting of 500 when using the 1-in. piston cell at the "high ratio" selection (*see* **Note 14**).
8. To remove cell debris, the extract is centrifuged for 5 min in a pre-cooled SS34 rotor (Sorvall) at 1954 g (5000 rpm) at 4°C.
9. The supernatant is collected and centrifuged again for 2 h at approx 150,000 g (40,000 rpm) in a Beckman 50.2Ti Rotor at 4°C to obtain a crude membrane pellet encompassing outer and inner membranes and ribosomes (*see* **Note 22**). The sticky pellets are carefully resuspended in buffer A by using a loosely fitting glass homogenizer (Fischer Scientific) to give a total volume of 8 mL. Crude membranes can be stored at −80°C after quick-freezing in liquid nitrogen.
10. Because of the different densities, the inner membrane vesicles can be separated from outer membranes and unbound ribosomes by sucrose gradient centrifugation. Prepare six sucrose gradients each one consisting of 12 mL 0.77 M, 12 mL 1.44 M and 10 mL 2.02 M sucrose solution in polyallomer centrifuge tubes (38.5 mL, 25 × 89 mm, Herolab centrifuge labware). Start with the 0.77 M sucrose cushion and always underlay the denser solutions by using a smoothly running syringe equipped with a horizontally cut, wide-bore needle. Equilibrate the gradients at 4°C for about 1 h. Finally, load 2–2.5 mL of the crude membranes on top of each gradient and spin at 4°C for at least 16 h at approx 81,500 g (25,000 rpm) in a swing-out rotor, type Sorvall AH 629/36.
11. After centrifugation the inner membrane fraction should be visible as a yellow layer at the interface between the 0.77 M and 1.44 M sucrose steps. Recover the membranes with a syringe by introducing the needle from the top of the

gradient or by carefully poking a hole into the tube wall at the height of the inner membrane layer. This is safely done (mind your fingers!) by use of a disposable hypodermic needle mounted on a syringe which is slowly turned clockwise and counter-clockwise between thumb and middle finger and thereby drilled across the tube wall. Push the vesicle suspension immediately into a tube placed on ice.

12. For subsequent collection, dilute the inner membrane vesicles with ice-cold 50 mM TeaOAc, pH 7.5 about fourfold (the fraction withdrawn from the gradients presumably stems to equal parts from both the 0.77 M and 1.44 M sucrose layers resulting in a calculated sucrose concentration of about 1.1 M).
13. Pellet the inner membranes by centrifugation for 2 h at approx 150,000 g (40,000 rpm) in a Beckman 50.2Ti Rotor at 4°C and carefully resuspend in INV buffer by using a loosely fitting glass homogenizer. The final desired volume of INV derived from 4 L of bacterial culture is about 1 mL. This will correspond to an absorption at 280 nm of about 30 U/mL (*see* **Note 23**).
14. Freeze the gradient-purified INV in small aliquots of about 15 μL in liquid nitrogen and store at −80°C (*see* **Note 24**).

3.3. Site-Specific Cross-Linking of In Vitro Synthesized E. coli *Precursor Proteins Using a pBpa-Specific Amber Suppressor tRNA and Its Cognate Amino Acyl-tRNA Synthetase*

3.3.1. In Vitro Transcription-Translation Reaction and Incorporation of pBpa via Amber Stop Codon Suppression

1. Plan the experiment according to the table in **Fig. 13.1**: synthesize a variant of the Tat-specific precursor protein pTorA-PhoA containing *p*Bpa at position Phe15 (pTorA-PhoA-F15) by use of an amber stop codon-suppressing S-135 cell extract. Synthesis is performed either in the absence of *p*Bpa (samples 1, 3, 5, and 7) or in the presence of *p*Bpa (samples 2, 4, 6, and 8), and at Mg^{2+}-concentrations of either 6 mM (samples 1 and 2), 7 mM (samples 3 and 4), 8 mM (samples 5 and 6), or 9 mM (samples 7, and 8). You will need 8 × 25 μL reactions. In order to provide enough material, plan for one additional reaction, i.e., a total of 9 × 25 μL reactions.

   Next, calculate the reaction mixture as exemplified in **Table 13.1**. The reaction mixture consists of all ingredients that are common to the eight individual reactions indicated in **Fig. 13.1**. The final volume of each reaction is 25 μL. Of those, 3.5 μL are used up by the individual additives listed in **Fig. 13.1**: H_2O, $Mg(OAc)_2$, and *p*Bpa. In the experiment outlined here 9 × 21.5 μL = 193.5 μL reaction mixture will be prepared and subsequently distributed in 21.5 μL aliquots onto the eight reaction tubes indicated in **Fig. 13.1**.
2. Prepare 500 μL Compensating Buffer (CB) for the transcription-translation reaction on ice (*see* **Note 25**).

Sample	1	2	3	4	5	6	7	8
H_2O	3.5μL	3μL	2.5μL	2μL	1.5μL	1μL	0.5μL	-
$Mg(OAc)_2$ [25m*M*]	-	-	1μL	1μL	2μL	2μL	3μL	3μL
*p*Bpa [2m*M*]	-	0.5μL	-	0.5μL	-	0.5μL	-	0.5μL
Reaction mixture (see Table 1)	21.5μL each							
Sum	25μL each							

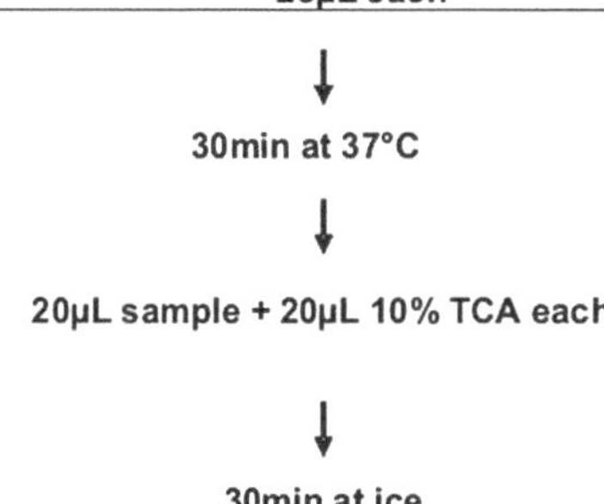

Fig. 13.1. In vitro incorporation of *p*Bpa into pTorA-PhoA-F15 via amber stop codon suppression. Pipetting scheme. The preparation of the reaction mix is detailed in **Table 13.1**. Mg^{2+}-solution and *p*Bpa solution are described in **Section 2.3.1**, items 5 and 16, respectively. TCA, trichloroacetic acid.

Efficient in vitro transcription of template DNA and translation of transcripts into protein needs defined reaction conditions. The following final concentration of ions have proven optimal for the wild type pTorA-PhoA DNA: 40 mM TeaOAc pH 7.5, 140 mM KOAc, 11 mM $Mg(OAc)_2$ (*see* **Note 26**). The optimal Mg^{2+}-concentration for suppression of the pTorA-PhoA-F15 amber stop codon mutant is determined in the experiment described here by titrating the Mg^{2+} concentration in mM-increments. The CB therefore is prepared with 6 mM Mg^{2+} as the lowest concentration intended (**Table 13.2**). In calculating the CB, the ionic contributions of the S-135 extract and plasmid DNA are taken into consideration. **Table 13.2** explains the calculation of CB for a 25 μL single reaction containing 3 μL of S-135 extract and 1 μL of plasmid DNA.

3. Thaw all required components. This is best done by placing small aliquots simply on ice and larger ones in a water bath at room temperature. Freshly prepare *p*Bpa solution (*see* **Section 2.3.1**, item 16) and protect from light. Set up a series of labelled 1.5-mL reaction tubes on ice and add H_2O, 25 mM $Mg(OAc)_2$, and *p*BPA as specified in **Fig. 13.1**.
4. Prepare the reaction mixture on ice according to **Table 13.1** strictly following the indicated order. Vortex before adding the first biological (creatine phosphokinase) and after the last addition, each time briefly spinning to collect all liquid at the bottom of the tube again.

Table 13.1
Calculation of the reaction mixture (Section 3.3.1)

	Concentration of stock solution	Final concentration	μL/25 μL	μL x 9
Compensating Buffer[a]	5x	1x	5	45
H_2O		up to 25 μL	5.4	48.6
Polyethylene glycol	40% (w/v)	3.2% (w/v)	2	18
18 amino acids	1 mM each	0.04 mM each	1	9
DTT	200 mM	2 mM	0.25	2.25
NTP mixture: (ATP GTP, UTP, CTP)	 50 mM 10 mM each	 2.5 mM 0.5 mM each	1.25	11.25
Phosphoenol pyruvate	200 mM	12 mM	1.5	13.5
Creatine phosphate	500 mM	8 mM	0.4	3.6
Creatine phosphokinase	10 mg/mL	40 μg/mL	0.1	0.9
DNA[b]	1 mg/mL	40 μg/mL	1	9
S-135			3	27
T7 RNA polymerase			0.1[c]	0.9
[^{35}S]-Met/Cys			0.5	4.5
Total (= Reaction mixture)			21.5 μL	193.5 μL
added separately (*see* **Fig. 13.1**):				
*p*Bpa	2 mM	0.04 mM	0.5	
$Mg(OAc)_2$	25 mM		up to 3	

[a]See **Table 13.2**.
[b]DNA used here is plasmid pET28a-TorA-PhoAF15 (28).
[c]Depends on activity; use 5–10U of a commercial enzyme.

5. Subdivide the reaction mixture onto the eight reaction tubes as indicated in **Fig. 13.1**. Vortex and briefly spin to collect all liquid at the bottom of the tubes.
6. Start the reactions by incubating all tubes at 37°C for 30 min (*see* **Note 27**).
7. Spin briefly and place tubes on ice to stop the synthesis reaction.
8. Add 20 μL of each reaction to 20 μL 10% trichloracetic acid each, mix, and let it precipitate on ice for at least 30 min (*see* **Note 28**).
9. Pellet precipitated proteins by centrifugation for 10 min in a tabletop microcentrifuge at room temperature. Carefully remove supernatant by aspiration into the radioactive waste.
10. Add 30 μL PAGE-loading buffer to each sample and shake vigorously at room temperature to dissolve the pellet com-

Table 13.2
Calculation of compensating buffer (CB)

	Tea[a]/Tris (nmol)	K^+ (nmol)	Mg^{2+} (nmol)	Spermidine (nmol)	H_2O
3 μL S135 (10 mM Tea, 60 mM K^+, 14 mM Mg^{2+})[b]	30	180	42		
1 μL DNA (10 mM Tris)[b]	10				
Total (1)	40	180	42		
Desired final concentration: 40 mM Tea, 140 mM K^+, 6 mM Mg^{2+}, 0.8 mM spermidine →nmol desired in 25-μL reaction (2)	1000	3500	150	20	
Difference (2)–(1) →nmol required for 25-μL reaction to be added via CB (3)	960	3320	108	20	
Required nmol (3) are added in 5 μL CB → required nmol/μL CB (4) (= mM concentration of CB)	192	664	21.6	4	
To prepare 1 mL of such CB from 1 M Tea, 4 M K^+, 1 M Mg^{2+}, 0.1 M spermidine stocks add	μL 192	μL 166	μL 21.6	μL 40	μL 580.4
To prepare 500 μL of such CB Add	96	83	10.8	20	290.2

[a]Tea, triethanolamine
[b]Components that contribute relevantly to the ionic composition of the reaction mixture

pletely. The colour of the loading buffer should remain dark blue. If it changes to yellow, add a few microliter of 1 M Tris base to neutralize residual trichloroacetic acid. Heat samples at 95°C for 5 min and analyse by SDS-PAGE and autoradiography.

3.3.2. SDS-Polyacrylamide Gel Electrophoresis (SDS-PAGE) and Autoradiography

1. Electrophoresis is carried out in large custom-made units. Dimensions of the gel are 35 cm × 25 cm × 1 cm (W × L × T). These gels are made from about 80 and 20 mL of separating and stacking gel solutions, respectively.
2. To prepare 100 mL of a 12% separating gel, add 40 mL acrylamide/bisacrylamide solution, 10 mL separating gel buffer, 0.4 mL 25% SDS to a measuring cylinder and adjust volume to 100mL with water. Add 0.04 mL TEMED and 0.6 mL ammonium peroxodisulfate to start polymerization, pour the solution into gel cassettes mounted in an upright position, and overlay with isobutanol.

3. After polymerization remove isobutanol, rinse with water and prepare 30 mL of stacking-gel solution by adding 5 mL acrylamide/bisacrylamide solution, 3.6 mL stacking-gel buffer, 0.12 mL 25% SDS to a measuring cylinder and adjusting the volume to 30 mL with water. Add 0.012 mL TEMED and 0.2 mL ammonium peroxodisulfate to start polymerization, pour the solution into the gel cassettes, and immediately insert a comb.
4. Prepare 2 L of running buffer by dilution from the 5x stock and addition of 8 mL 25% SDS; add to upper and lower chambers of the electrophoresis apparatus.
5. Load the samples completely into the wells of the gel and include one lane for prestained molecular weight markers. Electrophoresis is usually carried out overnight at a constant current of 20 mA until the bromophenol blue dye has run to the bottom of the gel. Avoid its running off the gel in order to retain any radioactive substance of similarly low molecular mass on the gel.
6. Remove the stacking gel and incubate the separating gel in fixing solution for 20 min on a shaking platform. Discard the fixing solution and incubate the gel in water three times for 10 min each.
7. Transfer the gel onto a prewetted Whatman 3MM paper, cover with plastic wrap, and dry at 70°C for 2 h on a vacuum dryer (Bio-Rad).
8. Expose the dried gel to a phosphorimaging screen overnight and analyse the autoradiogram on a PhosphorImager (e.g. Storm, GE Healthcare) using ImageQuant^{TM} software.
9. Print an image of the autoradiogram at magnification 1 and transfer the positions of the prestained molecular weight markers to the printout. The autoradiogram of the experiment described under **Section 3.3.1** is shown in **Fig. 13.2**.

3.3.3. In Vitro Cross-Linking of pBpa-Containing pTorA-PhoA with the TatC Subunit of the Tat Translocase of E. coli *Inner Membrane Vesicles*

1. Plan the experiment according to **Fig. 13.3**: synthesize pTorA-PhoA-F15 carrying a photo-reactive derivative of phenylalanine at position Phe15 by use of an amber stop codon-suppressing S-135 cell extract. Synthesis is performed either in the absence of *p*Bpa (sample 3) or in its presence (samples 1, 2 and 4). You will need 8 × 25 μL reactions (2 × 25 μL for each sample). In order to provide enough material, plan for two additional reactions, i.e., a total of 10 × 25 μL reactions. Calculate the reaction mixture according to **Table 13.3**. The reaction mixture contains all ingredients needed for in vitro synthesis of pTorA-PhoA-F15 except for *p*Bpa. Of the final volume of each 25 μL reaction, 0.5 μL is used up by the

[Mg^{2+}]	6m*M*		7m*M*		8m*M*		9m*M*	
*p*Bpa	-	+	-	+	-	+	-	+

pTorA-PhoA-F15 ►

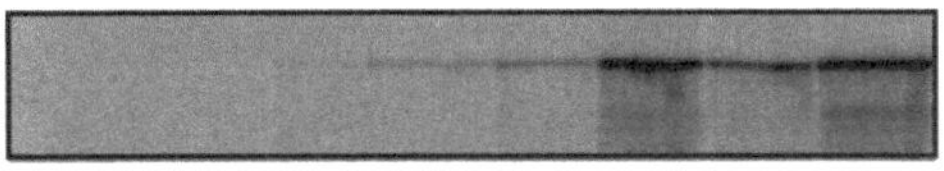

Fig. 13.2. In vitro suppression of the amber stop codon mutant pTorA-PhoAF15 by *p*-benzoyl-*L*-phenylalanine (*p*Bpa) and its dependence on the Mg^{2+}-concentration. Autoradiogram of the experiment outlined in **Fig. 13.1**. The precursor protein pTorA-PhoA-F15 was synthesized by a coupled in vitro transcription-translation system, which had been prepared from an *E. coli* strain transformed with plasmid pDULE-pBpa. This plasmid encodes a *p*Bpa-specific amber suppressor tRNA together with a cognate, *p*Bpa-specific amino acyl-tRNA-synthetase. The final concentration of Mg^{2+} and the addition of *p*Bpa are indicated. After synthesis samples were precipitated by trichloroacetic acid. Samples were separated by SDS-PAGE and visualized by phosphor imaging. Note the suppression of the amber stop codon in the presence of *p*Bpa leading to the synthesis of full-size pTorA-PhoA-F15 at Mg^{2+}-concentrations >6 mM, and some inevitable read-through of the stop codon, particularly at higher Mg^{2+}-concentrations.

subsequent addition of *p*Bpa. In the experiment outlined here 10 × 24.5 μL = 245 μL reaction mixture will be prepared and split into a 7 × 24.5 μL = 171.5 μL aliquot (*p*Bpa$^+$-mixture) and into a 49 μL aliquot (*p*Bpa$^-$-mixture) receiving 3.5 μL *p*Bpa solution and 1 μL H_2O, respectively.

2. Prepare 500 μL Compensating Buffer (CB). CB is composed as illustrated in **Table 13.2** except that it is calculated with an intended Mg^{2+}-concentration of 8 mM according to the results shown in **Fig. 13.2**.
3. Thaw all required components. Thawing is best done by placing small aliquots simply on ice and larger ones in a water bath at room temperature. Freshly prepare *p*Bpa solution (*see* **Section 2.3.1**, item 16) and protect from light.
4. Prepare the reaction mixture on ice according to **Table 13.3** strictly following the indicated order. Vortex before adding the first biological (creatine phosphokinase) and after the last addition, each time briefly spinning to collect all liquid at the bottom of the tube again. All subsequent steps should be performed in the dark to avoid untimely activation of the photo-probe.
5. Subdivide the reaction mixture according to **Fig. 13.3** onto two new reaction tubes (pBpa$^+$-mix and pBpa$^-$-mix) and add *p*Bpa solution or H_2O.
6. Start the reactions by incubating both tubes at 37°C for 30 min in the dark.
7. Stop synthesis by the addition of puromycin and incubation at 37°C for 10 min in the dark.

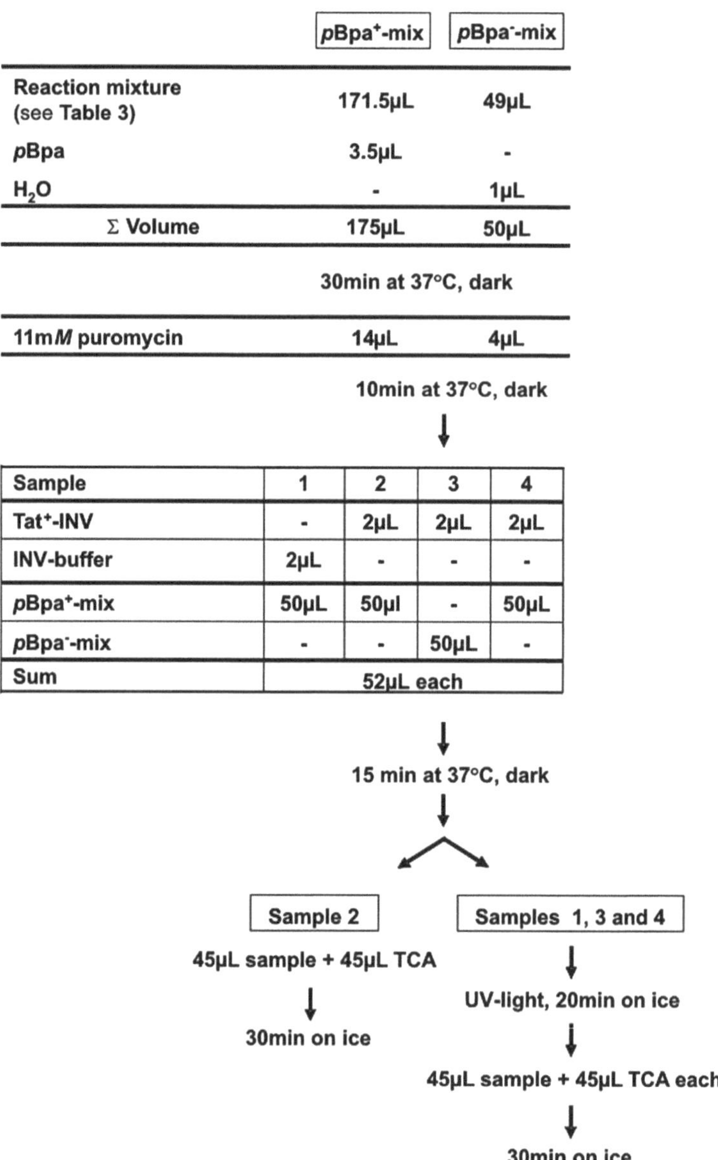

Fig. 13.3. In vitro cross-linking of *p*Bpa-containing pTorA-PhoA to the TatC subunit of the Tat translocase of *E. coli* inner membrane vesicles. The preparation of the reaction mixture is detailed in **Table 13.3**. To uncouple synthesis of pTorA-PhoA-F15 from its binding to membrane vesicles, protein synthesis is blocked by puromycin (*see* **Section 2.3.1**, item 17). Tat^+-INV were prepared from strain BL21(DE3) pLysS p8737. INV-buffer is described in **Section 2.2**, item 16. TCA, trichloroacetic acid.

8. Set up a series of labelled 1.5-mL reaction tubes on ice and add Tat^+-INV (*see* **Note 20**) or INV buffer as indicated in the lower table of **Fig. 13.3**.
9. Subdivide both reaction mixtures onto the four reaction tubes as indicated in the lower table of **Fig. 13.3**. Vor-

Table 13.3
Calculation of the reaction mixture (Section 3.3.3)

	Concentration of stock solution	Final concen-tration	μL/25μL	Reaction mix: μL x 10
Compensating Buffer[a]	5x	1x	5	50
H_2O		up to 25 μL	7.775	77.75
Polyethylene glycol	40% (w/v)	3.2% (w/v)	2	20
18 amino acids	1 mM each	0.04 mM each	1	10
DTT	200 mM	2 mM	0.25	2.5
NTP mixture			1.25	12.5
(ATP,	50 mM	2.5 mM		
GTP,UTP, CTP)	10 mM each	0.5 mM each		
Phosphoenol pyruvate	200 mM	12 mM	1.5	15
Creatine phosphate	500 mM	8 mM	0.4	4
Creatine phosphokinase	10 mg/mL	40 μg/mL	0.1	1
GSSG[b]	200 mM	5 mM	0.625	6.25
DNA[c]	1 mg/mL	40 μg/mL	1	10
S-135			3	30
T7 RNA polymerase			0.1[d]	1
[^{35}S]-Met/Cys			0.5	5
Total (= reaction mixture)			24.5 μL	245 μL
added separately (see **Fig.13.3**):				
*p*Bpa	2 mM	0.04 mM	0.5	

[a]The compensating buffer is calculated according to **Table 13.2** except that the final intended concentration of Mg^{2+} is now 8 mM.
[b]GSSG, oxidized Glutathion that is used to establish oxidative conditions during in vitro synthesis thereby allowing oxidative folding of the mature part of the TorA-PhoA precursor. Folding of pTorA-PhoA has been shown to be a prerequisite for a productive interaction of the signal peptide with TatC (28). This means that GSSG is a specific requirement of the substrate used in the experiment shown here and can therefore be omitted for other substrates.
[c]DNA used here is plasmid pET28aTorA-PhoAF15 (28).
[d]Depends on activity; use 5–10 U of a commercial enzyme.

tex and briefly spin to collect all liquid at the bottom of the tubes and incubate for additional 15 min on 37°C (*see* **Note 29**).

10. Add 45 μL of sample 2 to 45 μL 10% trichloracetic acid, mix, and let precipitate on ice for at least 30 min (*see* **Fig. 13.3**).

11. Spin the remaining samples 1, 3, and 4 and horizontally lay the closed tubes on ice. For UV irradiation of the samples, position UV lamp at the shortest possible distance right above the tubes (*see* **Note 30**). (UV radiation may

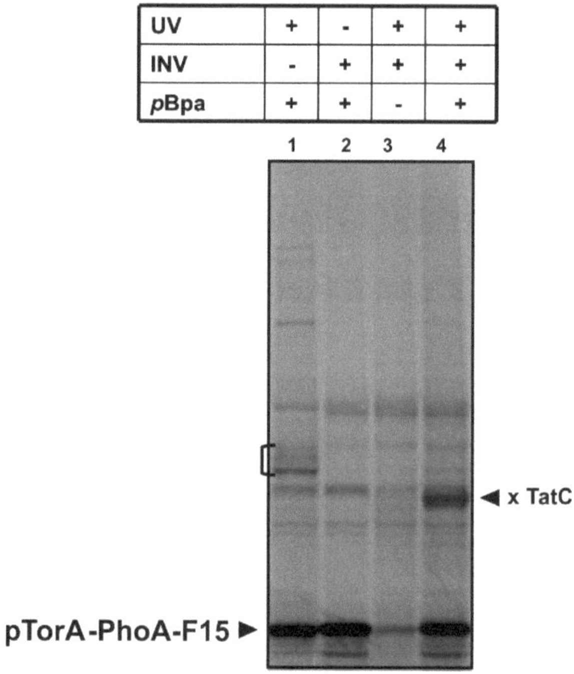

Fig. 13.4. In vitro synthesized pTorA-PhoA-F15 bearing *p*Bpa cross-links to TatC of inverted inner membrane vesicles (INV) from a TatABC-overproducing *E. coli* strain. Autoradiogram of the experiment described in the text (*see* **Section 3.3.3**). In vitro synthesis of pTorA-PhoA-F15 was performed using a coupled in vitro transcription-translation system from an *E. coli* strain transformed with plasmid pDULE. This plasmid encodes a *p*Bpa-specific amber suppressor tRNA together with a cognate, *p*Bpa-specific amino acyl-tRNA-synthetase. Efficient synthesis of pTorA-PhoA-F15 was obtained only if *p*Bpa was added (compare *lanes 3* to *lanes 1, 2*, and *4*). Following synthesis, reactions were supplemented with inverted inner membrane vesicles (INV) prepared from strain BL21(DE3) pLysS p8737 as indicated. For activation of *p*Bpa incorporated in pTorA-PhoA-F15, samples were irradiated with UV-light (365 nm) and subsequently precipitated with trichloracetic acid. Samples were separated by SDS-PAGE and visualized by phosphorimaging (× TatC, cross-link between pTorA-PhoA-F15 and TatC). UV-irradiation yields several radioactively labelled products that are larger in size than pTorA-PhoA-F15. In the absence of INV, adducts around 80 kDa are obtained (*brackets*) that have previously been identified as cross-links to the chaperones FkpA and TorD (28). In the presence of INV, however, activation of *p*Bpa by UV-light leads to an adduct of approx 70 kDa that by immunoprecipitation was previously identified to result from an interaction of pTorA-PhoA-F15 with TatC (28). Note that cross-linking of the *p*Bpa-containing precursor with TatC only occurs upon UV-irradiation (compare *lanes 2* and *4*) and moreover does not occur for pTorA-PhoA-F15 resulting from read-through of the stop codon rather than from incorporation of *p*Bpa (*lane 3*).

cause damages to skin and eyes. Always wear gloves and UV-protecting glasses when handling the UV source).

12. Add 45 μL each of sample 1, 3, and 4 to 45 μL 10% trichloracetic acid, mix, and let it precipitate on ice for at least 30 min (*see* **Fig. 13.3**).

13. Add 30μL PAGE-loading buffer to each pellet and shake vigorously at room temperature to dissolve it completely. The colour of the loading buffer should remain dark blue. If it changes to yellow, add a few microliter of 1 M Tris base to neutralize residual trichloroacetic acid. Incubate samples at 37°C for 10 min (*see* **Note 31**) and analyse by SDS-PAGE and autoradiography. The results from this experiment are shown in **Fig. 13.4**.

4. Notes

1. Media and solutions are prepared using deionized water.
2. Tetracyclin is used here for the growth of *E. coli* strains transformed with plasmid pDULE-*p*Bpa (19) encoding the orthogonal pair of an amber suppressor tRNA, which specifically accepts the photo-reactive derivative of Phe, *p*Bpa, as well as its cognate pBpa-specific amino acyl-tRNA synthetase (17). We have successfully used the alternative plasmid pSup-BpaRS-6TRN(D286R) (20).
3. Stock solutions are usually freed of microorganisms and particles by filtration through 0.22 μm mixed cellulose ester filters (Millipore). Solutions 4–7, 10, and 14 are also required for the in vitro transcription/translation reaction (*see* **Section 2.3.1**).
4. Alternatively use Pefabloc SC (Roche), a water-soluble inhibitor of serine proteases at a final concentration of 0.5 mg/mL.
5. Use gloves to touch the dialysis tubing. Perform the following treatment before use: boil the dialysis tubing in 1 L 2% $NaHCO_3$, 1 mM EDTA for 10 min. Rinse the dialysis tubing with water and boil it again in 1 L water for 10 min. Store dialysis tubing in water at 4°C.
6. To obtain sufficient synthesis in vitro, the gene of interest preferably should be under the control of the T7 promoter. We have successfully used vectors such as pKSM717 (21) and pET derivatives.
7. To minimize sterical perturbances due to the incorporation of the bulky and hydrophobic site chain of *p*Bpa, preferentially replace amino acids exhibiting similar properties, such as Trp, Phe, Tyr, Val, and Leu.
8. If the efficiency of synthesis unexpectedly drops, it often can be overcome by preparing fresh stocks of spermidine, DTT, and phosphoenol pyruvate.

9. Polyethylene glycol, 18 amino acids, DTT, NTP stock, phosphoenol pyruvate, and creatine phosphate can be combined according to the ratios indicated in **Table 13.1** and stored in 100- to 200-μL aliquots at −20°C. This should be done only after proof has been obtained that the individual solutions allow efficient protein synthesis in vitro.
10. Although *p*Bpa is considered to be stable at ambient light (22), *p*Bpa stocks, as well as every *p*Bpa-containing sample in subsequent steps should be protected from light by covering with aluminium foil.
11. *E. coli* strains suitable for the preparation of S-135 are MC4100 (23) and Top10 (Invitrogen, Carlsbad, CA, USA). Other strains can likely be used as long as they do not contain an endogenous amber suppressor tRNA. DH10B (Invitrogen), for example, has been used for *p*Bpa-specific photo-cross-linking of proteins in vivo (19).
12. Compared to the starter culture the concentration of tetracyclin is reduced in order to maximize protein expression (19).
13. Fast resuspension is achieved by repeatedly forcing the cell suspension through the medium-bore opening of a ball-equipped glass pipet harbouring a sufficiently large reservoir.
14. The best result is obtained by passing the cell suspension through the French pressure cell at a speed that allows a dropwise efflux. This requires more than one passage as the released DNA first causes high viscosity until it becomes fragmented by the applied shear forces.
15. Freeze the S-30 immediately in liquid nitrogen and store at −80°C or continue with the next step.
16. Use a volume ratio of S-30 to dialysis buffer of approx 1:100. One of the three steps can conveniently also be done overnight. After dialysis, the S-30 can be quick-frozen in liquid nitrogen and stored at −80°C.
17. High-speed centrifugation resulting in an S-135 extract is required to remove all membrane vesicles from the S-30 extract. High speed centrifugation also removes remaining polysomes from the extract.
18. Note that the time of spin and amount of supernatant withdrawn will have an influence on the performance of the S-135, the designation of which is an operational term rather than reflecting the actual g-force. Recovery of too much supernatant might still result in a contamination with endogenous membranes, whereas too little supernatant bears the risk of a shortage of monosomes. In the

latter case try to reduce time of ultracentrifugation when preparing the S-135 or add separately isolated ribosomes. Low translation activity of an S-135 preparation can also result from residual cold methionine added for the readout of endogenous polysomes. This is effectively removed by repeated passages of the S-135 through an Amicon ultra centrifugal filter unit (Millipore, molecular weight cut off of 10,000 Da), each time replacing the filtrate by fresh S-30 buffer.

19. Do not thaw and freeze the S-135 more than twice.
20. Efficient cross-linking of twin-arginine-containing precursor proteins to INV requires the preparation of INV from strains that overproduce the TatABC proteins, such as BL21(DE3) pLysS p8737TatABCD (24) or DADE (MC4100, Δ*tatABCD*Δ*tatE*) (25) transformed with plasmids pRep4 and pQE60-TatABC (16). In the BL21 derivative, in which *tatABC* is under T7 promoter control, IPTG induces expression of T7 RNA polymerase, whereas in the other strain, IPTG directly enhances expression of *tatABC* from the *lac* promoter. High levels of the TatABC proteins in the vesicles prepared from these strains are verified via Western blotting.
21. Cells destined for the preparation of INV must not be frozen before breakage in the French press. If necessary cell pellets can be kept on ice overnight.
22. If the protocol is to be directly continued beyond this step, prepare sucrose gradients during this 2-h centrifugation period.
23. For determining the absorbance of the vesicle suspension at 280 nm, prepare a 1:100 dilution in 2% SDS. With an absorbance of 30 or more, usually 1–2 μL of INV are sufficient to observe cross-links of a Tat precursor protein to the TatABC proteins in a 25-μL reaction.
24. Do not freeze and thaw INV more than two or three times.
25. To avoid contamination with proteases and RNases, always wear gloves and preferentially use sterile disposable reaction tubes and pipet tips.
26. With every new preparation of S-135 it is necessary to re-adjust the reaction conditions. The variable with the strongest impact on expression efficiency is the concentration of Mg^{2+} which even needs to be optimized for each particular DNA template. Sometimes inclusion of 8 mM putrescin into the reaction helps to improve expression. In this case the final Mg^{2+} concentration is usually lowered by about 3 mM and that of phosphoenol pyruvate by 6 mM.

If in vitro expression remains unsatisfactory, try to vary the amount of S-135 in the range of 2–4μL. Calculate and prepare a new CB for each experiment.

27. Incubation at 37°C is routinely used. In some cases (e.g. INV derived from *cs* mutants) it is necessary and possible to synthesize proteins also at lower temperatures.
28. Precipitation with trichloroacetic acid can be extended to an overnight incubation.
29. Incubation with INV enables the in vitro synthesized precursor to interact with the TatC protein. Omission of an energy-regenerating system largely prevents the H^+-gradient-dependent translocation of the precursor by the Tat translocase and thereby guarantees a prolonged interaction of the precursor with TatC, which is part of the receptor complex recognizing the precursor at the initial step of the translocational process (15, 26, 27).
30. The UV-lamp used here emits light of 365 nm. This wavelength is suitable for activation of *p*Bpa (activation at 350–360 nm (22)). Since we determined that approx 65–70% of the activating light is absorbed by the wall of the reaction tube, the time of irradiation might be shortened by placing the tube vertically on ice with the lid open. Using the same setup of irradiation, the sample size can be varied considerably, as long as the ratio between the irradiated surface of the sample and its volume is kept constant (19).
31. Because TatC is a largely hydrophobic membrane protein, denaturation should not be performed by boiling in SDS in order to avoid a smeary appearance on SDS-PAGE.

Acknowledgments

We gratefully acknowledge Dr. Peter Schultz, The Scripps Research Institute, La Jolla, for providing suppressor plasmids. This work was supported by grant LSHG-CT-2004-05257 of the European Union and grants from the Deutsche Forschungsgemeinschaft (Sonderforschungsbereich 388 and Graduiertenkolleg 434).

References

1. Krieg, U. C., Walter, P., and Johnson, A. E. (1986) Photocrosslinking of the signal sequence of nascent preprolactin to the 54-kilodalton polypeptide of the signal recognition particle. *Proc. Natl. Acad. Sci. USA* **83**, 8604–8608.
2. High, S., Martoglio, B., Gorlich, D., Andersen, S. S., Ashford, A. J., Giner,

A., Hartmann, E., Prehn, S., Rapoport, T. A., Dobberstein, B., and Brunner, J. (1993) Site-specific photocross-linking reveals that Sec61p and TRAM contact different regions of a membrane-inserted signal sequence. *J. Biol. Chem.* **268**, 26745–26751.

3. Martoglio, B., Hofmann, M. W., Brunner, J., and Dobberstein, B. (1995) The protein-conducting channel in the membrane of the endoplasmic reticulum is open laterally toward the lipid bilayer. *Cell* **81**, 207–214.
4. Do, H., Falcone, D., Lin, J., Andrews, D. W., and Johnson, A. E. (1996) The cotranslational integration of membrane proteins into the phospholipid bilayer is a multistep process. *Cell* **85**, 369–378.
5. Wiedmann, M., Kurzchalia, T. V., Bielka, H., and Rapoport, T. A. (1987) Direct probing of the interaction between the signal sequence of nascent preprolactin and the signal recognition particle by specific cross-linking. *J. Cell Biol.* **104**, 201–208.
6. Kurzchalia, T. V., Wiedmann, M., Girshovich, A. S., Bochkareva, E. S., Bielka, H., and Rapoport, T. A. (1986) The signal sequence of nascent preprolactin interacts with the 54 K polypeptide of the signal recognition particle. *Nature* **320**, 634–636.
7. Valent, Q. A., de Gier, J.-W. L., von Heijne, G., Kendall, D. A., ten Hagen-Jongman, C. M., Oudega, B., and Luirink, J. (1997) Nascent membrane and presecretory proteins synthesized in Escherichia coli associate with signal recognition particle and trigger factor. *Mol. Microbiol.* **25**, 53–64.
8. Houben, E. N., Urbanus, M. L., Van Der Laan, M., Ten Hagen-Jongman, C. M., Driessen, A. J., Brunner, J., Oudega, B., and Luirink, J. (2002) YidC and SecY mediate membrane insertion of a Type I transmembrane domain. *J. Biol. Chem.* **277**, 35880–35886.
9. Beck, K., Eisner, G., Trescher, D., Dalbey, R. E., Brunner, J., and Müller, M. (2001) YidC, an assembly site for polytopic Escherichia coli membrane proteins located in immediate proximity to the SecYE translocon and lipids. *EMBO Rep.* **2**, 709–714.
10. Beck, K., Wu, L. F., Brunner, J., and Müller, M. (2000) Discrimination between SRP- and SecA/SecB-dependent substrates involves selective recognition of nascent chains by SRP and trigger factor. *EMBO J.* **19**, 134–143.
11. Berks, B. C., Palmer, T., and Sargent, F. (2003) The Tat protein translocation pathway and its role in microbial physiology. *Adv. Microb. Physiol.* **47**, 187–254.
12. Lee, P. A., Tullman-Ercek, D., and Georgiou, G. (2006) The bacterial twin-arginine translocation pathway. *Annu. Rev. Microbiol.* **60**, 373–395.
13. Müller, M., and Klösgen, R. B. (2005) The Tat pathway in bacteria and chloroplasts. *Mol. Membr. Biol.* **22**, 113–121.
14. Robinson, C., and Bolhuis, A. (2004) Tat-dependent protein targeting in prokaryotes and chloroplasts. *Biochim. Biophys. Acta* **1694**, 135–147.
15. Alami, M., Lüke, I., Deitermann, S., Eisner, G., Koch, H. G., Brunner, J., and Müller, M. (2003) Differential interactions between a twin-arginine signal peptide and its translocase in *Escherichia coli. Mol. Cell* **12**, 937–946.
16. Holzapfel, E., Eisner, G., Alami, M., Barrett, C. M., Buchanan, G., Lüke, I., Betton, J. M., Robinson, C., Palmer, T., Moser, M., and Müller, M. (2007) The entire N-terminal half of TatC is involved in twin-arginine precursor binding. *Biochemistry* **46**, 2892–2898.
17. Chin, J. W., Martin, A. B., King, D. S., Wang, L., and Schultz, P. G. (2002) Addition of a photocrosslinking amino acid to the genetic code of *Escherichia coli. Proc. Natl. Acad. Sci. USA* **99**, 11020–11024.
18. Davanloo, P., Rosenberg, A. H., Dunn, J. J., and Studier, F. W. (1984) Cloning and expression of the gene for bacteriophage T7 RNA polymerase. *Proc. Natl. Acad. Sci. USA* **81**, 2035–2039.
19. Farrell, I. S., Toroney, R., Hazen, J. L., Mehl, R. A., and Chin, J. W. (2005) Photo-cross-linking interacting proteins with a genetically encoded benzophenone. *Nat. Methods* **2**, 377–384.
20. Ryu, Y., and Schultz, P. G. (2006) Efficient incorporation of unnatural amino acids into proteins in *Escherichia coli. Nat. Methods* **3**, 263–265.
21. Maneewannakul, S., Maneewannakul, K., and Ippen-Ihler, K. (1994) The pKSM710 vector cassette provides tightly regulated lac and T7lac promoters and strategies for manipulating N-terminal protein sequences. *Plasmid* **31**, 300–307.
22. Dorman, G., and Prestwich, G. D. (1994) Benzophenone photophores in biochemistry. *Biochemistry* **33**, 5661–5673.
23. Casadaban, M. J., and Cohen, S. N. (1979) Lactose genes fused to exogenous promoters

in one step using a Mu-lac bacteriophage: in vivo probe for transcriptional control sequences. *Proc. Natl. Acad. Sci. USA* **76**, 4530–4533.

24. Alami, M., Trescher, D., Wu, L. F., and Müller, M. (2002) Separate analysis of twin-arginine translocation (Tat)-specific membrane binding and translocation in Escherichia coli. *J. Biol. Chem.* **277**, 20499–20503.
25. Wexler, M., Sargent, F., Jack, R. L., Stanley, N. R., Bogsch, E. G., Robinson, C., Berks, B. C., and Palmer, T. (2000) TatD is a cytoplasmic protein with DNase activity. No requirement for TatD family proteins in Sec-independent protein export. *J. Biol. Chem.* **275**, 16717–16722.
26. Cline, K., and Mori, H. (2001) Thylakoid DeltapH-dependent precursor proteins bind to a cpTatC-Hcf106 complex before Tha4-dependent transport. *J. Cell Biol.* **154**, 719–729.
27. Gerard, F., and Cline, K. (2006) Efficient twin arginine translocation (Tat) pathway transport of a precursor protein covalently anchored to its initial cpTatC binding site. *J. Biol. Chem.* **281**, 6130–6135.
28. Panahandeh, S., Maurer, C., Moser, M., Delisa, M. P., and Müller, M. (2008) Following the path of a twin-arginine precursor along the TatABC translocase of *Escherichia coli*. *J. Biol. Chem.* 33267–33275.

Chapter 14

Tracking the Secretion of Fluorescently Labeled Type III Effectors from Single Bacteria in Real Time

Nandi Simpson, Laurent Audry, and Jost Enninga

Abstract

A large number of Gram negative pathogens use a specialized needle-like molecular machine known as Type III Secretion (T3S) system. This highly sophisticated molecular device consists of a basal body spanning the two bacterial membranes and a protruding needle structure that is connected to a distal translocator complex. The main features of the T3S system are (i) activation after host cellular membrane contact and (ii) the ability to "inject" effectors into host cells through the needle apparatus across three membranous structures—two bacterial and one host cellular—without effector leakage into the exterior space. The effector proteins execute multiple roles upon translocation including re-arranging the host cytoskeleton, manipulating signaling pathways and reprogramming the host immune response. We have established a novel approach to monitor the secretion of fluorescently labeled effectors through the T3S system of single living bacteria in real time. Our approach uses the tetracysteine-FlAsH labeling procedure. Here, we provide a detailed protocol and advice on its potential and experimental pitfalls. Using the entero-invasive pathogen *Shigella flexneri* for assay development, we have also successfully adapted our approach and developed procedures for T3S effector tracking for other pathogens such as Enteropathogenic *Escherichia coli* (EPEC).

Key words: Type III secretion system, translocation, *Shigella flexneri*, multidimensional imaging, tetracysteine labeling.

1. Introduction

The Type III Secretion (T3S) system or "injectisome" of Gram negative bacteria is an organelle with structural and functional similarity to flagellae (1). T3S enables (i) the secretion of bacterial protein "effectors" across both the inner and the outer bacterial membranes to the extra-bacterial environment and (ii) their

A. Economou (ed.), *Protein Secretion*, Methods in Molecular Biology 619,
DOI 10.1007/978-1-60327-412-8_14,

translocation into targeted cells in a single step. Since the first model proposed for the translocation of *Yersinia* proteins (2), T3S systems have been described in a wide range of Gram negative bacterial species, including pathogens, symbionts and commensals of mammals, plants, insects, and amoeba (3). Pathogenic bacteria use T3S to translocate effectors across the host cell membrane into the eukaryotic cytosol to undermine host cellular functions and to subvert the host immune system. Therefore, the study of T3S and their underlying mechanisms is very relevant in biomedical research.

Seven families of T3S systems have been described on the basis of phylogenetic studies. However, the genetic organization and structural components of the T3S machinery are generally conserved (3, 4). Construction of the T3S system requires approximately 25 proteins, including both structural components and ancillary proteins required for the assembly process (1). The T3S system is composed of a basal body or "needle complex," which spans the inner (IM) and outer (OM) bacterial membranes. During construction of the T3S machinery, the rings that span the IM are delivered to the bacterial envelope by the universal secretory pathway. These rings are anchored to the IM by N-terminal lipidation. A short rod traversing the periplasm joins IM rings to the OM component, which belongs to the secretin family of pore-forming proteins. The basal body terminates in a needle protruding from the bacterial surface, through which effectors are delivered to the extra-bacterial environment. Generally, the needle is a stiff helical polymer, and its length varies between species and even strains (5). In some instances, as exemplified by the EPEC T3S system, which traverses the thick glycocalyx of gastrointestinal epithelial cells, the needle is extended by a long flexible EspA filament. The needle tip complex is formed by a protein complex known as the "tranlocon," which is inserted into host cell membranes. Once the needle apparatus is activated by contact with host cells, translocation of effector proteins directly to the host cell cytoplasm occurs through the T3S apparatus without leakage to the extracellular medium. Together, these structures form a patent conduit from the bacterial to the host cell cytoplasm.

The process of secretion is tightly regulated and is triggered by contact between the needle tip and host cell, although the exact mechanism of activation is yet to be elucidated (6). Secretion of effectors through the channel, which has a diameter of around 25 to 30 Å requires that the majority of these proteins are unfolded. In *Yersinia*, the YscN family of ATPases is associated with the IM and functions both to unfold exported proteins and to remove their chaperones (7). The dynamic process of export is energized by ATP hydrolysis and a proton motive force (8). The majority of effectors translocated from the bacterial cytosol have a secretion signal within the first 20 to 30 amino acids; however there is evidence that for some effectors the "secretion signal" may be located in the mRNA and an additional level of control over effector secretion is derived from the activity of cytosolic chaperones (1). A wide range of T3S effectors have been correlated with pathogen virulence strategies. Effectors act by subverting host cell processes to promote bacterial adhesion or invasion, persistence and dissemination. Effectors often employ functional mimicry to manipulate host cell functions, affecting actin and tubulin cytoskeletal dynamics, host inflammatory responses, vesicular trafficking, cell cycle progression, gene expression and programmed cell death.

Classically, monitoring of effector translocation has involved the acquisition of a "static picture" by Western blot analysis, immunofluorescence of fixed samples or enzymatic assays. Although translational fusions between effectors and fluorescent proteins are unsuitable for translocation via the T3S apparatus, alternative detection strategies have been developed to demonstrate the accumulation of effectors in the eukaryotic cytosol. The first reporter enzyme assay was described in the early 1990s, involving translational fusions of the amino-terminal secretion signals of *Yersinia* effectors with the calmodulin-dependent adenylate cyclase (Cya) domain of the *B. pertussis* toxin cyclolysin, which converts ATP to cAMP in the presence of calmodulin (9). A later variation on this theme is a reporter system using translational fusion of effectors with the phosphorylable Elk peptide fused to the nuclear localization signal from large T antigen of SV40, which directs the fusion protein to the nucleus, where Elk is phosphorylated and can be detected by specific antibodies (10). More recently, a reporter system based on Cre-Lox recombination has been developed. Fusions between a bacteriophage P1 Cre recombinase and the signal sequence of the *Salmonella* effector SopE were created and eukaryotic cells previously transfected with constructs encoding Cre recombinase activity reporters (firefly luciferase or GFP) were infected with bacteria carrying the effector-Cre vector. Upon translocation of the Cre recombinase, Lox sites are cleaved and the reporter is transcribed, thereby, fluorescence activated cell sorting (FACS) analysis can be used to detect activation of the recombinase reporters upon effector translocation (11). A fluorescence-based translocation reporter system using translational fusion of T3S signal of the pathogenic *E. coli* effector Cif with mature TEM-1 β-lactamase, together with the β-lactamase substrate CCF2/AM was used to demonstrate translocation of Cif (12). In this assay, the CCF2/AM substrate enters the cell passively and is converted by cellular esterases to charged fluorescent CCF2. Excitation of CCF2 results in fluorescence energy transfer (FRET). In the presence of an effector fused to TEM-1, the CCF2 β-lactam ring is cleaved and FRET is disrupted. This technique was subsequently adapted for real time analysis of EPEC effector dynamics by fusing effectors to the β-lactamase gene and using a plate reader to measure the accumulation of the CCF2 hydrolysis product over time (13).

Two approaches using microscopy for the analysis of effector translocation at the single cell level have been described, one of which uses GFP-labeled chaperones as a probe for translocated effectors, the other uses the fluorescein-based biarsenical dye FlAsH to label effectors with a tetracysteine (4Cys) motif tag. The chaperone-based assay images effector accumulation in the host cell by detecting recruitment of a fluorescently labeled "probe." Complementary assays detecting the arrival of an effector in the eukaryotic cell cytosol and concurrent depletion from the bacterial cytosol were used to image secretion of the *Salmonella* effector SipA in real time (14). This system uses a pool of GFP-labeled chaperone as a specific probe to detect the effector in the eukaryotic cytosol. Fluorescence images were analyzed to measure GFP recruitment upon bacterial docking to the eukaryotic cell. Using time lapse microscopy to monitor how long each bacterium docked before fixation and subsequent confocal analysis of immunostained intrabacterial effector pools, depletion of the effector from the bacterial cytosol was correlated with accumulation in the eukaryotic cytosol. This assay using chaperones as specific probes monitors effector accumulation and chaperone recruitment rather than the direct kinetics of secretion. The FlAsH labeling method represents a two compound system for the fluorescent labeling of proteins: a genetically encoded sequence containing the 4Cys repeat sequence and the FlAsH compound (15, 16). This approach has been developed as an alternative to bulky fluorescent proteins such as GFP. Furthermore, the approach can be used with a number of similar metallo-organic compounds that fluoresce in different colors; at least one of which, named ReAsH has been used for correlative electron microscopy (17). Additionally, the 4Cys-FlAsH complexes remain stable during SDS-PAGE, so fluorescently labeled proteins can be detected via phospho-imager or simply using a UV box. Finally, FlAsH can also be linked to columns for protein purification. Importantly, the 4Cys-FlAsH method has been used for fluorescent labeling of proteins in living eukaryotic cells in reducing environments; however multiple groups have shown that it can also be applied for specific protein labeling inside living bacteria (18–21). The 4Cys-FlAsH method has been used to monitor the folding of proteins within living bacteria and to study the secretion of T3S effectors through T3S needle apparatuses. Here, we will provide a detailed protocol for the usage of the 4Cys-FlAsH labeling method to study T3S, presenting a method for fluorescent labeling of T3S effectors within living bacteria without functional loss.

2. Materials

2.1. Generation of Effector Proteins Fused to the Tetracysteine Sequence

2.1.1. Fusing Tetracysteine Sequences with Bacterial Effector Proteins

1. Annealing buffer: 100 mM potassium acetate, 30 mM HEPES-KOH pH=7.4, 2 mM magnesium-acetate
2. TAE buffer : 40 mM Tris-acetate, 1 mM EDTA
3. TAE loading buffer: 0.25%(w/v) bromophenol blue, 0.25%(w/v) xylene cyanol FF, 30% (v/v) glycerol in H_2O
4. Ethidium bromide

2.1.2. Bacteria

1. *E. coli strain*: DH5α
2. *Shigella* strains: M90T (22), *ipaB*(23), *ipaC*(23), *mxiD*(24)

2.1.3. Bacterial Culture

1. Trypticase casein soy broth medium (TCSB): casein (pancreatic digest) 17 g/L, soya peptone (papaic digest) 3 g/L, sodium chloride 5 g/L, dipotassium phosphate 2.5 g/L, dextrose 2.5 g/L
2. TCSB agar
3. LB medium : Bacto tryptone 1%, Bacto yeast extract 1%, NaCl 0.5%
4. LB-agar : Bacto tryptone 1%, Bacto yeast extract 1%, NaCl 0.5%, agar 1.5%

2.1.4. Kits

1. Rapid DNA Ligation kit (Roche)
2. Gel-extraction kit (Qiagen)
3. DNA miniprep (Qiagen)
4. PCR purification kit (Qiagen)

2.2. Labeling the 4Cys Effector Proteins in Living Shigella

1. FlAsH : Lumio Green (Invitrogen)
2. TAMRA : tetramethyl rhodamine (Molecular probes)
3. Arabinose
4. TCEP-HCl
5. Poly-(L-lysine)
6. Congo red
7. PBS

8. SDS-PAGE running buffer: 25 mM Tris, 192 mM glycine, 0.1% SDS, pH8.3
9. SDS-PAGE loading buffer: 250 mM TrisHCl pH6.8, 10% SDS, 30% glycerol, 20 mM TCEP, 0.02% bromophenol blue
10. 10% SDS-gel composition: 30% acrylamide mix, 1.5 m tris (pH 8.8), 10% SDS, 10% ammonium persulfate, temed 0.04%, 40% H_2O

2.3. Cell Culture, Infection with Bacteria Containing Fluorescently Labeled Effectors, and Microscopy

1. Dulbecco's Modified Eagle Medium (DMEM)
2. EM medium: 120 mM NaCl, 1.8 mM $CaCl_2$, 0.8 mM $MgCl_2$, 5 mM glucose, 7 mM KCl, 25 mM HEPES pH = 7.3
3. FBS
4. Cell line: HeLa ATCC CCL-2
5. Fixed samples are monitored using a Zeiss confocal microscope (Zeiss LSM510 Meta) with a 63x Apochromat 1.4 objective.
6. Secretion experiments are performed on an inverted Zeiss microscope attached to a spinning disc set-up and an AndorIQ EMCCD camera (see text and notes for further details). A 63x Apochromat 1.4 objective is used.

2.4. Data Processing and Analysis

1. ImageJ 1.40 (download: http://rsbweb.nih.gov/ij /)
2. Excel 2000 (Microsoft)

3. Methods

3.1. Generation of Effector Proteins Fused to the 4Cys Sequence

Generating chimeras between the 4Cys sequence and an effector protein of interest requires the addition of the 4Cys encoding sequence into a specific vector backbone. We have had good experience with the optimized 4Cys sequence that contains short flanking amino acid stretches (underlined below) to improve the 4Cys-FlAsH complex stability (25). This sequence consists of the amino acids AGS<u>FLN</u>**CCPGCC**<u>MEP</u>GGR encoded by the nucleotide sequence GCGGGCAGCTTTCTGAACTGCTGCCCGGGCTGCTGCATGGAACCGGGCGGCCGT. Multiple optimized flanking regions have been described (25). For the expression of *Shigella* effectors, we use the pBAD18 vector because of its tight regulation of protein expression (26). The 4Cys tag can also be multimerized to increase the fluorescence signals (*see* also **Note 1**).

1. The 4Cys sequence is inserted into the XbaI and HindIII sites of the pBAD18 vector to generate a vector named pBAD18-4Cys. Two complementary oligonucleotide sequences encoding the 4Cys sequence and containing the proper overhangs of XbaI and HindIII must be annealed, phosphorylated, and ligated into the linearized and dephosphorylated pBAD18 vector. The overhangs for the forward 4Cys sequence are 5′-CTAGA-3′ at the 5′-end, and TAAA at the 3′-end. For the reverse primer with the inverse complementary 4Cys sequence, the overhangs are 5′-AGCTTTTA-3′ at the 5′-end and 5′-T-3′ at the 3′-end.
2. 1 nmol/ml of each primer is heated in annealing buffer at 95°C for 5 min, then they are transferred into another heating block at 70°C for 10 min and subsequently the metal inlay of the heating block with the samples are removed to cool the samples slowly to room temperature. Following this, 1 μl of the annealed primers are phosphorylated in a 10 μl reaction volume using 1 μl T4 polynucleotide kinase (stock 10 U/μl) in the supplied phosphorylation buffer supplemented with 1 mM ATP at 37°C for 30 min. Afterwards, the enzyme is heat-inactivated at 70°C for 10 min, and the phosphorylated annealed oligonucleotides can be stored at −20°C.
3. 1 μg of pBAD18 vector is linearized with 4 μl of XbaI and 4 μl of HindIII (stock: 10 U/μl each) in the corresponding reaction buffer in a 100 μl reaction volume at 37°C for 3 h, and afterwards the cleaved vector is dephosphorylated with 1 μl shrimp alkaline phosphatase (SAP stock: 1U/μl) for 30 min. The SAP is heat-inactivated at 65°C for 10 min, and the entire reaction volume is then loaded on a 1% agarose gel (containing ethidium bromide as fluorescent marker for the migrating DNA) to separate the cleaved vector from the remaining short insert. The fluorescent band is cut out of the gel, and purified using a commercial gel-purification kit (Qiagen) following the manufacturer's instructions. 3 pmol of the cleaved dephosphorylated vector are ligated to 10 pmol of the phosphorylated 4Cys sequence stretch in a 20 μl volume using the rapid ligation kit (Roche) following the manufacturer's instructions. Subsequently, the ligated vector is transformed into chemically competent bacteria (*E. coli*, strain DH5α) by mixing 2 μl of the ligation reaction with 20 μl competent *E. coli*, keeping the reaction for 30 min at 4°C, performing a heat-shock at 42°C for 1 min, putting the reaction back on ice for 2 min, adding 1 ml of LB-medium at room temperature, and shaking the bacteria at 37°C for 90 min. 100 μl of the bacteria broth are then plated on LB-agarose plates containing ampicillin (final concentration

50 μg/ml) as selective marker. The following day, clones are selected and grown, plasmids are isolated by DNA preparation (Miniprep, Qiagen), and sequences are checked by sequencing.

4. Chimeras are generated by inserting *Shigella* effector sequences into the pBAD18-4Cys vector upstream of the 4Cys sequence. Particular care must be taken that these sequences are in frame with the 4Cys sequence (*see* **Note 2**). The molecular biology procedures are similar to those described above under point 3. The insert is not generated via primer annealing, but via PCR using a proofreading polymerase, such as Pfu (Stratagene) or Kod (Novagen) following the manufacturer's protocols (Stratagene or Novagen). The effector-4Cys fusions obtained are analyzed by DNA sequencing. The primers used for generating the effector PCR products contain restriction sites that can be used in conjunction with the sites present in the pBAD18-4Cys vector. For example, the primers used for amplification of the *Shigella* effector IpaB contain the sites NheI and XbaI. After the PCR, the PCR products are purified using a PCR purification kit (Qiagen), and then digested (see point 3 above). The plasmids encoding the chimeras should be checked by sequencing.

5. The 4Cys-effector chimeras are introduced into the corresponding *Shigella flexneri* strains via electroporation (voltage 2.2 V, resistance 200 Ω, capacity 25 μFd). Electrocompetent *Shigella* are obtained by growing a 1:100 dilution of an overnight culture in TCSB shaking vigorously at 37°C to an OD_{600}= 0.3. The volume depends on the number of competent bacteria that are required. 400 ml cultures can easily be handled by one person. At OD_{600}= 0.3, bacterial cultures are transferred on ice and are left for 20 min to cool down. Cultures are spun at 8000 g at 4°C for 8 min, and the bacterial pellets are washed twice in 100 ml ice-cold sterile, distilled H_2O with centrifuge spins as indicated above. Between the washes it is recommended to leave the bacterial cultures on ice for 20 min. Afterwards, the pellets are resuspended in 10 ml ice cold H_2O, and spun once. Finally, the bacterial pellets are resuspended in 2 ml H_2O supplemented with 10% glycerol, and stocks of 50 μl bacteria are stored at −80°C. These aliquots are used for electroporation (*see* **Note 3**).

3.2. Labeling the Effector-4Cys Proteins in Living Shigella

1. *Shigella* strains containing vectors with the effector-4Cys fusions are inoculated at 1:100 dilutions from overnight cultures in 8 ml TCSB supplemented with ampicillin (final concentration 50 μg/ml) and grown at 37°C in a heated shaker.

After 30 min of shaking, protein expression of the effector-4Cys chimeras is induced adding arabinose to a final concentration of 0.1% (*see* **Note 4**).

2. Bacteria are grown at 37°C to an OD_{600}=0.6, and 500 μl aliquots are centrifuged at 10,000 g at room temperature for 2 min (*see* **Note 5**). The supernatants are discarded, and replaced with 500 μl preheated (37°C) FlAsH staining solution consisting of TCSB plus 5 μM FlAsH, 10 μM TCEP-HCl, and 0.1% arabinose (around pH 7–7.5) (*see* **Note 6** and **Note 7**). Due to the photosensitivity of the reagents, the bacteria are grown in 14 ml Falcon tubes protected with aluminum foil at 37°C for 90 min.
3. The bacteria containing the fluorescently labeled effectors are centrifuged at 10,000 g, and then washed twice with PBS at room temperature. If bacterial invasion is required, *Shigella* are then coated with poly-(L-lysine) by incubating them at room temperature in 500 μl PBS supplemented with 10 μg/ml poly-(L-lysine). Afterwards, the bacteria are washed again two times in PBS and re-suspended in an appropriate volume of EM medium for host cell challenge. This volume can be calculated by the multiplicity of infection (MOI), and we recommend performing the experiments at a MOI of 50 bacteria per host cell.

3.3. Testing the Secretion Properties of Effector-4Cys Chimeras

It is very important to perform rigorous controls to test the capacity of the 4Cys-FlAsH labeled T3S effectors to be secreted via the T3S system to execute their functions within host cells. Certain regions of the T3S effectors contain functional domains, for example, the N-terminus is important for effector secretion (for more details, *see* **Note 8**).

1. Effector secretion must be checked for the effector-4Cys chimeras that have been fluorescently labeled with the FlAsH compound. Bacteria from overnight cultures are inoculated at 1:100 in TCSB. The effector-4Cys fusions are induced and fluorescently labeled as described in **Section 3.2**.
2. Subsequently, the bacteria are washed three times with PBS at room temperature in order to eliminate unbound FlAsH. Then, T3S is induced by the addition of congo red at a final concentration of 0.01% in the presence of 10 μM TCEP at 37°C for 60 min.
3. Secretion supernatants are separated from the bacteria by centrifugation at 13,000 g for 15 min (**Note 9**).
4. Subsequently, whole bacterial lysates and supernatants are prepared for SDS-PAGE using a sample buffer that contains 20 mM TCEP as reducing agent to protect the 4Cys-FlAsH complexes (*see* **Note 10**).

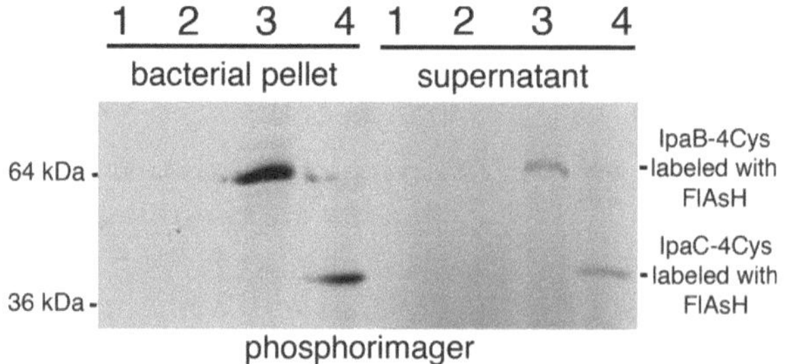

Fig. 14.1. Secretion of fluorescently labeled T3S effectors by phosphoimager analysis. Wild type *Shigella* (*lane 1*, M90T), a mutant that does not contain the virulence plasmid (*lane 2*, BS176), an *ipaB* mutant complemented with the pBAD18-IpaB-4Cys plasmid (*lane 3*), and an *ipaC* mutant complemented with the pBAD18-IpaC-4Cys (*lane 4*) are labeled with the FlAsH component after inducing protein expression with arabinose. The secretion of the effectors is induced by adding congo red to the growth medium, and supernatants are separated from the bacterial pellets after 60 min of effector secretion. Whole protein extracts are prepared for SDS-PAGE and the gels are scanned by a phosphoimager. The supernatants cannot be precipitated to concentrate the secreted effectors because this would disrupt the 4Cys-FlAsH complexes. Therefore, the fluorescent signals appear weaker in the supernatant. The specificity of the FlAsH labeling is illustrated by the absence of additional "contaminating" bands in the gel.

5. Gels are scanned with a multipurpose imager at 520 nm (λ_{Ex}= 488 nm). Typical results are shown in **Fig. 14.1**.

3.4. Cell Culture, Infection with Bacteria Containing Fluorescently Labeled Effectors, and Microscopy

1. Two days before the live cell experiment HeLa cells are seeded onto 35 mm glass-bottom dishes (Mattek) at a cell density of 2×10^5 cells per dish in 2 ml DMEM supplemented with 10% FBS.
2. Two days before the live cell experiment *Shigella flexneri* strains containing the 4Cys-effector chimeras are spread on TCSB plates supplemented with 0.05% congo red, and incubated overnight at 37°C. The next day clones are picked and transferred into 8 ml TCSB containing the appropriate antibiotic for overnight growth. This culture is then processed as described in **Section 3.2** (fluorescent labeling of 4Cys-effector chimeras). It is also possible to use other bacteria containing FlAsH labeled effector-4Cys fusions (*see* **Note 11**).
3. Before the experiment, HeLa cells are washed twice in PBS, and then incubated in EM buffer supplemented with (1 μg/ml) TAMRA for 10 min to label the cellular surface (*see* **Note 12**). Subsequently, the cells are washed twice with PBS, and 1 ml EM buffer (37°C) is added for the experiment (*see* **Note 13**). The cells are then placed in a temperature control box (TempControlII) on an inverted microscope (Zeiss 200 M) connected to a spinning disc confocal set up.

4. Imaging is performed using the following set-up. An inverted Zeiss 200 M microscope is connected to a Nipkow disc (Yokogawa CSU22) and a highly sensitive back illuminated CCD camera (DV885, Andor). As light source, the microscope is equipped with three diode pumped solid state lasers at wavelengths of 405 nm, 488 nm, and 560 nm. For tracking secretion in real time, only the 488 nm and 560 nm laser lines are required for monitoring the fluorescently labeled effetor-4Cys fusions and the host cellular membranes. Three-dimensional acquisition is achieved with a piezzo mono-objective. All elements of this set-up are driven by the AndorRevolution software for image acquistion. All images are taken through a 63x Apochromat 1.4 objective, and 1 × 1 binning (*see* **Note 14**).
4. Fluorescently labeled bacteria are added, and images are acquired in three dimensions in the red and green channel. For oversampling images are taken with a distance of 200 nm with the entire volume spanning 6–8 μm. Exposure is adjusted for each effector-4Cys fusion using secretion-deficient mxiD *Shigella* transformed with the effector-4Cys fusions to determine parameters with minimal photobleaching and photo-toxicity. Typically, three-dimensional stacks are measured every 20–40 s for an experimental length of 1 h. These parameters are controlled via the AndorRevolution software.

3.5. Data Analysis

1. The multi-dimensional data-sets obtained on the spinning disc confocal microscope are saved and converted into TIFF format for further processing using the program ImageJ (see **Fig. 14.2** for *Shigella* secreting FlAsH labeled IpaB-4Cys).
2. A simple analysis projects the 3D plus time image series to 2D plus time image series using the plugin "Grouped Z Projector" via maximum intensity projections.
3. Bacteria can be detected either automatically via the "Particle Tracking" function, or via following individual bacteria manually.
4. The normalized fluorescence intensities are calculated using the following equation: $F_{\text{t, normalized}} = 100 \times ((F_{\text{t, raw}} - F_{\text{background}})/(F_{\text{t}} = {}_{0,\text{ raw}} - F_{\text{background}})) \times$ bleachfactor, with the bleachfactor $= (F_{\text{t}} = 0,{}_{\text{no secretion}}/F_{\text{t, no secretion}})$.
5. The normalized fluorescence data-sets obtained are subsequently plotted using Excel (Microsoft). The position in the 3D volume of each bacterium that is classified as a secreting bacterium has to be controlled during the image acquisition. This can be done manually using the raw data sets (as has been done by us), or it can be followed using sophisticated

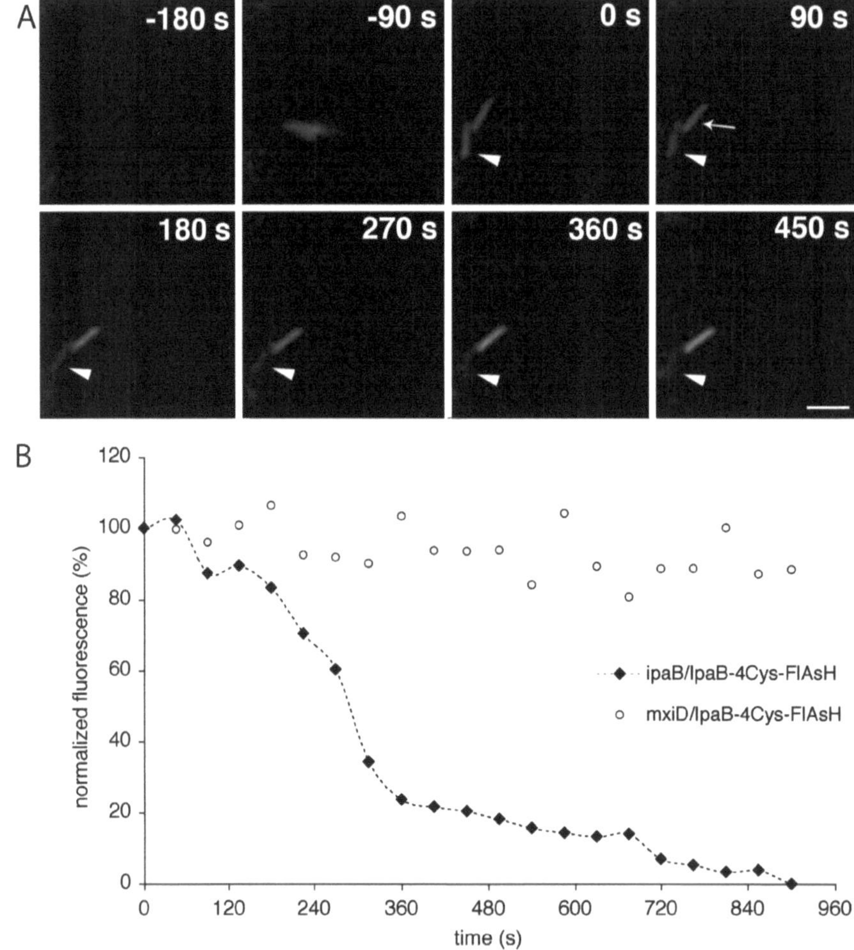

Fig. 14.2. Secretion of IpaB-4Cys labeled with FlAsH from single *Shigella* bacteria in real time. (**A**) Maximum intensity projections of 3D + time data-sets of the *ipaB* strain complemented with fluorescently labeled IpaB-4Cys. Two bacteria are shown; however only one of them (*arrowhead*) is in contact with a host cell. The intrabacterially stored fluorescent effectors are secreted and the bacterium looses its fluorescence. The bacterium that is not in contact with the host cell (*arrow*) remains fluorescent during the time course because it is not secreting its effectors. Scale bar: 5 μm. (**B**) Quantification of the effector secretion from single bacteria. The secretion-deficient strain *mxiD* cannot secrete fluorescently labeled effectors; therefore, the normalized fluorescence intensity remains unaltered during the measured time course (see text for details about the analysis). The secretion of FlAsH-labeled IpaB-4Cys occurs rapidly after host cellular contact. The entire pool of intrabacterially stored effectors is secreted in about 10 min after host cell contact.

image analysis algorithms (collaboration with Christophe Zimmer at Institut Pasteur, unpublished data).

6. Importantly, this analysis must be performed using secretion defective mutants that express effector-4Cys chimeras labeled with FlAsH. In the case of *Shigella*, this is the mxiD mutant expressing either IpaB-4Cys or IpaC-4Cys. The microscope settings must be adjusted to guarantee minimal or optimally, no bleaching. Consequently, these microscope settings can be used for all other experiments.

4. Notes

1. Multimerizing the 4Cys sequence has been described recently (21). This report employed a triple 4Cys sequence. Using this "enhanced 4Cys tag" for effector labeling, their detection was facilitated within the host cytoplasm after translocation. This approach is very promising to study effectors that are expressed at low levels and to study the function of the effectors after translocation.
2. Several aspects have to be considered generating the effector-4Cys fusions: (1) Most T3S effectors contain signal sequences at their N-terminus, therefore the 4Cys sequences should not be introduced here. (2) Linker sequences can be introduced between the 4Cys tag and the protein of interest to render the tag more flexible. This is important if the tagged terminus of the effector exerts a specific function. We have successfully used linkers made of GGGSGGG. (3) When effectors cannot be tagged at their C-terminus without functional loss, it is possible to introduce the 4Cys tag within the effector, for example we have used this strategy for IpaC (19).
3. A critical step for the generation of competent *Shigella* is the temperature. After cooling down the bacterial cultures to 4°C, it is very important to maintain this temperature. We recommend performing all subsequent steps on ice in a cold room.
4. The amount of arabinose for induction of protein expression using the pBAD vectors depends on the bacteria used and the protein that is to be expressed. This can be optimized for different experiments using different bacteria.
5. *Shigella* express the T3S system when growing at 37°C; however secretion is only triggered upon induction, for example, via contact with host cells. Importantly, bacteria loose their capacity to invade cells efficiently and to secrete their effectors when they are placed at temperatures below room temperature. Even short exposures to cool temperatures (e.g., 4°C) prevent efficient bacterial invasion, and should be avoided during the preparation of the bacteria. Therefore, all media and solutions used for staining and washing should be at a temperature between room temperature and 37°C.
6. Adding TCEP-HCl to the staining solution may change the pH. Therefore, it is important to adjust the pH of the staining solution before adding the bacteria. This can be

done by adding small amounts of NaOH until a neutral pH is reached.

7. The final concentration of the FlAsH component has to be adjusted for each effector protein. The concentration has to be chosen to minimize functional loss and to achieve strong labeling. We have labeled different effectors at FlAsH concentrations between 5 μM and 0.5 μM. Then, functional tests have to be performed for the different concentrations (see the following note).
8. We usually test secretion of *Shigella* effector/translocators IpaB and IpaC as described in the text. Furthermore, we control bacterial invasion, measuring actin foci and intracellular propagation by gentamicin protection assays (*see* (19) for details). For other effectors or different bacteria, other functional tests must be used.
9. An additional step can be introduced if the bacteria cannot be spun down completely. In such a case, the supernatants can be filtered (pore size below 100 nm) before continuing with the protocol.
10. Exchanging β-mercaptoethanol for TCEP is very important for the preparation of FlAsH labeled effectors for SDS-PAGE (17). Furthermore, the protein samples should be only heated at 70°C for 10 min before gel loading. Also, the 4Cys-FlAsH complexes are sensitive to low pH. Therefore, TCA precipitations cannot be used to concentrate the protein samples before SDS-PAGE.
11. Recently, the 4Cys-FlAsH labeling procedure has been described in *Salmonella* (21). We have also labeled effectors from Enteropathogenic *E. coli* (EPEC), and we have followed their secretion upon host cellular contact (unpublished data).
12. Obviously, it is possible to label the host cells with other fluorophores or to use host cells expressing fluorescently labeled proteins. We have combined the 4Cys-FlAsH labeling with fluorescent labeling of specific host factors to correlate effector secretion with the induced host responses (unpublished data).
13. It is very important that the washing buffers are heated to 37°C because changes in temperature affect the focus of the samples under the microscope. To avoid a focus drift, all media, samples, and bacterial cultures should be kept at 37°C before mounting them on the stage of the microscope.
14. Setting up the acquisition conditions properly is of highest importance to obtain images that can be quantified.

The major parameters are Laser power, exposure time, EM gain, and binning. These parameters must be set up for each machine. We recommend working with mutant strains with defunct T3S systems expressing fluorescently labeled effectors that remain within the bacteria during the time-course of acquisition. The parameters mentioned must be optimized to avoid bleaching and phototoxicity, but allow specific and robust detection of the 4Cys-FlAsH labeled effectors. For our microscope set-up, we use 24% power for the 488 nm laserline, 50 ms exposure, 140 EM gain, and a 1 × 1 binning and emission filters that allow parallel detection of the green and red fluorophores for measuring IpaB secretion. Of course, this has to be adjusted for other effectors.

Acknowledgements

We would like to thank Guy Tran Van Nhieu and Philippe Sansonetti for continuing support. Jost Enninga was supported by an HFSPO fellowship.

References

1. Cornelis, G. R. (2006). The type III secretion injectisome. *Nat Rev Microbiol*, **4**, 811–825.
2. Rosqvist, R., Magnusson, K. E., and Wolf-Watz, H. (1994). Target cell contact triggers expression and polarized transfer of Yersinia YopE cytotoxin into mammalian cells. *EMBO J*, **13**, 964–972.
3. Troisfontaines, P., and Cornelis, G. R. (2005). Type III secretion: more systems than you think. *Physiology (Bethesda)*, **20**, 326–339.
4. Pallen, M. J., Penn, C. W., and Chaudhuri, R. R. (2005). Bacterial flagellar diversity in the post-genomic era. *Trends Microbiol*, **13**, 143–149.
5. Journet, L., Agrain, C., Broz, P., and Cornelis, G. R. (2003). The needle length of bacterial injectisomes is determined by a molecular ruler. *Science*, **302**, 1757–1760.
6. Blocker, A. J., Deane, J. E., Veenendaal, A. K., Roversi, P., Hodgkinson, J. L., Johnson, S., and Lea, S. M. (2008). What's the point of the type III secretion system needle? *Proc Natl Acad Sci USA*, **105**, 6507–6513.
7. Woestyn, S., Allaoui, A., Wattiau, P., and Cornelis, G. R. (1994). YscN, the putative energizer of the Yersinia Yop secretion machinery. *J Bacteriol*, **176**, 1561–1569.
8. Galan, J. E. Energizing type III secretion machines: what is the fuel? *Nat Struct Mol Biol* 2008, 15, 127–128.
9. Sory, M. P., and Cornelis, G. R. (1994). Translocation of a hybrid YopE-adenylate cyclase from Yersinia enterocolitica into HeLa cells. *Mol Microbiol*, **14**, 583–594.
10. Day, J. B., Ferracci, F., and Plano, G. V. (2003). Translocation of YopE and YopN into eukaryotic cells by Yersinia pestis yopN, tyeA, sycN, yscB and lcrG deletion mutants measured using a phosphorylatable peptide tag and phosphospecific antibodies. *Mol Microbiol*, **47**, 807–823.
11. Briones, G., Hofreuter, D., and Galan, J. E. (2006). Cre reporter system to monitor the translocation of type III secreted proteins into host cells. *Infect Immun*, **74**, 1084–1090.
12. Charpentier, X., and Oswald, E. (2004). Identification of the secretion and translo-

cation domain of the enteropathogenic and enterohemorrhagic Escherichia coli effector Cif, using TEM-1 beta-lactamase as a new fluorescence-based reporter. *J Bacteriol*, **186**, 5486–5495.

13. Mills, E., Baruch, K., Charpentier, X., Kobi, S., and Rosenshine, I. (2008). Real-time analysis of effector translocation by the type III secretion system of enteropathogenic Escherichia coli. *Cell Host Microbe*, **3**, 104–113.
14. Schlumberger, M. C., Muller, A. J., Ehrbar, K., Winnen, B., Duss, I., Stecher, B., and Hardt, W. D. (2005). Real-time imaging of type III secretion: Salmonella SipA injection into host cells. *Proc Natl Acad Sci USA*, **102**, 12548–12553.
15. Griffin, B. A., Adams, S. R., and Tsien, R. Y. (1998). Specific covalent labeling of recombinant protein molecules inside live cells. *Science*, **281**, 269–272.
16. Gaietta, G., Deerinck, T. J., Adams, S. R., Bouwer, J., Tour, O., Laird, D. W., Sosinsky, G. E., Tsien, R. Y., and Ellisman, M. H. (2002). Multicolor and electron microscopic imaging of connexin trafficking. *Science*, **296**, 503–507.
17. Adams, S. R., Campbell, R. E., Gross, L. A., Martin, B. R., Walkup, G. K., Yao, Y., Llopis, J., and Tsien, R. Y. (2002). New biarsenical ligands and tetracysteine motifs for protein labeling in vitro and in vivo: synthesis and biological applications. *J Am Chem Soc*, **124**, 6063–6076.
18. Ignatova, Z., and Gierasch, L. M. (2004). Monitoring protein stability and aggregation in vivo by real-time fluorescent labeling. *Proc Natl Acad Sci USA*, **101**, 523–528.
19. Enninga, J., Mounier, J., Sansonetti, P., and Tran Van Nhieu, G. (2005). Secretion of type III effectors into host cells in real time. *Nat Methods*, **2**, 959–965.
20. Jaumouille, V., Francetic, O., Sansonetti, P. J., and Tran Van Nhieu, G. (2008). Cytoplasmic targeting of IpaC to the bacterial pole directs polar type III secretion in Shigella. *EMBO J*, **27**, 447–457.
21. Van Engelenburg, S. B., and Palmer, A. E. (2008). Quantification of real-time Salmonella effector type III secretion kinetics reveals differential secretion rates for SopE2 and SptP. *Chem Biol*, **15**, 619–628.
22. Clerc, P. L., Ryter, A., Mounier, J., and Sansonetti, P. J. (1987). Plasmid-mediated early killing of eucaryotic cells by Shigella flexneri as studied by infection of J774 macrophages. *Infect Immun*, **55**, 521–527.
23. Menard, R., Sansonetti, P. J., and Parsot, C. (1993). Nonpolar mutagenesis of the ipa genes defines IpaB, IpaC, and IpaD as effectors of Shigella flexneri entry into epithelial cells. *J Bacteriol*, **175**, 5899–5906.
24. Allaoui, A., Sansonetti, P. J., and Parsot, C. (1993). MxiD, an outer membrane protein necessary for the secretion of the Shigella flexneri lpa invasins. *Mol Microbiol*, **7**, 59–68.
25. Martin, B. R., Giepmans, B. N., Adams, S. R., and Tsien, R. Y. (2005). Mammalian cell-based optimization of the biarsenical-binding tetracysteine motif for improved fluorescence and affinity. *Nat Biotechnol*, **23**, 1308–1314.
26. Guzman, L. M., Belin, D., Carson, M. J., and Beckwith, J. (1995). Tight regulation, modulation, and high-level expression by vectors containing the arabinose PBAD promoter. *J Bacteriol*, **177**, 4121–4130.

Chapter 15

Comparative Analysis of Cytoplasmic Membrane Proteomes of *Escherichia coli* Using 2D Blue Native/SDS-PAGE

Susan Schlegel, Mirjam Klepsch, David Wickström, Samuel Wagner, and Jan-Willem de Gier

Abstract

Two-dimensional blue native (2D BN)/SDS-PAGE is the method of choice for the global analysis of the subunits of complexes in membrane proteomes. In the 1st dimension complexes are separated by BN-PAGE, and in the 2nd dimension their subunits are resolved by SDS-PAGE. The currently available protocols result in the distortion of the 1st dimension BN-gel lanes during their transfer to the 2nd dimension separation gels. This leads to low reproducibility and high variation of 2D BN/SDS-gels, making 2D BN/SDS-PAGE unsuitable for comparative analysis. Here, we present a 2D BN/SDS-PAGE protocol where the 1st dimension BN-gel is cast on a GelBond PAG film. Immobilization prevents distortion of BN-gel lanes when they are transferred to the 2nd dimension, which lowers variation and greatly improves reproducibility of 2D BN/SDS-gels. The use of 2D BN/SDS-PAGE with an immobilized first dimension is illustrated by the characterization of the cytoplasmic membrane proteome of *Escherichia coli* cells overexpressing cytochrome bo_3.

Key words: Membrane protein, membrane isolation, protein complex, two-dimensional blue native SDS-PAGE, comparative proteomics, *E. coli*.

1. Introduction

Two-dimensional blue native (2D BN)/SDS-PAGE was developed by Schägger and von Jagow (1) and has been widely used for the global analysis of the subunits of complexes in membrane proteomes [e.g., (2, 3)]. The 2D BN/SDS-PAGE technique in short: Isolated membranes are solubilized using a mild, non-ionic detergent. Protein complexes are negatively charged with coomassie

A. Economou (ed.), *Protein Secretion*, Methods in Molecular Biology 619,
DOI 10.1007/978-1-60327-412-8_15,

brilliant blue G-250 and separated according to their size in a gradient gel. The subunit composition of the complexes separated in the 1st dimension is subsequently analyzed in the 2nd dimension by SDS-PAGE. Unfortunately, the 2D-BN/SDS-PAGE protocols that are currently available are not suitable for comparative analysis since the first dimension BN-gel lanes get distorted during their transfer to the 2nd dimension SDS-gels (4). This distortion leads to low reproducibility and high variation between 2D BN/SDS-gels. Here, we present a 2D BN/SDS-PAGE protocol optimized for comparative analysis. In contrast to the protocols that have been published before [*see* e.g., (5, 6)], in our protocol the 1st dimension BN-gel is cast on a solid plastic support (i.e., a GelBond PAG film). Immobilization of the 1st dimension BN-gel prevents distortion during transfer of 1st dimension BN-gel lanes to the 2nd dimension SDS-gels. This greatly lowers variation between 2D BN/SDS-gels and allows using them for reliable comparative analysis. We have successfully used our 2D BN/SDS-PAGE protocol for the analysis of *Escherichia coli* inner membrane proteomes (7, 8, 9) and to create a reference map of the inner membrane proteome of *E. coli* BL21(DE3)pLysS (10). In this chapter we present our 2D BN/SDS-PAGE protocol where the 1st dimension is cast on a solid support. Its use is illustrated by the comparative analysis of the cytoplasmic membrane proteome from *E. coli* BL21(DE3)pLysS cells overexpressing the hetero-oligomeric membrane protein complex cytochrome *bo*3 (Cyt. *bo*3) with the one of control cells (i.e., harboring an empty expression vector).

2. Materials

2.1. Isolation of Cytoplasmic Membranes of E. coli and Sample Preparation

1. Pefabloc (100 mg/ml in ddH$_2$O) and DNase (5 mg/ml in ddH$_2$O), store in single use aliquots at −20°C and freshly add to buffer K.
2. Buffer K for cell resuspension: 50 mM triethanolamine (TEA), 250 mM sucrose, 1 mM ethylenediaminetetraacetic acid (EDTA), 1 mM 1,4-dithiotreitol (DTT), pH 7.5 with HAc. Pefabloc and DNase are added to final concentrations of 1 mg/ml and 5 μg/ml, respectively, before use.
3. Buffer 2x M: 100 mM TEA, 2 mM EDTA, 2 mM DTT, pH 7.5 with HAc.

4. 55% (w/w) sucrose solution in 2x M, all other sucrose solutions needed are subsequently prepared by diluting the 55% (w/w) stock solution in buffer 1x M.
5. Buffer L: 50 mM TEA, 250 mM sucrose, pH 7.5 with HAc.
6. Resuspension buffer: 750 mM 6-aminohexanoic acid (ACA), 50 mM Bis-Tris-HCl, pH 7.0.
7. BN Loading buffer (10x): 5% (w/v) coomassie brilliant blue G-250, 500 mM ACA, freshly prepared.
8. 10% (w/v) n-dodecyl-β-D-maltopyranoside (DDM) in ddH_2O, freshly prepared.

2.2. First Dimension BN-Gel Electrophoresis

1. GelBond PAG film (Lonza Rockland, Inc.).
2. Double sided scotch tape (tesa® Photo Film), Tesa AG (ordering number: 56663-02; *see* **Note 1**).
3. BN-gel buffer (3x): 1.5 M ACA, 150 mM Bis-Tris-HCl, pH 7.0, store at 4°C.
4. Duracrylamide: 30% T, 3% C (Genomic Solutions; *see* **Note 2**).
5. Ammonium persulfate (APS): 10% (w/v), freshly prepared.
6. Gel seal (e.g., from Amersham Biosciences)
7. Anode buffer (10x): 500 mM Bis-Tris-HCl, pH 7.0, store at 4°C.
8. Cathode buffer (10x): 500 mM tricine, 150 mM Bis-Tris, 0.2% (w/v) coomassie brilliant blue G-250. Do not adjust pH, store at 4°C.

2.3. Equilibration of 1st Dimension BN-Gel Strips

1. Equilibration buffer I: 2% (w/v) sodium dodecylsulfate (SDS), 5 mM tributyl-phosphine (TBP) (*see* **Note 3**).
2. Equilibration buffer II: 2% (w/v) SDS, 260 mM iodacetamide.

2.4. Second Dimension Gel Electrophoresis

1. Gel buffer: 3 M Tris-HCl, pH 8.45, 0.3% (w/v) SDS.
2. Duracrylamide: 30% T, 3% C (Genomic Solutions; *see* **Note 2**).
3. Glycerol (>98%).
4. Ammonium persulfate (APS): 10% (w/v), freshly prepared.
5. Anode buffer (10x): 2 M Tris-HCl (pH 8.9).
6. Cathode buffer (10x): 1 M Tris, 1 M tricine, 1% (w/v) SDS, do not adjust pH.
7. Low-melting agarose solution: 1% (w/v) low-melting agarose, 0.5% (w/v) SDS, few grains of bromphenol blue.

3. Methods

3.1. Isolation of Cytoplasmic Membranes from E. coli

1. Isolation of membranes is essentially done according to Wagner et al. (7). All steps during isolation (*see* **Section 3.1**) and sample preparation (*see* **Section 3.2**) should be carried out at 4°C or on ice and solutions should be chilled.
2. Harvest cells and wash the cell pellet with buffer K. We centrifuge the cells for 20 min at 12 krcf. Snap-freeze the cell pellet in liquid nitrogen and either thaw for immediate use or store at −80°C. Thawing and freezing facilitates breaking the *E. coli* cells in step 3.
3. Resuspend the pellet in buffer K (6 ml/1000 A_{600} of cells) freshly supplemented with 1 mg/ml Pefabloc and 5 μg/ml DNAse and break the cells by two cycles of French pressing (18,000 psi) (*see* **Note 4**).
4. Spin down the cell debris (16 krcf, 20 min).
5. Prepare a two-step sucrose gradient (bottom: 0.8 ml 55% (w/w) sucrose solution, top: 5 ml 8.8% (w/w) sucrose in buffer 1x M) in a 14 × 89 mm Ultra Clear^{TM} Centrifuge Tube (Beckman) and load the supernatant on top. If other tubes are used, adjust the volumes accordingly.
6. Spin for 2.5 h, 210 krcf. The membrane fraction (outer and inner membranes) will accumulate on top of the 55% sucrose solution.
7. Collect the membranes with a syringe and adjust the volume to 3.3 ml with buffer 1x M (*see* **Note 5**).
8. Prepare a six-step sucrose gradient (from the bottom to the top: 0.7 ml 55% (w/w) sucrose, 1.4 ml 50% (w/w) sucrose, 1.5 ml 45% (w/w) sucrose, 2.2 ml 40% (w/w) sucrose, 1.8 ml 35% (w/w) sucrose, 0.9 ml 30% (w/w) sucrose). Load the membrane fraction on top of the gradient and spin for 16 h at 220 krcf. Set the centrifuge to slow acceleration and do not use the brake.
9. Collect the cytoplasmic membrane fraction (*see* **Note 6**) and add buffer 1x M to double the volume. Transfer the fraction to 1.5 ml ultracentrifuge tubes and spin for 30 min at 265 krcf. Discard the supernatant.
10. Resuspend the membrane pellet in 100–200 μl of buffer L and determine the protein concentration. For protein concentration determinations, we use the BCA-Assay from Pierce. Adjust the volume with buffer L to the desired concentration (0.5–1 mg/ml of protein), add DTT to a final

concentration of 1 mM, and aliquot the samples. If not used immediately, the aliquots can be snap-frozen in liquid nitrogen and stored at −80°C (*see* **Note 7**).

3.2. Sample Preparation

1. Spin down the desired membrane quantities (corresponding to 100–150 μg of protein) at 186 krcf for 45 min and discard the supernatant. The amount of total protein needed may vary depending on the method you choose to visualize the protein spots after the 2nd dimension run (*see* **Note 8**).
2. Carefully resuspend the membrane pellet in resuspension buffer (85 μl per 100 μg of protein) by gentle pipetting.
3. Add 10% (w/v) DDM to a concentration of 0.56% (w/v) (5 μl per 100 μg of protein, final concentration including loading dye: 0.5% (w/v)) and incubate on ice for 15 min to solubilize the membranes. Gently vortex every 5 min (*see* **Note 9**).
4. Centrifuge at 100 krcf for 30 min to remove all non-resuspended material. Transfer the supernatant to a new tube containing 10 μl of BN loading buffer per 100 μg of protein.

3.3. Casting of the 1st Dimension BN-Gel

1. The following quantities/volumes apply to using a GE Healthcare SE660 standard vertical electrophoresis unit with 24 cm long glass plates and 1.0 mm thick spacers. Clean the glass plates and spacers carefully prior to use. Always wear gloves and keep equipment and solutions SDS-free.
2. Cut the GelBond PAG film (backing) approximately 1.5 mm smaller in width and 0.5 mm smaller in length than the glass plates (*see* **Note 10**).
3. Place the backing on one glass plate. Note that the hydrophilic size should be facing up (*see* **Note 11**).
4. Fix the bottom of the film using double-sided scotch tape (**Fig. 15.1A**). To improve adhesion of the backing to the glass plate, place a few droplets of water right above the scotch tape between the film and the glass. Use a soft, lint-free tissue to wipe carefully from the bottom to the top of the backing, thereby evenly distributing the water and removing air bubbles. This will result in tight adherence of the film to the glass plate.
5. Slightly grease the spacers with gel seal on the side that is facing the backing and assemble the gel cassette (**Fig. 15.1B**). You may also slightly grease the bottom of the glass plates to prevent leakage while casting the gel (*see* **step 7**).

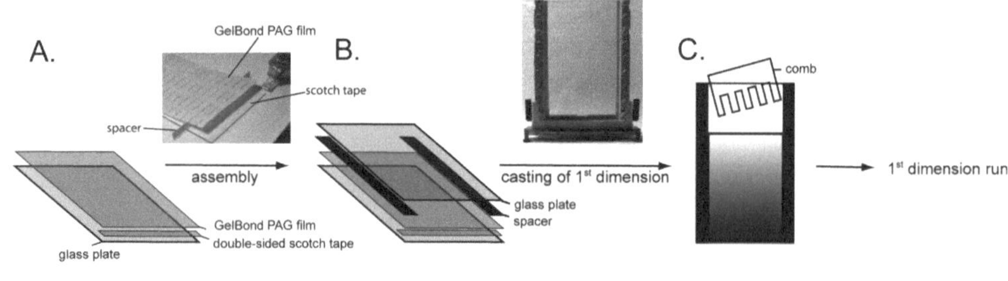

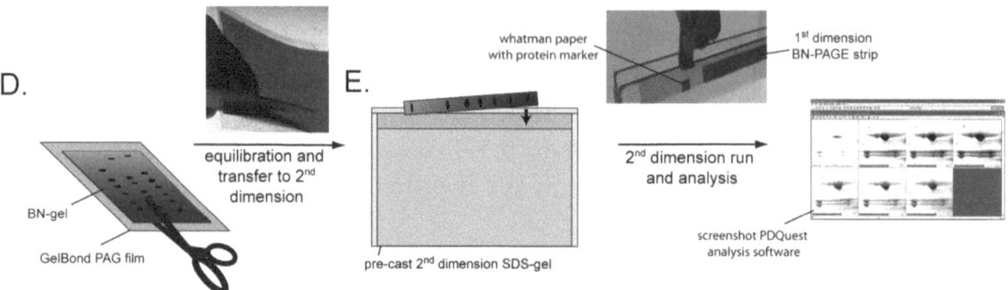

Fig. 15.1. Casting of the 1st dimension BN-gel on a GelBond PAG film, excising BN-gel lanes, and transferring the 1st dimension to the 2nd dimension. (**A**) Fixing the GelBond PAG film with double-sided scotch tape to the glass plate, (**B**) assembly of the gel cassette for the 1st dimension BN-gel, (**C**) to prevent the formation of air bubbles assemble the comb already in the cassette slightly tilted before casting the stacking gel, (**D**) BN-gel lanes are excised with scissors, (**E**) transfer of the equilibrated 1st dimension BN-PAGE lanes to the 2nd dimension SDS-gels. After the 2nd dimension run, proteins are visualized and gels can be used for comparative analysis.

6. Prepare the solutions for the 1st dimension BN gradient gel according to **Table 15.1**. A 5–14% separation gel is suitable to separate protein complexes between 66 and 1000 kDa (*see* **Note 12**).
7. Pour the separation gel and overlay with ddH_2O. Let the gel polymerize at room temperature. Complete polymerization will take approximately 6 h, preferably the gel should be allowed to polymerize for 10 h.
8. Prepare the solution for the stacking gel.
9. Carefully remove the water from the top of the gradient gel (for the removal of residual water use Whatman paper) and pour the stacking gel. To avoid occurrence of air bubbles in the bottom of the slots, place the comb slightly tilted between the glass plates and cast the gel (**Fig. 15.1C**). Allow the gel about 2 h to polymerize at room temperature. After polymerization, the gels can be stored up to 1 week if they are kept cold (4°C) and moist.

3.4. First Dimension Run

1. The 1st dimension run is performed in the cold room (4°C) with additional water cooling. Prepare 1x cathode buffer and 1x anode buffer prior to the run and cool them at 4°C.

Table 15.1
Recipe for the 1st dimension BN gradient gel

Ingredients	Separation gel 5%	Separation gel 14%	Stacking gel
ddH_2O	7.3 ml	0.5 ml	4.8 ml
3x BN-gel buffer	4.9 ml	4.9 ml	3.0 ml
Duracrylamide (30% T, 3% C)	2.4 ml	6.8 ml	1.2 ml
80% (w/v) Glycerol	–	2.4 ml	–
TEMED	7.5 µl	7.5 µl	4 µl
10% APS	73 µl	73 µl	40 µl

2. Fill the anode buffer compartment and assemble the gel into the tank (*see* **Note 13**).
3. Add DDM to a final concentration of 0.03% (w/v) to the cathode buffer. Fill only the wells of the gel with cathode buffer. This way you can easily load the protein samples.
4. Load the samples. Fill the remaining empty wells according to the amount of protein in the previously loaded samples. Per 100 µg of protein use a mixture of 75 µl resuspension buffer, 15 µl 10% (w/v) DDM, and 15 µl of BN loading buffer. For other protein amounts adjust the volumes accordingly. If you want to determine the approximate molecular weight of the protein complexes in the 1st dimension load an appropriate high molecular weight marker in one of the wells.
5. Carefully pour the remaining cathode buffer in the cathode buffer compartment.
6. Perform electrophoresis as follows: 5 mA/gel, start: 100 V for 2 h, o/n 250 V, finish run at 450 V.
7. Stop the run when about 1 cm of the dark blue coomassie front still remains in the gel and disassemble the gel cassette.
8. Remove the stacking gel and the dark blue coomassie front with a scraper. This will create short overhangs of the backing which later will facilitate handling of the gel strips. If necessary, overhangs can be trimmed to 4 mm.
9. Excise single BN-gel lanes from the BN-PAGE gel with scissors (**Fig. 15.1D**). You can seal the lanes in plastic for later use. They can be stored at –20°C or –80°C for up to 1 week or 2 months, respectively.

3.5. Casting of the 2nd Dimension Tricine-SDS-Gels

1. For the 2nd dimension run we use an EttanDalt Twelve Gel system (GE Healthcare) with glass cassettes for 1.5 mm thick gels (*see* **Note 14**). Clean the glass plates carefully and pre-cool solutions on ice.
2. Assemble the glass cassettes into the gel tank.
3. Mix 400 ml duracrylamide, 400 ml 2nd dimension gel buffer, 160 ml glycerol (>98%), and 240 ml ddH_2O, and stir the solution gently on ice for 0.5 h (avoid air bubbles).
4. Add 360 μl TEMED and 2.4 ml freshly prepared 10% (w/v) APS and cast the gels.
5. Overlay with EtOH and wrap in moist paper tissues. Allow polymerization at room temperature.
6. Disassemble the gel caster and remove polymerized acrylamide from the outside of the glass plates before use.

3.6. Equilibration of the 1st Dimension BN-Gel Strips and Transfer to the 2nd Dimension

1. Equilibrate the gel strips in equilibration buffer I for 15 min. The tributyl-phosphine will lead to reduction of cysteines. Subsequently, transfer the strips to Equilibration buffer II and incubate for another 15 min. Iodacetamide will cause alkylation of cysteine residues and thus prevent re-oxidation. Do not prolong the incubation times since this may cause small proteins to diffuse out of the gel.
2. Apply 5 μl of protein marker to small pieces of Whatman paper (4 × 4 mm) and allow them to dry. These will later be used in the 2nd dimension run (*see* **step 6**).
3. Pre-warm the Tricine-SDS-gels for about 5 min in a 56°C incubator and carefully remove residual water from the top using Whatman paper.
4. Heat the low-melting agarose solution in the microwave until it boils. Subsequently, leave it at room temperature for 10 min. Overlay the top of the Tricine-SDS-gels with the hot low-melting agarose solution.
5. Place the 1st dimension BN-PAGE strips on top of the agarose solution and carefully push them on top of the Tricine-SDS-gels using a spatula (**Fig. 15.1E**). Only push on the backing, never on the gel. Make sure that no air bubbles form underneath the 1st dimension strips, and always place the strips in the same orientation with respect to the silicon sealed side of the gel cassette. This way gels run in parallel can later be used for comparative analysis (*see* **Note 15**).
6. Place one piece of Whatman paper with marker (*see* **step 2**) next to the 1st dimension gel strip and let the agarose solidify.

3.7. Second Dimension Run and Gel Analysis

1. Assemble the gels in the EttanDalt Tank and fill up the respective buffer compartments with 1x anode buffer and 1x cathode buffer (*see* **Note 16**).
2. Run the gels at 4°C. Start electrophoresis at 30 V until the proteins have completely entered the separation gel. Then switch to 80 V or 130 V. In total the electrophoresis will take approximately 2 days (80 V) or 30 h (130 V). Note that 130 V is only recommended when running not more than 8 gels at the same time.

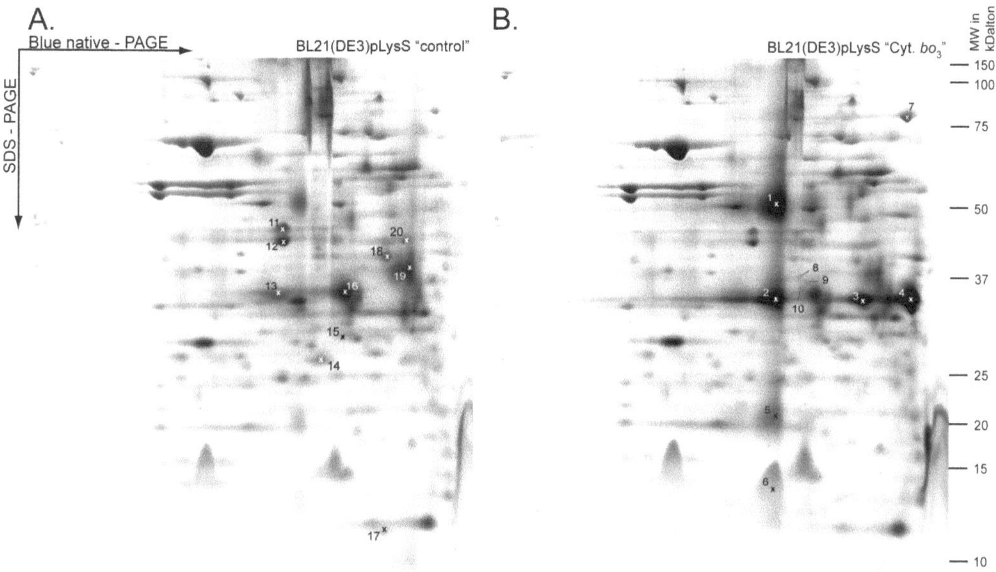

Fig. 15.2. Comparative analysis of cytoplasmic membrane proteomes of *E. coli* by 2D BN/SDS-PAGE with an immobilized 1st dimension. The hetero-tetrameric Cyt. bo_3 complex is one of the two terminal oxidases in *E. coli.* It is encoded by the *cyoABCDE* operon, where *cyoA*, *cyoB*, *cyoC*, and *cyoD* encode the 4 structural subunits of the Cyt. bo_3 complex. The protein CyoE is assumed to assist in the assembly of Cyt. bo_3 and is not part of the complex. From here on, the overexpression of the *cyoABCDE* operon will be referred to as overexpression of Cyt. bo_3. Cyt. bo_3 was overexpressed in BL21(DE3)pLysS as described before [*see* (7, 15)]. Cytoplasmic membranes of cells overexpressing Cyt. bo_3 and of control cells (i.e., cells with an empty expression vector) were isolated as described in this protocol. In the 1st dimension, 150 μg of protein were separated. Independent triplicates of cytoplasmic membrane samples were analyzed by 2D BN/SDS-PAGE. Proteins were visualized by colloidal coomassie staining, and differences between replicate groups were analyzed using PDQuest (Bio-Rad) essentially as described before [*see* (7)]. Differential protein accumulation levels were analyzed using the Student's t-test with a 98% level of confidence and a threshold for quantitative differences of at least twofold (including "on" and "off" responses) upon Cyt. bo_3 overexpression. Proteins were identified by MS and/or matching with a reference map [for the reference map *see* (10)]. In total 123 protein spots could be identified, representing 104 different proteins. Of the 123 protein spots, 8 contained more than 1 protein and were therefore excluded from the analysis. Proteins with an at least twofold increased accumulation level upon Cyt. bo_3 overexpression are indicated in (**B**): CyoB (1), CyoA (2;3;4), CyoC (5), CyoD (6), DnaK (7), DegS (8), HldD (9) and YadG (10). Proteins 1, 2, 5, and 6 are subunits of the tetrameric Cyt. bo_3complex. Proteins with an at least twofold decreased accumulation level upon Cyt. bo_3overexpression are indicated in (**A**): MalK (11), MalF (12), ArnC (13), OppB (14), FrdB (15), PtkC (16), YjcH (17), ArtP (18), PutP (19) and DacA/C (20). As for the overexpression of single membrane proteins (*see* (7)] the main effects upon overexpression of Cyt. bo_3 were an increased accumulation level of DnaK and lowered accumulation levels of several cytoplasmic membrane proteins. Overall, the effects observed upon the overexpression of Cyt. bo_3 did not notably differ from the effects of the overexpression of single membrane proteins [(7, 9)].

3. Stop electrophoresis when the bromphenol blue front (upper blue front) is about 1–2 cm away from the bottom of the gels. A typical example of the described process is illustrated by the comparative analysis of the cytoplasmic membrane proteome from *E. coli* BL21(DE3)pLysS cells overexpressing the hetero-oligomeric membrane protein complex cytochrome bo_3 (Cyt. bo_3) (**Fig. 15.2**).
4. Disassemble the gel cassettes. Visualize the protein spots and analyze the gels with the methods of your choice (*see* **Note 17**).

4. Notes

1. To fix the backing to the glass plate (*see* **Section 3.3.4.**) we use double sided scotch tape from Tesa (http://www.tesatape.com/). Other tapes tested were either hard to remove after the 1st dimension run and/or did not endure the electrophoresis.
2. We use duracrylamide rather than acrylamide for both the 1st and the 2nd dimension gels. Duracrylamide based gels are less prone to breaking and therefore easier to handle (11). To obtain 3% crosslinking add 0.12 g of N′N ′-Methylene-bis-acrylamide per 100 ml of duracrylamide (30% T, 2.6% C).
3. TBP can ignite spontaneously when exposed to oxygen and the resulting gas as well as the chemical itself are both toxic. Therefore, extreme care should be taken when preparing, storing, and using the 200 mM TBP solution. Always wear gloves, a lab coat, and safety goggles, and work in a fume hood on a glassplate (never use paper to cover the bottom) when you work with TBP solution. Both the TBP solution and the isopropanol used to prepare the 200 mM TBP solution should be overlaid with an inert gas (we use nitrogen) using a sealing rubber cap and a syringe system. Spilled TBP-droplets should be wiped away immediately using a paper towel soaked in water. We frequently exchange the gas on top of the 200 mM TBP solution for new nitrogen to prevent enrichment of oxygen and avoid underpressure.
4. Usually two cycles of French pressing are sufficient to break *E. coli* cells, especially when using strains that express T7 lysozyme (e.g., BL21(DE3)pLysS/E. However, depending on the strain additional cycles and/or

addition of lysozyme prior to French pressing may be necessary.

5. Be careful to collect the membranes without imbibing too much of the 55% sucrose solution. Otherwise the density of the membrane containing sucrose solution in the end may be too high and lead to improper separation of the membranes in the second gradient.
6. The cytoplasmic membrane fractions accumulate between 30 and 35% (w/w) sucrose, whereas the outer membrane fractions accumulate on top of the 55% (w/w) sucrose solution. In case cells produce aggregates, these will accumulate just below the outer membrane fraction. In the gradient the cytoplasmic membrane fraction can easily be distinguished from the outer membrane fraction (white color) due to their yellow/orange shade. If you have problems to see the cytoplasmic membrane fraction enhance the contrast by positioning a piece of black paper behind the gradient. We strongly recommend to mark the different sucrose layers on the tube before centrifugation
7. To make sure that the amounts of protein in the different membrane protein samples are properly adjusted, i.e., there are equal amounts of protein in each sample, you can separate approximately 15–30 μg of proteins (after determining and adjusting the protein content in your sample, *see* **Section 3.1.10.**) using standard SDS-gel electrophoresis. Stain the gel after electrophoresis and compare the total intensities in the different lanes.
8. We visualize proteins in a 2D-BN/SDS-PAGE gel using colloidal coomassie (14). If you use another staining technique, keep in mind that the sensitivity of the method of your choice can differ from the sensitivity of colloidal coomassie. Therefore, you may have to adopt the amount of protein used in the 1st dimension BN-PAGE. If you want to identify your protein spots using mass spectrometry (MS) make sure your staining method is MS compatible.
9. The non-ionic detergent DDM at a final concentration of 0.5% (w/v) gives the best results in our hands. However, depending on, e.g., the type of membrane or proteins/complexes to be analyzed optimization of solubilization, i.e., screening different detergents and concentrations, will be necessary [*see* e.g., (12)].
10. If the film is cut too small, this will lead to leakage of buffer from the anode buffer compartment to the cathode buffer compartment during the electrophoresis run. If the size of the film exceeds the size of the glass plate, in addition to causing leakage of buffer from the anode buffer

compartment the film will bulge, thereby leading to irregularities in gel thickness.

11. To distinguish between the hydrophilic and hydrophobic side you may place a few droplets of water on a part of each side of the backing that is not used for crosslinking. Before you assemble the gel cassette carefully remove the droplets with a lint-free, soft tissue.
12. By varying the gradient, BN-PAGE can be used to separate proteins/protein complexes with masses between 10 and 10,000 kDa [*see* e.g., (1)].
13. Since Bis-Tris is rather expensive we re-use the anode buffer up to 3 times. Store it at 4°C and keep it SDS-free.
14. If you use another system make sure that the spacers you use for the 1st dimension BN-gel are thinner than the ones used for the 2nd dimension SDS-gels. This will enable you to fit the 1st dimension BN-PAGE-strips between the glass plates of the 2nd dimension SDS-gels.
15. After placing the 1st dimension BN-PAGE-strip on the 2nd dimension gel you may observe air bubbles on top of the strip. These may cause disturbances of the current flow thereby leading to irregularities in protein separation. To remove air bubbles, carefully overlay the top of the gel with hot low melting agarose.
16. We use 1x concentrated cathode buffer to run up to eight gels in parallel. It is our experience that when more than eight 1.5 mm thick gels are run at the same time, proteins in the upper part of the 2nd dimension gel are not properly resolved any longer. This is most likely due to depletion of ions in the cathode buffer (13). We therefore use 2x concentrated cathode buffer when running more than eight gels (1.5 mm thick) at the same time.
17. We routinely stain the proteins in the gels with colloidal coomassie (14), scan the gels with a GS-800 densitometer from Bio-Rad, and analyze the gels with the PDQuest software from Bio-Rad. Of course one may use other staining/visualization techniques, scanning equipment, and software for the analysis.

Acknowledgements

This research was supported by grants from the Swedish Research Council and The Center for Biomembrane Research, which is supported by the Swedish Foundation for Strategic Research.

References

1. Schägger, H., and von Jagow, G. (1991) Blue native electrophoresis for isolation of membrane protein complexes in enzymatically active form. *Anal Biochem* **10**, 223–231.
2. McKenzie, M., Lazarou, M., Thorburn, D.R., and Ryan, M. T. (2007) Analysis of mitochondrial subunit assembly into respiratory chain complexes using Blue Native polyacrylamide gel electrophoresis. *Anal Biochem* **364**, 128–137.
3. Stenberg, F., Chovanec, P., Maslen, S. L., Robinson, C. V., Ilag, L. L., von Heijne, G., Daley, D. O. (2005) Protein Complexes of the *Escherichia coli* Cell Envelope. *J Biol Chem* **280**, 34409–34419.
4. Brookes, P. S., Pinner, A., Ramachandran, A., Coward, L., Barnes, S., Kim, H., and Darley-Usmar, V.M. (2002) High throughput two-dimensional blue native electrophoresis: a tool for functional proteomics of mitochondria and signaling complexes. *Proteomics* **2**, 969–977.
5. Wittig, I., Braun, H. P., and Schägger, H. (2006) Blue native PAGE. *Nat Protoc* **1**, 418–428.
6. Schamel, W. W. A. (2008) Two-dimensional blue native polyacrylamide gel elektrophoresis. *Curr Protoc Cell Biol*, Chapter 6:Unit 6.10.
7. Wagner, S., Baars, L., Ytterberg, A. J., Klussmeier, A., Wagner, C. S., Nord, O., Nygren, P. A., van Wijk, K. J., and de Gier, J.-W. (2007) Consequences of membrane protein overexpression in *Escherichia coli*. *Mol Cell Proteomics* **6**, 1527–1550.
8. Baars, L., Wagner, S., Wickström, D., Klepsch, M., Ytterberg, A.J., van Wijk, K. J., and de Gier, J.-W. (2007) Effects of SecE depletion on the inner and outer membrane proteomes of *Escherichia coli*. *J Bacteriol* **190**, 3505–3525.
9. Wagner, S., Klepsch, M., Schlegel, S., Appel, A., Draheim, R., Tarry, M., Högbom, M., van Wijk, K. J., Slotboom, D. J., Persson, J. O., and de Gier, J. W. (2008) Tuning *Escherichia coli* for membrane protein overexpression. *PNAS* **105**, 14371–14376.
10. Klepsch, M. M., Schlegel, S., Wickström, D., Friso, G., van Wijk, K. J., Persson, J.-O., de Gier, J.-W., and Wagner, S. (2008) Immobilization of the first dimension in 2D Blue Native/SDS-PAGE allows the relative quantification of membrane proteomes. *Methods* **46**, 48–53.
11. Patton, W. F., Lopez, M. F., Barry, P., and Skea, W. M. (1992) A mechanically strong matrix for protein electrophoresis with enhanced silver staining properties. *Biotechniques* **12**, 580–585.
12. Reisinger, V., and Eichacker, L. A. (2008) Solubilization of membrane protein complexes for blue native PAGE. *J Proteomics* **171**, 277–283.
13. Werner, W. E. (2003) Run parameters affecting protein patterns from second dimension electrophoresis gels. *Anal Biochem 317*, 280–283.
14. Neuhoff, V., Arold, N., Taube, D., and Ehrhardt, W. (1988) Improved staining of proteins in polyacrylamide gels including isoelectric focusing gels with clear background at nanogram sensitivity using Coomassie Brilliant Blue G-250 and R-250. *Electrophoresis* **6**, 255–262.
15. Stenberg, F., von Heijne, G., and Daley, D. O. (2007) Assembly of the Cytochrome *bo3* Complex. *J Mol Biol* **371**, 765–773.

Chapter 16

Using Hidden Markov Models to Discover New Protein Transport Machines

Vladimir A. Likic, Pavel Dolezal, Nermin Celik, Michael Dagley, and Trevor Lithgow

Abstract

Protein import and export pathways are driven by protein translocases, often comprised of multiple subunits, and usually conserved across a range of organisms. Protein import into mitochondria is fundamental to eukaryotic organisms and is initiated when substrate proteins are translocated across the mitochondrial outer membrane by the TOM complex. The essential subunit of this complex is a protein called Tom40, which is probably a β-barrel in structure and serves as the translocation pore. We describe a hidden Markov model search designed to find the Tom40 sequence in the amoeba *Entamoeba histolytica*. This organism has a highly reduced "mitosome", an organelle whose relationship to mitochondria has been the subject of controversy. The Tom40 sequence could not be found with BLAST-based searches, but a hidden Markov model search identified a likely candidate to form the protein import pore in the outer mitosomal membrane in *E. histolytica*.

Key words: Hidden Markov models, sequence database search, protein translocases, mitochondria, TOM complexes, Tom40.

1. Introduction

Computational approaches are first line tools for the assignment of function to protein sequences identified from genome sequence data. Widely available tools such as BLAST (1) allow searches of this type to be widely deployed by non-specialists in conjunction with comprehensive sequence databases such as EMBL (2) or UniProt (3). This allows for biological insights, such as how proteins are translocated across membranes, to

A. Economou (ed.), *Protein Secretion*, Methods in Molecular Biology 619,
DOI 10.1007/978-1-60327-412-8_16, © Springer Science+Business Media, LLC 2010

be extended from knowledge gained in experimentally tractable systems such as the yeast *Saccharomyces cerevisiae* or bacterium *Escherichia coli*.

Underpinning the sequence database search is the problem of sequence comparison: to find similar sequences in the database one needs to compare every sequence in the database to the sequence at hand. Sequence comparison, in turn, implies the sequence alignment problem. In order to compare two sequences in a consistent and a non-biased way one first needs to best align the two sequences. The best alignment of two sequences throughout their entire length is achieved with the Needleman-Wunsch algorithm (4) (global alignment), while the best alignment of two sequences involving only portions of the two sequences is achieved with the Smith-Waterman algorithm (5) (local alignment).

The biological complexity associated with proteins is high and databases therefore contain a wide variety of sequences in terms of their properties, total length, etc. The Smith-Waterman algorithm, which is the appropriate means for such a database search, is relatively computationally expensive. For this reason routine searches of large sequence databases (such as GenBank at NCBI) typically rely on BLAST, a heuristic algorithm that approximates the results of the Smith-Waterman algorithm. In contrast to the Smith-Waterman algorithm BLAST does not guarantee to find the best match; however in most practical situations BLAST is accurate enough while it provides significant computational savings relative to the Smith-Waterman algorithm.

Similarly as the application of the Smith-Waterman algorithm for database search, BLAST is inherently a single sequence search: a given query sequence is compared to each of the sequences in the database with the aim of finding similar sequences. In practice one may have a family of related sequences at hand, and would like to identify additional homologous sequences within a given database. A typical example of this is searching a newly sequenced genome for a member of a known family of proteins. While BLAST can potentially be used to find the new member, its effectiveness is limited by the degree of pairwise sequence similarity between a single given query sequence and the target sought. In many biological scenarios, such as searches across large phylogenetic distances or searches involving organisms from cryptic environments where evolutionary pressure is high, this can be a critical limitation.

Here we demonstrate the use of hidden Markov models to search the *E. histolytica* genome for proteins from the Tom40 family. Tom40 is the channel through which imported proteins cross the mitochondrial outer membrane (6–8). Tom40 is predicted to be a β-barrel protein (9, 10), and thereby likely to be derived from an ancestral protein of bacterial origins. Tom40 has been found in a vast range of eukaryotes, leading to the suggestion that it was a fundamental component of the original protein import system installed in protomitochondria (11). The amoeba *E. histolytica* has a highly reduced compartment called a mitosome, an organelle whose relationship to mitochondria has been the subject of some controversy (12, 13). For some time, *E. histolytica* had been considered an "amitochondriate" organism, in large part because BLAST-based searches do not yield homologues of typical mitochondrial proteins like Tom40 from the genome sequence data of *E. histolytica*.

Figure 16.1 shows a portion of the multiple sequence alignment of known Tom40 proteins. The alignment shows islands of conservation among the non-conserved regions and insertions. This is typical in a family of divergent but homologous sequences: due to evolutionary pressures, mutations in functionally critical regions are not tolerated very well, while mutations in regions not essential for function may abound given sufficient evolutionary distance. A set of homologous sequences, such as the one shown in **Fig. 16.1**, provides significantly more information about inherent features of the protein family compared to any single sequence from the family taken in isolation. For example, the multiple sequence alignment shows which amino acid positions are relatively conserved and the positions where insertions and deletions are more frequent. Leveraging this information can be critical for detecting evolutionary distant members of the family. This is because the alignment of a protein family conveys the information as to which amino acid positions should match for a highly significant hits (e.g. the position 190 in **Fig. 16.1** where G is highly conserved), and also positions that are probably not significant, that is where sequence variations or gaps in the alignment are evident. For conserved positions, one can estimate both the degree and the nature of conservation from a multiple sequence alignment: for example, in the position 190 in **Fig. 16.1**, the amino acid G would be highly significant, increasing the likelihood that the sequence is a true member of the family. However, one can also see that a true member of the family can tolerate S, F, or I in this position.

The information inherent in the family of sequences can be captured by the so-called "'profile'" methods that convert a multiple sequence alignment into a position specific scoring matrix, which in turn allows a more sensitive database search (14). A more advanced method for capturing features inherent in a family of sequences is based on hidden Markov models (15). Hidden Markov models (HMMs) are general statistical models for certain types of pattern recognition problems, widely used in speech recognition for example. The HMM variants used to capture the inherent properties of a family of protein sequences are called profile HMMs. A profile HMM is "trained" on an aligned set of

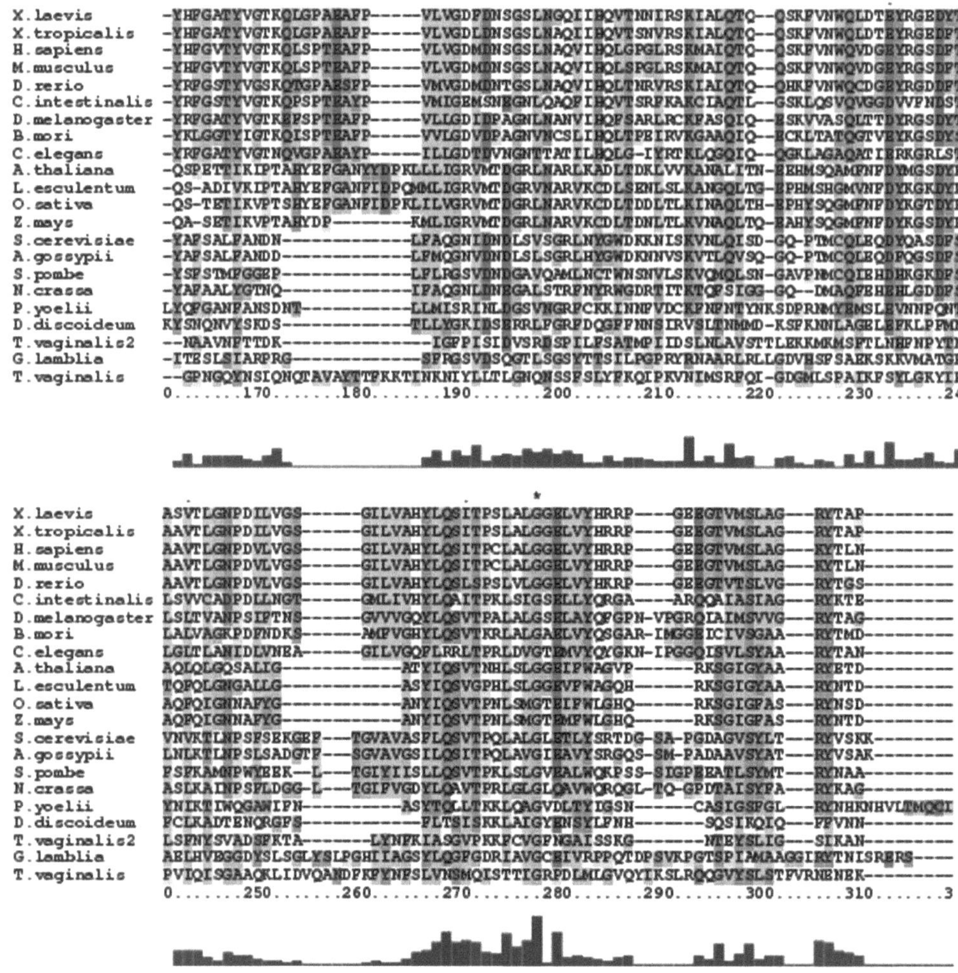

Fig. 16.1. A portion of the multiple sequence alignment of Tom40 sequences used in this work (22 known Tom40 sequences were used in the alignment). The segment shows residue positions 160–318.

sequences, and subsequently any sequence from the database can be scored against the HMM to give the probability of it belonging to the family.

Figure 16.2 shows a simplified example of a hidden Markov model illustrating how a HMM is set up and trained on a family of sequences (in this case hypothetical DNA sequences are used for simplicity). Given an arbitrary sequence the trained HMM can give the probability that a sequence belongs to the family. The sequence of states shown in **Fig. 16.2b** is a Markov chain in mathematical terminology because the probability of the next state depends only on the present state and does not depend on the past states. The Markov model is "hidden" because the states are not observed directly, only the residues that the states generate are observed directly (*see* the caption of **Fig. 16.2**).

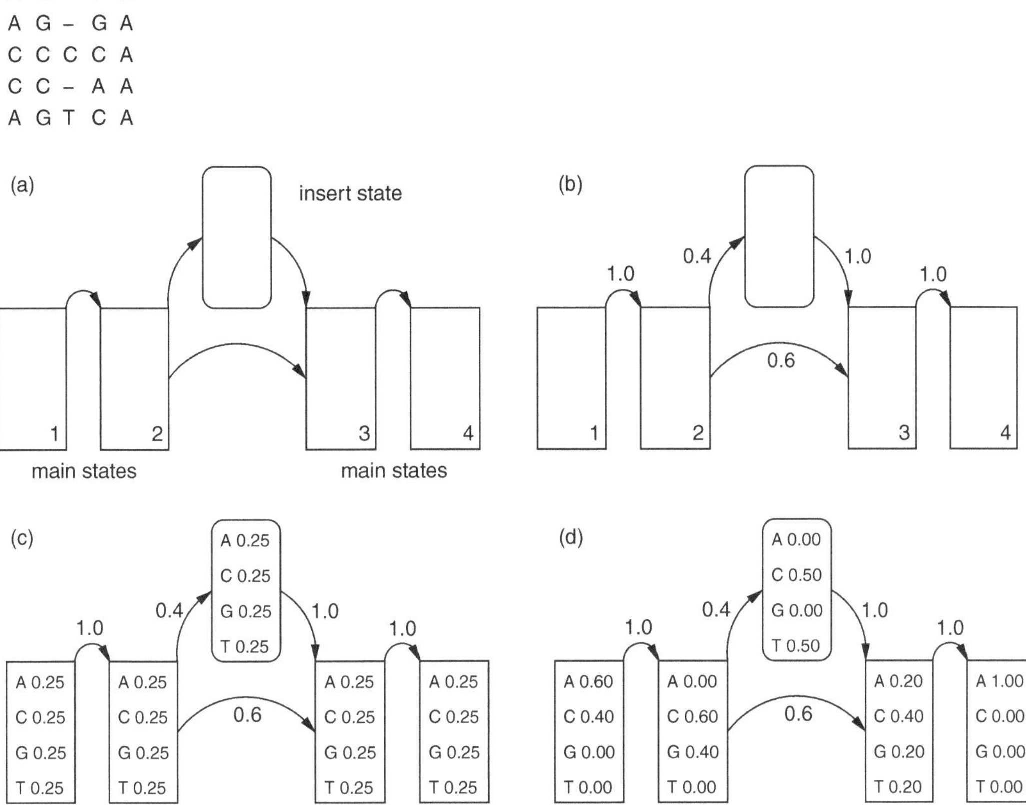

Fig. 16.2. A simplified explanation of how hidden Markov models are set up. For simplicity the DNA sequence "family" was used, where each sequence has either 4 or 5 residues. The best alignment of sequences is shown in the top left corner. **Panel (a)**: From the alignment the states of the HMM are constructed, which involve four main states and one insert state that models the residue insertion at position three. **Panel (b)**: The probabilities for state transitions are set from the multiple sequence alignment. The probability for an insertion of a residue at position 3, modeled by an insert state, is set to 0.4. This is deduced from the multiple sequence alignment, where there is two residue inserted in five sequences ($2/5 = 0.4$). Therefore the probability of direct transition from the main state 2 to the main state 3 is $1–0.4 = 0.6$. **Panel (c)**: The residue probabilities are initialized. There are four residues (A, T, G and C) and the initial probability for each is set to $1/4 = 0.25$. **Panel (d)**: The residue probabilities are trained based on the multiple sequence alignment and the HMM topology. The residue probabilities are set based on the residue counts in each position of the sequence alignment. For example, in position 1 the residue A occurs three times out of five; therefore the probability for A is $3/5 = 0.60$. The model in panel (**d**) can be used to predict the probability of *any* 4 or 5 letter DNA sequence. For example, the sequence 'AGTA' would have the probability of 0.6 (residue A in state 1) $\times$ 0.4 (residue G in state 2) $\times$ 0.6 (direct transition to state 3) $\times$ 0.2 (residue T in state 3) $\times$ 1.0 (residue A in state 4) $= 0.028$ (*see* **Note 3**). The sequence 'CCTTA' would have the probability of 0.0096. Therefore the sequence 'AGTA' is a better match to this model than the sequence 'CCTTA'.

The example shown in **Fig. 16.2** is grossly simplified, and profile HMMs useful in practice are significantly more complex. However freely available software for the application of HMMs can be used to shield the user from many details and one does not need to master the theory behind HMMs in full detail to use them effectively. In the example presented here we consider the family of 23 known Tom40 protein sequences, and ask the question: Is there a Tom40 in the *E. histolytica* genome (*Eh*Tom40)? A BLAST search of the *E. histolytica* genome with any individual sequence from the Tom40 training set failed to return a reasonable Tom40 candidate. Here we describe step-by-step the protocol used to build the Tom40 HMM, and demonstrate the ability of the resulting model to delineate the *Eh*Tom40 candidate protein import channel in *E. histolytica*.

2. Materials

2.1. Setup and Notation

1. *The computer system.* The HMM searches described here was performed on the computer running Red Hat Linux 5. The Red Hat 5 installation was default, with specific bioinformatics programs installed in addition, as described below.
2. *Conventions.* Program names, file and directory (folder) names are written in single quotes. The computer terminal outputs are written in `Courier font`. In Unix, folders are commonly called directories; throughout the text the term "folders" will be used. In the main text, folder names will be appended with a forward slash (but not always in the computer terminal output, which is copied verbatim). For each set of commands executed on the computer screen it is assumed that the starting folder is 'workspace/'.
3. *Folder for the project.* For the purpose of the search examples a folder was created named 'workspace/', with the absolute path '/home/workspace/' (*Note: in all examples below, replace this with your own path*). Four additional sub-folders were created in the folder 'workspace/':

```
$ cd workspace
$ mkdir clustalw hmmer search ehist
```

The purpose of these folders is as follows:

- 'workspace/' – overall workspace for this project
- 'clustalw/' – installation folder for the multiple sequence alignment program Clustal-w
- 'hmmer/' – installation folder for the HMMER software package
- 'search/' – to contain Tom40 sequences, HMMs and search outputs
- 'ehist/' – folder for the *E. histolytica* predicted hypothetical proteins

2.2. Additional Bioinformatics Programs

1. Clustal-w installation (the program for multiple sequence alignment). The program Clustal-w was installed in the folder '/home/workspace/clustalw/. Clustal-w was downloaded from the FTP site 'ftp://ftp.ebi.ac.uk/pub/software/clustalw2' as the file 'clustalw-2.0.9-linux-i386-libcppstatic.tar.gz'. This file was placed in the folder 'clustalw/', and unpacked as follows:

```
$ cd /home/workspace/clustalw
$ ls
clustalw-2.0.9-linux-i386-libcppstatic.tar.gz
$ tar xvfz clustalw-2.0.9-linux-i386-libcppstatic.tar.gz
```

   The last command created the folder 'clustalw-2.0.9-linux-i386-libcppstatic/' which contained the executable 'clustalw2'. For convenience this file was moved into the folder 'workspace/clustalw', and the original 'clustalw-2.0.9-linux-i386-libcppstatic/' folder was removed:

```
$ mv clustalw-2.0.9-linux-i386-libcppstatic/* .
$ rm -rf clustalw-2.0.9-linux-i386-libcppstatic
$ ls
clustalw2 clustalw_help
```

2. HMMER installation (Sean Eddy's program for Hidden Markov Models search (15)). The software package HMMER was downloaded from 'http://hmmer.janelia.org/'; as the file 'hmmer-2.3.2.tar.gz'. This file was placed in the folder '/home/workspace/hmmer/'. The installation:

```
$ cd workspace/hmmer
$ ls
hmmer-2.3.2.tar.gz
$ tar xvfz hmmer-2.3.2.tar.gz
$ ls -CF
hmmer-2.3.2/ hmmer-2.3.2.tar.gz
$ cd hmmer-2.3.2
$ ./configure --prefix=/home/workspace/hmmer
[---output deleted---]
$ make
[---output deleted--]
$ make install
```

   With this several HMMER components were installed in '/home/workspace/hmmer/':

```
$ cd /home/workspace/hmmer
$ ls -CF
bin/ hmmer-2.3.2/ hmmer-2.3.2.tar.gz man/
```

The HMMER executables were installed in 'bin/' (note that HMMER consists of several programs):

```
$ ls -CF bin
hmmalign* hmmcalibrate* hmmemit* hmmindex* hmmsearch*
hmmbuild* hmmconvert* hmmfetch* hmmpfam*
```

2.3. Downloading the E. histolytica Predicted Proteins

The *E. histolytica* conceptual proteome was downloaded from 'ftp.tigr.org' as follows:

```
$ cd /home/workspace/ehist
$ ftp ftp.tigr.org
Connected to www.tigr.org.
220 JCVI FTP Server
Name: anonymous
331 Anonymous login ok, send your complete email
   address as your password.
Password:**********
[---output deleted---]
ftp> cd pub/data/Eukaryotic_Projects/e_histolytica/
  annotation_dbs
ftp> get EHA1.pep
ftp> quit
$ ls
EHA1.pep
```

3. Methods

All commands described in this section are performed in the folder '/home/workspace/search/'.

3.1. Preparation of Sequences for Hidden Markov Model Search

1. *Preparation of Tom40 model sequences.* Known Tom40 protein sequences were collected into a single file in preparations for building of the Tom40 HMM. The sequence file was named 'Tom40.fas', and contained sequences in the FASTA format. In FASTA format each sequence starts with the comment, designated with '>' as the first character, and followed by the comment until the end of the line. The actual sequence starts in the line following the comment. The sequence is given in one letter code, until the new sequence comments is reached (indicated by another '>' character, as first in the line) or the end of file. Our Tom40 training set contained 23 sequences of known Tom40 proteins from different organisms. The snippet of this file, showing only the first three sequences, is given below (*see* **Note 1**):

```
>C.intestinalis Tom40
MGNAHAASWGWSSSTPAETAATPPPVEAPPPVVPVEPLPPSSPVDATPVHSKTATNSVGT
FEEIHKPCKDIALQPFEGLRFIVNKGLSSHFQAQHTVHLNNEGSSYRFGSTYVGTKQPSP
TEAYPVMIGEMSNEGNLQAQFIHQVTSRFKAKCIAQTLGSKLQSVQVGGDVVFNDSTLSV
VCADPDLLNGTGMLIVHYLQAITPKLSIGSELLYQRGAARQQAIASIAGRYKTENWQAAG
TIAAGGMHASFYRKANENVQVGVELEASLKNKESVTTFAYQMDLPKMNLLFKGMLTSEWT
IGSALEKRLQPLPITLNLTGTYNIKKDKVAVGIGAVLG
>A.thaliana Tom40
MADLLPPLTAAQVDAKTKVDEKVDYSNLPSPVPYEELHREALMSLKSDNFEGLRFDFTRA
LNQKFSLSHSVMMGPTEVPAQSPETTIKIPTAHYEFGANYYDPKLLLIGRVMTDGRLNAR
LKADLTDKLVVKANALITNEEHMSQAMFNFDYMGSDYRAQLQLGQSALIGATYIQSVTNH
LSLGGEIFWAGVPRKSGIGYAARYETDKMVASGQVASTGAVVMNYVQKISDKVSLATDFM
YNYFSRDVTASVGYDYMLRQARVRGKIDSNGVASALLEERLSMGLNFLLSAELDHKKKDY
KFGFGLTVG
>O.sativa Tom40
MGSAASAAAPPPPPTAQPHMAAPPYGAGLAGILPPKPDGEEEGKKKEVEKVDYLNLPCPV
PFEEIQREALMSLKPELFEGLRFDFTKGLNQKFSLSHSVFMGSLEVPSQSTETIKVPTSH
YEFGANFIDPKLILVGRVMTDGRLNARVKCDLTDDLTLKINAQLTHEPHYSQGMFNFDYK
GTDYRAQFQIGNNAFYGANYIQSVTPNLSMGTEIFWLGHQRKSGIGFASRYNSDKMVGTL
QVASTGIVALSYVQKVSEKVSLASDFMYNHMSRDVTSSFGYDYMLRQCRLRGKFDSNGVV
AAYLEERLNMGVNFLLSAEIDHSKKNYKFGFGMTVGE
[---other entries deleted---]
```

2. Building the Tom40 multiple sequence alignment. The building of the hidden Markov model requires sequences to be aligned in the regions of similarity. The sequence alignment can be achieved with different programs, such as Clustal-w and T-COFFEE. In this example we use Clustal-w. The input to the program Clustal-w is the set of unaligned sequences (in this case the 'Tom40.fas' FASTA file), and the output is multiple sequence alignment:

```
$ cd /home/workspace/search
$ ls
Tom40.fas
$ ../clustalw/clustalw2 -outfile=Tom40.gcg
     -output=gcg -infile=Tom40.fas
  CLUSTAL 2.0.9 Multiple Sequence Alignments
Sequence format is Pearson
Sequence 1: C.intestinalis  338 aa
Sequence 2: A.thaliana      309 aa
Sequence 3: O.sativa        337 aa
[---output deleted---]
$ ls
Tom40.dnd Tom40.fas Tom40.gcg
```

The above command executed the program 'clustalw2' (installed in the **section 2.2** step 1 above) in a non-interactive mode. In the command line we have specified that 'Tom40.fas' is the input file, 'Tom40.gcg' will be the output file with multiple sequence alignment (to be created), and that the output file should be in the GCG format. The above command has produced the alignment file 'Tom40.gcg' and also the dendrogram file 'Tom40.dnd'. The latter will not be used and can be deleted:

```
$ rm Tom40.dnd
$ ls
Tom40.fas Tom40.gcg
```

3.2. Building the Tom40 Hidden Markov Model

All commands described in this section are performed in the folder '/home/workspace/search/'.

1. *Building the hidden Markov model.* Building the Tom40 hidden Markov model requires two steps: creating the raw HMM and calibrating the HMM. The first step used the multiple sequence alignment in the input file ('Tom40.gcg') and produced a raw hidden Markov model file ('Tom40g.hmm'):

```
$ ../hmmer/bin/hmmbuild -n Tom40g Tom40g.hmm Tom40.gcg
hmmbuild - build a hidden Markov model from an alignment
HMMER 2.3.2 (Oct 2003)
Copyright (C) 1992-2003 HHMI/Washington University School of
  Medicine
Freely distributed under the GNU General Public License (GPL)
- - - - - - - - - - - - - - - - - - - - - - - - - - - - -
Alignment file:                 Tom40.gcg
File format:                    MSF
Search algorithm configuration: Multiple domain (hmmls)
Model construction strategy:    MAP (gapmax hint: 0.50)
Null model used:                (default)
Prior used:                     (default)
Sequence weighting method:      G/S/C tree weights
New HMM file:                   Tom40g.hmm
- - - - - - - - - - - - - - - - - - - - - - - - - - - - - -


Alignment:           #1
Number of sequences: 23
Number of columns:   478
Determining effective sequence number ... done. [14]
Weighting sequences heuristically     ... done.
Constructing model architecture       ... done.
Converting counts to probabilities    ... done.
Setting model name, etc.              ... done. [Tom40g]
Constructed a profile HMM (length 398)
Average score:       500.08 bits
Minimum score:       282.32 bits
Maximum score:       608.28 bits
Std. deviation:       99.39 bits
```

```
Finalizing model configuration ... done.
Saving model to file           ... done.
//
$ ls
Tom40.fas Tom40.gcg Tom40g.hmm
```

In the above command the argument '-n Tom40g' specified that this hidden Markov model will be called 'Tom40g' (this information is recorded internally in the model). The two arguments Tom40g.hmm and Tom40.gcg are the hidden Markov model file (to be produced) and the input multiple sequence alignment (*see* **Note 2**). This command takes a few seconds to execute on a modern computer.

2. *Calibration of the hidden Markov model.* The next step is to calibrate the hidden Markov model 'Tom40g.hmm'. This step is important to optimize the sensitivity of the hidden Markov model search. The empirical calibration is performed by fitting a distribution to the scores obtained from a Monte Carlo simulation. The calibration is performed with the HMMER program 'hmmcalibrate':

```
$ ../hmmer/bin/hmmcalibrate --num 10000 Tom40g.hmm
hmmcalibrate -- calibrate HMM search statistics
HMMER 2.3.2 (Oct 2003)
Copyright (C) 1992-2003 HHMI/Washington University School of Medicine
Freely distributed under the GNU General Public License (GPL)
- - - - - - - - - - - - - - - - - - - - - - - - - - - - - - - - - - --
HMM file:                 Tom40g.hmm
Length distribution mean: 325
Length distribution s.d.: 200
Number of samples:        10000
random seed:              1215572532
histogram(s) saved to:    [not saved]
- - - - - - - - - - - - - - - - - - - - - - - - - - - - - - - --
HMM    : Tom40g
mu     : -194.942474
lambda : 0.136655
max    : -146.145996//
```

The argument '- -num 10000' specifies the number of samples: 5000 is the default in HMMER version 2.3.2. This step is computationally expensive, and calibration of a single hidden Markov model may require a few minutes on a modern computer.

3.3. Running the Tom40 Hidden Markov Model Search

1. Tom40 hidden Markov model search of *E. histolytica* predicted proteins. The HMM search for a single hidden

Markov model is straightforward, performed with the program 'hmmsearch':

```
$ ../hmmer/bin/hmmsearch -E 0.1 Tom40g.hmm ../ehist/EHA1.pep > Tom40g.OUT
```

The command 'hmmsearch' can take several arguments, including the name of the hidden Markov model file ('Tom40g.hmm'), and the name of the sequence database for the search ('EHA1.pep' in this case). The optional argument '-E 0.1' specifies the E-value cutoff for reporting hits (the E-value is interpreted similarly as in BLAST searches). The lower E-value the more significant is the hit, and typically hits with E-values of 1 or larger are not significant. In the command above the output of the program 'hmmsearch' has been collected in the file 'Tom40g.OUT'. The *E. histolytica* conceptual proteome contains 9772 sequences. To run the Tom40 hidden Markov model search against this database on a Xeon 3.2 GHz CPU required one minute (*see* **Note 4**).

Inspection of the output file Tom40g.OUT shows one hit, with the E-value of 0.0039:

```
Sequence  Description                                        Score  E-value N
--------  -----------                                        ------ -------- --
38.m00236 hypothetical protein 38.t00034 AAFB01000158 74.0   0.0039  1
Parsed for domains:
Sequence   Domain   seq-f   seq-t   hmm-f    hmm-t   score   E-value
--------   -------  -----   -----   ------   -----   ------  -------
38.m00236  1/1      22      304 ..  1        398 [] -74.0    0.0039
```

This is the best Tom40 candidate in *E. histolytica* based on the training set of Tom40 sequences used in this study. The E-value shows that the similarity of the sequence '38.m00236' to the Tom40 model is well in the grey zone (a closely related sequence would have E-value of $<10^{-100}$), yet the observed level of similarity is not very likely to occur by chance. A visual inspection of the sequence '38.m00236' and a comparison with the training sequences showed some similarity typical to Tom40 throughout the entire sequence length, with some shortening relative to typical Tom40 sequences. Protein shortening is a general feature seen in many parasites, as a means of overall reduction in genome size, and makes pairwise (e.g. BLAST) sequence searches even less likely to succeed in scenarios such as identifying protein transport components in organisms like *E. histolytica*. Based on the clues provided by HMM searches such as the one documented here, the protein import machinery in the mitosomes of *E. histolytica* is being fully evaluated (16).

4. Notes

1. The full set of Tom40 sequences used in this work is available from the authors on request.
2. By default, the program 'hmmbuild' builds a model optimized for local comparison with respect to the sequence and global comparison with respect to the HMM. This is akin to the Needleman-Wunsch type sequence alignment, rather than the Smith-Waterman type sequence alignment. To build a model which is local with respect to the sequence and with respect to the HMM, use the -f switch (i.e. 'hmmbuild -f').
3. An immediately apparent limitation of the model shown in **Fig. 16.2** is that any residue in state 4 other than 'A' will generate zero probability for the entire sequence. In practice residue probabilities are not set to zero even for residues that do not feature in a given position, but to a small background value deduced from statistical reasoning.
4. The example chosen for this "Methods" paper demonstrates the use of HMM searches against a single genome dataset as such searches can be easily performed on modern computers in a short time period. It is also possible to interrogate far larger databases with HMMs, for example the entire UniProt database. Searches on very large databases can take days to complete on desktop workstations. As the rate of released genome sequence data continues to increase rapidly and exceeds the rate of advances in computational power, routine searches of the largest databases increasingly require access to specialized supercomputing facilities.

References

1. Altschul, S. F., Gish, W., Miller, W., Myers, E. W., and Lipman, D. J. (1990) Basic local alignment search tool. *J Mol Biol* **215**, 403–410.
2. Kulikova, T., Akhtar, R., Aldebert, P., Althorpe, N., Andersson, M., Baldwin, A., Bates, K., Bhattacharyya, S., Bower, L., Browne, P., Castro, M., Cochrane, G., Duggan, K., Eberhardt, R., Faruque, N., Hoad, G., Kanz, C., Lee, C., Leinonen, R., Lin, Q., Lombard, V., Lopez, R., Lorenc, D., McWilliam, H., Mukherjee, G., Nardone, F., Pastor, M. P., Plaister, S., Sobhany, S., Stoehr, P., Vaughan, R., Wu, D., Zhu, W., and Apweiler, R. (2007) EMBL Nucleotide Sequence Database in 2006. *Nucleic Acids Res* **35**, D16–20.
3. Wu, C. H., Apweiler, R., Bairoch, A., Natale, D. A., Barker, W. C., Boeckmann, B., Ferro, S., Gasteiger, E., Huang, H., Lopez, R., Magrane, M., Martin, M. J., Mazumder, R., O'Donovan, C., Redaschi, N., and Suzek, B. (2006) The Universal Protein Resource (UniProt): an expanding universe of protein information. *Nucleic Acids Res* **34**, D187–191.
4. Needleman, S. B., and Wunsch, C. D. (1970) A general method applicable to the search for similarities in the amino acid sequence of two proteins. *J Mol Biol* **48**, 443–453.

5. Smith, T. F., and Waterman, M. S. (1981) Identification of common molecular subsequences. *J Mol Biol* **147**, 195–197.
6. Vestweber, D., Brunner, J., Baker, A., and Schatz, G. (1989) A 42 K outer-membrane protein is a component of the yeast mitochondrial protein import site. *Nature* **341**, 205–209.
7. Model, K., Prinz, T., Ruiz, T., Radermacher, M., Krimmer, T., Kuhlbrandt, W., Pfanner, N., and Meisinger, C. (2002) Protein translocase of the outer mitochondrial membrane: role of import receptors in the structural organization of the TOM complex. *J Mol Biol* **316**, 657–666.
8. Kiebler, M., Pfaller, R., Sollner, T., Griffiths, G., Horstmann, H., Pfanner, N., and Neupert, W. (1990) Identification of a mitochondrial receptor complex required for recognition and membrane insertion of precursor proteins. *Nature* **348**, 610–616.
9. Gabriel, K., Buchanan, S. K., and Lithgow, T. (2001) The alpha and the beta: protein translocation across mitochondrial and plastid outer membranes. *Trends Biochem Sci* **26**, 36–40.
10. Hill, K., Model, K., Ryan, M. T., Dietmeier, K., Martin, F., Wagner, R., and Pfanner, N. (1998) Tom40 forms the hydrophilic channel of the mitochondrial import pore for preproteins. *Nature* **395**, 516–521.
11. Dolezal, P., Likic, V., Tachezy, J., and Lithgow, T. (2006) Evolution of the molecular machines for protein import into mitochondria. *Science* **313**, 314–318.
12. Tovar, J., Fischer, A., and Clark, C. G. (1999) The mitosome, a novel organelle related to mitochondria in the amitochondrial parasite Entamoeba histolytica. *Mol Microbiol* **32**, 1013–1021.
13. Mai, Z., Ghosh, S., Frisardi, M., Rosenthal, B., Rogers, R., and Samuelson, J. (1999) Hsp60 is targeted to a cryptic mitochondrion-derived organelle ("crypton") in the microaerophilic protozoan parasite Entamoeba histolytica. *Mol Cell Biol* **19**, 2198–2205.
14. Gribskov, M., McLachlan, A. D., and Eisenberg, D. (1987) Profile analysis: detection of distantly related proteins. *Proc Natl Acad Sci USA* **84**, 4355–4358.
15. Eddy, S. R. (1998) Profile hidden Markov models. *Bioinformatics* **14**, 755–763.
16. Dolezal, P., Dagley, M., Kono, M., Wolynec, P., Likic, V., Foo, J.H., Sedinova, M. Tachezy, J., Bachmann, A., Bruchhaus, I., and Lithgow, T. (under revision) The essentials of protein import in the degenerate mitochondrion of Entamoeba histolytica. *PLoS Pathogens* (under revision).

Chapter 17

Bioinformatics Predictions of Localization and Targeting

Shruti Rastogi and Burkhard Rost

Abstract

One of the major challenges in the post-genomic era with hundreds of genomes sequenced is the annotation of protein structure and function. Computational predictions of subcellular localization are an important step toward this end. The development of computational tools that predict targeting and localization has, therefore, been a very active area of research, in particular since the first release of the groundbreaking program PSORT in 1991. The most reliable means of annotating protein structure and function remains homology-based inference, i.e. the transfer of experimental annotations from one protein to its homologs. However, annotations about localization demonstrate how much can be gained from advanced machine learning: more proteins can be annotated more reliably. Contemporary computational tools for the annotation of protein targeting include automatic methods that mine the textual information from the biological literature and molecular biology databases. Some machine learning-based methods that accurately predict features of sorting signals and that use sequence-derived features to predict localization have reached remarkable levels of performance. Sustained prediction accuracy has increased by more than 30 percentage points over the last decade. Here, we review some of the most recent methods for the prediction of subcellular localization and protein targeting that contributed toward this breakthrough.

Key words: Protein subcellular localization prediction, sorting signals, neural networks, support vector machines, hidden Markov models, amino acid composition, text analysis.

Abbreviations

NLS	nuclear localization signal
NN	neural networks
HMM	hidden Markov model
SP	signal peptide
SVM	support vector machine.

A. Economou (ed.), *Protein Secretion*, Methods in Molecular Biology 619,
DOI 10.1007/978-1-60327-412-8_17,

1. Introduction

Proteins are the gene-products that constitute the fundamental functional components for the machinery of life. Almost one thousand organisms have been completely sequenced; the burning challenge now becomes the annotation of function and structure for most of those gene-products. Even for organisms as well studied as yeast or *E. coli*, most proteins remain experimentally uncharacterized, and it is only through computational biology that some relevant annotations can be obtained for almost half of all those proteins (1–7).

Knowing the native subcellular localization of a protein is often instrumental to the experimental characterization of its cellular function (8). Several high-throughput experiments aim at the characterization of localization for large chunks of the proteome (9–13). These large datasets provide information about protein function and more generally global cellular processes. However, they currently remain significantly below 100% coverage (only some fraction of all proteins is experimentally characterized), and they also do not achieve levels of accuracy comparable to high-accuracy low-throughput experiments (8, 14, 15). This is the background against which computational methods that cover all proteins at high levels of accuracy become important.

Many studies have reviewed different methods that predict protein targeting and localization either based on their background methodologies or based on organisms specificity (16–18). In this chapter, we focused on a few state-of-the-art methods that predict protein targeting and localization based on organelle and organism specificity. One challenge for users of recent tools pertains to the reliability of the tool: publishing over-fitted machine learning-based methods that appear to work but do not stand up to their claims is particularly easy and prominent in this field. Conversely, many groups who have generated semi-automated analyses of localization fall prey to overestimates of the accuracy for homology-based inference and of some experimental annotations. The degree of the challenge for users is raised further by the fact that performance is likely to increase further in the future: the major limitation in developing better tools today appears to be the shortage of reliable experimental data. In other words, the best tools have yet to be developed but many tools that will be developed in the future will not work as well as tools that are with us since over a decade.

Good state-of-the-art methods suffice to arrive at surprising findings when analyzing large data sets. For instance, it came as a surprise that most eukaryotic proteins with experimental structures are secreted (19) and that almost 35% of all non-membrane human proteins are native to the nucleus (20). Many publications on comparative genomics rooted in the application of methods that predict targeting and localization (8, 17).

2. Methods Using Sorting Signals

Bacterial cells generally consist of a single intracellular compartment surrounded by the plasma membrane. Gram-positive bacteria have one membrane, and Gram-negative bacteria have an additional outer membrane. In contrast, eukaryotic cells are subdivided into functionally distinct, membrane-bound compartments and each compartment harbors a variety of proteins that carry out various biochemical reactions necessary for viability of the cell. The major constituents of eukaryotic cells are extracellular space, cytoplasm, nucleus, mitochondria, Golgi apparatus (Golgi), endoplasmic reticulum (ER), chloroplasts (plants only), peroxisome, vacuoles, cytoskeleton, and the ribosome; other subdivisions include nucleoplasm, nucleolus, and nuclear matrix (21). A significant portion of the proteins in a eukaryotic cell is targeted to one specific organelle. Since proteins are usually synthesized on the ribosomes in the cytosol, the question becomes how this sorting is regulated.

Guenter Blobel found that "protein sorting" or "protein tracking," requires information encoded in the protein sequence itself in the form of topogenic signals, as well as the cellular machinery that decodes this information and delivers the protein to its correct location (22). A wide variety of localization signals target proteins to their functional site. Some signals are contained in peptides that are ultimately cleaved. Such pre-sequences are typically 15–70 residues long, but have a common overall structure (**Fig. 17.1**): the N-terminal part (n-region) has a variable length and generally carries a net positive charge which is followed

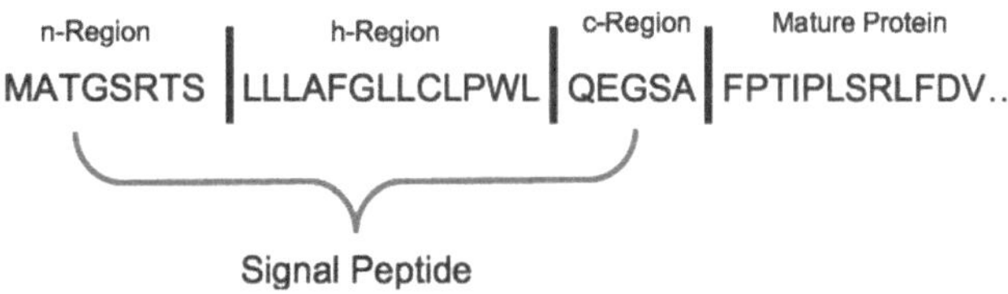

Fig. 17.1. Sketch of typical signal peptides. *n-region*: Positively-charged amino acids (arginine or lysine) are prevalent at the N-terminus around residues 2–15; *h-region*: hydrophobic amino acids dominate the following region over typically at least 8 residues (note that this region can be much longer and can be inserted as one or more integral transmembrane helices); *c-region*: less hydrophobic residues follow in front of the recognition site for the signal peptidase (typically the c-region is only around 6 residues long). All these three regions are cleaved from the mature, native protein.

by a central hydrophobic core (h-region) of 6–15 amino acids, which is essential for signal peptide function. Often helix-breakers such as proline, glycine or serine mark the transition between the hydrophobic h-region and the c-region with the cleavage site. Other signals (in particular signal peptides that target secretion) have the potential to form amphiphilic alpha helices that might be inserted as integral transmembrane helices.

Targeting to chloroplasts, mitochondria, nucleus, and the extracellular space are typical examples for sorting signal-based determination of protein localization. However, even for these compartments, the signals differ substantially (**Table 17.1**). Signals that determine the import into and the export from the nucleus (NLS: nuclear localization signals) are particularly diverse (23–27). Most NLS remain unknown (20). Nuclear localization signals are also special in that some NLS imply particular aspects of function following three step functional cycle (25): (1) NLS recognized by shuttle protein such as importin, (2) once in the nucleus the importin leaves the NLS and the same site is used to bind DNA or RNA, (3) after completion of the nucleotide-related function, exportins bind to the freed NLS and shuttle the protein outside again thereby readying it for another cycle. Unfortunately, only a limited number of such "functional" NLS can be identified today (8, 26) although we now have methods that reliably predict DNA-binding to regions that have no recognizable similarity to known DNA-binding regions (28).

Table 17.1
Some typical signal sequences

Sorting signal	Example for sequence *
ER signal peptide	+H3N-M-M-S-F-V-S-**L-L-L-V-G-I-L-F-W-A**-T-*E*-A-*E*-Q-L-Thr-*K*-C-*E*-V-F-Q-
Mitochondria transit peptide (mTP)	+H3N-M-L-S-L-R-Q-S-I-R-F-F-K-P-A-T-R-T-L-C-S-S-R-Y-L-L-
Nuclear Localization Signal (NLS)	-P-P-K-K-K-R-K-V-
Twin-arginine signal peptide	+H3N-M-N-N-*E*-*E*-T-F-Y-Q-A-M-R-R-Q-G-V-T-R-R-S-F-L-K-Y-C-S-L-A-A-T-S-L-G-L-G-A-G-M-A-P-K-I-A-W-A-
Chloroplast transit Peptide (cTP)	+H3N-M-A-M-A-M-R-S-T-F-A-A-R-V-G-A-K-P-A-V-R-G-A-R-P-A-S-R-M-S-C-M-A-
Lipoprotein signal peptide	+H3N-M-K-R-Q-A-L-A-A-M-I-A-S-**L-F-A-L-A-A**-

* *Positively* charged amino acids are underlined, and *negatively* charged amino acids are in italics. An extended block of hydrophobic amino acids is shown in bold. +H3N indicates the amino terminus of a protein; COO- indicates the carboxyl terminus.

Proteins translated on eukaryotic ribosomes are often classified according to the presence or absence of an N-terminal signal peptide. Proteins without a signal peptide are translated on "free" ribosomes and may remain in the cytosol and those transported to the nucleus have a distinct signal for it to be transported through the nuclear pore by gated transport or transported to mitochondria or chloroplasts by transmembrane transport. Proteins with signal peptides and complete translation on ER-attached ribosomes (or "bound" ribosomes) either stay in the ER, the Golgi, vacuoles, or are secreted to the plasma membrane, cell wall or extracellular matrix by vesicular transport (**Fig. 17.2**). While signal peptides in eukaryotes are usually at the N-terminus, in prokaryotes such as *E. coli* SRP-dependent signal peptides may be within integral transmembrane helices.

All bacterial cells consist of a cytoplasm surrounded by a lipid bilayer called the cytoplasmic membrane. With some exceptions, Gram-positive bacteria are further surrounded by a thick peptidoglycan cell wall whereas Gram-negative bacteria are surrounded by a different structure, the cell envelope. The cell envelope consists of a peptidoglycan layer, a lipid bilayer known as the outer membrane and the space between the membranes, which is called the periplasm. Following its synthesis in the cytoplasm, a protein

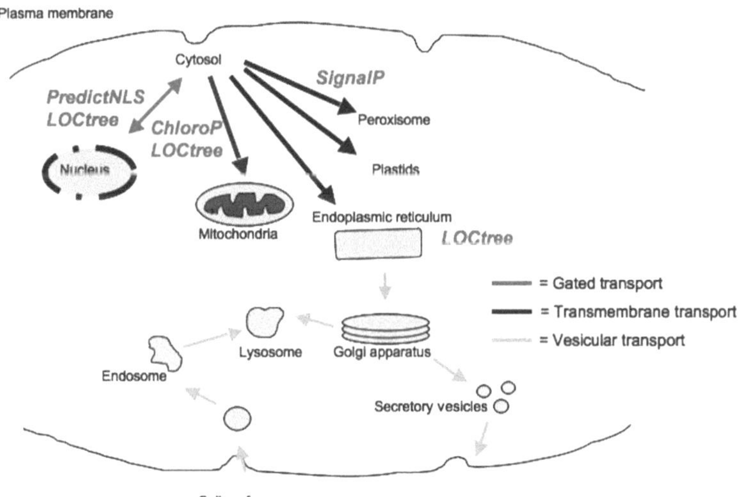

Fig. 17.2. Sketch of protein targeting in eukaryotes. Proteins can move from one compartment to another by gated transport (*dark gray*), transmembrane transport (*black*), or vesicular transport (*light gray*). Proteins are synthesized in the cytosol from where they are sorted to their respective localizations. The signals that direct a given protein's movement through the system, and thereby determine its eventual location in the cell, are contained in each protein's amino acid sequence. At each compartment, a decision is made as to whether the protein is to be retained in that compartment or transported further. In principle, a signal could be required for either retention in or exit from a compartment.

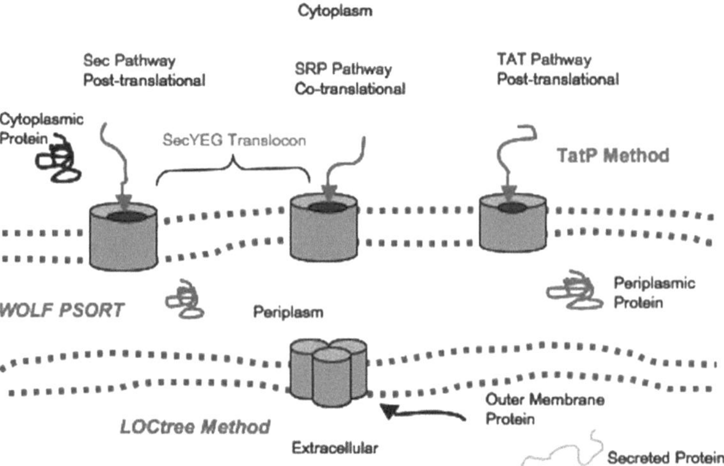

Fig. 17.3. Sketch of protein targeting in bacteria. All bacterial proteins are synthesized in the cytoplasm, and most remain there to carry out their unique functions. Other proteins contain export signals that direct them to other cellular locations. In Gram-negative bacteria, they include the cytoplasmic membrane, the periplasm, the outer membrane and the extracellular space. In Gram-positive bacteria, these include the cytoplasmic membrane, cell wall, and extracellular space. In most cases, the whole protein is located in a single compartment; however, proteins can also span multiple localization sites.

can either remain in cytoplasm or be targeted to one or more sites (**Fig. 17.3**) through one of several different transport systems. One of these is the type I secretion system (29), which shuttles proteins directly from the cytoplasm into the extracellular space. The type II secretion system, also called the general secretory pathway (30), is a two-step process involving insertion into, or translocation across, the cytoplasmic membrane by either the Sec-dependent pathway, the signal recognition particle (SRP)-dependent pathway or the twin-arginine translocation (Tat) pathway (31) systems. In Gram-negative bacteria this is followed by insertion into or translocation across the outer membrane. The other transport systems include the type III (32) and type IV (33) secretion systems, which directly inject products into the cytoplasm of a neighboring cell, and the type V secretion system (34), or auto-transporter system, which self-transports a passenger domain using a C-terminal pore domain.

Secreted proteins and integral plasma membrane proteins are of special interest since they play key roles in important biological processes, e.g. signal transduction and transmission, and cellular differentiation. Moreover, these protein families are comparably easily accessible by drug molecules, due to their localization and have therefore constituted the majority of the proteins targeted by drugs in the past (35). Proteins in special organelles of parasites might be better candidates to develop novel anti-infective agents (36, 37).

Unlike NLS, signal peptides (secretory pathway, SP), transit peptides (chloroplasts, cTP), and targeting peptides (mitochondria, mTP) have many generic features that lend themselves to the development of generic tools that identify such features from sequence (38–43). The first advanced prediction methods yielded levels around 70–80% accuracy for secretory proteins (39), current techniques reach up to 90% accuracy (often also referred to as specificity, i.e. correctly predicted as secreted/correctly predicted as secreted) at 90% coverage (often also referred to as sensitivity, i.e. correctly predicted as secreted/observed as secreted) (44). Early methods of predicting signal peptides were essentially based on consensus signals, using discriminate functions with weight matrices (41, 45). Modern machine-learning techniques such as neural networks (NNs) and hidden Markov models (HMMs) (46, 47) learn to discriminate automatically from the data, using experimentally verified examples as input. Such methods successfully predict secretory signal peptides (SPs) (46, 48), mitochondrial targeting peptides (mTPs) (49, 50), chloroplast targeting peptides (cTPs) (51), lipoprotein signal peptides in Gram-negative bacteria (52), and twin-arginine signal peptide cleavage sites in bacteria (53). The most advanced methods simultaneously combine several sequence features for prediction; not all methods function for all organisms (**Table 17.2**). Here, we focused on four prominent methods: TargetP (44), LipoP (52), TatP (54), and PredictNLS (25, 26).

On the one hand, signal peptides do not cover all secreted proteins, instead some proteins are secreted pass through an alternative mechanism (54). Some methods explicitly target proteins that are secreted through non-classical paths (53, 54); others address this issue indirectly [e.g., LOCtree correctly identifies all classical cases through its hierarchy of methods (20, 55)].

On the other hand, signal peptides are not conclusive evidence for secretion, because some proteins with signal peptides that enter the secretory pathway are not secreted. Many are retained in the Golgi apparatus, the ER, and some in vesicles. For Golgi and ER, several retention signals are known that are typically C-terminal (e.g., KDEL for ER). The Golgi/ER retention signals are so far also the only instance of experimentally verified conformational signals, i.e. signals that will only be present in fully folded proteins as opposed to unfolded fragments (56) (note: KDEL is a linear not a conformational signal). Unfortunately, very few of the retention signals are experimentally known and not a single one of the signals is as accurate as are the known NLS, i.e. proteins that have these signals are usually retained but they might not be (56).

Table 17.2
Selected prediction tools and databases for signal sequence analysis.*

Name	URL	Organisms	Method
Sequence homology-based localization annotations			
LOChom	www.rostlab.org/db/LOChom/	Eukaryotes	SVM
Methods based on N-terminal sorting signals			
SignalP	www.cbs.dtu.dk/services/SignalP/	Eykaryotes, bacteria	ANN, HMM
ChloroP	www.cbs.dtu.dk/services/ChloroP/	Plants	ANN
TatP	www.cbs.dtu.dk/services/TatP /	Bacteria	ANN
LipoP	www.cbs.dtu.dk/services/LipoP/	Bacteria	HMM
TargetP	www.cbs.dtu.dk/services/TargetP/	Non-plant, plant	ANN, HMM
iPSORT	hc.ims.u-tokyo.ac.jp/iPSORT/	Non-plant, plant	Rule-based classifier
Predotar	urgi.versailles.inra.fr/predotar/ predotar.html	Multicellular eukaryotes	ANN
Prediction and analysis of nuclear localization signals			
PredictNLS	www.rostlab.org/predictNLS/	Multicellular eukaryotes	Local database searching
NucPred	www.sbc.su.se/~maccallr/nucpred	Eukaryotes	GP
Methods based on amino acid composition			
LOCnet	www.rostlab.org/services/LOCnet/P	Eukaryotes, prokaryotes	Local database searching
SubLoc	www.bioinfo.tsinghua.edu.cn/SubLoc/P	Eukaryotes, prokaryotes	SVM
Methods using multiple features for predicting localizations			
WoLFPsort	wolfpsort.org/	Animal, fungi, plants	Rule-based classifier
PSORT-B	www.psort.org/psortb/	Bacteria	Rule-based classifier
LOCtarget	www.rostlab.org/services/LOCtarget/	Eukaryotes, prokaryotes	SVM, Local database searching
MultiLoc	www-bs.informatik.uni-tuebingen.de/ Services/MultiLoc/	Non-plant, plant	SVM
LOCtree	www.rostlab.org/services/LOCtreeP	Animal, plant, prokaryotes	SVM
BaCelLo	gpcr.biocomp.unibo.it/bacello/	Animal, fungi, plants	SVM

2.1. TargetP: Secretory Pathway, Mitochondria, and Chloroplasts

Basics: TargetP (44) is a NN and HMM-based tool for predicting the subcellular location of eukaryotic proteins. The localizations are assigned to proteins based on the predicted presence of any of the N-terminal signal peptides, chloroplast transit peptide (cTP), mitochondrial targeting peptide (mTP) or secretory pathway signal peptide (SP), and also based on predicted presence of a potential cleavage site. TargetP uses ChloroP (51) and SignalP (39, 57) to predict cleavage sites for cTP and SP, respectively.

Performance: The method can discriminate between proteins destined for the secretory pathway, mitochondria, chloroplast, and other localizations with an accuracy of 96% (plant) or 94% (non-plant). TargetP predicts reaches 95% accuracy for secretory signal peptides and 70–80% accuracy for pre-sequences in chloroplasts and mitochondria.

Availability: http://www.cbs.dtu.dk/services/TargetP/. Response time: typically seconds to minutes; TargetP 1.1 is available as a stand-alone software package for download for academic users and others can get it on request.

2.2. LipoP: Lipoprotein Motifs

Basics: LipoP (52) predicts signal peptides in lipoproteins of Gram-negative bacteria. It is based on a hidden Markov model (HMM) that identifies lipoproteins and discriminates between lipoprotein signal peptides, other signal peptides, and N-terminal membrane helices.

Performance: LipoP is reported to reach around 96% accuracy and to significantly outperform similar methods. Although Gram-positive lipoprotein signal peptides differ from those in Gram-negatives, LipoP also correctly identified over 90% of the lipoproteins included in a Gram-positive test set.

Availability: http://www.cbs.dtu.dk/services/LipoP/; response time: typically seconds to minutes; LipoP 1.0 is available as a stand-alone software package for download for academic users and others can get it on request.

2.3. TatP: Bacterial Tat Signal Peptides

Basics: TatP (53, 54) predicts the presence of bacterial Tat signal peptides through a combination of regular expressions and neural networks. The Tat pathway has recently been discovered in bacteria (58–60). It operates in parallel to the well-characterized Sec export pathway. Substrates of the Tat pathway are often redox cofactor binding proteins, which acquire their cofactors, and therefore fold in the cytoplasm. Thus, in contrast to protein export via the Sec pathway, Tat substrates are folded prior to export (58, 61). Proteins entering the Tat pathway have signal peptides with a tripartite structure that is much like classical Sec signal peptides and indeed are probably also cleaved by leader peptidase (62). However, in contrast to Sec signal peptides, a striking twin-arginine motif is found at the border between the n- and h-regions of the Tat signal peptide. A consensus sequence for this twin-arginine motif has previously been defined as (S/T)RRxFLK (58), where the two consecutive arginines are invariant. With an average length of 37 amino acids (generally found in the n-region), Tat signal peptides are significantly longer than classical Sec signal peptides. In addition, the h-region of Tat signal peptides has a lower average hydrophobicity than classical Sec signal peptides (63). A complex consensus pattern, which covers 97% of putative Tat signal peptides in the training set, is implemented.

Performance: TatP is reported to reach about 96% accuracy at 91% sensitivity.

Availability: Web server at: http://www.cbs.dtu.dk/services/TatP/; response time: typically seconds to minutes; TatP 1.0 will soon be available as a package on a commercial license.

2.4. PredictNLS: Nuclear Localization Signals (NLS)

Basics: Many distinct NLSs have been experimentally implicated in nuclear transport (64–67). NLSdb (26) is the largest publicly available database of experimental NLSs. However, known experimental NLSs can account for fewer than 10% of known nuclear proteins. To remedy this, PredictNLS (25) uses a procedure of *in silico* mutagenesis to discover new NLSs by changing or removing some residues from the experimentally characterized NLS motifs and monitoring the resulting true (nuclear) and false (non-nuclear) matches, then discarding any potential NLSs that are found in known non-nuclear proteins as false matches. By using PredictNLS, 194 potential NLSs are discovered which increased the coverage of known nuclear proteins to 43%. All proteins in the PDB (68) and UniProt/KB (43) databases were annotated by using the full list of experimental and potential NLSs. NLSdb contains over 6,000 predicted nuclear proteins and their targeting signals from the PDB and Swiss-Prot databases.

Performance: PredictNLS is designed to only contain motifs that identify no single protein known not to be nuclear (100%). This stringency is paid for by low coverage (~10%). Users can assess the accuracy and coverage for particular motifs that they want to probe.

Availability: Web server: http://www.rostlab.org/services/predictNLS ; response time: seconds to minutes; standalone program available upon request. Input: sequence or motif; Output: NLS with reliability and functional annotations if available; if queried with a motif, the server returns the degree to which this motif maps only in nuclear proteins, i.e. to which it is a good candidate for a new NLS.

3. Methods Directly Using Other Experimental Annotations

3.1. Localization Inferred Through Sequence Similarity

Proteins with similar sequences have similar function (69–74), and they are native to the same subcellular location (75). In fact, the inference of annotations based on sequence similarity to proteins with experimental annotations is arguably the most important reason to begin any experiment with database searches. Often, but not always, the reason for the connection between sequence similarity and feature similarity (same localization) is related to the evolutionary connection between the protein for which some experimental data is available and that for which the inference is made. Even Swiss-Prot (43), the arguably best expert curated comprehensive protein database is riddled with such homology-based inferences that are frequently confused for facts by users.

Many studies have explored the relationship of sequence and structure similarity to conservation of various aspects of protein function (76–79). These studies observed that thresholds of 50–60% sequence identity could be used for transferring annotations. Recent studies suggested that these levels of sequence similarity might not be sufficient to accurately inferring function (75, 78, 79). Moreover, for many protein chains, similarity-based annotation cannot be carried due to unavailability of experimental data. In spite of this, homology-based approaches continue to be the most reliable for annotating subcellular localization (56, 78, 80).

Given the importance of homology-based inference, it is surprising that very few methods explicitly relate sequence similarity to annotations of localization along with estimates for accuracy. As often, the logic appears to be as follows: similar enough in sequence implies identical localization; the user knows what "similar enough" means. The logic is double-flawed. Firstly, the level of similarity needed depends on the particular protein family as well as the particular location. In the most extreme example, an NLS can be as short as five residues that if attached to any protein will lead to the import of that protein into the nucleus. Secondly, the transition from the semi-safe zone of inference to the twilight zone constitutes what physicists refer to as phase transitions, i.e. the drop from where homology-based inferences almost always succeed to where they are most of the time is extremely sharp (75, 78, 81–84). Together these two realities clearly challenge those who want to infer localization through sequence signals without guidance by specific tools developed toward this end.

3.1.1. LOChom: Homology-Based Inference of Location

Basics: LOChom (75) is a comprehensive database containing homology-based subcellular localization annotations for less than 50% of all proteins in the UniProt/KB database (43) and around 20% of sequences from five entirely sequences eukaryotic genomes. In this method sequence homologs are identified using pairwise BLAST (85) and PSI-BLAST (86). The subcellular localization is assigned based on three measures of sequence similarity: (i) pairwise sequence identity (PIDE); (ii) BLAST e-values (EVAL) (85); and (iii) HSSP-values that measure the distances from the HSSP-threshold as a function of alignment length and PIDE (HVAL) (82, 87). Of the three measures, the HVAL is considered to be most successful in annotating location. LOChom infers location from the HVAL closest homolog and provides estimate of accuracy/coverage for the given transfer.

Performance: LOChom is a part of a web server and database that predicts and annotates sub-cellular localization for structural genomics targets. The decisions made by LOChom for all compartments are significantly more accurate than de novo predictions.

Availability: Web server: http://www.rostlab.org/db/LOC hom; response time: minutes to hours (if queue overloaded); standalone program available upon request. Input: sequence; Output: localization with reliability and functional annotations.

3.2. Localization Inferred Through Text Analysis

Automatic text mining methods try to close the sequence–annotation gap by extracting the wealth of knowledge contained in the literature or in controlled vocabularies of curated databases. One crucial bottleneck remains the mapping of gene/protein names (88–91). This problem is solved and the challenge for the text mining reduced when applying methods to controlled vocabularies in databases. Many methods explore the functional annotations in UniProt (89, 92–97). Both fully automated and semi-automated methods have been applied to predict subcellular localization from database keywords. The fully automatic methods extract rules from keywords by using statistical learning methods like probabilistic Bayesian models (98), symbolic rule learning (99), and M-ary (multiple category) classi?ers such as the k-Nearest neighbor algorithm (100). Among the major methods in this category are LOCkey (93), Proteome Analyst (94), and Spearmint (101, 102). The semi-automated methods are based on building dictionaries of rules, examples are EUCLID (92), Meta_A (95), and RuleBase (103).

3.2.1. LocKey: Using Swiss-Prot Keywords

Basics: LocKey (93) infers localization by using experimental descriptions of protein function as contained in the controlled vocabulary of Swiss-Prot keywords (43). First, the target sequence is aligned to sequences in Swiss-Prot using pairwise BLAST (86). Then all Swiss-Prot keywords for all sequence homologs that meet specified thresholds in terms of sequence similarity and the content of these keywords are extracted. LOCkey (93) then infers subcellular localization through an automated lexical analysis of the extracted Swiss-Prot keywords. In contrast to dictionary-based approaches, LOCkey is fully automated and the rule libraries used to infer localization from keywords are generated dynamically.

Performance: The method is very accurate when several keywords are known (~90%); accuracy remains high (>82% accuracy) as long as any informative keyword is identified in Swiss-Prot.

Availability: Web server: http://www.rostlab.org/services/LOCkey ; response time: minutes to hours (if queue overloaded); standalone program available upon request.

3.2.2. Proteosome Analyst: Using Swiss-Prot Keywords

Basics: Proteome Analyst (94) predicts subcellular localization and the molecular function according to the GeneOntology (GO) classification. It uses a precise, annotation keyword-based approach comprising two steps. First, the query blasted (86) against Swiss-Prot (43). Secondly, the keywords in the annotation that might be indicative of a particular localization site are extracted from the Swiss-Prot records and are passed onto a Bayesian classifier. This classifier then uses the extracted keywords to assign the query protein to one of three Gram-positive (cytoplasm, cytoplasmic membrane, or extracellular space) or five Gram-negative localization sites. The program can also predict location for animals, plants, archaea, and fungi proteins. Proteome Analyst was trained and tested using sequences with annotated localization information extracted from Swiss-Prot (43).

Performance: The authors report precision of 96.9% and 97.2% and recall of 95.3% and 95.6% for the Gram-negative and Gram-positive classifiers, respectively. Given that this exceeds what can be expected from careful low-throughput experiments, these numbers might constitute over-estimates.

Availability: Web server: http://pickardville.cs.ualberta.ca:8080/pa ; response time: 30 sec/input sequence.

4. De Novo Prediction Methods

The overall amino acid composition correlates with the native compartment (104–106). This observation has led to the development of a variety of prediction methods based solely on composition (107–112). Higher order correlations (residues i and $(i + n)$, $n = 2,3,4$) have been accounted for by using pseudo-amino acid composition (113, 114). With the availability of many completely sequenced genomes, phylogenetic profiles have been employed to identify subcellular localization (115). So far, this approach has been much less accurate than methods based solely on composition. PSORT II is a knowledge-based system that integrates rules based on amino acid composition with known sequence motifs (108, 116, 117), and also uses other methods such as NNPSL (110). Thus, the accuracy of PSORT II somehow depends on the accuracy of the underlying original methods. Drawid & Gerstein have proposed a Bayesian system based on a diverse range of 30 different features (118). They applied their system to predicting localization of yeast proteins. Another approach using optimally weighted fuzzy k-nearest neighbors (OWFKNN) algorithm has been used to predict subcellular locations of proteins based on their amino acid composition (119). The k-nearest neighbor (k-NN) rule is most often used for classification of a query pattern and is classified according to the classification of the nearest neighbor from a database of known pattern classes (120). Authors reported accuracy of about 88.5% for prokaryotic sequences and 86.2% for eukaryotic sequences in a jackknife test.

Amino acid composition correlates with localization (104, 105). This basic observation is intuitive because the environments in compartments differ; hence, the proteins should adopt their amino acid composition somehow to the environment. Only the protein surface "sees" the environment. Therefore, the correlation between amino acid composition and localization should largely originate from surface residues. This is indeed the case (121), and has led to the development of methods that use surface composition and related features (19, 20, 122, 123). LOCtree, appears to be one of the best methods for the prediction of localization in five classes for eukaryotes and in three classes for prokaryotes; it combines information about surfaces and secondary structure and implements a hierarchy of support vector machines (SVMs) that mimic the mechanism of cellular sorting (20).

4.1. WoLF PSORT: Third Improvement of a Standard

Basics: WoLF PSORT (124) is an extension and improvement over the long successful line of PSORT programs. It converts protein amino acid sequences into numerical localization features, based on sorting signals, amino acid composition, and functional motifs such as DNA-binding motifs. WoLF PSORT uses a wrapper method to select and use only the most relevant features. This reduces the amount of information, which needs to be considered (and displayed) for the user to interpret individual predictions and may also make the predictor less prone to over learning. After conversion, a simple k-nearest neighbor classifier is used for prediction. WoLF PSORT classifies proteins into more than 10 localizations, including dual localization such as proteins, which shuttle between the cytosol and the nucleus. WoLF PSORT not only provides subcellular localization prediction with competitive accuracy, but also provides detailed information relevant to protein localization to help users to form their own hypotheses. Each prediction can be shown in two ways: (i) a list of proteins of known localization with the most similar localization features to the query and (ii) tables with detailed information about individual localization features. For convenience, sequence alignments of the query to similar proteins and provided with links to UniProt and Gene Ontology (125).

Performance: The overall prediction accuracy has been estimated to be very high, although a head-to-head comparison with other methods based on identical data sets remains to be performed. WoLF PSORT also displays some information about detected sorting signals which is useful in helping users determine the reliability of the prediction in specific cases.

Availability: Web server at http://wolfpsort.org/; response time: 30 sec /input sequence; standalone program free for academia.

4.2. LOCtree: Hierarchy of SVMs

Basics: LOCtree (20) uses a hierarchical system combining SVMs and other prediction methods; it exploits a variety of sequence and predicted structural features in its input. One important feature contributing essentially to the reliability of the prediction is evolutionary information. Currently LOCtree does not predict localization for membrane proteins, since the compositional properties of membrane proteins significantly differ from those of non-membrane proteins. Information about function can be used by the system if available.

Performance: When evaluated on a non-redundant test set, LOCtree achieved sustained levels of 74% accuracy for non-plant eukaryotes, 70% for plants, and 84% for prokaryotes. We rigorously benchmarked LOCtree in comparison to the best alternative methods for localization prediction. LOCtree outperformed all other methods in nearly all benchmarks. Localization assignments using LOCtree agreed quite well with data from recent large-scale experiments.

Availability: Web server: http://www.rostlab.org/services/LOCtree; response time is typically between minutes and hours (the bottleneck is the generation of the evolutionary information (database searches) used as input; standalone program available upon request. Input: sequence; Output: localization with reliability and functional annotations as well as alignments and other predictions.

5. Conclusions

The field of predicting protein targeting and subcellular localization is a perfect example for the strengths and difficulties of computational biology. Firstly, the quality of methods remains limited by the sparseness of reliable experimental data. This is true for many fields, but it is particularly true for localization prediction. The most successful methods make use of modern machine learning algorithms. Such methods when, e.g. applied to the prediction of inter-residue contacts (126) can use over 50 million of reliable experimental data points. In contrast, localization is experimentally established reliably for fewer than 50 thousand proteins. The more reliable data, the better the methods will become.

Secondly, machine-learning methods are extremely powerful; however, if not applied carefully they can lead to over-optimizations that render useless methods. Methods that are based on simpler optimizations also have a long history for parameter over-fitting in the localization prediction field. It is up to the user to separate the chaff from the wheat. The TargetP program series stands out as a beacon of carefully assessing performance and avoiding over-fitting.

Overall, the field has improved significantly over the last decade, thanks to more data and better algorithms. Secreted proteins remain the class that is predicted best, due to the presence of signal peptides and compositional differences, as the relative abundance in cysteines. Methods such as SignalP/TargetP, WoLF PSORT, and LOCtree perform best. Good performance is also achieved in recognizing that proteins are either nuclear or cytoplasmic. However, the further distinction into which of those two remains problematic. Similarly, in plants it is quite easy to predict proteins native to the mitochondria or chloroplasts, but it is very hard to discriminate the two classes. Computational biology is amazingly successful in the prediction of targeting: the best methods reach levels of accuracy in coarse-grained prediction of localization that reach the performance of less careful high-throughput experiments. This clearly suffices for coarse-grained annotations of entire proteomes.

Acknowledgements

The work of SR and BR was supported by the grant R01-GM079767 from the National Institute of General Medical Sciences (NIGMS) at the NIH. Last but not least, thanks to Amos Bairoch (SIB, Geneva), Rolf Apweiler (EBI, Hinxton), Phil Bourne (San Diego Univ.), and their crews for maintaining excellent databases and to all experimentalists who enabled this analysis by making their data publicly available.

References

1. Rost, B., Liu, J., Nair, R. et al. (2003) Automatic prediction of protein function, *Cel Mol Life Sci*, **60**, 2637–2650.
2. Sharan, R., Ulitsky, I., Shamir, R. (2007) Network-based prediction of protein function, *Mol Syst Biol*, **3**, 88.
3. Smith, T.F. (1998) Functional genomics–bioinformatics is ready for the challenge, *Trends Genet*, **14**, 291–293.
4. Koonin, E. V. (2005) Orthologs, paralogs, and evolutionary genomics, *Annu Rev Genet*, **39**, 309–338.
5. Koonin, E. V., Wolf, Y. I. (2008) Genomics of bacteria and archaea: the emerging dynamic view of the prokaryotic world, *Nucleic Acids Res*, **36**, 6688–6719.
6. Ashburner, M., Ball, C. A. , Blake, J. A. et al. (2000) Gene ontology: tool for the unification of biology. The Gene Ontology Consortium, *Nature Genet*, **25**, 25–29.
7. Camon, E., Barrell, D., Lee, V. et al. (2004) The Gene Ontology Annotation (GOA) Database–an integrated resource of GO annotations to the UniProt Knowledgebase, *In Silico Biol*, **4**, 5–6.
8. Nair, R., Rost, B. (2008) Predicting protein subcellular localization using intelligent systems, *Methods Mol Biol*, **484**, 435–463.
9. Marion, J., Bach, L., Bellec, Y. et al. (2008) Systematic analysis of protein subcellular localization and interaction using high-throughput transient transformation of Arabidopsis seedlings, *Plant J*, **56**, 169–179.
10. Hu, Y.H., Vanhecke, D., Lehrach, H. et al. (2005) High-throughput subcellular protein localization using cell arrays, *Biochem Soc Trans*, **33**, 1407–1408.
11. Barrios-Rodiles, M., Brown, K.R., Ozdamar, B. et al. (2005) High-throughput mapping of a dynamic signaling network in mammalian cells, *Science*, **307**, 1621–1625.
12. Kumar, A., Agarwal, S., Heyman, J. A. et al. (2002) Subcellular localization of the yeast proteome, *Genes Dev*, **16**, 707–719.
13. Huh, W. K., Falvo, J. V., Gerke L. C. et al. (2003) Global analysis of protein localization in budding yeast, *Nature*, **425**, 686–691.
14. Rey, S., Gardy, J. L., Brinkman, F. S. (2005) Assessing the precision of high-throughput computational and laboratory approaches for the genome-wide identification of protein subcellular localization in bacteria, *BMC Genomics*, **6**, 162.
15. Davis, T. N. (2004) Protein localization in proteomics, *Curr Opin Chem Biol*, **8**, 49–53.
16. Schneider, G., Fechner, U. (2004) Advances in the prediction of protein targeting signals, *Proteomics*, **4**, 1571–1580.
17. Gardy, J. L., Brinkman, F. S. (2006) Methods for predicting bacterial protein subcellular localization, *Nat Rev Microbiol*, **4**, 741–751.
18. Casadio, R., Martelli, P. L., Pierleoni, A. (2008) The prediction of protein subcellular localization from sequence: a shortcut to functional genome annotation, *Brief Funct Genomic Proteomic*, **7**, 63–73.
19. Nair, R., Rost, B. (2003) LOC3D: annotate sub-cellular localization for protein structures, *Nucleic Acids Res*, **31**, 3337–3340.
20. Nair, R., Rost, B. (2005) Mimicking cellular sorting improves prediction of subcellular localization, *J Mol Biol*, **348**, 85–100.
21. Lodish, H. (2004) *Mol Cell Biol*, 5th ed., WH Freeman, New York.
22. Blobel, G., Dobberstein, B. (1975) Transfer of proteins across membranes. II. Reconstitution of functional rough microsomes from heterologous components, *J Cell Biol*, **67**, 852–862.
23. Boulikas, T. (1993) Nuclear localization signals (NLS), *Crit Rev Eukaryot Gene Expr*, **3**, 193–227.
24. Moroianu, J. (1999) Nuclear import and export: transport factors, mechanisms and regulation, *Crit Rev Eukaryot Gene Expr*, **9**, 89–106.
25. Cokol, M., Nair, R., Rost, B. (2000) Finding nuclear localization signals, *EMBO Rep*, **1**, 411–415.
26. Nair, R., Carter, P., Rost, B. (2003) NLSdb: database of nuclear localization signals, *Nucleic Acids Res*, **31**, 397–399.
27. La Cour, T., Gupta, R., Rapacki, K. et al. (2003) NESbase version 1.0: a database of nuclear export signals, *Nucleic Acids Res*, **31**, 393–396.
28. Ofran, Y., Mysore, V., Rost, B. (2007) Prediction of DNA-binding residues from sequence, *Bioinformatics*, **23**, i347–353.
29. Holland, I. B., Schmitt, L., Young, J. (2005) Type 1 protein secretion in bacteria, the ABC-transporter dependent pathway (review), *Mol Membr Biol*, **22**, 29–39.
30. Pugsley, A. P. (1993) The complete general secretory pathway in gram-negative bacteria, *Microbiol Rev*, **57**, 50–108.
31. Muller, M., Klosgen, R. B. (2005) The Tat pathway in bacteria and chloroplasts (review), *Mol Membr Biol*, **22**, 113–121.
32. Journet, L., Hughes, K. T., Cornelis, G.R. (2005) Type III secretion: a secretory

pathway serving both motility and virulence (review), *Mol Membr Biol,* **22**, 41–50.
33. Christie, P. J., Cascales, E. (2005) Structural and dynamic properties of bacterial type IV secretion systems (review), *Mol Membr Biol,* **22**, 51–61.
34. Thanassi, D. G., Stathopoulos, C., Karkal, A. et al. (2005) Protein secretion in the absence of ATP: the autotransporter, two-partner secretion and chaperone/usher pathways of gram-negative bacteria (review), *Mol Membr Biol,* **22**, 63–72.
35. Ofran, Y., Punta, M., Schneider, R. et al. (2005) Beyond annotation transfer by homology: novel protein-function prediction methods to assist drug discovery, *Drug DiscovToday,* **10**, 1475–1482.
36. Gierasch, L. M. (1989) Signal sequences, *Biochemistry,* **28**, 923–930.
37. Zheng, N., Gierasch L.M. (1996) Signal sequences: the same yet different, *Cell,* **86**, 849–852.
38. Nielsen, H., Engelbrecht, J., Brunak, S. et al. (1997) A neural network method for identification of prokaryotic and eukaryotic signal peptides and prediction of their cleavage sites, *Int J Neural Syst,* **8**, 581–599.
39. Nielsen, H., Engelbrecht, J., Brunak, S. et al. (1997) Identification of prokaryotic and eukaryotic signal peptides and prediction of their cleavage sites, *Protein Eng,* **10**, 1–6.
40. Nielsen, H., Engelbrecht, J., von Heijne, G. et al. (1996) Defining a similarity threshold for a functional protein sequence pattern: the signal peptide cleavage site, *Proteins,* **24**, 165–177.
41. von Heijne, G. (1983) Patterns of amino acids near signal-sequence cleavage sites, *Eur J Biochem,* **133**, 17–21.
42. Claros, M. G., Brunak, S., von Heijne, G. (1997) Prediction of N-terminal protein sorting signals, *Curr Opin Struct Biol,* **7**, 394–398.
43. Boeckmann, B., Bairoch, A., Apweiler, R. et al. (2003) The SWISS-PROT protein knowledgebase and its supplement TrEMBL in 2003, *Nucleic Acids Res,* **31**, 365–370.
44. Emanuelsson, O., Brunak, S., von Heijne, G. et al. (2007) Locating proteins in the cell using TargetP, SignalP and related tools, *Nat Protoc,* **2**, 953–971.
45. Laforet, G. A., Kendall, D. A. (1991) Functional limits of conformation, hydrophobicity, and steric constraints in prokaryotic signal peptide cleavage regions. Wild type transport by a simple polymeric signal sequence, *J Biol Chem,* **266**, 1326–1334.
46. Nielsen, H., Brunak, S., von Heijne, G. (1999) Machine learning approaches for the prediction of signal peptides and other protein sorting signals, *Protein Eng,* **12**, 3–9.
47. Emanuelsson, O., Nielsen, H., Brunak, S. et al. (2000) Predicting subcellular localization of proteins based on their N-terminal amino acid sequence, *J Mol Biol,* **300**, 1005–1016.
48. Kall, L., Krogh, A., Sonnhammer, E.L. (2004) A combined transmembrane topology and signal peptide prediction method, *J Mol Biol,* **338**, 1027–1036.
49. Fujiwara, Y., Asogawa, M., Nakai, K. (1997) Prediction of Mitochondrial Targeting Signals Using Hidden Markov Model, *Genome Inform Ser Workshop Genome Inform,* **8**, 53–60.
50. Emanuelsson, O., von Heijne, G. (2001) Prediction of organellar targeting signals, *Biochim Biophys Acta,* **1541**, 114–119.
51. Emanuelsson, O., Nielsen, H., von Heijne, G. (1999) ChloroP, a neural network-based method for predicting chloroplast transit peptides and their cleavage sites, *Protein Sci,* **8**, 978–984.
52. Juncker, A. S., Willenbrock, H., Von Heijne, G. et al. (2003) Prediction of lipoprotein signal peptides in Gram-negative bacteria, *Protein Sci,* **12**, 1652–1662.
53. Bendtsen, J. D., Nielsen, H., Widdick, D. et al. (2005) Prediction of twin-arginine signal peptides, BMC *Bioinformatics,* **6**, 167.
54. Bendtsen, J.D., Kiemer, L., Fausboll, A. et al. (2005) Non-classical protein secretion in bacteria, *BMC Microbiol,* **5**, 58.
55. Nair, R., Rost, B. (2008) Protein subcellular localization prediction using artificial intelligence technology, *Methods Mol Biol,* **484**, 435–463.
56. Wrzeszczynski, K. O., Rost, B. (2004) Annotating proteins from endoplasmic reticulum and Golgi apparatus in eukaryotic proteomes, *Cell Mol Life Sci,* **61**, 1341–1353.
57. Bendtsen, J. D., Nielsen, H., von Heijne, G. et al. (2004) Improved prediction of signal peptides: SignalP 3.0, *J Mol Biol,* **340**, 783–795.
58. Berks, B. C. (1996) A common export pathway for proteins binding complex redox cofactors?, *Mol Microbiol,* **22**, 393–404.
59. Sargent, F., Bogsch, E. G., Stanley, N. R. et al. (1998) Overlapping functions of components of a bacterial Sec-independent protein export pathway, *EMBO J,* **17**, 3640–3650.
60. Weiner, J. H., Bilous, P. T., Shaw, G. M. et al. (1998) A novel and ubiquitous system for membrane targeting and secretion of cofactor-containing proteins, *Cell,* **93**, 93–101.

61. Berks, B. C., Palmer, T., Sargent, F. (2003) The Tat protein translocation pathway and its role in microbial physiology, *Adv Microb Physiol*, **47**, 187–254.
62. Yahr, T. L., Wickner, W.T. (2001) Functional reconstitution of bacterial Tat translocation in vitro, *EMBO J*, **20**, 2472–2479.
63. Cristobal, S., de Gier, J.W., Nielsen, H. et al. (1999) Competition between Sec- and TAT-dependent protein translocation in Escherichia coli, *EMBO J*, **18**, 2982–2990.
64. Mattaj, I. W., Englmeier, L. (1998) Nucleocytoplasmic transport: the soluble phase, *Annu Rev Biochem*, **67**, 265–306.
65. Tinland, B., Koukolikova-Nicola, Z., Hall, M. N. et al. (1992) The T-DNA-linked VirD2 protein contains two distinct functional nuclear localization signals, *Proc Natl Acad Sci U S A*, **89**, 7442–7446.
66. Moede, T., Leibiger, B., Pour, H. G. et al. (1999) Identification of a nuclear localization signal, RRMKWKK, in the homeodomain transcription factor PDX-1, *FEBS Lett*, **461**, 229–234.
67. Jans, D. A., Xiao, C. Y., Lam, M.H. (2000) Nuclear targeting signal recognition: a key control point in nuclear transport?, *Bioessays*, **22**, 532–544.
68. Berman, H. M., Westbrook, J., Feng, Z. et al. (2000) The Protein Data Bank, *Nucleic Acids Res*, **28**, 235–242.
69. Thornton, J. W., DeSalle, R. (2000) Gene family evolution and homology: genomics meets phylogenetics, *Annu Rev Genomics Hum Genet*, **1**, 41–73.
70. Whisstock, J. C., Lesk, A. M. (2003) Prediction of protein function from protein sequence and structure, *Quart Rev Biophys*, **36**, 307–340.
71. Baxter, S. M., Fetrow, J. S. (2001) Sequence- and structure-based protein function prediction from genomic information, *Curr Opin Drug Discov Devel*, **4**, 291–295.
72. Rost, B., Liu, J., Nair, R. et al. (2003) Automatic prediction of protein function, *Cell Mol Life Sci*, **60**, 2637–2650.
73. Wass, M. N., Sternberg, M. J. (2008) ConFunc–functional annotation in the twilight zone, *Bioinformatics*, **24**, 798–806.
74. Ng, P., Nagarajan, N., Jones, N. et al. (2006) Apples to apples: improving the performance of motif finders and their significance analysis in the Twilight Zone, *Bioinformatics*, **22**, e393–401.
75. Nair, R., Rost, B. (2002) Sequence conserved for subcellular localization, *Protein Sci*, **11**, 2836–2847.
76. Orengo, C. A., Todd, A. E., Thornton, J.M. (1999) From protein structure to function, *Curr Opin Struct Biol*, **9**, 374–382.
77. Wilson, C. A., Kreychman, J., Gerstein, M. (2000) Assessing annotation transfer for genomics: quantifying the relations between protein sequence, structure and function through traditional and probabilistic scores, *J Mol Biol*, **297**, 233–249.
78. Rost, B. (2002) Enzyme function less conserved than anticipated, *J Mol Biol*, **318**, 595–608.
79. Pawlowski, K., Godzik, A. (2001) Surface map comparison: studying function diversity of homologous proteins, *J Mol Biol*, **309**, 793–806.
80. Gardy, J. L., Spencer, C., Wang, K. et al. (2003) PSORT-B: Improving protein subcellular localization prediction for Gram-negative bacteria, *Nucleic Acids Res*, **31**, 3613–3617.
81. Alexandrov, N. N., Soloveyev, V. V. (1998), Statistical significance of ungapped sequence alignments, in Altman R.B., Dunker A.K., Hunter L. et al. Eds., *HICCS' 98: Pacific Symposium on Biocomputing' 98*, World Scientific, Maui, Hawaii, U.S.A., pp. 463–472.
82. Rost, B. (1999) Twilight zone of protein sequence alignments, *Protein Eng*, **12**, 85–94.
83. Wrzeszczynski, K. O., Rost, B. (2004) Annotating proteins from Endoplasmic reticulum and Golgi apparatus in eukaryotic proteomes, *Cel Mol Life Sci*, **61**, 1341–1353.
84. Pawlowski, K., Jaroszewski, L., Rychlewski, L. et al. (2000) Sensitive sequence comparison as protein function predictor, *Pac Symp Biocomput*, **8**, 42–53.
85. Altschul, S. F., Gish, W., Miller, W. et al. (1990) Basic local alignment search tool, *J Mol Biol*, **215**, 403–410.
86. Altschul, S. F., Madden, T. L., Schaffer, A. A. et al. (1997) Gapped BLAST and PSI-BLAST: a new generation of protein database search programs, *Nucleic Acids Res*, **25**, 3389–3402.
87. Sander, C., Schneider, R. (1991) Database of homology-derived protein structures and the structural meaning of sequence alignment, *Proteins*, **9**, 56–68.
88. Valencia, A., Pazos, F. (2002) Computational methods for the prediction of protein interactions, *Curr Opin Struct Biol*, **12**, 368–373.
89. Stapley, B. J., Kelley, L. A., Sternberg, M. J. (2002) Predicting the sub-cellular location of proteins from text using support vector machines, *Pac Symp Biocomput*, 374–385.

90. Mika, S., Rost B. (2004) Protein names precisely peeled off free text, *Bioinformatics,* **20**, I241–I247.
91. Mika, S., Rost, B. (2004) NLProt: extracting protein names and sequences from papers, *Nucleic Acids Res,* **32**, W634–W637.
92. Tamames, J., Ouzounis, C., Casari, G. et al. (1998) EUCLID: automatic classification of proteins in functional classes by their database annotations, *Bioinformatics,* **14**, 542–543.
93. Nair, R., Rost, B. (2002) Inferring subcellular localization through automated lexical analysis, *Bioinformatics,* **18** Suppl 1, S78–86.
94. Lu, Z., Szafron, D., Greiner, R. et al. (2004) Predicting subcellular localization of proteins using machine-learned classifiers, *Bioinformatics,* **20**, 547–556.
95. Eisenhaber, F., Bork, P. (1999) Evaluation of human-readable annotation in biomolecular sequence databases with biological rule libraries, *Bioinformatics,* **15**, 528–535.
96. Hatzivassiloglou, V., Duboue, P. A., Rzhetsky, A. (2001) Disambiguating proteins, genes, and RNA in text: a machine learning approach, *Bioinformatics,* **17** Suppl 1, S97–106.
97. Luscombe, N. M., Greenbaum, D., Gerstein, M. (2001) What is bioinformatics? A proposed definition and overview of the field, *Methods Inf Med,* **40**, 346–358.
98. Lewis, D. D., Ringuitte, M. (1994) Comparison of two learning algorithms for text characterization, In proceeding of the Third Annual Symposium on Document Analysis and Information Retrival (SDAIR' 94), 81–93.
99. Apte, C., Damerau, F., Weiss, S. (1994) Towards language independent automated learning of text categorization models, In proceedings of The 17th Annual ACM/SIGIR Conference, 23–30.
100. Dasarathy, B. V. (1991) *Nearest Neighbor (NN) Norms: NN Pattern Classification Techniques,* Los Alamitos: IEEE Computer Society Press.
101. Kretschmann, E., Fleischmann, W., Apweiler, R. (2001) Automatic rule generation for protein annotation with the C4.5 data mining algorithm applied on SWISS-PROT, *Bioinformatics,* **17**, 920–926.
102. Bazzan, A. L., Engel, P. M., Schroeder, L. F. et al. (2002) Automated annotation of keywords for proteins related to mycoplasmataceae using machine learning techniques, *Bioinformatics,* **18** Suppl 2, S35–43.
103. Fleischmann, W., Moller, S., Gateau, A. et al. (1999) A novel method for automatic functional annotation of proteins, *Bioinformatics,* **15**, 228–233.
104. Nishikawa, K., Kubota, Y., Ooi, T. (1983) Classification of proteins into groups based on amino acid composition and other characters. II. Grouping into four types, *J Biochem,* **94**, 997–1007.
105. Nishikawa, K., Kubota, Y., Ooi, T. (1983) Classification of proteins into groups based on amino acid composition and other characters. I. Angular distribution, *J Biochem,* **94**, 981–995.
106. Nakashima, H., Nishikawa, K. (1994) Discrimination of intracellular and extracellular proteins using amino acid composition and residue-pair frequencies, *J Mol Biol,* **238**, 54–61.
107. Horton, P., Nakai, K. (1997) Better prediction of protein cellular localization sites with the k nearest neighbors classifier, *Proc Int Conf Intell Syst Mol Biol,* **5**, 147–152.
108. Nakai, K., Horton, P. (1999) PSORT: a program for detecting sorting signals in proteins and predicting their subcellular localization, *Trends Biochem Sci,* **24**, 34–36.
109. Horton, P., Nakai, K. (1996) A probabilistic classification system for predicting the cellular localization sites of proteins, *Proc Int Conf Intell Syst Mol Biol,* **4**, 109–115.
110. Reinhardt, A., Hubbard, T. (1998) Using neural networks for prediction of the subcellular location of proteins, *Nucleic Acids Res,* **26**, 2230–2236.
111. Yuan, Z. (1999) Prediction of protein subcellular locations using Markov chain models, *FEBS Lett,* **451**, 23–26.
112. Cedano, J., Aloy, P., Perez-Pons, J. A. et al. (1997) Relation between amino acid composition and cellular location of proteins, *J Mol Biol,* **266**, 594–600.
113. Chou, K. C. (2000) Prediction of protein subcellular locations by incorporating quasi-sequence-order effect, *Biochem Biophys Res Commun,* **278**, 477–483.
114. Chou, K. C. (2001) Prediction of protein cellular attributes using pseudo-amino acid composition, *Proteins,* **43**, 246–255.
115. Marcotte, E. M., Xenarios, I., van Der Bliek, A. M. et al. (2000) Localizing proteins in the cell from their phylogenetic profiles, *Proc Natl Acad Sci USA,* **97**, 12115–12120.
116. Nakai, K., Kidera, A., Kanehisa, M. (1988) Cluster analysis of amino acid indices for prediction of protein structure and function, *Protein Eng,* **2**, 93–100.
117. Nakai, K., Kanehisa, M. (1992) A knowledge base for predicting protein localization sites in eukaryotic cells, *Genomics,* **14**, 897–911.

118. Drawid, A., Gerstein M. A. (2000) Bayesian system integrating expression data with sequence patterns for localizing proteins: comprehensive application to the yeast genome, *J Mol Biol,* **301**, 1059–1075.
119. Nasibov, E., Kandemir-Cavas, C. (2008) Protein subcellular location prediction using optimally weighted fuzzy k-NN algorithm, *Comput Biol Chem,* **32**, 448–451.
120. Larose, D.T. (2005) *Discovering Knowledge in Data: An Introduction to Data Mining,* John Wiley and Sons, Inc, Hoboken, NJ.
121. Andrade, M. A., O'Donoghue, S. I., Rost, B. (1998) Adaptation of protein surfaces to subcellular location, *J Molr Biol,* **276**, 517–525.
122. Nair, R., Rost, B. (2004) LOCnet and LOCtarget: sub-cellular localization for structural genomics targets, *Nucleic Acids Res,* **32**, W517–521.
123. Nair, R., Rost, B. (2003) Better prediction of sub-cellular localization by combining evolutionary and structural information, *Proteins,* **53**, 917–930.
124. Horton, P., Park, K. J., Obayashi, T. et al. (2007) WoLF PSORT: protein localization predictor, *Nucleic Acids Res,* **35**, W585–587.
125. Ashburner, M., Ball C. A., Blake, J. A. et al. (2000) Gene ontology: tool for the unification of biology. The Gene Ontology Consortium, *Nat Genet,* **25**, 25–29.
126. Punta, M., Rost, B. (2005) PROFcon: novel prediction of long-range contacts, *Bioinformatics,* **21**, 2960–2968.

Chapter 18

The Chloroplast Protein Import Machinery: A Review

Penelope Strittmatter, Jürgen Soll, and Bettina Bölter

Abstract

Plastids are a heterogeneous family of organelles found ubiquitously in plants and algal cells. Most prominent are the chloroplasts, which carry out such essential processes as photosynthesis and the biosynthesis of fatty acids as well as of amino acids. As mitochondria, chloroplasts are derived from a single endosymbiotic event. They are believed to have evolved from an ancient cyanobacterium, which was engulfed by an early eukaryotic ancestor. During evolution the plastid genome has been greatly reduced and most of the genes have been transferred to the host nucleus. Consequently, more than 98% of all plastid proteins are translated on cytosolic ribosomes. They have to be posttranslationally targeted to and imported into the organelle. Targeting is assisted by cytosolic proteins which interact with proteins destined for plastids and thereby keep them in an import competent state. After reaching the target organelle, many proteins have to conquer the barrier of the chloroplast outer and inner envelope. This process is mediated by complex molecular machines in the outer (Toc complex) and inner (Tic complex) envelope of chloroplasts, respectively. Most proteins destined for the compartments inside the chloroplast contain a cleavable N-terminal transit peptide, whereas most of the outer envelope components insert into the membrane without such a targeting peptide.

Key words: Chloroplasts, protein targeting, in vivo import, translocation machinery, Toc/Tic, sorting.

1. What Happens in the Cytosol?

The proteins destined for the chloroplast are synthesized as precursor proteins with a transit sequence that is necessary and sufficient for correct targeting. There are two types of targeting signals. The first class of targeting signals involves internal noncleavable signals, mostly found in outer envelope proteins (OEP), such as OEP7, OEP16, OEP21, OEP24, as well as in the Toc34 (translocon at the outer envelope of chloroplasts) and Toc159

A. Economou (ed.), *Protein Secretion*, Methods in Molecular Biology 619,
DOI 10.1007/978-1-60327-412-8_18, © Springer Science+Business Media, LLC 2010

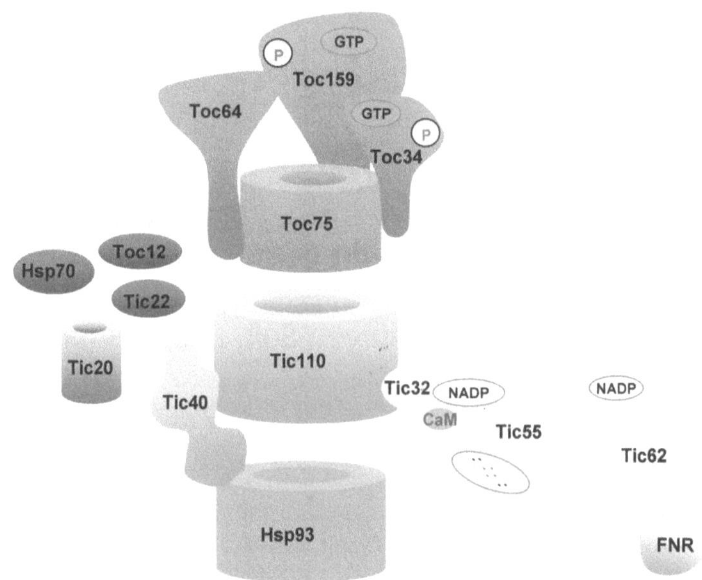

Fig. 18.1. Model of the chloroplast envelope translocation machinery The translocon at the outer envelope of chloroplasts (Toc) consists of five components: The receptor proteins Toc159 and Toc34 and the translocation pore Toc75 comprise the so-called core complex, whereas Toc64 and Toc12 are more loosely associated. Seven members of the translocon at the inner envelope (Tic) have been identified so far. Tic110 and Tic20 are discussed as being part of the import channel. Tic62, Tic55, and Tic32 exhibit features typical for redox-regulated proteins and could therefore function in the regulation of the import process. Tic40 has been shown to bind molecular chaperones. Tic22 forms an intermembrane complex with Toc12 and Toc64.

(1–6) (**Fig. 18.1**). Some OEPs can insert spontaneously into the outer envelope. Although the process of insertion is not energy dependent, a step in the insertion was shown to be stimulated by nucleotides (5, 6).

Besides some exceptions the majority of proteins that are targeted to internal chloroplast compartments are synthesized as precursor proteins with a cleavable N-terminal presequence. Targeting sequences reveal only little similarity at the level of primary sequence or length (7), but all contain predominantly positively charged and hydroxylated amino acids, such as threonine and serine (8). In vitro, serine and threonine can be phosphorylated by a cytosolic protein kinase that exclusively phosphorylates chloroplast but not mitochondrial targeting signals (9). Phosphorylation of the presequence is important not only for the binding to a cytosolic guidance complex but also for the interaction with the import receptors Toc34 and Toc159 (10, 11).

The lack of a secondary structure (12) makes preproteins good candidates for interaction with heat shock proteins (Hsp). For mitochondrial protein import the interaction of precursor proteins with chaperones is well described (13). In chloroplast protein import experiments using preLHCP purified cytosolic Hsp70 could partially substitute for leaf extract (14). No stimulation of import could be observed for soluble stroma proteins as preFd and preSSU (15, 16). But May and Soll (2000) could show an interaction of preSSU as well as preOE23 with an Hsp70 homologue after translation (17). Nevertheless, precursors with altered affinity for Hsp70 in their transit peptides are efficiently imported into chloroplasts (18). Thus, other organelle-specific factors are needed. Besides a putative Hsp70-binding site the presequence contains a kinase phosphorylation motif as mentioned above. This motif shows strong similarities to phosphopeptide binding motifs for 14-3-3 proteins, implicating that 14-3-3 proteins might be the organelle-specific factor. May and Soll presented evidence that cytosolic 14-3-3 proteins interact with phosphorylated preproteins. Together with an Hsp70 isoform and maybe other components a guidance complex is formed. This guidance complex is supposed to dock the preprotein to the chloroplast surface component prior to translocation.

At the outer envelope of chloroplasts Toc34 recognizes both, the transit sequence (19) and the 14-3-3 protein of the guidance complex via a GTP-regulated cytosolic domain (20). Through heterodimerization of Toc34 with Toc159, the precursor protein is handed from the initial receptor Toc34 to the motor protein Toc159.

Furthermore, another route for preprotein recognition at the chloroplast surface is known. The import receptor Toc64 (**Fig. 18.1**) is dynamically associated with the Toc core complex. It contains three tetratricopeptide repeats (TPR) (21). TPR motifs are domains for protein–protein interaction (22) and often mediate the interaction of proteins with molecular chaperones such as Hsp70 and Hsp90 (23). Import kinetics revealed the importance of Toc64 as initial receptor at the chloroplast surface. It was demonstrated that the C-terminal TPR domain of Toc64 recognizes precursor proteins through their interaction with Hsp90. Afterward, Toc64 itself associates with the GTP-charged Toc complex by interaction of its TPR domain with Toc34. The precursor proteins further dissociate from Hsp90 and are then recognized by Toc34. At last, delivery of the preproteins from Toc64 to the core complex leads to the dissociation of Toc64 (24).

2. How to Pass the Outer Envelope?

The core of the translocon in the outer envelope of chloroplasts comprises three proteins called Toc34, Toc75, and Toc159 (**Fig. 18.1**). They are supposed to build a complex of around

500 kDa in which four copies of Toc34, four copies of Toc75, and one copy of Toc159 are found (25).

Toc75 is the most abundant protein of the outer envelope membrane and it forms the import pore of the Toc translocon. When reconstituted into liposomes Toc75 forms a cation-selective, voltage-gated channel with a pore width of 14–26 Å (26). Resistance to high concentrations of protease indicates that Toc75 is deeply embedded in the outer envelope membrane (27). Before entering the translocation channel dephosphorylation of the preprotein has to occur since phosphorylated precursors are unable to pass Toc75 (28). Toc75 contains a precursor binding site distinct from that on the chloroplast surface and does not recognize the mature part of the preprotein. In *Arabidopsis thaliana*, four homologues of Toc75 are found, namely atToc75-I, atToc75-III, atToc75-IV, and atToc75-V, the latter also known as OEP80 (29, 30). AtToc75-I was shown to be a pseudogene, due to an insertion of a retrotransposon (31). AtToc75-IV is only expressed at very low levels and has a significant N-terminal truncation.

In contrast atToc75-III encodes a full-length protein which is expressed in all tissues. It even shows an expression pattern similar to that in pea (27, 31). Knockout mutants of atToc75-III are embryo lethal (31–33). Therefore it is very likely that Toc75-III is the general import pore in *A. thaliana*.

The strong similarities of atToc75-V to a class of bacterial transport proteins, the Omp85-family (34), suggest that it represents the most ancestral or earliest form of a Toc75-like proteins which was further modified during evolution. It has been proposed that the Toc75-like channel is derived from an ancient prokaryotic channel of smaller size and evolved by partial gene duplication in the amino terminal region of the protein (35). The fairly abundant presence of atToc75-V (about 5–10% of atToc 75-III) indicates that it has retained special functional properties that are required for the import (or export?) of its substrate(s). Alternatively, atToc75-V could be involved in membrane insertion of ß-barrel proteins, as has been shown for proteins of the Omp85-family. Recently it was shown that atToc75-V is essential for viability, but the function still remains elusive (36).

Toc34 and Toc159 have both been characterized as receptor proteins. They share the presence of a GTP-binding domain and it was shown that they function as GTP-dependent precursor protein receptors. For Toc159 a second possible function was described, which is to provide the driving force needed for traversing the membrane. The driving force for the unidirectional movement through the translocon seems to be generated on the cytosolic side by the Toc159 GTPase in a pushing mechanism. This mechanism is visualized as a sewing-machine-like process: Precursor binding causes the activation of the endogenous

Toc159 GTPase activity and a conformational change of the proteins. This activity causes movement of a preprotein binding domain toward the translocation channel. Through several rounds of preprotein binding and GTP hydrolysis the precursor protein is pushed into the channel and then across the membrane (27).

Toc159 can be divided into three parts: the A-domain, which is rich in acidic amino acids; the G-domain, which contains the GTP-binding domain; and the M-domain, which is the carboxy-terminal membrane domain. Four homologues of Toc159 can be found in *A. thaliana*: atToc159, atToc132, atToc120, and atToc90 (37). The proteins differ mainly in the size of the A-domain, which is completely absent in atToc90. The expression of atToc159 seems to be essential for correct chloroplast development. The analysis of T-DNA insertion mutants of atToc159 (*ppi2*) led to the conclusion that Toc159 is the receptor for photosynthetic proteins, while atToc132 and atToc120 seem to be primarily involved in import of non-photosynthetic proteins and act as receptors of other members in the plastidic family (38).

During protein import Toc34 is in close proximity to the preprotein as could be shown by cross-linking (39) and co-immunoprecipitation with αToc34 antibodies (40). Additionally, Toc34 can be connected covalently with the translocation pore Toc75 via a disulfide bridge (6). The close physical proximity ensures efficient translocation initiation and delivery of the preproteins to the translocation channel. Toc34 is anchored by a single transmembrane domain at the carboxy terminus in a way that the amino terminus is facing the cytosol (6,39,41). It is a preprotein receptor that is regulated both by GTP/GDP-binding and by phosphorylation in an elaborate mechanism: The affinity of the specific interaction of Toc34 with the transit peptide of the preprotein is drastically increased in the GTP-bound state. Bound precursor is released upon hydrolysis of GTP. The GDP can be replaced by GTP and either Toc34 enters a new round of transit peptide recognition or Toc34 is phosphorylated and in this way inactivated because the phosphorylated Toc34 can neither recognize preproteins nor bind GTP. Dephosphorylation and at the same time activation of Toc34 is carried out by an ATP-dependent phosphatase (40). In addition, Toc34 has a higher affinity for phosphorylated precursors than for non-phosphorylated ones. After dephosphorylation, Toc34 binds GTP and subsequently forms a high-affinity complex with phosphorylated precursors (40). A homodimerization (42) of Toc34 and a heterodimerization with the GTP-binding domain of Toc159 might cause the GDP/GTP exchange of the intrinsic GTPase. On the other hand, the preprotein can act as a GTPase-activating factor for Toc34 which causes a 10–50 fold increase in GTPase activity. The resulting GDP-Toc34 precursor complex has a lower affinity for the preprotein than the GTP-bound form and allows the preprotein to dissociate from the receptor and continues its passage through the translocon (43). Dimerization with Toc159 occurs in the GDP-bound form of Toc34.

In *A. thaliana* two Toc34 homologues could be identified: atToc33 and atToc34, both of which are expressed in vivo (44). While atToc34 is almost equally expressed in all tissues and all stages of development, the expression of atToc33 seems to be upregulated in photosynthetic and meristematic tissue (41). Another difference between both proteins is that only atToc33 is regulated by phosphorylation. T-DNA insertion lines of atToc33 (*ppi1*) show a pale green phenotype and retarded chloroplast development (44, 45). In later stages of development, however, plants recover and are able to grow on soil. This observation is probably due to the fact that atToc34 can partially take over the function of atToc33. Nevertheless, both proteins show clear preferences for different classes of preproteins.

The discovery of the two very similar Toc components atToc33 and atToc34; the small protein family of atToc159, atToc132, and atToc120; and two atToc75 isoforms could be an indication for the existence of at least two distinct translocon complexes in *Arabidopsis.* The capacity of plastids to import proteins is regulated developmentally and peaks during the early days of organ expansion (27, 46). The reason for changing import efficiencies could be the presence of translocon complexes made of different subunit combinations.

In addition to the core components, there are two auxiliary components: Toc64 and Toc12 (21, 47). Toc64 could be cross-linked to several Toc components. It is an integral membrane protein that seems to be built from two independent modules. One module exhibits homologies to amidases. No amidase activity could be measured either from isolated envelope membranes or from overexpressed Toc64 since the amidase function seems to be inactivated by a point mutation in the active site. As already mentioned, the second module is a threefold repeated TPR motif, which is exposed to the cytosol (21).

There are three homologues of atToc64 (atToc64-III, atToc64-V, and atToc64-I) in the genome of *A. thaliana*, which are all expressed in vivo (39, 48). They are reported to be localized in chloroplasts, mitochondria, and the cytosol, respectively (49). Because atToc64-III (localized in the chloroplast outer membrane) and atToc64-V (localized in mitochondria) show high similarity to psToc64, it was proposed that Toc64 may play a role in recognition of dual-targeted proteins (50).

In chloroplasts, the proposed role of Toc64 is to function early in preprotein translocation not only as a recognition site for chaperone-bound preproteins (21, 24) but also as a major component of an intermembrane space complex, together with Toc12, Tic22, and an Hsp70 (**Fig. 18.1**). This complex was suggested to coordinate the interaction of Toc and Tic translocons, allowing efficient and direct translocation of preproteins through the intermembrane space. The intermembrane space domain of Toc64 is involved in preprotein recognition and association with the Toc complex independent of the cytosolic domain of the Toc64 receptor (51).

Toc12 contains a J-domain that is common to a family of Dna-J proteins (47). This J-domain is required for the interaction of these proteins with Hsp70, as it interacts with an Hsp70 in the intermembrane space (47). How the precursor engages the Tic complex is unknown, but the association of Toc and Tic complexes has been shown (52, 53) and indicates that a de novo formation of joint translocation sites is not absolutely required. The precursor proteins thereby engage both the Toc and the Tic complexes simultaneously during translocation.

3. What About the Inner Envelope?

Unlike the Toc complex, knowledge about composition and function of the Tic (translocon at the inner envelope of chloroplasts) complex is rather scarce. Until now seven components have been described: Tic110, Tic62, Tic55, Tic40, Tic32, Tic22, and Tic20 (**Fig. 18.1**). Recently an additional component was proposed, namely Tic21 (54), but the function as part of the Tic complex is questioned as another study identified the same protein as a metal permease (55).

Tic110 was the first component to be found. After cross-linking of preproteins, Tic110 and Toc75 could be co-immunoprecipitated with antibodies against preSSU (56) implicating a close proximity of Tic110 to the Toc complex. Additionally, Tic110 is supposed to recruit chaperones from the stroma such as Cpn60 (Hsp60) or the Hsp100 homologue Hsp93/ClpC (caseinolytic protease, subunit C) (53, 57). ClpC is supposed to play a role in driving preprotein translocation, while Cpn60 is believed to be responsible for folding newly imported proteins. Heins and co-workers could show by electrophysiological measurements and binding studies with preSSU that Tic110 is a central part of the import pore of the Tic complex, forming a cation-selective channel similar to Toc75 (58). In *Arabidopsis* only one homologue of Tic110 is present (48), and T-DNA insertion lines of atTic110 are embryo lethal, pointing at a general role for atTic110 the translocation of plastidic precursor proteins across the inner envelope membrane (59).

Tic40 was originally suggested to be involved in preprotein translocation at both membranes (60). However, Stahl and co-workers demonstrated by cross-linking and immunoprecipitation that it is an exclusive component of the Tic complex (61). Tic40 is predicted to be integrated into the inner membrane by a single N-terminal transmembrane domain.

Tic40 most likely forms a ternary complex together with Tic110 and Hsp93/ClpC (62–64). The C-terminus of Tic40 shares similarity with the C-terminal Sti1 domains of the mammalian Hsp70-interacting protein (Hip) and Hsp70/Hsp90-organizing protein (Hop) co-chaperones. In addition, Tic40 may possess a tetratricopeptide repeat (TPR) protein–protein interaction domain, another characteristic feature of Hip/Hop co-chaperones (65). Moreover, it was recently demonstrated that *tic40*-null mutants accumulate soluble intermediates of inner envelope proteins in the stroma due to a slower rate of insertion (66). Therefore, Tic40 is now believed to act as a co-chaperone, increasing the efficiency of precursor processing and translocation (67).

Three components of the Tic complex, namely Tic32, Tic55, and Tic62, contain structural motifs that could act as redox-active regulatory components in the import process at the stage of the inner envelope (68).

Tic32 was found by screening for interaction partners of the N-terminus of Tic110 (69). It is a member of the short-chain dehydrogenases/reductase (SDR) family and does indeed show dehydrogenase activity which was shown to be NADPH dependent (70). Furthermore, it was demonstrated that Tic32 is a calmodulin-binding protein and that its dehydrogenase activity is affected by calmodulin, suggesting that redox regulation might be modulated by calcium signals (70, 71).

Using blue-native gel electrophoresis, Tic55 and later Tic62 were found in a translocation complex together with Tic110 (72, 73). Intriguingly, both proteins show features from known redox proteins: Tic55 contains a Rieske-type iron–sulfur cluster and a mononuclear iron-binding site. Usually Rieske-iron clusters are known to be involved in electron transfer chains as, e.g., photosynthesis and the mitochondrial respiration chain. The cytochrome b_6f complex, for example, contains a Rieske-type iron–sulfur cluster. There is also some evidence that iron–sulfur proteins function as sensors. The SoxR-system in *E. coli,* for example, is composed of a protein iron–sulfur center to regulate gene expression. The oxidation state of the iron–sulfur center in the redox-sensing SoxR protein controls its own activity as a transcription activator independent of DNA-binding ability. Thus, iron–sulfur centers link cellular oxidative stress to the expression of defense genes (74, 75). Another clue for the participation of iron–sulfur proteins in plastidic protein translocation came from import studies in which the chloroplasts had been pretreated with diethyl pyrocarbonate (DEPC). DEPC changes histidine residues in iron–sulfur centers. Import into pretreated chloroplasts was inhibited at the level of the inner envelope translocon (72), possibly because the function of Tic55 was disturbed. Recently, it could be demonstrated that Tic55 is a target for thioredoxins. Thioredoxins contain a redox-active disulfide bridge and seem to interact with the redox-active CxxC motif of Tic55. Therefore, it was proposed that thioredoxins play a role in dark/light regulation of protein translocation (76).

Tic62 is a member of the extended family of SDRs with an active dehydrogenase domain at its N-terminus, while the C-terminal region contains a repetitive module rich in serine and proline residues, which acts as a specific ferredoxin-$NADP^+$-oxidoreductase (FNR)-binding motif (68,73,77). Whereas the N-terminal module of Tic62 is highly conserved among all oxyphototrophs, the C-terminal region is only found in vascular plants (77). Therefore, the ability to interact with FNR seems to be an evolutionary new feature of Tic62. The $NADP^+$/NADPH ratio in the stroma has a strong influence on Tic62. Changes in the $NADP^+$/NADPH pool result in a shuttling behavior of Tic62 between the membrane and the stroma. Not only the localization of Tic62 within the chloroplast but also the interaction of Tic62 with FNR and the Tic complex is affected (68). These results indicate that Tic62 possesses redox-dependent properties that would allow it to fulfil a role as redox-sensor protein in the chloroplast.

Thus Tic32, Tic55, and Tic62 seem to be perfect candidates to fulfil the role of a regulatory system at the inner envelope.

The two smallest components, Tic22 and Tic20, were identified by label transfer cross-link experiments indicating that both proteins are in close proximity to the preprotein during translocation (78).

Tic22 is supposed to be located in the intermembrane space, where it is a part of the intermembrane space complex together with Toc64, Toc12, and Hsp70 (47, 51). In the *Arabidopsis* genome there are two homologues of Tic22, namely atTic22-III and atTic22-VI (48).

Tic20 was shown to be an integral membrane protein with four predicted α-helical membrane spans. In the *Arabidopsis* genome there are four homologues for Tic20 (54, 79), namely atTic20-I (most closely related to the pea protein), atTic20-II, atTic20-IV, and atTic20-V. The importance of Tic20 for chloroplast viability was revealed in *Arabidopsis* antisense lines where the expression of atTic20-I was altered. The plants exhibited chloroplast defects illustrated by pale leaves, reduced accumulation of plastid proteins, and significant growth defects (79). Subsequently, the atTic20-I knockout mutant revealed an albino phenotype (54).

Tic20 shows weak similarities to prokaryotic branched-chain amino acid transporters and to the mitochondrial channel proteins Tim23 (translocase of the inner mitochondrial membrane), Tim22, and Tim17 (80). Therefore it was favored to be part of the translocation pore. However, no in vitro data support this idea so far. Therefore the exact function of Tic20 needs further investigation.

Although Tic110 and Tic20 are both proposed to form preprotein translocation channels, they were never shown to interact with each other directly. One explanation could be that they might work autonomously, possibly in different subcomplexes. In mitochondria, a plurality of translocons has been described in protein import as there are at least two Tim complexes (81). The Tim23 complex imports matrix proteins with the typical N-terminal targeting sequence and the Tim22 complex is in charge of targeting integral inner membrane proteins. Considering the mitochondrial import system, the existence of two independent channels also in chloroplasts is conceivable. Moreover, different import pathways for chloroplastidic proteins have been described. The analysis of different inner envelope proteins, e.g., already revealed several possibilities for the sorting and insertion:

Proteins using the conservative sorting pathway, e.g., Tic110 and Tic40, are first fully translocated into the stroma and then re-targeted into the membrane (82–85).

In contrast, proteins using the stop-transfer mechanism are arrested within the translocon and laterally discharged into the membrane. This was proposed for Arc6 (84) and for the triose phosphate-3-phosphoglycerate-phosphate translocator (TPT) from spinach (86).

Only recently, import of a number of hydrophobic inner envelope proteins comprising a transit peptide has been investigated (87). It was shown that they all use the stop-transfer pathway, but that they diverge in their processing to the mature form. In this respect, they can be classified into two groups. Members of the first group possess a typical transit peptide, which is processed by the stromal processing peptidase (SPP) after passage through the Toc and the Tic complex, probably still arrested in the translocon with only the transit peptide protruding into the stroma. After processing the mature form is integrated directly from the translocon into the plane of the membrane. Proteins of the second subfamily have a bipartite transit peptide, of which the processing steps are not yet clear, and are integrated into the inner envelope even before the second cleavage (87).

A completely different pathway is used by ceQORH and Tic32 (88, 89). These proteins are synthesized without cleavable transit peptides and use a route alternative from the general import pathway via Toc and Tic.

Taken together, the import into the inner envelope and the assembly of the Tic complex/complexes is much more multi-faceted and diverse than has been assumed so far.

4. Arrival in the Stroma

Like Hsp70 in mitochondria the stromal Hsp100-like protein Hsp93/ClpC is supposed to pull precursor proteins across the envelope membranes in an ATP-dependent manner (90). Hsp93/ClpC was found to be permanently associated with the Tic complex, whereas the stromal chaperone Cpn60 seems to interact with the Tic complex only in the presence of a translocating precursor (91). Cpn60 is most likely involved in folding and assembly of the translocated protein (92). After translocation into the stroma the SPP completes import by the removal of the transit sequence (93). The proteins are then either folded in an active conformation or further sorted toward the thylakoid membrane or lumen. Generally there are four different sorting pathways into the thylakoid compartment. Thylakoid lumen proteins are either targeted via the Tat (twin-arginine translocation) or the Sec pathways, while proteins destined for the thylakoid membrane are inserted through the SRP (signal recognition particle) pathway or are inserted spontaneously (94).

The pathways have been characterized in some detail and are reviewed elsewhere (95, 96).

References

1. Bölter, B., Soll, J., Hill, K., Hemmler, R., and Wagner, R. (1999) A rectifying ATP-regulated solute channel in the chloroplastic outer envelope from pea. *EMBO J* **18**, 5505–5516.
2. Li, H. M. and Chen, L. J. (1996) Protein targeting and integration signal for the chloroplastic outer envelope membrane. *Plant Cell* **8**, 2117–2126.
3. Pohlmeyer, K., Soll, J., Steinkamp, T., Hinnah, S., and Wagner, R. (1997) Isolation and characterization of an amino acid-selective channel protein present in the chloroplastic outer envelope membrane. *PNAS* **94**, 9504–9509.
4. Pohlmeyer, K., Soll, J., Grimm, R., Hill, K., and Wagner, R. (1998) A high-conductance solute channel in the chloroplastic outer envelope from Pea. *Plant Cell* **10**, 1207–1216.
5. Salomon, M., Fischer, K., Flugge, U. I., and Soll, J. (1990) Sequence analysis and protein import studies of an outer chloroplast envelope polypeptide. *PNAS* **87**, 5778–5782.
6. Seedorf, M., Waegemann, K., and Soll, J. (1995) A constituent of the chloroplast import complex represents a new type of GTP-binding protein. *Plant J* 7, 401–411.
7. von Heijne, G., Steppuhn, J., and Herrmann, R. G. (1989) Domain structure of mitochondrial and chloroplast targeting peptides. *Eur J Biochem* **180**, 535–545.
8. Cline, K. (2000) Gateway to the chloroplast. *Nature* **403**, 148–149.
9. Waegemann, K. and Soll, J. (1996) Phosphorylation of the transit sequence of chloroplast precursor proteins. *J Biol Chem* **271**, 6545–6554.
10. Fulgosi, H., Soll, J., de Faria, M. S., Korthout, H. A., Wang, M., and

Testerink, C. (2002) 14-3-3 proteins and plant development. *Plant Mol Biol* **50**, 1019–1029.
11. Oreb, M., Hofle, A., Mirus, O., and Schleiff, E. (2008) Phosphorylation regulates the assembly of chloroplast import machinery. *J Exp Bot* **59**, 2309–2316.
12. vonHeijne, G. (1989) Control of topology and mode of assembly of a polytopic membrane protein by positively charged residues. *Nature* **341**, 456–458.
13. Pfanner, N., Craig, E. A., and Honlinger, A. (1997) Mitochondrial preprotein translocase. *Annu Rev Cell Dev Biol* **13**, 25–51.
14. Waegemann, K., Paulsen, H., and Soll, J. (1990) Translocation of proteins into chloroplasts requires cytosolic factors to obtain import competence. *FEBS Lett* **261**, 89–92.
15. Pilon, M., de Boer, A. D., Knols, S. L., Koppelman, M. H., van der Graaf, R. M., de Kruijff, B., and Weisbeek, P. J. (1990) Expression in Escherichia coli and purification of a translocation-competent precursor of the chloroplast protein ferredoxin. *J Biol Chem* **265**, 3358–3361.
16. Pilon, M., Rietveld, A. G., Weisbeek, P. J., and de Kruijff, B. (1992) Secondary structure and folding of a functional chloroplast precursor protein. *J Biol Chem* **267**, 19907–19913.
17. May, T. and Soll, J. (2000) 14-3-3 proteins form a guidance complex with chloroplast precursor proteins in plants. *Plant Cell* **12**, 53–64.
18. Rial, D. V., Ottado, J., and Ceccarelli, E. A. (2003) Precursors with altered affinity for Hsp70 in their transit peptides are efficiently imported into chloroplasts. *J Biol Chem* **278**, 46473–46481.
19. Schleiff, E., Soll, J., Sveshnikova, N., Tien, R., Wright, S., Dabney-Smith, C., Subramanian, C., and Bruce, B. D. (2002) Structural and guanosine triphosphate/diphosphate requirements for transit peptide recognition by the cytosolic domain of the chloroplast outer envelope receptor, Toc34. *Biochemistry* **41**, 1934–1946.
20. Schleiff, E., Jelic, M., and Soll, J. (2003) A GTP-driven motor moves proteins across the outer envelope of chloroplasts. *Proc Natl Acad Sci U S A* **100**, 4604–4609.
21. Sohrt, K. and Soll, J. (2000) Toc64, a new component of the protein translocon of chloroplasts. *J Cell Biol* **148**, 1213–1221.
22. Lamb, J. R., Tugendreich, S., and Hieter, P. (1995) Tetratrico peptide repeat interactions: to TPR or not to TPR? *Trends Biochem Sci* **20**, 257–259.
23. Frydman, J. and Hohfeld, J. (1997) Chaperones get in touch: the Hip-Hop connection. *Trends Biochem Sci* **22**, 87–92.
24. Qbadou, S., Becker, T., Mirus, O., Tews, I., Soll, J., and Schleiff, E. (2006) The molecular chaperone Hsp90 delivers precursor proteins to the chloroplast import receptor Toc64. *EMBO J* **25**, 1836–1847.
25. Schleiff, E., Soll, J., Kuchler, M., Kuhlbrandt, W., and Harrer, R. (2003) Characterization of the translocon of the outer envelope of chloroplasts. *J Cell Biol* **160**, 541–551.
26. Hinnah, S. C., Wagner, R., Sveshnikova, N., Harrer, R., and Soll, J. (2002) The chloroplast protein import channel Toc75: pore properties and interaction with transit peptides. *Biophys J* **83**, 899–911.
27. Tranel, P. J., Froehlich, J., Goyal, A., and Keegstra, K. (1995) A component of the chloroplastic protein import apparatus is targeted to the outer envelope membrane via a novel pathway. *EMBO J* **14**, 2436–2446.
28. Waegemann, K. and Soll, J. (1996) Phosphorylation of the transit sequence of chloroplast precursor proteins. *J Biol Chem* **271**, 6545–6554.
29. Eckart, K., Eichacker, L., Sohrt, K., Schleiff, E., Heins, L., and Soll, J. (2002) A Toc75-like protein import channel is abundant in chloroplasts. *EMBO Rep* **3**, 557–562.
30. Jackson-Constan, D. and Keegstra, K. (2001) Arabidopsis genes encoding components of the chloroplastic protein import apparatus. *Plant Physiol* **125**, 1567–1576.
31. Baldwin, A., Wardle, A., Patel, R., Dudley, P., Park, S. K., Twell, D., Inoue, K., and Jarvis, P. (2005) A molecular-genetic study of the Arabidopsis Toc75 gene family. *Plant Physiol* **138**, 715–733.
32. Constan, D., Patel, R., Keegstra, K., and Jarvis, P. (2004) An outer envelope membrane component of the plastid protein import apparatus plays an essential role in Arabidopsis. *Plant J* **38**, 93–106.
33. Hust, B. and Gutensohn, M. (2006) Deletion of core components of the plastid protein import machinery causes differential arrest of embryo development in Arabidopsis thaliana. *Plant Biol* 18–30.
34. Schleiff, E. and Soll, J. (2005) Membrane protein insertion: mixing eukaryotic and prokaryotic concepts. *EMBO Rep* **6**, 1023–1027.
35. Reumann, S. and Keegstra, K. (1999) The endosymbiotic origin of the protein import machinery of chloroplastic envelope membranes. *Trends Plant Sci* **4**, 302–307.
36. Patel, R., Hsu, S. C., Bedard, J., Inoue, K., and Jarvis, P. (2008) The omp85-related

chloroplast outer envelope protein OEP80 is essential for viability in Arabidopsis. *Plant Physiol* **148**, 235–245.
37. Hiltbrunner, A., Bauer, J., Alvarez-Huerta, M., and Kessler, F. (2001) Protein translocon at the Arabidopsis outer chloroplast membrane. *Biochem Cell Biol* 79, 629–635.
38. Kubis, S., Patel, R., Combe, J., Bedard, J., Kovacheva, S., Lilley, K., Biehl, A., Leister, D., Rios, G., Koncz, C., and Jarvis, P. (2004) Functional specialization amongst the Arabidopsis Toc159 family of chloroplast protein import receptors. *The Plant Cell* **16**, 2059–2077.
39. Kessler, F., Blobel, G., Patel, H. A., and Schnell, D. J. (1994) Identification of two GTP-binding proteins in the chloroplast protein import machinery. *Science* **266**, 1035–1039.
40. Sveshnikova, N., Soll, J., and Schleiff, E. (2000) Toc34 is a preprotein receptor regulated by GTP and phosphorylation. *PNAS* **97**, 4973–4978.
41. Hirsch, S., Muckel, E., Heemeyer, F., von Heijne, G., and Soll, J. (1994) A receptor component of the chloroplast protein translocation machinery. *Science* **266**, 1989–1992.
42. Sun, Y. J., Forouhar, F., Li Hm, H. M., Tu, S. L., Yeh, Y. H., Kao, S., Shr, H. L., Chou, C. C., Chen, C., and Hsiao, C. D. (2002) Crystal structure of pea Toc34, a novel GTPase of the chloroplast protein translocon. *Nat Struct Biol* 9, 95–100.
43. Jelic, M., Sveshnikova, N., Motzkus, M., Hörth, P., Soll, J., and Schleiff, E. (2002) The chloroplast import receptor Toc34 functions as preprotein regulated GTPase. *J Biol Chem* **12**, 1875–1883
44. Jarvis, P., Chen, L. J., Li, H., Peto, C. A., Fankhauser, C., and Chory, J. (1998) An Arabidopsis mutant defective in the plastid general protein import apparatus. *Science* **282**, 100–103.
45. Gutensohn, M., Schulz, B., Nicolay, P., and Flugge, U. I. (2000) Functional analysis of the two Arabidopsis homologues of Toc34, a component of the chloroplast protein import apparatus. *Plant J* **23**, 771–783.
46. Dahlin, C. and Cline, K. (1991) Developmental regulation of the plastid protein import apparatus. *Plant Cell* **3**, 1131–1140.
47. Becker, T., Hritz, J., Vogel, M., Caliebe, A., Bukau, B., Soll, J., and Schleiff, E. (2004) Toc12, a novel subunit of the intermembrane space preprotein translocon of chloroplasts. *Mol Biol Cell* **15**, 5130–5144.
48. Jackson-Constan, D. and Keegstra, K. (2001) Arabidopsis genes encoding components of the chloroplastic protein import apparatus. *Plant Physiol* **125**, 1567–1576.
49. Aronsson, H., Boij, P., Patel, R., Wardle, A., Töpel, M., and Jarvis, P. (2007) Toc64/OEP64 is not essential for the efficient import of proteins into chloroplasts in Arabidopsis thaliana. *Plant J* **52**, 53–68.
50. Chew, O. and Whelan, J. (2004) Just read the message: a model for sorting of proteins between mitochondria and chloroplasts. *Trends Plant Sci* **9**, 318–319.
51. Qbadou, S., Becker, T., Bionda, T., Reger, K., Ruprecht, M., Soll, J., and Schleiff, E. (2007) Toc64 - A Preprotein-receptor at the outer membrane with bipartide function. *J Mol Biol* **367**, 1330–1346.
52. Akita, M., Nielsen, E., and Keegstra, K. (1997) Identification of protein transport complexes in the chloroplastic envelope membranes via chemical cross-linking. *J Cell Biol* **136**, 983–994.
53. Nielsen, E., Akita, M., Davila-Aponte, J., and Keegstra, K. (1997) Stable association of chloroplastic precursors with protein translocation complexes that contain proteins from both envelope membranes and a stromal Hsp100 molecular chaperone. *EMBO J* **16**, 935–946.
54. Teng, Y. S., Su, Y. s., Chen, L. J., Lee, Y. J., Hwang, I., and Li, H. M. (2006) Tic21 is an essential translocon component for protein translocation across the chloroplast inner envelope membrane. *Plant Cell* **18**, 2247–2257.
55. Duy, D., Wanner, G., Meda, A. R., von Wiren, N., Soll, J., and Philippar, K. (2007) PIC1, an ancient permease in Arabidopsis chloroplasts, mediates iron transport. *Plant Cell* **19**, 986–1006.
56. Lübeck, J., Soll, J., Akita, M., Nielsen, E., and Keegstra, K. (1996) Topology of IEP110, a component of the chloroplastic protein import machinery present in the inner envelope membrane. *EMBO J* **15**, 4230–4238.
57. Kessler, F. and Blobel, G. (1996) Interaction of the protein import and folding machineries of the chloroplast. *PNAS* **93**, 7684–7689.
58. Heins, L., Mehrle, A., Hemmler, R., Wagner, R., Küchler, M., Hörmann, F., Sveshnikov, D., and Soll, J. (2002) The preprotein conducting channel at the inner envelope membrane of plastids. *EMBO J* **21**, 2616–2625.
59. Inaba, T., Alvarez-Huerta, M., Li, M., Bauer, J., Ewers, C., Kessler, F., and Schnell, D. J. (2005) Arabidopsis tic110 is essential for the assembly and function of the protein import machinery of plastids. *Plant Cell* **17**, 1482–1496.

60. Ko, K., Budd, D., Wu, C., Seibert, F., Kourtz, L., and Ko, Z. W. (1995) Isolation and characterization of a cDNA clone encoding a member of the Com44/Cim44 envelope components of the chloroplast protein import apparatus. *J Biol Chem* **270**, 28601–28608.
61. Stahl, T., Glockmann, C., Soll, J., and Heins, L. (1999) Tic40, a new "old" subunit of the chloroplast protein import translocon. *J Biol Chem* **274**, 37467–37472.
62. Chou, M. L., Fitzpatrick, L. M., Tu, S. L., Budziszewski, G., Potter-Lewis, S., Akita, M., Levin, J. Z., Keegstra, K., and Li, H. M. (2003) Tic40, a membrane-anchored co-chaperone homolog in the chloroplast protein translocon. *EMBO J* **22**, 2970–2980.
63. Inaba, T., Li, M., Alvarez-Huerta, M., Kessler, F., and Schnell, D. J. (2003) AtTic110 functions as a scaffold for coordinating the stromal events of protein import into chloroplasts. *J Biol Chem* **278**, 38617–38627.
64. Kovacheva, S., Bedard, J., Patel, R., Dudley, P., Twell, D., Rios, G., Koncz, C., and Jarvis, P. (2005) In vivo studies on the roles of Tic110, Tic40 and Hsp93 during chloroplast protein import. *Plant J* **41**, 412–428.
65. Bedard, J., Kubis, S., Bimanadham, S., and Jarvis, P. (2007) Functional similarity between the chloroplast translocon component, Tic40, and the human co-chaperone, Hsp70-interacting protein (Hip). *J Bio Chem* **282**, 21404–21414.
66. Chiu, C. C. and Li, H. M. (2008) Tic40 is important for reinsertion of proteins from the chloroplast stroma into the inner membrane. *Plant J* Epub ahead of print.
67. Chou, M. L., Chu, C. C., Chen, L. J., Akita, M., and Li, H. M. (2006) Stimulation of transit-peptide release and ATP hydrolysis by a cochaperone during protein import into chloroplasts. *J Cell Biol* **175**, 893–900.
68. Stengel, A., Benz, J. P., Balsera, M., and Soll, J. A. B. B. (2008) Tic62 – Redox regulated translocon composition and dynamics. *J Biol Chem* **283**, 6656–6667.
69. Hormann, F., Kuchler, M., Sveshnikov, D., Oppermann, U., Li, Y., and Soll, J. (2004) Tic32, an essential component in chloroplast biogenesis. *J Biol Chem* **279**, 34756–34762.
70. Chigri, F., Soll, J., and Vothknecht, U. C. (2005) Calcium regulation of chloroplast protein import. *Plant J* **42**, 821–831.
71. Chigri, F., Hormann, F., Stamp, A., Stammers, D. K., Bolter, B., Soll, J., and Vothknecht, U. C. (2006) Calcium regulation of chloroplast protein translocation is mediated by calmodulin binding to Tic32. *Proc Natl Acad Sci U S A* **103**, 16051–16056.
72. Caliebe, A., Grimm, R., Kaiser, G., Lubeck, J., Soll, J., and Heins, L. (1997) The chloroplastic protein import machinery contains a Rieske-type iron-sulfur cluster and a mononuclear iron-binding protein. *EMBO J* **16**, 7342–7350.
73. Kuchler, M., Decker, S., Hormann, F., Soll, J., and Heins, L. (2002) Protein import into chloroplasts involves redox-regulated proteins. *EMBO J* **21**, 6136–6145.
74. Ding, H., Hidalgo, E., and Demple, B. (1996) The redox state of the [2Fe-2S] clusters in SoxR protein regulates its activity as a transcription factor. *J Biol Chem* **271**, 33173–33175.
75. Hidalgo, E., Ding, H., and Demple, B. (1997) Redox signal transduction via iron-sulfur clusters in the SoxR transcription activator. *Trends Biochem Sci* **22**, 207–210.
76. Bartsch, S., Monnet, J., Selbach, K., Quigley, F., Gray, J., von Wettstein, D., Reinbothe, S., and Reinbothe, C. (2008) Three thioredoxin targets in the inner envelope membrane of chloroplasts function in protein import and chlorophyll metabolism. *Proc Nat Acad Sci* **105**, 4933–4938.
77. Balsera, M., Stengel, A., Soll, J., and Bolter, B. (2007) Tic62: a protein family from metabolism to protein translocation. *BMC Evol Biol* **7**, 43.
78. Kouranov, A., Chen, X., Fuks, B., and Schnell, D. J. (1998) Tic20 and Tic22 are new components of the protein import apparatus at the chloroplast inner envelope membrane. *J Cell Biol* **143**, 991–1002.
79. Chen, X., Smith, M. D., Fitzpatrick, L., and Schnell, D. J. (2002) In vivo analysis of the role of atTic20 in protein import into chloroplasts. *Plant Cell* **14**, 641–654.
80. Rassow, J., Dekker, P. J., van Wilpe, S., Meijer, M., and Soll, J. (1999) The preprotein translocase of the mitochondrial inner membrane: function and evolution. *J Mol Biol* **286**, 105–120.
81. Koehler, C. M. (2000) Protein translocation pathways of the mitochondrion. *FEBS Lett* **476**, 27–31.
82. Li, M. and Schnell, D. J. (2006) Reconstitution of protein targeting to the inner envelope membrane of chloroplasts. *J Cell Biol* **175**, 249–259.
83. Lubeck, J., Heins, L., and Soll, J. (1997) A nuclear-coded chloroplastic inner envelope membrane protein uses a soluble sorting intermediate upon import into the organelle. *J Cell Biol* **137**, 1279–1286.

84. Tripp, J., Inoue, K., Keegstra, K., and Froehlich, J. E. (2007) A novel serine/proline-rich domain in combination with a transmembrane domain is required for the insertion of AtTic40 into the inner envelope membrane of chloroplasts. *Plant J* **52** (5), 824–838.
85. Vojta, L., Soll, J., and Bolter, B. (2007) Requirements for a conservative protein translocation pathway in chloroplasts. *FEBS Lett* **581**, 2621–2624.
86. Flügge, U., Fischer, K., Gross, A., Sebald, W., Lottspeich, F., and Eckerskorn, C. (1989) The triose phosphate-3-phosphoglycerate-phosphate translocator from spinach chloroplasts: nucleotide sequence of a full-length cDNA clone and import of the in vitro synthesized precursor protein into chloroplasts. *EMBO J* **8**, 39–46.
87. Firlej-Kwoka, E., Strittmatter, P., Soll, J., and Bölter, B. (2008) Import of preproteins into the chloroplast inner envelope membrane. *Plant Mol Biol* **68**, 505–519.
88. Miras, S., Salvi, D., Ferro, M., Grunwald, D., Garin, J., Joyard, J., and Rolland, N. (2002) Non-canonical transit peptide for import into the chloroplast. *J Biol Chem* **277**, 47770–47778.
89. Nada, A. and Soll, J. (2004) Inner envelope protein 32 is imported into chloroplasts by a novel pathway. *J Cell Sci* **117**, 3975–3982.
90. Keegstra, K. and Cline, K. (1999) Protein import and routing systems of chloroplasts. *Plant Cell* **11**, 557–570.
91. Nielsen, E., Akita, M., Davila-Aponte, J., and Keegstra, K. (1997) Stable association of chloroplastic precursors with protein translocation complexes that contain proteins from both envelope membranes and a stromal Hsp100 molecular chaperone. *EMBO J* **16**, 935–946.
92. Tsugeki, R. and Nishimura, M. (1993) Interaction of homologues of Hsp70 and Cpn60 with ferredoxin-NADP+ reductase upon its import into chloroplasts. *FEBS Lett* **320**, 198–202.
93. Richter, S. and Lamppa, G. K. (1998) A chloroplast processing enzyme functions as the general stromal processing peptidase. *PNAS* **95**, 7463–7468.
94. Di Cola, A., Klostermann, E., Robinson, C. (2005) The complexity of pathways for proteins import into thylakoids: it's not easy being green. *Biochem Soc Trans* **33**, 1024–1027
95. Gutensohn, M., Fan, E., Frielingsdorf, S., Hanner, P., Hou, B., Hust, B., and Klosgen, R. B. (2006) Toc, Tic, Tat et al.: structure and function of protein transport machineries in chloroplasts. *J Plant Physiol* **163**, 333–347.
96. Robinson, C., Thompson, S. J., and Woolhead, C. (2001) Multiple pathways used for the targeting of thylakoid proteins in chloroplasts. *Traffic* **2**, 245–251.

Chapter 19

Measurement of the Energetics of Protein Transport Across the Chloroplast Thylakoid Membrane

Steven M. Theg

Abstract

Protein transport across cellular membranes represents an unknown, possibly significant drain on the total energy pool. Many protein transport systems utilize a mixture of energetic inputs, with contributions from both NTP hydrolysis and transmembrane electrochemical gradients. Both of these parameters will have to be measured before we can know the cost to the cell of its considerable protein transport activities. We describe here methods to evaluate the magnitude of the ΔpH across the thylakoid membrane, which serves as the driving force for protein transport on the cpTat pathway, and to determine how much energy is drained therefrom per protein translocated. The methods derive from spectroscopic techniques, well known in the field of thylakoid energetics, to monitor the light-dependent ΔpH across the membrane and the rate of proton flux through the thylakoid lumen, combined with those to measure the rate of protein transport across the thylakoid membrane.

Key words: Protein transport, energetics, thylakoid, membrane, ΔpH, proton pump.

1. Introduction

Protein traffic across biological membranes represents a considerable cellular activity, as it has been estimated that as many as 50% of the proteins made on cytoplasmic ribosomes are translocated, either cotranslationally or posttranslationally, across one or more membranes (1). Almost every protein translocation mechanism described requires the expenditure of energy to affect transport, and many studies have been directed at elucidating the nature of that energetic input (some of which are reviewed in (2–4)).

A. Economou (ed.), *Protein Secretion*, Methods in Molecular Biology 619,
DOI 10.1007/978-1-60327-412-8_19,

Accordingly, for any given translocation apparatus it is generally known whether it requires, for instance, ATP for its transport activity. Many such systems display complex energy inputs, utilizing the energy available in both NTP hydrolysis and the transmembrane electrochemical potential. Examples of these latter systems include the Sec transporter in *E. coli* (5) and the cpSec- and cpSRP transporters in chloroplasts (6), each of which has a strict NTP requirement, but can also be assisted by a membrane electrochemical gradient. The complex quality of the energetics is further underscored by the facts that new information about the nature of the energy inputs is still being discovered (7, 8), and that some otherwise well-characterized translocation systems remain for which the energy input has eluded detection altogether (9).

In contrast to the large body of work on the *nature* of the energy required for protein transport, relatively few studies have addressed the *amount* of energy expended during these processes. The first papers directed to this topic studied the so-called protein translocation ATPase of the *E. coli* Sec transporter (10, 11), in which the ATP hydrolyzed during protein transport into inverted plasma membrane vesicles was assayed. Driessen (11) found that the transport of a single proOmpA precursor was accompanied by the hydrolysis of approximately 1,000 molecules of ATP in the presence of an assisting proton-motive force (pmf), and that this number increased to around 5,000 when the pmf was eliminated. More recently, this group calculated a rate of ATP hydrolysis during the transport of proteins of differing sizes as approximately five ATP per amino acid, putting the cost of proOmpA transport at ~1,700 ATP (12). However, the total amount of energy expended to transport this protein was not precisely determined because the contribution of the pmf was not assessed.

The first complete accounting of the free energy of protein transport, the so-called $\Delta G_{\mathrm{transport}}$, was performed in our laboratory using the chloroplast Tat (cpTat) system (13). We chose that particular transporter because the pmf is the only required energy input (6, 7, 14), and the techniques to quantitate this parameter in thylakoids are known. In order to assess the drain by protein transport on total cellular energy pools, it is necessary that other protein transporters be examined. In plastids, the in vitro import of proteins across the envelope membranes is also governed by a single energy input, in this case, ATP hydrolysis, and we are currently working to develop a translocation ATPase assay for this system. Determination of the $\Delta G_{\mathrm{transport}}$ for the cpSec- and cpSRP transporters will require an accounting of both the contributions of the required NTP hydrolysis and the pmf. We report herein the procedure for quantitating the contribution of the chloroplast pmf to protein transport across the thylakoid membrane on the cpTat pathway.

The free energy change associated with any process powered by a pmf is calculated as the product of two measurable parameters:

$$\Delta G = \Delta\tilde{\mu}_{H^+} \times n_{H^+}, \quad (1)$$

where $\Delta\tilde{\mu}_H+$ is the thermodynamic threshold below which the energy contained in the pmf is insufficient to drive the process and $n_{\text{H}}+$ is the number of protons released from the pmf per unit of work, in this case, per protein transported. Accordingly, both of these parameters must be experimentally measured. The thermodynamic threshold for protein transport is determined by plotting the rate of protein transport as a function of the ΔpH developed across the thylakoid membrane in the light. The rate of protein transport is measured as the rate of uptake of radiolabeled protein (with or without the envelope transit peptide, but retaining the thylakoid signal peptide) into isolated thylakoids (c.f. ref. (15) from this series). The ΔpH is determined from the light-induced quenching of 9-aminoacridine (9-AA) fluorescence and can be varied either by changing the actinic light intensity or by titration with nigericin. The stoichiometry of proton release to protein transport can be measured by a number of independent procedures, some direct and some indirect, but all of which lead to the same answer (13). The least technically challenging of these is to measure the decrease in the ΔpH caused by draining the gradient through protein transport and calibrating this change by mapping the ΔpH into the rate of proton pumping determined as the rate of light-dependent alkalinization of the medium in which the thylakoids are suspended (16).

In order for these measurements to be meaningful, it is important that the protein transport and energetic assays be conducted simultaneously in the same cuvette. Since the energetics depends critically on the actinic light intensity, and this in turn is a function of the geometry of the illumination apparatus, it is not possible to ensure that the same energetic parameters are established when thylakoids are illuminated in differently configured locations in the lab or with different chlorophyll concentrations.

2. Materials

Unless otherwise noted, all chemicals are prepared in distilled water.

2.1. Plant Growth

1. Seeds from *Pisum sativa* var. Little Marvel (Seedway LLC, Hall, NY, USA).
2. Vermiculite.

2.2. Thylakoid Isolation

1. Grinding buffer: 330 mM sorbitol, 50 mM K-Tricine pH 8.0, 1 mM $MgCl_2$, 1 mM $MnCl_2$, 2 mM Na_2EDTA, 0.1% BSA.
2. Import buffer: 330 mM sorbitol, 50 mM K-Tricine pH 8.0, 5 mM $MgCl_2$.
3. Breaking buffer: 10 mM K-Hepes pH 8.0, 5 mM $MgCl_2$.
4. Percoll (Sigma-Aldrich).
5. Miracloth (Calbiochem).
6. Isoascorbaic acid.
7. Glutathione.
8. 80% acetone.

2.3. Preparation of Radiolabeled iOE17

2.3.1. In Vitro Transcription/Translation

1. SP6 RNA polymerase transcription (Promega).
2. Wheat germ translation kit (Promega).
3. [^{3}H]leucine (NEN).

2.3.2. Expression in Bacteria and Isolation Therefrom

1. [^{3}H]leucine or [^{35}S]methionine (NEN).
2. Isopropyl β-DAD-1-thiogalactopyranoside (IPTG).
3. CelLytic Express (Sigma-Aldrich).
4. Ni-NTA agarose beads (Qiagen).
5. Urea.
6. Bovine serum albumin (BSA).

2.4. SDS-PAGE and Fluography

1. Resolving gel: ~50 ml prepared as follows: 16.2 ml H_2O, 12.6 ml 1.5 M Tris-HCl pH 8.8, 0.51 ml 10% SDS, 21.1 ml 30% acrylamide/bis, 0.25 ml 10% ammonium persulfate (APS), 26 μl TEMED.
2. Stacking gel: ~23 ml prepared as follows: 13.7 ml H_2O, 5.6 ml 0.5 M Tris-HCl pH 6.8, 0.225 ml 10% SDS, 2.9 ml 30% acrylamide/bis, 0.112 ml 10% APS, 22.5 μl TEMED.
3. Running buffer: 25 mM Tris, 192 mM glycine, 0.1% SDS.
4. SDS-sample buffer: 0.125 M Tris-HCl pH 6.8, 20% glycerol, 10% β-mercaptoethanol, 0.05% bromophenol blue.
5. Coomassie stain: 0.25% Coomassie brilliant blue, 40% methanol, 10% acetic acid.
6. Destain: 7.5% acetic acid, 1% glycerol, 40% ethanol.
7. 2,5-Diphenyloxazole (PPO): 1 mM in glacial acetic acid.
8. High-range Rainbow molecular weight markers (Amersham).

2.5. Measurement of $\Delta\tilde{\mu}_{H^+}$

1. 9-AA (stocks made in ethanol).
2. Methyl viologen.

2.6. Measurement of n_{H^+}

1. 9-AA (stocks made in ethanol).
2. Methyl viologen.
3. Valinomycin (stocks made in ethanol).
4. Phenol red.
5. Import buffer (*see* **Section 2.2**).
6. Proton pump buffer: 0.1 mM K-Tricine pH 8.0, 50 mM KCl, 50 mM sorbitol, and 6 mM $MgCl_2$.
7. KOH.

3. Methods

3.1. Plant Growth

1. Approximately 100 ml of pea seeds are soaked for a few hours in deionized water.
2. Seeds are sown onto a wet bed of vermiculite in a flat plastic pan (35 cm × 20 cm × 6 cm) with holes in the bottom blocked by a layer of paper towels. The seeds are covered by a shallow layer of wet vermiculite.
3. Plants are grown in a controlled environment chamber at 20°C in a 12/12 h light/dark cycle.
4. Seedling leaves are harvested at 10–14 days.

3.2. Thylakoid Isolation

Where possible, all procedures are carried out at 4°C.

1. Mix in a 30 ml Corex tube 15 ml Percoll, 15 ml 2x grinding buffer, and 10 mg each of isoascorbate and glutathione. Centrifuge for 30 min at 37,000 *g* (max) to form a continuous Percoll density gradient. Place tube on ice and use within a few hours.
2. Briefly grind leaves from a flat in approximately 200 ml grinding buffer in a Waring blender fitted with sharpened blades. Filter the slurry through a layer of Miracloth and divide the filtrate among four 50 ml centrifuge tubes.
3. Pellet the chloroplasts in the slurry by centrifuging the tubes in a swing-out rotor at 3,000 *g* (max) for 5 min.
4. Resuspend the pellet in 2–4 ml import buffer and carefully layer on top of the pre-formed Percoll gradient. Spin in a swing-out rotor at 8,000 *g* (max) for 10 min. Intact chloroplasts will collect in the lower of the two resulting chlorophyll (Chl)-containing bands. Aspirate the liquid above this

band and remove the chloroplasts to a single fresh 50 ml tube. Wash the chloroplasts in 40 ml IB, pelleting at 1,500 *g* (max) for 5 min.

5. Resuspend chloroplasts in 20 ml low osmotic strength breaking buffer and incubate on ice for 5 min. Double the volume with 2x IB, pellet the thylakoids as above, and resuspend in 1–2 ml IB.
6. Measure the Chl content of the thylakoids by mixing 10 μl thylakoids with 5 ml 80% acetone, filtering, and then reading the absorbance of the filtrate at 645, 663, and 720 nm. Chl (in mg/ml) is calculated as $4.02 \times (A_{663} - A_{720}) + 10.14 \times (A_{645} - A_{720})$.
7. Dilute the thylakoids to 1 mg Chl/ml and keep in a covered ice bucket until use.

3.3. Preparation of Radiolabeled iOE17

3.3.1. In Vitro Transcription/Translation

We have successfully produced transport-competent proteins using a variety of protocols and products, including sequential in vitro transcription and translation using wheat germ or reticulocyte lysate, and the TNT-coupled translation system (Promega). Our procedures rarely deviate from the manufacturer's instructions. We generally prefer wheat germ to translate mRNA produced using the SP6 polymerase, but T3 and T7 polymerases are also suitable for mRNA synthesis. Our cDNA clone for pea iOE17 (containing the thylakoid-targeting signal sequence but not the stromal-targeting transit peptide) is placed behind the SP6 promoter in a pGem7zf vector, and so we use SP6 polymerase to make mRNA for this protein, followed by translation in wheat germ lysate with the addition of [^{3}H]leucine (*see* **Note 1**). The translation products are then aliquoted and stored at –80°C until use (*see* **Note 2**).

3.3.2. Expression in Bacteria and Isolation Therefrom

In vitro translation from cDNA clones yields extremely low amounts of high specific activity proteins. For those experiments in which we wish to observe changes in any number of chloroplast parameters upon transport of an added protein, it is necessary to express "chemical quantities" of transport substrate in bacteria. We generally do this using the pET expression system in *E. coli* (Novagen). To this end we cloned iOE17 into pET21a, placing it behind the inducible T7 promoter. For convenience we also included at its C terminus a sequence containing four methionines (for ^{35}S labeling), a cysteine (for cross-linking) and a six histidine tag (for purification). This protein is induced for overexpression by IPTG in the presence of radiolabeled [^{35}S]Met or [^{3}H]leucine according to the manufacturer's instructions. Inclusion bodies containing iOE17 are solubilized in 8 M urea and then purified using a Ni column. Urea is included in the wash and elution buffers. The specific activity of the protein is determined by scintillation counting and by comparison to BSA in a Coomassie-stained gel containing a dilution series of both iOE17 and BSA.

3.4. SDS-PAGE and Fluography

We generally assess the transport of OE17 into thylakoids using a 12% resolving gel in the Mini-PROTEAN II electrophoresis apparatus (Bio-Rad).

1. Using the multi-gel casting stand, prepare nine gels at a time. Prepare 50 ml of resolving gel solution as described in **Section 2.4**, adding APS and TEMED last. Fill the gels in the casting chamber, leaving room for a stacking gel. Overlay each gel with isopropanol. When they have solidified, rinse the isopropanol off the gels with distilled water and pipette the stacking gel solution on top of the running gels. Insert combs, making sure to avoid air bubbles. When solidified, gels are removed from the casting chamber and stored in running buffer for up to a week at 4°C.
2. For use, remove the comb from the stacking gel and rinse the wells out with H_2O. Load the gels into the Mini-PROTEAN II apparatus and add running buffer to the appropriate chambers in the apparatus.
3. Samples in SDS-sample buffer are prepared for electrophoresis by solubilizing any sample pellets using a bath sonicator (Laboratory Supplies Co. Inc, Hicksville, NY, USA) or by vigorously scraping the microfuge tubes over the holes in a plastic 16-position microfuge tube holder. When the pellet can no longer be seen, the samples are placed into a near-boiling water bath for 3 min (*see* **Note 3**).
4. Load 8–12 μl of the samples into the wells of the stacking gel using a pipettor or Hamilton syringe. Include a lane containing an aliquot of transport substrate alone that you would expect to produce bands of a similar intensity to those from the experiment. This is often 20% of the protein added to each sample for those experiments performed with in vitro translated protein and 2–5% of the protein added per sample for bacterially expressed protein. If quantitation is critical for the experiment, a dilution series of the protein substrate is loaded into adjacent wells in the same gel with the samples. Include a lane of pre-stained molecular weight markers, which can be useful to monitor the progress of the migration of samples in the gel.
5. Run the gel at 170 V, constant voltage, for approximately 1 h.
6. When the electrophoresis is completed, separate the plates, cut off the stacking gel, and cut a corner of the resolving gel in a distinctive manner. From here, the gel is either dried on a gel drier and then placed on a phosphorimager screen (for

[^{35}S]-labeled protein) or soaked in PPO, dried, and placed on x-ray film (for [^{3}H]-labeled protein).

3.5. Measurement of $\Delta\tilde{\mu}_{H}+$

1. Transport reactions are carried out in a fluorimeter (*see* **Note 4)** in a stirred cuvette with a 500 μl reaction volume. The reaction mixture contains import buffer (IB), thylakoids at 20 μg Chl, 2 μl of in vitro translated radiolabeled iOE17, and 20 μM 9-AA and methyl viologen, respectively (*see* **Note 5**). The fluorescence measurement should commence quickly after all the additions are in the cuvette. A useful timing sequence is $t = -20$ s, start the recording (to establish a no fluorescence baseline, Fo); $t = -15$ s, turn on the measuring beam (to excite the fluorescence without initiating electron flow, to establish the maximal fluorescence, Fm); $t = 0$, turn on the actinic light (to start electron flow and proton pumping and to initiate both the protein transport reaction and 9-AA fluorescence quenching to the level, Fq); $t = 6$ min, remove the sample, dilute with 1.0 ml of ice-cold IB and place on ice in the dark until all samples are collected (*see* **Note 6**). This sequence is repeated with decreasing actinic light intensities until the 9-AA quenching becomes negligible (*see* **Note 7**). A typical recording of 9-AA quenching fluorescence, which in this case includes a return of the ΔpH to zero after the actinic light has been extinguished, is shown in **Fig. 19.1**.

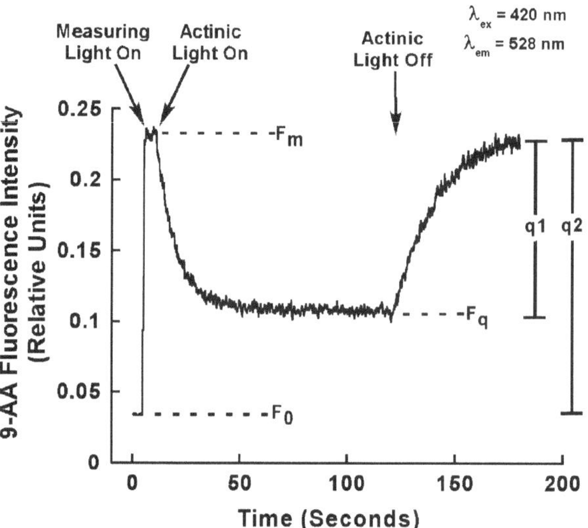

Fig. 19.1. A representative 9-AA fluorescence quenching recording. The trace shows the quenching of 9-AA fluorescence in response to the light-induced development of a pH gradient across the thylakoid membrane. Fluorescence levels corresponding to Fo, Fm, and Fq are indicated.

2. Determine Fo, Fm, and Fq for each fluorescence trace. Calculate the ΔpH according to the equation:

$$\Delta\text{pH} = \log\left(\frac{Q}{1-Q}\right) - \log(V), \tag{2}$$

where $Q = \left(\frac{\text{Fm}-\text{Fq}}{\text{Fm}-\text{Fo}}\right)$ and V = ratio of the volume of the thylakoid lumen to the total volume (in this experiment = $\left(\frac{20\mu\text{l/mg Chl} \times 0.02\ \text{mg Chl}}{500\mu\text{l}}\right) = 8 \times 10^{-4}$)(17).

3. Centrifuge the samples held on ice from Step 1 above in a microfuge at full speed for 2 min. Remove the (radioactive) supernatant and resuspend the pellet in 40 µl SDS-sample buffer.

4. Subject the samples to SDS-PAGE and phosphorimaging or fluography (*see* **Section 3.4**). Quantitate the amount of protein transport relative to the input using a suitable software package such as NIH Image, Scion Image, or ImageQuant (*see* **Note 8**).

5. Plot the rate of protein transport determined in Step 4 above (since the reaction was terminated during the linear portion of the transport reaction, the amount of protein transport is proportional to the rate) versus the ΔpH developed in the light during the reaction as determined in Step 2. This will produce a straight line with an extrapolated abscissa intercept equal to the ΔpH that corresponds to $\Delta\tilde{\mu}_{H^+}$ (*see* **Notes 9** and **10**). One such plot generated in this manner is shown in **Fig. 19.2**.

3.6. Measurement of n_{H^+}

Measurements with 9-AA yield the ΔpH resulting from the pumping of protons from the external medium into the thylakoid lumen. However, without a detailed knowledge of the lumen buffer capacity, this number cannot be related to the number of protons pumped or, more germane to this discussion, the number of protons released from the lumen per protein transported. Described below is the method for determining n_{H^+} that we find to be the least technically challenging. It involves mapping the rate of protons pumped in the light and measured as a pH change in the thylakoid suspension as a function of light intensity against the ΔpH developed under identical conditions. This is then used to calculate the number of protons lost from the transmembrane gradient as a protein is transported, which is seen as a change in the ΔpH induced by protein transport under light-limiting conditions.

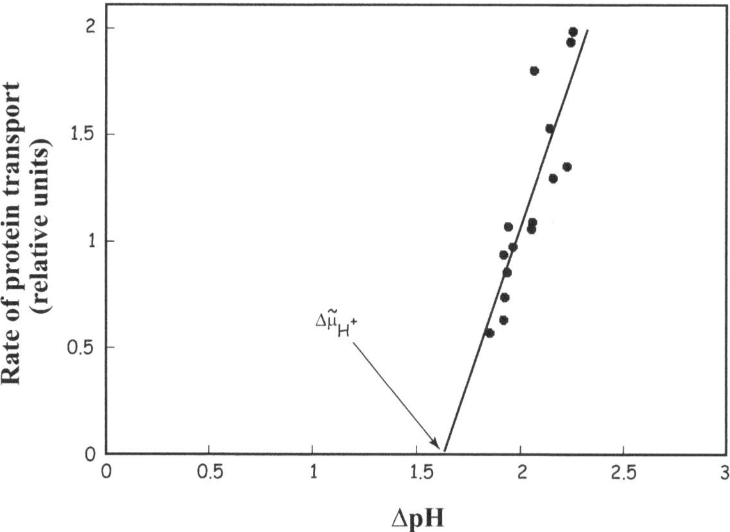

Fig. 19.2. Determination of the threshold ΔpH for protein transport. Values of the ΔpH and corresponding protein transport rates were measured as described in the text. A least-squares line was fitted to the points and extended to the abscissa. Conversion of the intercept value into the thermodynamic threshold energy for protein transport is described in **Note 10**.

3.6.1. Mapping ΔpH onto the Number of Protons Pumped into the Thylakoid Lumen

1. Generate a curve showing the ΔpH generated in the light as a function of light intensity as in Steps 1 and 2 in **Section 3**.5, except omitting the radiolabeled protein transport substrate.
2. Proton pump measurements are performed by monitoring the change in absorbance at 540 nm of the pH indicating dye phenol red in a spectrophotometer setup for kinetics readout and actinic illumination at 90° to the path of the measuring beam (*see* **Note 11**). Prepare a 500 μl reaction containing proton pump buffer, thylakoids at 20 μg Chl/ml, 40 μM methyl viologen, 2 μM valinomycin, and 20 μM phenol red. Record a baseline corresponding to the initial dye absorbance at pH 8.0 and then turn on the actinic light and monitor the absorbance change accompanying alkalinization of the medium as protons are pumped into the lumen. The absorbance reaches a steady-state level when the rate of proton pumping into the lumen exactly balances the rate of proton return across the membrane, either by leak or a productive process (13). When this steady state is reached, turn the actinic light off and record the return of the dye's absorbance to its initial level. You are interested in the initial decay of the absorbance signal from its steady-state level the moment the light was extinguished. This corresponds to the rate of proton deposition just before the light was

switched off (16). When the absorbance has returned to its initial level, calibrate the total absorbance change by repeatedly adding OH^- to the cuvette using a Hamilton syringe in 10 nmol aliquots until the absorbance has surpassed the maximum deflection reached during the illumination period. Using this calibration, calculate the number of protons that were pumped from the external medium into the lumen during the illumination period and the rate of proton deposition the moment the actinic light was extinguished. Repeat this procedure, lowering the intensity of the actinic illumination to match those used in Step 1 above, generating a plot of the rate of the steady-state proton pumping as a function of light intensity. An example of this measurement is shown in **Fig. 19.3**.

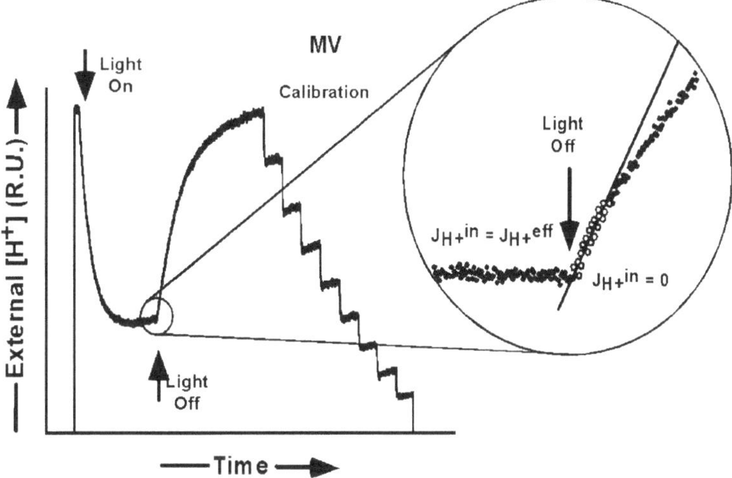

Fig. 19.3. Measurement of the proton influx rate. The trace shows the response of phenol red absorbance at 540 nm to actinic illumination and calibrating additions of KOH. The inset shows a blow-up of the light on-to-off transition, from which the rate of proton influx in the steady state is determined.

3. For each light intensity examined, plot the values of the ΔpH and the rate of the proton pump measured in Steps 1 and 2 on the ordinate and abscissa, respectively. You now have a curve that can be used to calculate a change in the rate of proton flux across the thylakoid membrane from an observed change in ΔpH (*see* **Note 12**).

3.6.2. Measuring the Drain on the ΔpH by Transport of iOE17 Under Light-Limiting Conditions

1. Set up a 9-AA fluorescence quenching experiment as in Step 1 in **Section 3.5**. Adjust the intensity of the actinic light to give a fluorescence quench of approximately 75% of that measured in saturating intensities, i.e., that allows the fluorescence quench to be 75% of the maximum (see **Note 13**).

2. Record 9-AA fluorescence during actinic illumination until it has reached a steady-state quenched level, say for 3 min. At this point inject into the cuvette 400–800 nmol of radiolabeled iOE17 produced by overexpression in bacteria(13). This will cause the ΔpH to decrease as it is utilized to transport the newly introduced substrate protein, which will be reflected as a decreased quenching of the fluorescence. Record the new level of 9-AA fluorescence reached in the steady state after addition of iOE17.
3. Three minutes after injection of the transport substrate, remove the sample from the fluorimeter and evaluate the amount of protein that was transported as in Steps 1–4 in **Section 3.5**.
4. Using the relationship between the rate of proton pumping and the ΔpH developed in Step 3 in **Section 3.6.1**, determine the change in proton flux that corresponds to the change in ΔpH observed upon introduction of the transport substrate in Step 2 above.
5. Dividing the change in the rate of proton pumping induced by protein transport (expressed as the number of protons pumped per minute per 20 μg Chl) by the rate of protein transport (the number of proteins transported per minute per 20 μg Chl) gives the desired parameter, n_{H+}.

The $\Delta G_{transport}$ is calculated according to Equation 1 using the $\Delta\tilde{\mu}_{H+}$ (expressed as kJ/mol H^+) and n_{H+} (mol H^+ expended from the gradient/mol protein transported) determined as described above. This yields a value of $\Delta G_{transport}$ in kJ/mol iOE17 protein transported; in our hands it was 6.5×10^5 kJ/mol OE17 transported. This very large number only took on meaning for us when we converted it to an energy equivalence of >10,000 moles of ATP. This is done using a value of 50 kJ/mol ATP, which reflects the prevailing levels of ATP, ADP, and Pi found in a typical cell. This calculation is tantamount to saying that although the reaction examined does not utilize energy derived from ATP hydrolysis, if it did, it would utilize >10,000 moles of ATP per mole of protein transported.

4. Notes

1. This protein, as well as another well-studied cpTAT pathway substrate, OE23, contains no methionine beyond the start codon. This necessitates the use of the less energetic [^{3}H] instead of the preferred [^{35}S].

2. We have had varying success in storing precursors for later use in transport experiments. Although sometimes they retain their activity for many months at –80°C, we generally prefer to use precursors relatively quickly after synthesis, and on the same day if they are not taken up efficiently into chloroplasts or thylakoids.
3. Some chloroplast proteins, notably PsbW, TatC, and others, appear to run abnormally on SDS gels after they have been boiled. OE17 is not among that group, and we boil these samples before electrophoresis.
4. The instrument we use most often for this measurement is a homemade device that can be arranged for either fluorescence of absorbance measurements (13). However, many commercial fluorimeters are set up for side actinic illumination and work well. 9-AA is excited at 420 nm and fluorescence is detected at 520 nm. An important point to consider is that the actinic light must be kept from scattering into the detector, and this can be an annoying problem. This is accomplished with filters placed at the actinic light source to select long-wavelength red light, say > 670 nm, and another filter set placed before the detector that does not transmit this long-wavelength light. We have had the best results using color glass filters in combination with narrow band-pass interference filters; the latter still transmit enough light far from their transmission peaks to ruin the measurement. A working setup for us is to place a Corion S10-520 narrow band-pass filter (Newport Corp., Franklin, MA, USA) with a Corning CS4-96 color glass filter (Kopp Glass, Pittsburgh, PA, USA) in front of the detector and a Corion 650 high band-pass filter with a Corning CS3-67 color glass filter in line with the actinic light.
5. In our experience, an electron acceptor must be added to isolated thylakoids in order for them to transport electrons at rates leading to the development of a significant pH gradient. However, we have found that methyl viologen at concentrations normally used to measure photosynthetic oxygen evolution (i.e., 0.1 mM) is inhibitory to the thylakoid protein transport reaction. Accordingly, the concentration suggested here, 10 μM, represents a compromise between these two competing activities.
6. Six minutes is within the linear range of iOE17 transport across the thylakoid membrane. Thus, the 6-min point can be taken to represent the rate of protein transport.
7. The actinic light intensity should ideally be adjusted with neutral density filters as opposed to lowering the voltage on the lamp, which could, in principle, change the spectral

output of the lamp. In our instrument a useful range of light intensities is between 0.5 and 135 μE/m^2·s.

8. NIH Image and Scion Image are freeware programs for the Macintosh and PC platforms, respectively; both work well for the purpose of quantitating the densities of bands in gels. We currently use ImageQuant, originally marketed by Molecular Dynamics for use with phosphorimager files.
9. As is to be expected, it is difficult to accurately measure low rates of protein transport. In addition, 9-AA does not respond to pH gradients less than approximately 1.6 pH units. Accordingly, the abscissa intercept in these experiments must be determined by extrapolation of the best-fit line to data points well above the abscissa, as seen in **Fig. 19.2**.
10. ΔpH is related to $\Delta\tilde{\mu}_{\mathrm{H}}+$ through the equation$\Delta\tilde{\mu}_{\mathrm{H}}+ = 2.303RT\left(\Delta\mathrm{pH}\right)$, where $R =$ the gas constant and $T =$ temperature in kelvin. At 25°C, a ΔpH of one pH unit = 5.7 kJ/mol. Thylakoids also maintain a rather small but significant electrical potential under steady-state illumination, a good estimate of which is 30 mV (7). This should be added into the calculation of $\Delta\tilde{\mu}_{\mathrm{H}}+$, increasing the term by $F\Delta\psi$, where $F = 96.49$ kJ/V·mol or 2.9 kJ/mol.
11. For this measurement we use the homebuilt instrument described in **Note 4** and configured as a spectrophotometer. Many laboratory spectrophotometers are not set up for side illumination, but it is often the case that illumination from above could be achieved using a fiber optic lamp such as those often employed in microscopy. The same light filtering considerations described in **Note 4** apply here as well, and the same filter combination described above work for this measurement, except the Corion S10-520 filter is replaced by a Corion S10-540 narrow band-pass filter, allowing the measurement to be made at 540 nm.
12. This curve is shown in (13) in Figure 4a.
13. This measurement must be made at sub-saturating actinic light intensities. The electron transport chain is subject to feedback control by the ΔpH across the thylakoid membrane. Under saturating conditions when this control is in force, a drain on the pH gradient would be compensated by an increase in the rate of electron transport, effectively restoring the ΔpH to its pre-drain level. When the steady-state ΔpH is already limited by the light intensity, any drain on the pH gradient will result in a further lowering of the ΔpH, which can be measured as a decrease in 9-AA quenching. See **Fig. 2b** in (13) for an example of this.

Acknowledgments

This work is supported by US Department of Energy Grant DE-FG02-03ER15405.

References

1. Schatz, G., and Dobberstein, B. (1996) Common principles of protein translocation across membranes. *Science* **271,** 1519–1525.
2. Alder, N. N., and Theg, S. M. (2003) Energy use by biological protein transport pathways. *Trends Biochem Sci* **28,** 442–451.
3. Mokranjac, D., and Neupert, W. (2008) Energetics of protein translocation into mitochondria. *Biochim Biophys Acta* **1777,** 758–762.
4. Wickner, W., and Schekman, R. (2005) Protein translocation across biological membranes. *Science* **310,** 1452–1456.
5. Schiebel, E., Driessen, A. M., Hartl, F.-U., and Wickner, W. (1991) Delta mu H+ and ATP function at different steps of the catalytic cycle of preprotein translocase. *Cell* **64,** 927–939.
6. Cline, K., Ettinger, W. F., and Theg, S. M. (1992) Protein-specific energy requirements for protein transport across or into thylakoid membranes. Two lumenal proteins are transported in the absence of ATP. *J Biol Chem* **267,** 2688–2696.
7. Braun, N. A., Davis, A. W., and Theg, S. M. (2007) The chloroplast tat pathway utilizes the transmembrane electric potential as an energy source. *Biophys J* **93,** 1993–1998.
8. Bageshwar, U. K., and Musser, S. M. (2007) Two electrical potential-dependent steps are required for transport by the Escherichia coli Tat machinery. *J Cell Biol* **179,** 87–99.
9. Diekert, K., Kispal, G., Guiard, B., and Lill, R. (1999) An internal targeting signal directing proteins into the mitochondrial intermembrane space. *Proc Natl Acad Sci USA* **96,** 11752–11757.
10. Lill, R., Cunningham, K., Brundage, L. A., Ito, K., Oliver, D., and Wickner, W. (1989) SecA protein hydrolyzes ATP and is an essential component of the protein translocation ATPase of Escherichia coli. *EMBO J* **8,** 961–966.
11. Driessen, A. J. M. (1992) Precursor protein translocation by the Escherichia coli translocase is directed by the protonmotive force. *EMBO J* **11,** 847–853.
12. Tomkiewicz, D., Nouwen, N., van Leeuwen, R., Tans, S., and Driessen, A. J. (2006) SecA supports a constant rate of preprotein translocation. *J Biol Chem* **281,** 15709–15713.
13. Alder, N. N., and Theg, S. M. (2003) Energetics of protein transport across biological membranes: A study of the thylakoid DeltapH-dependent/cpTat pathway. *Cell* **112,** 231–242.
14. Hulford, A., Hazell, L., Mould, R. M., and Robinson, C. (1994) Two distinct mechanisms for the translocation of proteins across the thylakoid membrane, one requiring the presence of a stromal protein factor and nucleotide triphosphates. *J Biol Chem* **269,** 3251–3256.
15. Bolter, B., and Soll, J. (2007) Import of plastid precursor proteins into pea chloroplasts. *Methods Mol Biol* **390,** 195–206.
16. Berry, S., and Rumberg, B. (1996) H+/ATP coupling ratio at the unmodulated CF0CF1-ATP synthase determined by proton flux measurements. *Biochim Biophys Acta* **1276,** 51–56.
17. Schuldiner, S., Rottenberg, H., and Avron, M. (1972) Determination of delta pH in chloroplasts: 2. Fluorescent amines as a probe for the determination of delta pH in chloroplasts. *Eur J Biochem* **25,** 64–70.

Chapter 20

In Vitro Dissection of Protein Translocation into the Mammalian Endoplasmic Reticulum

Ajay Sharma, Malaiyalam Mariappan, Suhila Appathurai, and Ramanujan S. Hegde

Abstract

In eukaryotic cells, roughly one-fourth of all mRNAs code for secretory and membrane proteins. This class of proteins must first be segregated to the endoplasmic reticulum, where they are either translocated into the lumen or inserted into the lipid bilayer. The study of these processes has long relied on their successful reconstitution in cell-free systems. The high manipulability of such in vitro systems has allowed the identification of key machinery, elucidation of their functional roles in translocation, and dissection of their mechanisms of action. Here, we provide the basic methodology for (i) setting up robust mammalian-based in vitro translation and translocation systems, (ii) assays for protein translocation, insertion, and topology, and (iii) methods to solubilize, fractionate, and reconstitute ER membranes. Variations of these methods should be applicable not only to forward protein translocation systems but also for dissecting other poorly understood membrane-associated processes such as retrotranslocation.

Key words: in vitro translation, microsomes, membrane proteins, reconstitution, proteoliposomes, protease protection, protein topology.

1. Introduction

The endoplasmic reticulum (ER) is the major site for the biosynthesis, maturation, quality control, and degradation of secretory and membrane proteins. Each of these basic processes employs multiple distinct pathways that operate in parallel to provide the cell considerable flexibility in handling the tremendous diversity of proteins that transit through the ER. A major goal of cell biology has long been to identify and dissect the mechanism of action of the machinery that define these pathways of secretory and membrane protein metabolism.

A. Economou (ed.), *Protein Secretion*, Methods in Molecular Biology 619,
DOI 10.1007/978-1-60327-412-8_20, © Springer Science+Business Media, LLC 2010

One of the principal approaches that has been applied to this problem is the reconstitution of key pathways or sub-reactions in a cell-free system. The tremendous manipulability of such systems affords a direct window into the biochemical and mechanistic dissection of any of these pathways. By understanding the basic features of the standard cell-free translation and translocation system, it can be sensibly customized in various ways to address any of a number of processes that occur at or within the ER and cytosol. A few of the major ER-associated processes that are amenable to dissection using this or similar in vitro systems are shown in **Fig. 20.1**.

In co-translational translocation (**Fig. 20.1A**), secretory and membrane proteins are recognized as they are being synthesized. The signal recognition particle (SRP) binds to a hydrophobic

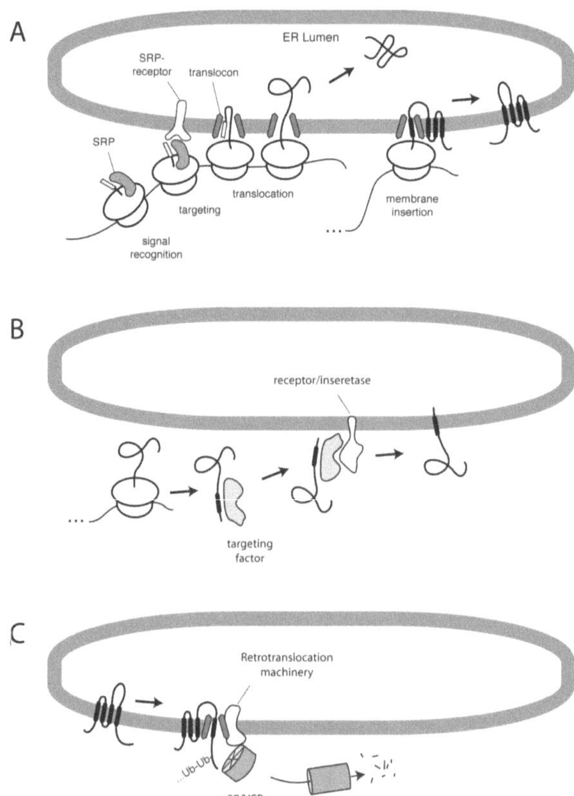

Fig. 20.1. Examples of ER-associated pathways amenable to in vitro reconstitution. (**A**) The SRP-dependent co-translational translocation pathway. (**B**) A post-translational translocation pathway for tail-anchored membrane protein insertion. (**C**) Post-translocational pathway of membrane protein metabolism involving ubiquitination, retrotranslocation, and proteasomal degradation. In each of these instances, the substrate is synthesized in vitro, making it the only protein that becomes radiolabeled. The other components of the system can be manipulated to analyze the requirements for substrate translocation, maturation, degradation, etc.

domain (either a signal sequence or transmembrane domain) in the nascent polypeptide as it emerges from the ribosome. This complex of ribosome–nascent chain–SRP is then targeted to the membrane (via the SRP receptor) and transferred to a translocon whose central component is the Sec61 complex. Secretory proteins are translocated through the Sec61 complex, while membrane proteins are laterally released by the Sec61 complex into the membrane bilayer. Because everything occurs co-translationally, the study of these events typically depends on translating the protein of interest in a cytosolic extract *in the presence of* a source of ER membrane [typically rough microsomes (RM) isolated from canine pancreas (1) or other tissues/cells (2, 3)]. If the ER membranes are added to the reaction after translation is completed, translocation will not occur.

In post-translational translocation pathways (**Fig. 20.1B**), the protein remains competent for translocation even after it has been fully synthesized and released from the ribosome. Such pathways can be employed by a subset of secretory proteins, some small proteins, and certain types of membrane proteins. These pathways, particularly in higher eukaryotes, are not nearly as well understood as the SRP-dependent co-translational translocation pathway. For example, a novel and well-conserved pathway for tail-anchored membrane proteins was only recently discovered (4, 5) and whose full complement of machinery remains to be clarified. From a practical standpoint, these post-translational pathways are in many ways easier to study because the translation reaction can be uncoupled from the translocation reaction. Thus, proteins can be translated, after which various manipulations can be applied (e.g., removal of energy, addition of inhibitors, change in conditions) before initiating the translocation reaction by adding a source of ER membranes. This provides greater flexibility than co-translational reactions where conditions must be maintained within the narrow range that is compatible with efficient protein synthesis.

And finally, reactions after both synthesis and translocation have been completed can also be studied using these same in vitro systems (**Fig. 20.1C**). Examples of such processes include maturation events in the ER lumen (6), quality control of misfolded proteins, retrotranslocation, ubiquitination, and degradation (7). Again, many of these pathways are still relatively poorly understood, particularly from a mechanistic point of view. As with post-translational translocation, these events can often be uncoupled from protein synthesis (and in some cases, even translocation), allowing experimental flexibility.

The study of all of these processes in vitro depends on three basic tools. An in vitro translation (IVT) system, clear and definitive assays for translocation and topology, and, in the case of events occurring at the ER, methods to manipulate the composition of the membrane. These basic tools can be applied in a wide range of ways. The IVT system allows one to produce in a physiologic system a protein radiolabeled with very high specific activity that can be followed. By scaling up these reactions, biochemical amounts can also be generated to identify interacting partners (4). Highly specific assays for translocation can be used to identify protein or lipid requirements (4, 8, 9) and analyze the action of small molecule inhibitors (10). Manipulation of the membrane provides access to the requirements at this usually inaccessible compartment (8, 9, 11). And the ability to isolate the membrane after insertion provides the ability to study subsequent events (such as degradation) in isolation (7).

2. Materials

2.1. Preparation of the Transcription Mix (T1)

For general advice regarding these reagents, see the following:

1. 1 M HEPES, pH 7.6: Prepare a solution of 1 M HEPES (free acid), titrated with 0.45 M NaOH. This will be the correct pH when diluted in the buffers below. Filter and store at 4°C (*see* **Notes 1, 2, 8,** and **9**).
2. 2 M $MgCl_2$, store at RT or 4°C.
3. 100 mM spermidine (Sigma); very hygroscopic. Dissolve 145 mg spermidine in 10 mL water. Freeze in nitrogen and store at –20°C.
4. 1 M DTT (1,4-dithiothreitol; Roche). Dissolve 1.54 g DTT in 10 mL water. Aliquot and freeze in nitrogen. Store at –80°C. Do not freeze-thaw more than twice to prevent oxidation.
5. 10X NTPs. 5 mM each of ATP, UTP, and CTP and 1 mM of GTP in water. Adjust to pH ~7 with NaOH as needed. Aliquot and freeze in nitrogen. Store at –80°C. Do not freeze-thaw more than five times.
6. 10X Cap: 7-methyl diguanosine triphosphate cap structure analog (New England Biolabs). Each vial contains 25 A_{260} units. Add 300 μL water directly to the vial (to make ~5 mM solution), mix well to dissolve, aliquot, and freeze in nitrogen. Store at –80°C. Do not freeze-thaw more than five times. (*see* **Note 3**).

2.2. Preparation of the Translation Mix (T2)

1. Crude rabbit reticulocyte lysate (RRL). This can be prepared in-house (12), although very few labs currently do this due to practical limitations. Purchase from Green Hectares. They offer two products. We buy the more dilute material which they say is for 'purification of factors.' The

other product 'for in vitro translation' is somewhat more concentrated (and expensive), but we have not found this to be a significant advantage. The more dilute product typically comes in rather large aliquots (~50 mL), which should be stored at –80°C. Do not freeze-thaw more than twice. The first time you thaw a 50 mL aliquot, make 10 mL aliquots in 15 mL polypropylene tubes. To one of these 10 mL aliquots, add hemin, treat with micrococcal nuclease, and further sub-aliquot it (*see* **Section 3.2.**). The other 10 mL aliquots can be frozen directly in nitrogen and stored at –80 until needed. Crude lysate is stable for up to 10 years. Nuclease-treated RRL may be less stable, which is why we do not nuclease everything at once.

2. Micrococcal nuclease (Calbiochem; 15,000 units/vial). Dissolve in 1 mL 50 mM HEPES buffer pH 7.4 (i.e., 15 units per μL); aliquot and freeze in liquid nitrogen and store at –20°C.
3. Hemin (Sigma). Prepare 10 mL of a 100 μM stock by mixing the following *in this exact order* (to avoid problems with precipitation): 6.5 mg hemin, 250 μL 1 N KOH, 500 μL 200 mM Tris, pH 8.0, 8.9 mL ethylene glycol, 190 μL 1 N HCl, 50 μL water. Mix by vortexing and store at –20°C; will not freeze due to ethylene glycol.
4. 100 mM $CaCl_2$: Dissolve in water. Keep at 4°C.
5. 200 mM EGTA. Prepare as follows: To 760 mg EGTA powder, add ~7 or 8 mL water and vortex to resuspend the powder (it will not go into solution). Add 950 μL 5 N NaOH (at which point the EGTA will go into solution). Adjust to 10 mL with additional water. Store at 4°C.
6. Calf liver tRNA (Novagen). This is typically supplied as a 10 mg/mL stock. Freeze in nitrogen and store at –80°C. Stable to multiple freeze-thaws as long as it is kept on ice when thawed and frozen in liquid nitrogen immediately after use (*see* **Note 4**).
7. 1.2 M creatine phosphate (Roche). Dissolve in water, freeze in nitrogen, and store at –80°C.
8. 20 mg/mL creatine kinase (Roche). Dissolve 100 mg in 5 mL of 10 mM HEPES, pH 7.5, 50% glycerol. Store at –20°C.
9. 5 M KOAc stock (for each 500 mL add 2 mL of 12 N HCl to bring pH to ~ 7). Store at RT.
10. Amino acid stocks and mixes – each one is made up individually as a 20 mM stock, and these are mixed to prepare an amino acid mix of 19 amino acids (1 mM each) without methionine (which will be supplied as a ^{35}S-methionine to

label translated proteins). Prepare stock solutions of each amino acid (purchased as powders from Sigma) at 20 mM in either 0.01 N HCl (Trp, Val, Ile, Asn, Phe, Asp, Glu, and Lys), 0.1 N HCl (Tyr), or water (the remaining ones). To make the 19 amino acid mix (1 mM each) simply mix 0.5 mL of each of the 19 with 0.5 mL of H_2O. Mixes lacking different amino acids for labeling with residues other than methionine can also be made. Freeze in aliquots and store both the stock amino acids and the mixes at –80°C (*see* **Note 5**).

11. 10X Emix: for 1 mL, mix 305 μL water, 400 μL 19 amino acid mix (1 mM each; step 10 above), 10 μL 1 M HEPES (not pH adjusted), 1.87 μL 8 N KOH, 83.3 μL 1.2 M creatine phosphate (step 7 above), 100 μL 0.1 M ATP, 100 μL 0.1 M GTP. Freeze in nitrogen and store at –80°C.
12. CB20X: 627.5 μL water, 240 μL 1 M HEPES (not pH adjusted), 22.5 μL 8 N KOH, 100 μL 5 M KOAc pH 7 (step 9 above), 10 μL 2 M $MgCl_2$. Freeze in nitrogen and store at –80°C (*see* **Note 10**).

2.3. Linked Transcription and Translation

1. Reagents for PCR amplification of the desired cDNA.
2. Qiagen PCR purification kit.
3. RNAsin (Promega). *Do not freeze in nitrogen*; keep at –20°C. This comes in 50% glycerol. Stable for at least 2 years.
4. T7 or SP6 polymerase (New England Biolabs). *Do not freeze in nitrogen*; keep at –20°C. This comes in 50% glycerol. Stable for at least 2 years.
5. ^{35}S-methionine (from PerkinElmer; 1,000 Ci/mmol, in aqueous solution). Store in aliquots of 100 μL or less at –80°C. Suitable precautions should be taken when working with radioisotopes.
6. Canine pancreatic rough microsomes (RM). RMs can be obtained in small amounts (50–200 μL) from commercial sources (e.g., Promega). However, for large amounts as needed for fractionation and purification of membrane proteins, purchasing RMs becomes prohibitively expensive (*see* **Notes 6** and **11**).

2.4. Assays for Translocation and Topology

1. Physiologic salt buffer (PSB): 100 mM KOAc, 2 mM $Mg(OAc)_2$, 50 mM HEPES, pH 7.4. It is often convenient to prepare a 10X PSB stock, which is diluted with various other components (e.g., sucrose) as needed.
2. Proteinase K (PK; Roche): 10 mg/ mL dissolved in 20 mM HEPES, pH 7.4. Prepare single-use 20 μL aliquots, freeze in nitrogen, and store at –80. After thawing and use, discard

remainder. We usually prepare ~100 aliquots at a time (*see* **Note 7**).

3. 10% Triton X-100 solution. Store at 4°C.
4. PK-kill buffer (PKB): 1% SDS, 0.1 M Tris, pH 8.0.

2.5. Reconstitution of Membrane Proteins into Proteoliposomes

1. Bovine liver phosphatidylcholine (PC, supplied in organic solvent, usually chloroform or chloroform:methanol; from Avanti Polar Lipids) (*see* **Note 12**).
2. Bovine liver phosphatidylethanolamine (PE, supplied in organic solvent, usually chloroform or chloroform: methanol; from Avanti Polar Lipids).
3. Lissamine–rhodamine–dipalmotyl–PE (DPPE, supplied in organic solvent, usually chloroform or chloroform: methanol; from Avanti Polar Lipids).
4. Lipid hydration buffer: 50 mM HEPES, pH 7.4, 10 mM DTT (added from a freshly prepared 1 M DTT stock solution).
5. Bio-Beads SM2 (Bio-Rad). Two grades are sold. We buy standard grade product (less expensive) as it seems to behave identically to the more expensive 'biotechnology' grade Bio-Beads.
6. RM (*see* **Note 6**).
7. 10% DeoxyBigChap (DBC; Calbiochem). Add ~9.5 mL water to a 1 g vial of DBC to bring the total volume to 10 mL. Dissolve by gentle agitation at room temperature. Store at 4°C for short-term use (6 months or less); otherwise freeze and store at either –20 or –80°C (*see* **Note 13**).
8. Pre-extraction buffer: 50 mM HEPES, pH 7.4, 250 mM sucrose, 2 mM $MgCl_2$, 0.2% DBC.
9. Extraction buffer: 400 mM KOAc, 5 mM MgCl2, 50 mM HEPES, pH 7.4, 15% glycerol, 1 mM DTT (added just before use from a freshly prepared 1 M stock solution).

3. Methods

In our lab, we have simplified the transcription and translation reactions to essentially be in-house generated 'kits' that are easy to use. The kit components consist of a T1 mix (for transcription) and a RRL-based T2 mix (for translation). Because a single (meticulous and responsible) person can be charged with preparing and maintaining aliquots of T1 and T2, the experiments become highly reproducible and sufficiently straightforward that they are amenable to even the most inexperienced lab members. Furthermore, the use of this approach is far more economical and allows for much larger scale applications of in vitro translation reactions than is possible from commercial kits. This scalability actually makes it reasonable to purify the translated proteins [and identify interesting co-associating factors (4)], something that is typically not considered an option with mammalian in vitro systems.

The stock solutions in Steps 1 and 2 of **Section 2.1** are all quite stable, and we have one primary person responsible for preparing and maintaining them. They are not used for any other purpose and are therefore kept in a separate place for use only in preparing transcription and translation reagents. This is to maintain reliability and prevent RNAse contamination. The T1 and T2 mixes are also quite stable, but less so than individual reagents. We therefore make it from the stock solutions from time to time. If you are doing lots of in vitro transcription and translation reactions, you can make 7.6 mL and 17.1 mL T1 and T2 mixes, respectively, at a time (as we usually do). Otherwise, 760 μL and 1.71 mL at a time is ample, as is described here.

3.1. Preparation of the Transcription Mix (T1)

1. To prepare 760 μL T1 Mix, put into a microcentrifuge tube on ice the following components, in order, with gentle mixing after each addition (to avoid problems with precipitation): 487 μL water, 40 μL 1 M HEPES, pH 7.6, 3 μL 2 M $MgCl_2$, 20 μL 100 mM spermidine, 10 μL 1 M DTT, 100 μL 10X NTPs, 100 μL 10X Cap. Mix well (*see* **Note 8**).
2. Prepare aliquots (100 μL) into pre-chilled microcentrifuge tubes on ice, freeze in nitrogen, and store at –80°C. Do not freeze-thaw more than four times. Stable for at least 2 years.

3.2. Preparation of the Translation Mix (T2)

Before using the RRL, it must be supplemented with hemin [to prevent translational inhibition due to eIF2-α phosphorylation by the heme-regulated kinase (13)] and treated with micrococcal nuclease (to digest endogenous mRNAs, primary coding for globin). This is done on 10 mL at a time and sub-aliquoted for later use to make the T2. Steps 1–5 below describe how to nuclease the RRL, and steps 7–9 describe the preparation of T2.

1. Thaw a 10 mL aliquot of crude reticulocyte lysate quickly and put immediately on ice.
2. Add 400 μL of hemin solution (final concentration will be 4 μM), 100 μL of 100 mM $CaCl_2$, and 100 μL of micrococcal nuclease (15 U/μL stock). Mix gently but thoroughly (by repeated inversion).
3. Incubate in 25°C water bath for 12 min, making sure the entire sample is immersed in the water to ensure even warming. Mix gently by inversion after ~3 or 4 min of incubation.

4. Transfer to ice, immediately add 100 μL of 200 mM EGTA, and mix gently but thoroughly by repeated inversion.
5. Dispense 1 mL aliquots into pre-chilled *2 mL microcentrifuge tubes* on ice, thereby leaving enough room to add other components to prepare the T2 mix below.
6. Freeze in liquid nitrogen and store at –80°C. Nucleased RRL should generally be stable for at least 1–2 years.
7. To prepare T2 mix, thaw quickly and put immediately on ice a 1 mL aliquot of hemin/nuclease-treated RRL from step 6 above.
8. Add the following, in order: 6 μL creatine kinase (20 mg/ mL), 30 μL tRNA (10 mg/mL), 224 μL water, 300 μL Emix, and 150 μL CB 20X. The total volume will be 1.71 mL . Mix gently but thoroughly (*see* **Note 8**).
9. Dispense 200 μL aliquots into pre-chilled tubes on ice, freeze in nitrogen, and store at –80°C. The T2 mix is stable for up to a year at –80°C and will tolerate around four or five freeze-thaws if handled properly.

3.3. Linked Transcription and Translation Reactions

The translation mixture is optimized such that the products of the transcription reaction should be used directly in the translation reaction without any purification of the transcript. This 'linked' system works because the composition of the translation mix is adjusted to account for the Mg^{+2}, DTT, and spermidine contributed from the transcription. Thus, it is critical to use the transcription reaction directly (i.e., not purified transcript or mRNA) because it is providing Mg^{+2}, DTT, and spermidine to the translation, all of which are important. If purified RNA is used, then you need to add Mg^{+2} to 1.2 mM, spermidine to 0.4 mM, and DTT to at least 0.5 mM (to maintain a reducing environment in the translation reaction). The protocol below is an example of a typical linked transcription–translation reaction. It can be scaled as necessary.

1. Design and obtain oligos to PCR amplify the coding region of interest (typically from a plasmid-containing cDNA of interest). The PCR product should contain a 5′ T7 promoter for transcription. The 3′ primer should anneal at or beyond the stop codon. The following 5′ oligo contains the T7 promoter (italics) followed by a few linker nucleotides, a Kozak's site, the start codon (bold), and the nucleotides that should be chosen to anneal to the coding region to be amplified (indicated here by underlined Ns) 5′-*TAATACGACTCACTATAGGG*AGACC**ATG** <u>NNNNNNNNNNNNNNNNNN</u>-3′.
2. Use the above oligos and desired template plasmid to PCR amplify the coding region of interest. Any of several

thermostable polymerases can be used, following the manufacturer's supplied reagents and protocol. As an example, a typical 100 μL PCR reaction for Taq polymerase (New England Biolabs) contains 1X PCR buffer (final concentration; supplied as a 10X stock with the polymerase), 20 ng template plasmid, 1 μM (final concentration) of each primer, 200 μM (final concentration) each dNTP, 5 units Taq polymerase, water to 100 μL. We usually perform 30 cycles.

3. Use the Qiagen PCR purification kit to purify the PCR product as described by the manufacturer. Elute the product from the spin column with 50 μL water (not TE). Check an aliquot (usually 1 or 2 μL) on an agarose gel to confirm and estimate the quantity of amplification. Typically, 50 ng/μL concentration is normal for an average-sized product (~750 bp). Anywhere from 25 to 200 ng/μL should suffice for use as template in the transcription reaction below (*see* **Notes 14** and **15**).
4. *For each DNA template to be tested, set up on ice the following:* 7.6 μL T1 mix, 0.2 μL RNAsin, 0.2 μL RNA polymerase (either SP6 or T7, depending on the promoter of your template DNA), and 2 μL template DNA (*see* **Note 16**).
5. Incubate for 60 min at 40°C for SP6 and at 37°C for T7, then transfer to ice (*see* **Note 17**).
6. *To translate the products of the transcription reaction, add the following directly to the completed transcription reaction on ice:* 28.5 μL T2 mix, 5 μL ^{35} *S*-Met, and water/other reagents as desired to a final reaction volume of 50 μL. If co-translational translocation is being carried out, rough microsomes (RMs), proteoliposomes, liposomes, etc. are included, typically at ~2 to 5 μL per 50 μL reaction (*see* **Note 18**).
7. Incubate the translation reaction at 32°C for 30 min (for the typical 25–50 kD product). Longer for larger products, shorter for smaller products (*see* **Note 19**).
8. Following the translation reaction, transfer the tubes to ice. Remove 1 μL to a separate tube containing 19 μL SDS-PAGE sample buffer for direct analysis, keeping the remaining 49 μL on ice for downstream analysis (*see* **Section 3.2.**). Typically, we run half of this (saving the other 10 μL in case of technical problems with the gel) on a 0.75 mm thick minigel, which is then fixed, Coomassie stained (to confirm equal loading of all lanes), dried, and applied to film (Kodak-MR single emulsion film). Such direct analysis of the total translation products is helpful in troubleshooting downstream assays (*see* **Section 3.2.**) in case they yield confusing results because you will know exactly what you started with before any additional manipulations were performed (*see* **Note 20**).

3.4. Assays for Translocation and Topology

There are several ways to assay the segregation of proteins to the ER membrane. The three main ones are depicted in **Fig. 20.2**. Other assays have also been employed, but are not discussed further here. The first is to exploit an ER-specific modification such as glycosylation or signal sequence cleavage (**Fig. 20.2A**). Although this is very straightforward (typically detected as a

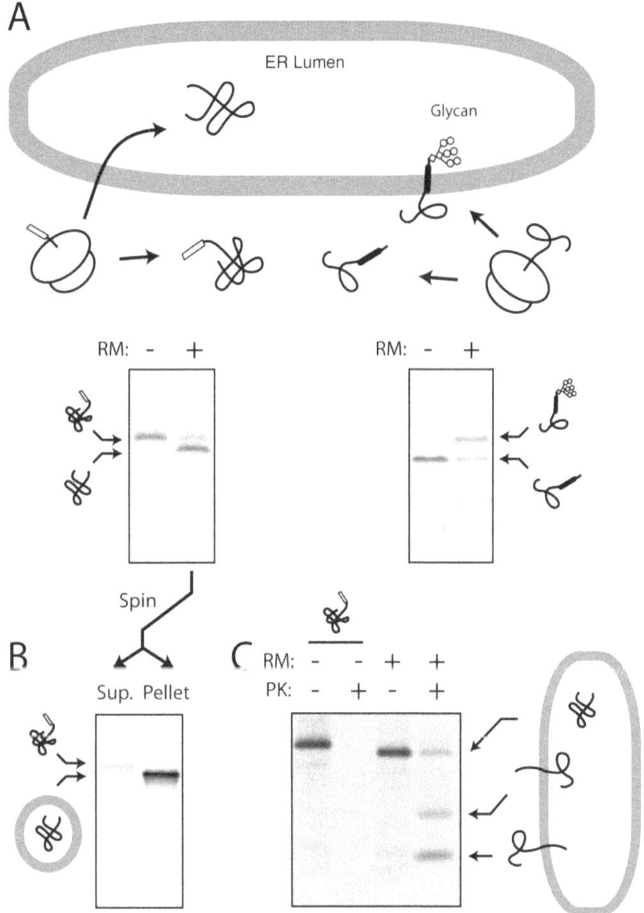

Fig. 20.2. Assays for protein segregation to the ER. (**A**) ER translocation-dependent modification. In the example on the *left*, a precursor becomes processed by signal peptidase only upon its translocation into the lumen of rough microsomes (RM), an event that can be monitored by a change in migration on SDS-PAGE. In the example on the right, a glycosylation site becomes modified upon successful translocation (8). (**B**) Co-fractionation assays. A sample similar to that from panel **A** can be separated by centrifugation into a cytosolic supernatant and membrane pellet to assess successful translocation. (**C**) Protease protection assay. Upon addition of proteinase K (PK) to the products of a translocation reaction, proteins that are either fully or partially translocated into the lumen of RMs are protected. Even a protein that generates multiple topological forms (such as mammalian prion protein; *see* ref. 20) can be resolved by this assay. By contrast, lack of translocation leads to complete digestion upon PK treatment.

change in migration by SDS-PAGE), it is not always applicable, may need the introduction of artificial glycosylation sites or epitopes into the substrate (8), and depends on the presence of functional and topologically restricted enzymes with very high activity and efficiency. This generally precludes its use if the ER components are fractionated. The second approach is to separate ER membranes from cytosol and determine whether the substrate co-fractionates with the ER (**Fig. 20.2B**). Varying levels of stringency can be applied during the fractionation (e.g., high salt, pH 11.5, urea) to increase the specificity of substrate association and reduce background. However, even with these precautions, this approach cannot be used to definitively illustrate translocation (versus peripheral association) and cannot reliably provide information about protein topology. Nonetheless, such methods are very useful to re-isolate the membranes for post-translocational assays (e.g., retrotranslocation). And finally, protease protection assays can be employed to assay both translocation and topology (**Fig. 20.2C**). In this strategy, proteins (or portions of proteins) that are translocated or membrane inserted are shielded from proteases added to the cytosolic side of the membrane. By employing very high concentrations of an aggressive and relatively non-specific protease, essentially all cytosolically exposed protein can be digested to leave only the specific translocated population. By combining this approach with immunoprecipitation, the background can be markedly minimized to allow detection of even very low efficiency translocation events. Below is a typical protocol for both fractionation (by both sedimentation and floatation) and protease protection assays from the 50 μL translation reaction produced in **Section 3.3**.

3.4.1. Isolation of Microsomes from Translation Reactions by Sedimentation

1. Remove 10 μL of the translation reaction and dilute with 90 μL 1X PSB on ice.
2. Layer this onto 100 μL of the sucrose cushion in an ultracentrifuge tube (either a tube for the TLA120.1 rotor or a micro-test tube for the TL100.3 rotor).
3. Spin for 5 min at 200,000 *g* (e.g., 70,000 rpms in the TL100.3 rotor). This spin time is suitable for traditional pancreatic RMs. If semi-permeabilized cells are used as source of ER, 5 min in a microcentrifuge may be sufficient. If proteoliposomes or smooth ER is used, longer spin times may be needed (e.g., 200,000 *g* for 30 min).
4. Remove the supernatant to a separate tube and resuspend the pellet in 1X PSB. It can be analyzed further if desired (e.g., by a protease protection assay as in **Section 3.4.3.**) or directly prepared for SDS-PAGE.

5. Analysis of the fractionation can be assessed by running equivalent amounts of the total (saved in step 8 of **Section 3.3.**), supernatant, and pellet fractions.

3.4.2. Isolation of Microsomes from Translation Reactions by Floatation

1. Remove 10 μL of the translation reaction and dilute with 90 μL of 2.2 M sucrose in 1X PSB. Considerable mixing will be required to ensure homogeneity.
2. Put into the bottom of a TLA120.1 tube. Layer with 100 μL of 1.8 M sucrose in 1X PSB followed by 25 μL of 1X PSB.
3. Centrifuge at 350,000 *g* for 1 h at 4°C.
4. Carefully remove the top 60 μL (the membranes will have floated to the top of the 1.8 M sucrose step). If the vesicles in this sample need to be recovered, the sample can be diluted in 1X PSB to 200 μL and centrifuged at 200,000 *g* for 20 min.
5. The floated vesicles can be analyzed by SDS-PAGE relative to the starting sample to assess the extent of membrane association.

3.4.3. Protease Protection Assay

1. Divide the translation reaction into three aliquots of 9 μL each on ice (A, B, and C).
2. Add 1 μL PSB to A, 0.5 μL PK and 0.5 μL PSB to B, and 0.5 μL PK and 0.5 μL Triton X-100 to C. Mix well and incubate on ice for 60 min.
3. To terminate the protease digest, both a protease inhibitor and rapid transfer to denaturating conditions are used. Either step alone is not fully sufficient to avoid artifacts. Approximately 10 min before the digestion reaction above is completed, start a boiling water bath.
4. Aliquot 100 μL of PKB into an appropriate number of empty tubes corresponding to the proteolysis reactions.
5. Dissolve a small amount of PMSF (2–5 mg) in DMSO to 0.25 M at room temperature.
6. Dip a P-2 pipette to draw up a very small amount (~0.1–0.2 μL) into the tip by capillary action and expel this into each proteolysis tube, mix, and put on ice.
7. After all tubes are completed, put the tubes containing the PKB into boiling water bath to pre-heat them to 100°C.
8. After tubes heated up (~1–2 min), transfer each proteolysis reaction directly into the boiling SDS solution and mix by rapidly pipetting up/down several times. Continue

boiling sample for additional 1–2 min and remove to room temperature.

9. Analyze an aliquot (5 or 10 μL) of the three reactions (A–C) by SDS-PAGE and autoradiography. The remainder can be subjected to immunopreciptation if desired.

3.5. Reconstitution of Membrane Proteins into Proteoliposomes

The basic steps in preparing reconstituted proteoliposomes are to prepare a mixture of membrane proteins, detergent, and lipid. The detergent is then slowly removed, during which the detergent–protein–lipid micelles will assemble into lipid vesicles containing the membrane proteins (**Fig. 20.3**). Removal of the detergent can be accomplished in several ways including sim-

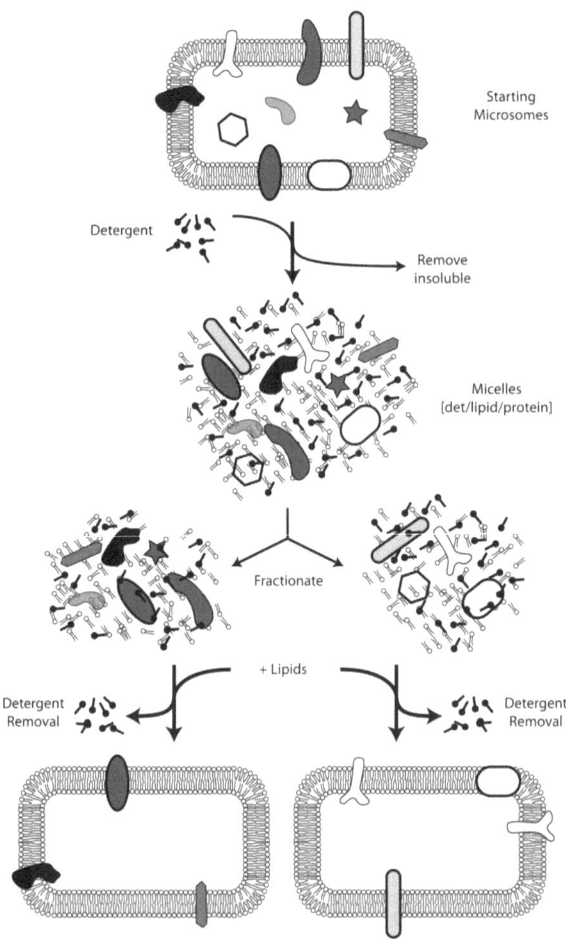

Fig. 20.3. Schematic depiction of membrane protein reconstitution. Crude microsomes are solubilized with detergent, fractionated, and reconstituted into proteoliposomes by removal of detergent in the presence of phospholipids. Note that not all proteins are successfully reconstituted, and the orientation achieved after reconstitution must be checked empirically.

ple dilution, dialysis, or adsorption. The method below is based on adsorption of the detergent to hydrophobic beads (Bio-Beads SM2, from Bio-Rad). The optimal conditions for reconstitution are difficult to predict and must be determined on a case-by-case basis. Furthermore, the orientation acquired by the membrane protein in the proteoliposomes is also not stochastic. Although it is often assumed that the orientation will be random, direct analysis shows this is not the case for many individual proteins.

From a practical standpoint, the protein, detergent, and lipids come from different sources and are mixed together just before reconstitution. The membrane protein(s), either a crude mixture or purified proteins, are already in detergent solutions to make them soluble. In addition, crude detergent extracts of membranes will also contain lipids. If the detergent and lipid contents need to be changed, the easiest method is to bind the membrane proteins to a chromatography resin, wash extensively with buffers containing the desired detergent, and elute in a buffer containing this new detergent. This is typically known as 'detergent exchange' and is often used if the detergent used for solubilization or purification is different from the one that proves to be best for reconstitution. A common example is the use of digitonin for solubilization and purification (due to its very gentle properties in maintaining membrane protein complexes), but exchange to another detergent just prior to reconstitution (because digitonin is very difficult to remove by dialysis or binding to Bio-Beads). *See* ref. (11) for an example.

The lipids are often provided separately, allowing the investigator to control the composition of the resulting proteoliposomes. They are usually easiest to prepare and handle as liposomes, but need to be added to the membrane protein as detergent-solubilized micelles. This way, the membrane proteins, lipid, and detergent will form mixed micelles at the start of the reconstitution.

3.5.1. Preparation of Lipids and Bio-Beads

Lipids are essential to making proteoliposomes containing solubilized membrane proteins. Although the precise composition of lipids may be largely irrelevant for reconstitution per se, they can significantly influence the activities of the reconstituted proteins. A simple mixture of phosphatidylcholine (PC) and phosphatidylethanolamine (PE) from a natural source (liver) in a 4:1 ratio is used here. However, other lipids, as well as cholesterol, can also be included if these are deemed important. Lipids are prone to oxidation and are therefore typically supplied in sealed ampules containing an inert gas (such as argon). With further manipulations, a reducing agent (e.g., DTT) is often included to prevent oxidation. In addition, lipids in organic solvents are typically not handled with plastic pipettes or put into plastic tubes. This is because the lipids often stick to the plastic, and the organic solvent can extract various contaminants from the plastic into your sample. Once lipids are hydrated in aqueous solutions (in which they form liposomes) or solubilized in detergent (in which they form micelles), they can be handled with the usual plastic pipettes and tubes.

1. Using glass measuring tools (e.g., glass pipettes), transfer 20 mg PC and 5 mg PE to a glass test tube or glass vial. If a tracer is desired to follow lipid recovery, include 0.5 mg rhodamine–DPPE (with accordingly less PE).
2. Dry the lipids under high vacuum. A conventional SpeedVac can be used. We usually leave it overnight to ensure complete removal of solvent. Alternatively, the organic solvent can be evaporated and the lipids dried to a film on the side of the tube using a stream of nitrogen (do this in the hood to avoid breathing chloroform vapors). After the bulk solvent is removed, the lipids can be lyophilized under high vacuum to remove any traces of organic solvent.
3. Hydrate the lipids by adding 0.4 mL lipid hydration buffer to the dried lipids and resuspend by agitation, vortexing, and/or sonication in a bath sonicator. Resist the temptation to mechanically resuspend the lipids (e.g., with a pipette) as they can be quite sticky at this stage. Be patient, and they will be fully resuspended with vortexing, at which point you will have a homogeneous cloudy/milky suspension of liposomes. Once resuspended, they can be handled with plastic pipettes and put into plastic tubes.
4. Transfer the lipids to an Eppendorf tube and adjust to 0.5 mL with additional lipid hydration buffer as necessary to make a 50 mg/mL suspension. This can be divided into 100 μL aliquots, frozen in liquid nitrogen, and stored at –80°C (*see* **Note 22**).
5. Put ~20–30 mL of dry Bio-Beads into a 50 mL polypropylene tube. Fill with MeOH, mix, and let the beads settle. Pour off the MeOH and fill with distilled water. Mix, let the beads settle (or if you are impatient, brief centrifugation), and pour off the water. Repeat this extensively (20 or more times) until all traces of MeOH are removed. Alternatively, pour the beads out into a disposable filter flask and use a vacuum to extensively wash the beads with distilled water. Put the beads back into a 50 mL polypropylene tube, fill with water, and store at 4°C. They are stable indefinitely.

3.5.2. Solubilization of ER Membranes

The solubilization of membrane proteins with detergent is influenced by many parameters. The most important are (i) the choice of detergent (*see* **Note 21**), (ii) the detergent concentration (more specifically, the relative ratio of detergent:protein:lipid),

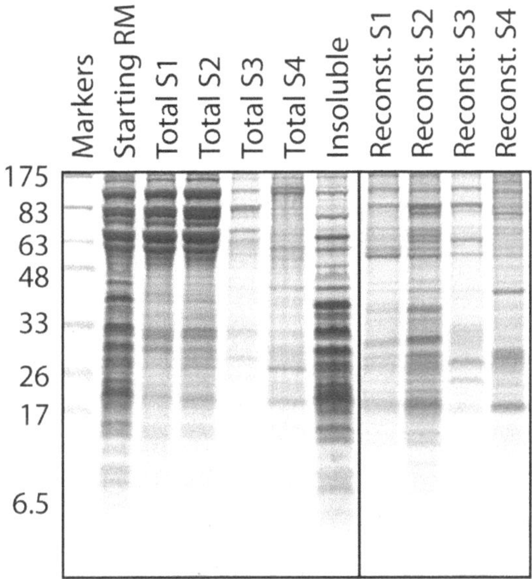

Fig. 20.4. Example of differential solubilization and reconstitution of membrane proteins. Crude RMs (*lane* 1) were sequentially extracted by four buffers containing different amounts of detergent and salt to generate four supernatant fractions (S1–S4) and an insoluble pellet (primarily containing ribosomes). Each of these four fractions was then reconstituted in the presence of phospholipids by detergent removal, and the resulting proteoliposomes were also analyzed on the gel. Note that the very abundant high molecular weight proteins in S1 and S2 (primarily lumenal proteins) are not reconstituted. Note also the different protein profiles of the different proteoliposomes illustrating the utility of differential membrane protein extraction as a purification step. The detergent in this case was DeoxyBigCHAP, although similar results can be obtained with other detergents.

and (iii) the concentration and type of salt present during solubilization. The fact that different proteins are solubilized under different conditions can actually be used to effect some degree of purification via sequential selective solubilization (*see* **Fig. 20.4**). In addition to solubilization of proteins, maintaining its functionality further constrains the conditions. Hence, conditions that most efficiently solubilize a membrane protein may also result in its irreversible denaturation. While there are general guidelines, the solubilization conditions for any given protein or activity must be determined emperically. A general protein stabilizing agent such as 10–15% glycerol is often helpful. Some of the other less common parameters that influence stability of certain types of membrane proteins include divalent cations (e.g., Mg+2), nucleotides, free phospholipids, or specific co-factors. Below is a generic protocol for solubilizing and enriching for most ER membrane proteins under conditions that allow their relatively straightforward reconstitution. Although not described, the solubilized proteins can of course be fractionated (e.g., by ion exchange or lectin chromatography) prior to reconstitution of individual fractions by similar methods.

1. 1 mL pancreatic rough microsomes (RM) at a concentration of 50 A_{260} units is put into a 3 mL thickwall polycarbonate centrifuge tube (Beckman) on ice. Add an equal volume of ice-cold pre-extraction buffer and mix well.
2. Centrifuge at 400,000 *g* in a TL100.3 rotor for 15 min at 4°C. Remove the supernatant (which will contain primarily lumenal proteins and some lipids that have been extracted by the low concentration of detergent making holes in the membrane). This can be saved for other uses if desired.
3. Resuspend the pellet in 0.9 mL ice-cold extraction buffer. This can be done by either gentle repeated pipetting or transfer to a small smooth glass homogenizer and manual homogenization with either a teflon or glass pestle. After resuspension, the solution should be homogeneous and turbid.
4. Add 100 μL of the 10% DBC on ice to solubilize the membrane. The solution should become more clear. Let sit on ice for 5–15 min to allow complete solubilization.
5. Centrifuge at 400,000 *g* in a TL100.3 rotor for 30 min at 4°C. Transfer the supernatant (the detergent extract, containing most of the membrane proteins and lipids) to a fresh tube on ice for use in the reconstitutions (*see* step 3 of **Section 3.3.**). The pellet will contain ribosomes, any tightly associated membrane proteins, and other membrane proteins that were not solubilized in step 4. If these remaining membrane proteins are desired, they can be solublized in a subsequent step using harsher conditions (e.g., a different detergent, higher salt). Otherwise, it can be discarded.

3.5.3. Reconstitution of Solubilized Membrane Proteins

The amount of Bio-Beads needed to effectively remove the detergent without significant removal of the lipids or protein, while accomplishing this slowly enough to allow formation of proteoliposomes without aggregation, needs to be determined empirically. The optimum varies depending on the detergent being used, its concentration, and the salt conditions. Below is a sample protocol for how we usually do this for any given detergent extract or sample (*see* **Note 23**).

1. Mix 40 μL of the liposome stock (prepared in step 1 of **Section 3.3.**) with 10 μL of the 10% DBC. If your samples to be reconstituted were prepared in another detergent, use that to solubilize the liposomes. Thus, the sample to be reconstituted typically contains only a single detergent.
2. Add the solubilized lipids from step 1 to the 1 mL of detergent extract (prepared in step 2 of **Section 3.3.**) on ice.

3. Dispense different amounts of the Bio-Beads (prepared in step 1 of **Section 3.3.**) into 0.5 mL microcentrifuge tubes and remove the water. A good range to start is between 30 and 200 μL of packed bead volume. An alternative is to use round-bottomed 2 mL microcentrifuge tubes.
4. Add 105 μL of the detergent–lipid mixture (from step 2) to each of the tubes containing the Bio-Beads.
5. Incubate with gentle overhead mixing for ~12–16 h at 4°C. If using the 2 mL tubes, use orbital shaking to mix (e.g., an Eppendorf Thermomixer place in the cold room, set at 700–800 rmps).
6. Separate the fluid from the beads. If using the 0.5 mL tubes, we briefly spin down the sample (a pulse in the microcentrifuge), cut off the lid, puncture the bottom with a 26-gauge needle, put the tube into a larger 1.5 mL microcentrifuge tube, and spin for ~1 min to recover the fluid into the larger tube (while retaining the beads in the smaller tubes). Alternatively, remove the fluid carefully with a thin long pipette tip (e.g., a gel-loading tip) and transfer to 1.5 mL tubes.
7. Add five volumes (500 μL) of ice-cold water (or 1X PSB). This serves to dilute any residual detergent and reduce the concentration of glycerol to reduce the density of the solution.
8. Centrifuge the samples in a TL100.3 rotor with micro-test tubes and adaptors at 200,000 *g* for 30 min.
9. Remove the supernatant and resuspend the pellet in 25 μL 1X PSB by careful repeated pipetting.
10. The efficiency of lipid recovery can be monitored by including rhodamine DPPE in the lipid mixture (*see* step 1 of **Section 3.3.**). The amount of lipid in the starting sample (step 2 above) can then be compared to the amount in the final proteoliposomes (step 9 above) by measuring absorbance at 560 nm.
11. The efficiency of protein reconstitution can be assessed by comparing the starting extract and final proteoliposomes by SDS-PAGE and staining for total proteins or immunoblotting for individual membrane proteins.

4. Notes

1. We use several precautions for handling all of the reagents for the T1 and T2 mixes. Use de-ionized, clean,

RNAse-free water for all of the solutions. In general (unless otherwise indicated), we quick-freeze items by immersion in liquid nitrogen and thaw quickly using either the warmth of your hands or a room temperature water bath. The rapid freezing and thawing are important for a few reasons. First, it minimizes/avoids precipitation of various reagents that can occur upon slow freezing/thawing. Second, it minimizes oxidation. Third, proteins are better protected from damage/denaturation by ice crystals if freezing/thawing is rapid. Immediately after thawing, reagents are generally kept on ice unless otherwise noted and immediately put away after use.

2. The pH of solutions is important. To avoid differences in the ways different people measure pH (different brands of pH paper, pH meter, etc.), we have provided the exact amounts of acid/base to add for most solutions that require accurate buffer/pH conditions. Other items can be adjusted to approximately the indicated pH using pH paper as a rough guide. We also typically favor HEPES over Tris buffers because the former has a p*K*a closer to 7, and its pH does not vary with temperature.
3. Cap can be omitted from the transcription reaction if cost is an issue, and if an approximately twofold lower efficiency of translation can be tolerated. In this case, use 5 mM instead of 1 mM GTP in the 10X NTPs stock above.
4. Supplementing RRL translations with tRNA is not absolutely necessary and can be omitted if cost is an issue. It does, however, stimulate the translation of some proteins (presumably because the endogenous levels of some tRNAs in the RRL are limiting), and it is possible that some RRL preparations contain relatively low amounts of tRNA and thus would benefit from supplementation.
5. Endogenous RRL usually contains reasonable amounts of all of the amino acids, making supplementation optional. However, some batches of RRL may contain more or less of particular amino acids, so we always supplement with a complete mixture to avoid variability.
6. The most active and well-characterized source of ER-derived microsomes is from canine pancreas. Preparation of canine pancreas RMs in-house follows a well-established protocol (1). Microsomes have also been prepared from many other sources including rat liver (2) and cultured cells. In addition, semi-permeabilized cells have also been used and are a viable alternative (3). However, for the purposes of fractionation and reconstitution studies, rather large amounts of microsomes are needed and can realisti-

cally only be isolated from tissue since such large-scale cell culture is prohibitive. It is worth noting that microsomes from different sources may have different functional properties even though they are all ER-derived.

7. PK should be of the highest purity available commercially. Because it is purified from natural sources, contamination with other enzymes that could affect reliability or disrupt membranes (e.g., lipases) is a potential concern. Historically, preparations of PK were 'pre-digested' for 10 min at 25°C to proteolytically destroy any possible contaminants (PK is stable to its auto-digestion), but this has proven unnecessary in our experience.

8. Mix solutions gently but thoroughly. Keep in mind that many components are of considerably different density and do not always mix together with just a little tap of the tube. When mixing protein-containing solutions, avoid making bubbles or frothing the sample to minimize the possibility of denaturation.

9. RNAses are a potential problem because they are ubiquitous and will obviously preclude the transcription and translation reactions. We have found that simple cleanliness and care are the best strategies. The use of DEPC-treated water for the various reagents is helpful. Keeping your pipettes and tips clean and free from dust is helpful. Do not touch your tubes and tips with bare hands. The inclusion of RNAsin is helpful, but only inhibits certain RNAses, and is not really necessary. In general, we can do all of this without RNAse inhibitor by following the above precautions, and we do not have problems with degradation.

10. Optimal conditions for translation can vary somewhat between different transcripts. The most important parameters are Mg^{+2} (typically between 1.5 and 3 mM) and K^{+} (typically between 50 and 150 mM). Note that because a crude reticulocyte lysate is used, it contains endogenous Mg^{+2} and K^{+} (estimated to be ~1.7 and ~40 mM, respectively), probably at around one-third the levels typical for the cytosol. Hence the need to supplement these salts, especially because the added nucleotides can chelate Mg^{+2}.

11. Microsomes subjected to multiple freeze-thaws will lose lumenal contents by leakage. Presumably, the integrity of the vesicles is sensitive to freeze-thawing. If the presence of these proteins in the reaction might pose a problem for your assays, it is helpful to re-sediment the microsomes and resuspend them in PSB just before use.

12. Cheaper alternatives to purified liver phospholipids are egg or soy-derived PC and PE. However, the acyl chain compositions can be quite different and may influence your proteins and/or activities. This needs to be tested emperically. In addition synthetic lipids of defined acyl chain composition are also available (e.g., DOPC, POPC) if you wish to precisely control lipid composition of the membrane. Other fluorescent lipids are also available to suit your requirements.
13. DBC is a detergent of the bile salt family. Others in the same family are Cholate, CHAPS, and BigChap. Among these, Cholate is the least expensive, but is also anionic and would therefore interfere with ion exchange fractionation. CHAPS is a good alternative that is relatively inexpensive and zwitterionic, making it more suitable for use in fractionation. Another commonly used detergent for reconstitution is octyl-glucoside because it is uncharged and is easily removed by either dialysis or Bio-Beads. A more thorough discussion of detergents is presented in ref. (14).
14. Although circular or linearized plasmids can be used for transcription/translation, PCR products are usually easiest and work well. In either case, you will need an SP6 or T7 promoter and a consensus Kozak's sequence at the start codon. A poly-A tail is not necessary for efficient in vitro translation. Even if there is no promoter in the plasmid containing the coding region of interest, you can simply encode the SP6 or T7 promoter and Kozak's sequence into the 5′ primer used for PCR (as is described in step 1 of **Section 3.3.**). Note that the 5′ primer can be designed to anneal internally to generate an N-terminally truncated protein. Similarly, 3′ primers containing a stop codon can be designed within the coding region to generate a C-terminally truncated protein. A DNA template lacking a stop codon will result in a translation product that remains tethered to the ribosome via the last amino-acyl tRNA (15). Such truncated translation intermediates have many applications (for example, *see* ref. 16–19), but are not considered further here.
15. If you have numerous constructs in a vector that already has an SP6 or T7 promoter (as we do), you can use the same primers to amplify them all for translation: just use a 5′ primer that anneals to the SP6 or T7 site and a 3′ primer that anneals to the vector sequence beyond the open reading frame. The circular plasmids can also be used directly in the transcription reaction. For this purpose, use DNA that is free of RNAse contamination and is at between 100

and 1000 ng/μL concentration (as is typical for standard Qiagen minipreps of high-copy plasmids).

16. In instances where many transcription reactions are being performed simultaneously, a master mix lacking the DNA can be prepared, aliquoted, and supplemented with the different DNAs. This can easily be scaled up or down as needed. A control reaction with water instead of DNA can be performed to determine the extent of background in the translation reactions.
17. Transcription reactions can be incubated for longer times (up to 2 or 3 h), but there is no significant increase in yield under these conditions (the nucleotides become limiting and the pyrophosphate that is generated inhibits the polymerase). After transcription, the sample can be kept on ice for some time (an hour or two) without any concern of degradation. If frozen in nitrogen and stored at –80°C, the transcript can be used again. However, we almost always make it fresh for each experiment and discard any leftover material (to avoid the risk of degradation with storage).
18. The amount of ^{35} *S*-Met can be decreased depending on the amount of labeling needed. If unlabeled translations are being performed, cold Met (at a stock concentration of 0.4 mM) should be added instead (to 40 μM final concentration).
19. Incubations are generally at ~30–32°C, but translation works at anywhere from 23 to 37°C. Typical translation times are 30–60 min, depending on size of expected products. In general, it takes ~5 min to complete a 100 residue protein (~10 kD) at 32°C. Thus, we generally incubate for ~10 min per 10 kD of expected size product. Longer incubations generally do not produce more product since translational activity of the lysate declines. Translation is slower at lower temperatures.
20. We find that for a small- to average-sized protein (~25–30 kD) containing an average proportion of methionines, a band corresponding to the translation product can be detected in as little as 5 or 10 min for optimally translating proteins. Preprolactin and prion protein are often used in our lab, and both express comparably well. Note that hemoglobin, present in the RRL at ~50 mg/mL, migrates at 14 kD. Although it is not radiolabeled, it distorts this region of the gel and can cause artifacts if your translated product is of the same size. This can be avoided by loading 10-fold less of the translation product and exposing the gel for longer times. Alternatively, the translation products can be immunoprecipitated.

21. There have been many instances where detergents vary sufficiently from lot to lot to affect the results of functional reconstitution (e.g., ref. 11). It is worth keeping this in mind if you have issues with reproducibility. Some detergents are more variable than others, and ones from natural sources (e.g., digitonin) are especially difficult in terms of reliability. Cost can also be an issue in detergent choice. Thus, some degree of screening can be important in initial studies. The time invested is usually very worthwhile and pays off in the long run.
22. Although freezing and thawing liposomes are not recommended in many protocols, this is not really relevant here (as long as oxidation is avoided), since they will be subsequently solubilized in detergent. Thus, alteration of liposome size or maintaining them as unilamellar vesicles is not an issue.
23. Reconstitution of purified membrane proteins in detergent solution can be accomplished using basically the same methodology. However, it may be necessary to re-optimize conditions to maximize recovery, which can be expected to be at least 50%. Generally speaking, one tries to use the least amount of lipids needed to fully incorporate the protein of interest into vesicles (to maximize density in the membrane).

Acknowledgments

Work in the Hegde lab is supported by the NICHD Intramural Research Program of the National Institutes of Health.

References

1. Walter, P., and Blobel, G. (1983) Preparation of microsomal membranes for cotranslational protein translocation. *Methods Enzymol* **96**, 84–93.
2. Adelman, M.R., Blobel, G., and Sabatini, D.D. (1973) An improved cell fractionation procedure for the preparation of rat liver membrane-bound ribosomes. *J Cell Biol* **56**, 191–205.
3. Wilson, R., Allen, A.J., Oliver, J., Brookman, J.L., High, S., and Bulleid, N.J. (1995) The translocation, folding, assembly and redox-dependent degradation of secretory and membrane proteins in semi-permeabilized mammalian cells. *Biochem J* **307**, 679–687.
4. Stefanovic, S., and Hegde, R.S. (2007) Identification of a targeting factor for posttranslational membrane protein insertion into the ER. *Cell* **128**, 1147–1159.
5. Schuldiner, M., Metz, J., Schmid, V., Denic, V., Rakwalska, M., Schmitt, H.D., Schwappach, B., and Weissman, J.S. (2008) The GET complex mediates insertion of tail-anchored proteins into the ER membrane. *Cell* **134**, 634–645.
6. Daniels, R., Kurowski, B., Johnson, A.E., and Hebert, D.N. (2003) N-linked glycans direct the cotranslational folding pathway of influenza hemagglutinin. *Mol Cell* **11**, 79–90.

7. Oberdorf, J., and Skach, W.R. (2002) In vitro reconstitution of CFTR biogenesis and degradation. *Methods Mol Med* **70**, 295–310.
8. Brambillasca, S., Yabal, M., Soffientini, P., Stefanovic, S., Makarow, M., Hegde, R.S., and Borgese, N. (2005) Transmembrane topogenesis of a tail-anchored protein is modulated by membrane lipid composition. *EMBO J* **24**, 2533–2542.
9. Fons, R.D., Bogert, B.A., and Hegde, R.S. (2003) Substrate-specific function of the translocon-associated protein complex during translocation across the ER membrane. *J Cell Biol* **160**, 529–539.
10. Garrison, J.L., Kunkel, E.J., Hegde, R.S., and Taunton J. (2005) A substrate-specific inhibitor of protein translocation into the endoplasmic reticulum. *Nature* **436**, 285–289.
11. Görlich, D., and Rapoport, T.A. (1993) Protein translocation into proteoliposomes reconstituted from purified components of the endoplasmic reticulum membrane. *Cell* **75**, 615–630.
12. Jackson, R.J., and Hunt, T. (1983) Preparation and use of nuclease-treated rabbit reticulocyte lysates for the translation of eukaryotic messenger RNA. *Methods Enzymol* **96**, 50–74.
13. Trachsel, H., Ranu, R.S., and London, I.M. (1978) Regulation of protein synthesis in rabbit reticulocyte lysates: purification and characterization of heme-reversible translational inhibitor. *Proc Natl Acad Sci USA* **75**, 3654–3658.
14. Helenius, A., and Simons, K. (1975) Solubilization of membranes by detergents. *Biochim Biophys Acta* **415**, 29–79.
15. Perara, E., Rothman, R.E., and Lingappa, V.R. (1986) Uncoupling translocation from translation: implications for transport of proteins across membranes. *Science* **232**, 348–352.
16. Wiedmann, M., Kurzchalia, T.V., Hartmann, E., and Rapoport, T.A. (1987) A signal sequence receptor in the endoplasmic reticulum membrane. *Nature* **328**, 830–833.
17. Görlich, D., Hartmann, E., Prehn, S., and Rapoport, T.A. (1992) A protein of the endoplasmic reticulum involved early in polypeptide translocation. *Nature* **357**, 47–52.
18. Crowley, K.S., Reinhart, G.D., and Johnson, A.E. (1993) The signal sequence moves through a ribosomal tunnel into a noncytoplasmic aqueous environment at the ER membrane early in translocation. *Cell* **73**, 1101–1115.
19. Hegde, R.S., and Lingappa, V.R. (1996) Sequence-specific alteration of the ribosome-membrane junction exposes nascent secretory proteins to the cytosol. *Cell* **85**, 217–228.
20. Hegde, R.S., Mastrianni, J.A., Scott, M.R., DeFea, K.A., Tremblay, P., Torchia, M., DeArmond, S.J., Prusiner, S.B., and Lingappa, V.R. (1998) A transmembrane form of the prion protein in neurodegenerative disease. *Science* **279**, 827–834.

Chapter 21

In Vitro Reconstitution of the Selection, Ubiquitination, and Membrane Extraction of a Polytopic ERAD Substrate

Kunio Nakatsukasa and Jeffrey L. Brodsky

Abstract

Secretory and membrane proteins that are destined for intracellular organelles in eukaryotes are first synthesized at the endoplasmic reticulum (ER) and are then delivered to their final destinations. The ER contains high concentrations of molecular chaperones and folding enzymes that assist substrates to acquire their native conformations. However, protein misfolding is an inevitable event especially when cells are exposed to stress or during development or aging. ER-associated degradation (ERAD) is a major mechanism to eliminate misfolded proteins from the secretory pathway. The importance of ERAD is underscored by the fact that mutations in secretory and membrane proteins or corruption of the ERAD machinery have been linked to human diseases. Many components involved in ERAD have been identified by a genetic analysis using the yeast *Saccharomyces cerevisiae*, and it now appears that most of these factors are conserved in higher eukaryotes. In this chapter, we describe a method to recapitulate the ubiquitination and extraction of misfolded polytopic membrane proteins in vitro using materials prepared from yeast. These techniques provide a powerful tool to further dissect the ERAD pathway into elementary steps.

Key words: Endoplasmic reticulum, ER-associated degradation, ATP, proteasome, Ufd2, Cdc48/p97, microsomes, yeast.

1. Introduction

Newly synthesized secretory and membrane proteins that fail to achieve their native conformations are retained in the endoplasmic reticulum (ER) and may be degraded. This process is referred to as ER-associated degradation (ERAD). From studies over the past 13 years, it is now clear that ERAD substrates are first recognized in the ER and are then retrotranslocated back to the cytoplasm

A. Economou (ed.), *Protein Secretion*, Methods in Molecular Biology 619,
DOI 10.1007/978-1-60327-412-8_21,

where they are ubiquitinated and degraded by the proteasome (1–5). Earlier genetic studies using the yeast *Saccharomyces cerevisiae* have identified core components required for ERAD, including membrane-associated E2/E3 ubiquitination enzymes, cytoplasmic and luminal chaperones, and the proteasome. Although the detailed mechanism for substrate recognition and retrotranslocation is not yet clear, current evidence suggests that, depending on the location of the misfolded lesion, molecular chaperones and chaperone-like lectins either in the ER or in the cytoplasm help select ERAD substrate (6–9). To further dissect the ERAD reaction into elementary steps and to characterize the functions of known and novel components, it is vital to biochemically reconstitute ERAD.

Ste6p is a yeast **a**-factor mating pheromone transporter that is synthesized in the ER and is delivered to and functions at the plasma membrane. A mutant form of Ste6p, which is called Ste6p*, is retained in the ER and is degraded by the proteasome via ERAD (10). Ste6p* has 12 transmembrane domains and is structurally similar to the cystic fibrosis transmembrane conductance regulator (CFTR), which is also an ERAD substrate and which when mutated results in cystic fibrosis. Genetic analysis has shown that Ste6p* degradation is slowed when specific E2 ubiquitin-conjugating enzymes (Ubc6p and Ubc7p), E3 ubiquitin ligases (Doa10p and Hrd1p), cytoplasmic Hsp70 and Hsp40 chaperones (Ssa1p and Ydj1p/Hlj1p), and a AAA-ATPase Cdc48p are disabled (6, 11). Although the ERAD pathway for Ste6p* is relatively well-defined, until recently it was not clear how this substrate is selected for ubiquitination and whether it is degraded in the cytoplasm or at the ER membrane.

We recently reconstituted the ubiquitination and extraction of Ste6p* using materials prepared from yeast (12). This assay has proven that Ssa1p is essential for ubiquitination. Moreover, ubiquitinated Ste6p* is extracted from the ER membrane to the cytosol in an ATP- and Cdc48p-dependent manner. We also discovered that Ufd2p, an E4 polyubiquitin chain-extending enzyme, elongates ubiquitin chains. Theoretically, this assay can be applied to any misfolded membrane protein that can be expressed in yeast. This assay also has the potential to further dissect the pathway of these ERAD substrates using yeast genetic mutants.

2. Materials

2.1. Preparation of ER-Derived Microsomes

1. Plasmids that encode misfolded polytopic membrane substrates: Ste6p*-3HA is encoded by pSM1082 (2μ *URA3* p_{ste6}*ste6-166::HA*) or pSM1911 (2μ *URA3* p_{PGK}ste6-166::*HA*) and CFTR-3HA is encoded by pSM1152

(2μ *URA3* p_{PGK}*CFTR*::*HA*) (11, 13). Yeast cells are transformed with one of these plasmids and grown in a selective medium using established methods (14). Filter-sterilized medium is used for the protocol in **Section 3.1.2**.

2. Lyticase: It was obtained commercially or has been produced using a heterologous expression system (15, 16).
3. Lyticase buffer: 0.7 M sorbitol, 0.75% (w/v) yeast extract, 1.5% (w/v) bacto peptone, 0.5% glucose, 10 mM tris(hydroxymethyl)aminomethane (Tris)–hydrochloric acid (HCl), pH 7.4
4. Lysis buffer: 0.1 M sorbitol, 50 mM potassium acetate (KOAc), 2 mM ethylenediaminetetraacetic acid (EDTA), 20 mM *N*-(2-hydroxyethyl)piperazine-*N*′(2ethanesulfonic acid) (HEPES)–NaOH, pH 7.4. The following reagents were added immediately prior to use: 1 mM dithiothreitol (DTT), 1 mM phenylmethylsulfonyl fluoride (PMSF), 1 μg/mL leupeptin, and 0.5 μg/mL pepstatin A.
5. Cushion 1: 0.8 M sucrose, 1.5% Ficoll 400, 20 mM HEPES–NaOH, pH 7.4
6. Cushion 2: 1.0 M sucrose, 50 mM KOAc, 20 mM HEPES–NaOH, pH 7.4. DTT (1 mM) is added immediately prior to use.
7. Buffer 88 (B88): HEPES–NaOH, pH 6.8, 150 mM KOAc, 5 mM magnesium acetate (MgOAc), 250 mM sorbitol in double-deionized water (ddH_2O). The solution should be filter sterilized and stored at 4°C.

2.2. Preparation of Yeast Cytosol

1 Liquid nitrogen (~4 L)
2. Stainless steel blender
3. B88 (*see* **Section 2.1**, item 7)

2.3. Ubiquitination of Ste6p* -3HA

1. B88 (*see* **Section 2.1**., item 7).
2. Microsomal membranes (*see* **Section 3.1**).
3. Yeast cytosol (*see* **Section 3.2**).
4. 10X ATP-regenerating system: 10 mM ATP, 500 mM creatine phosphate, 2 mg/mL of creatine phosphokinase in B88. We typically store this solution in aliquots of approximately 100 μL at –80°C and use only once (i.e., do not re-freeze).
5. ^{125}I-labeled ubiquitin: Bovine ubiquitin (Sigma) is dissolved in phosphate-buffered saline at a concentration of 10 μg/μL and labeled with ^{125}I (NEN Research, BioRad) using the ICl method (17, 18). The labeled ubiquitin is enriched with a D-salt Excellulose Desalting

column (Pierce) and is stored at a final concentration of 0.2 μg/μL (~1.0 × 10^6 cpm/μL) (*see* **Notes 1** and 7).

6. Apyrase (Sigma).
7. Methylated ubiquitin (Boston Biochem).
8. 1.25% SDS stop solution: 50 mM Tris–Cl, pH 7.4, 150 mM NaCl, 5 mM EDTA, 1.25% sodium dodecyl sulfate (SDS). The following reagents are added immediately prior to use: 1 mM PMSF, 1 μg/mL leupeptin, 0.5 μg/mL pepstatin A, and 10 mM *N*-ethylmaleimide (NEM).
9. 2 or 1% Triton X-100 solution: 50 mM Tris–Cl, pH 7.4, 150 mM NaCl, 5 mM EDTA, 2 or 1% Triton X-100. The following reagents are added immediately prior to use: 1 mM PMSF, 1 μg/mL leupeptin, 0.5 μg/mL pepstatin A, and 10 mM NEM.
10. 2X SDS–PAGE sample buffer: 4% β-mercaptoethanol, 4% SDS, 130 mM Tris–Cl, pH 6.8, 20% glycerol, 10 mg/mL bromophenol blue.
11. Trichloroacetic acid (TCA) sample buffer: 80 mM Tris–Cl, pH 8.0, 8 mM EDTA, 0.25 M DTT, 3.5% SDS, 15% glycerol, 0.08% Tris-base, 0.01% bromophenol blue.
12. Anti-HA antibody: 5 mg/mL (Roche).
13. Protein A-Sepharose: Sepharose 50% (v/v) (GE Health care) is equilibrated with a buffer (50 mM Tris–Cl, pH 7.4, 150 mM NaCl, 5 mM EDTA, 1 mM azide) and is stored at 4°C.
14. IP wash buffer: 50 mM Tris–Cl, pH 7.4, 150 mM NaCl, 5 mM EDTA, 1% Triton X-100, 0.2% SDS. 10 mM NEM (1 M stock in dimethyl sulfoxide) is added immediately prior to use.
15. SDS–PAGE fixative: 25% (v/v) isopropanol, 10% (v/v) glacial acetic acid in ddH_2O.

3. Methods

3.1. Preparation of ER-Derived Microsomes

In vitro ubiquitination of Ste6p* and CFTR depends on relevant ubiquitination enzymes (e.g., Ubc6p/7p, Hrd1p/Doa10p, and Ufd2p) and Hsp70 and Hsp40 molecular chaperones (e.g., Ssa1p and Ydj1p/Hlj1p). To assay the effects of these agents, yeast microsomes are prepared in one of three different ways from mutant cells and isogenic wild-type cells expressing Ste6p* or CFTR. When microsomes are prepared from deletion mutant cells (e.g., *ubc6Δubc7Δ*, *hrd1Δ*, *doa10Δ*, *hrd1Δdoa10Δ*, and *ufd2Δ*) and isogenic wild-type strains, the cell walls are first digested with lyticase at room temperature or at 30°C for <1 h before the preparation of cell homogenate (*see* **Section 3.1.1**). However, during this incubation at the permissive temperature, the temperature-sensitive defect may be lost. Therefore, when microsomes are instead prepared from temperature-sensitive mutants (e.g., *ssa1-45*, a mutant form of *SSA1*, and the *ydj1-151/hlj1Δ* strains) and isogenic wild-type strains, cells are grown at a permissive temperature of 26°C and then are shifted to a restrictive temperature of 37°C. Cells are then collected on ice and are physically disrupted with glass beads by keeping them on ice to strictly control the temperature (*see* **Section 3.1.2** or **3.1.3**) (*see* **Note 4**).

3.1.1. Preparation of Microsomes from Homogenates After Spheroplast Formation (Large Scale)

The following procedure, used routinely in our laboratory, is based on a protocol previously described (19–21).

1. Yeast microsomes are prepared from cells expressing the desired substrate (Ste6p* or CFTR) grown to log to late-log phase (optical density at 600 nm [OD_{600}] of 2–3). Typically yeast cells are grown in 1–2 L of selective medium.
2. The cell walls are digested with lyticase, and the resulting spheroplasts are collected by centrifugation through Cushion 1. The plasma membrane is then broken with a Teflon-glass motor-driven homogenizer.
3. Lysates are layered onto Cushion 2 and centrifugation is used to obtain a crude microsomal fraction, which is then concentrated and washed with B88 by centrifugation at approximately 15,000 *g* for 10 min.
4. The concentration of microsomes is adjusted to approximately 10 mg protein/mL (OD_{280} = 40 in 2% SDS) with B88. Microsomes should be stored in single-use aliquots (~50 μL), which are stable indefinitely at –80°C and should be thawed on ice immediately before use.

3.1.2. Preparation of Microsomes From Homogenates After Glass Bead Disruption (Small Scale)

1. Cells are grown to log phase (OD_{600}= 0.7–1.5) at a permissive temperature (e.g., *ssa1-45* and *ydj1-151/hlj1Δ* at room temperature) and are shifted to 37°C for approximately 1 h. A shaking water bath is used to strictly control the temperature.
2. Approximately 20–30 OD_{600} equivalents of cells are collected by centrifugation at 4,300 *g* for 5 min at 4°C and are washed once with 20–30 mL of ice-cold distilled water. The pelleted cells are resuspended in 1 mL of ice-cold water, transferred to a new microcentrifuge tube, and recentrifuged and the remaining water is removed.

The cells are then frozen in liquid nitrogen and stored at –80°C.

3. To prepare a crude membrane fraction, add 250 μL of lysis buffer to the frozen cells and disperse the cells quickly by agitation. Transfer the cell suspension to 13 × 100 mm glass test tube (VWR International) and add acid-washed 0.5-mm glass beads (Scientific Industries) to the meniscus.
4. The cells are broken by vigorous agitation on a vortex mixer for 30 s 10 times, with 30-s intervals on ice between each treatment.
5. Add 500 μL of ice-cold B88 to the homogenate and agitate for approximately 1 s.
6. Transfer cell suspension to a pre-cooled microcentrifuge tube on ice.
7. Wash the glass beads with 500 μL of ice-cold B88 by brief agitation and pool the wash with the homogenate above (step 6).
8. Remove unbroken cells by two rounds of centrifugation at 830 *g* for 5 min at 4°C in a refrigerated microcentrifuge.
9. To obtain the subcellular membrane fraction, the resulting supernatant is centrifuged at 18,000 *g* for 20 min at 4°C.
10. Wash the membrane fraction with 1 mL of ice-cold B88 and re-collect the membranes as above (step 9).
11. Adjust the protein concentration and store at –80°C as in **Section 3.1.1**, step 4.

3.1.3. Preparation of Microsomes from Homogenate After Glass Beads Disruption (Medium Scale)

1. Cells are grown to log phase (OD_{600} = 0.7–1.5) at a permissive temperature and are shifted to 37°C for approximately 1 h (*see* **Section 3.1.2**, step 1).
2. Approximately 200 OD_{600} equivalents of cells are collected by centrifugation at 4,300 *g* for 5 min at 4°C and are washed and re-collected by centrifugation two times with 20–30 mL of ice-cold distilled water. Cells can be placed in a polycarbonate centrifugation tube and frozen in liquid nitrogen and stored at –80°C.
3. To prepare a crude membrane fraction, add 2 mL of lysis buffer to the frozen cells and glass beads to the meniscus.
4. The cells are broken as in **Section 3.1.2**, step 4.
5. Add 5 mL of ice-cold B88 to the homogenate and agitate for approximately 1 s.
6. Transfer the cell homogenate to a new pre-cooled tube on ice.

7. Wash the glass beads with 5 mL of ice-cold B88 by brief agitation and pool this wash with the homogenate above (step 6).
8. Overlay the cell homogenate (~15 mL) onto 15 mL of Cushion 2 in a polycarbonate centrifugation tube and centrifuge in a swinging-bucket HB-6 rotor at 7,000 *g* for 10 min at 4°C.
9. Transfer the upper layer to a new polycarbonate tube and centrifuge at 15,000 *g* for 10 min at 4°C.
10. Wash the pelleted membrane fraction with 20 mL of B88 and centrifuge again as above (step 9).
11. Adjust the concentration and store at –80°C as in **Section 3.1.1**, step 4.

3.2. Preparation of Yeast Cytosol

The following procedure, used routinely in our laboratory, is based on a protocol previously described (21).

1. Grow yeast cells in rich medium to log phase (OD_{600} = ~2.0) at 30°C. When the *cdc48-3* mutant and isogenic wild-type cells are being used, the yeast are propagated at room temperature and shifted to a restrictive temperature (37°C) for 5 h (22).
2. Collect the cells and wash with distilled water.
3. Resuspend the cells in a minimal amount of B88 to form a thick yeast slurry (e.g., <6 mL of B88 per 6 L of initial yeast culture).
4. Freeze the cells by drop-wise addition to 500 mL liquid nitrogen in a plastic beaker. After the excess liquid nitrogen evaporates, store these "popcorn" like particles at 80°C.
5. Add the particles to approx 500 mL of liquid nitrogen and blend at high speed for 8–10 min in a stainless-steel blender. Maintain the volume of liquid nitrogen above the rotating blades by periodic addition of liquid nitrogen during blending.
6. After the liquid nitrogen evaporates, transfer the powder containing broken cells to a 50 mL Falcon tube, which can be stored at –80°C.
7. Place the tube on ice and add a minimal amount of B88 (e.g., ~0.5 mL/40 mL of broken yeast slurry) containing 1 mM DTT. Then, thaw the cells in a room temperature water bath.
8. After thawing, centrifuge the lysate at 10,000 *g* for 10 min at 4°C. The supernatant is then collected and centrifuged at 300,000 *g* for 1 h at 4°C to remove membranes/aggregated protein.

7. The supernatant from this final spin is aliquoted (~100–200 μL), snap-frozen in liquid nitrogen, and stored at –80°C. The concentration of resultant cytosol is usually 20–30 mg/mL (*see* **Note 2**). Avoid repeated freeze and thaw cycles as the activity of the lysate diminishes upon each cycle.

3.3. Ubiquitination of Ste6p*

The in vitro ubiquitinated Ste6p* and the presence of unmodified Ste6p* can be detected by autoradiography and by western blotting, respectively. The following procedure results in sample volumes of approximately 28 μL, but 12-μL samples are sufficient for autoradiography or western blotting. The same protocol can be used to detect in vitro ubiquitinated CFTR.

1. Combine the reagents in the following order on ice: B88 (sufficient amount for an initial reaction volume of 18 μL), 2 μL of microsomes, 2 μL of 10× ATP-regenerating system, and the appropriate final concentration of cytosol (typically 1–4 mg/mL). As a negative control, microsomes prepared from the strain lacking the Ste6p* expression vector and B88 instead of cytosol can be used. Add reaction inhibitors such as apyrase (ATP control) or methylated ubiquitin (inhibitor of ubiquitin extension; see **Fig. 21.1**) at this point (*see* **Note 5**).
2. Pre-incubate the reaction at 23°C for 10 min.

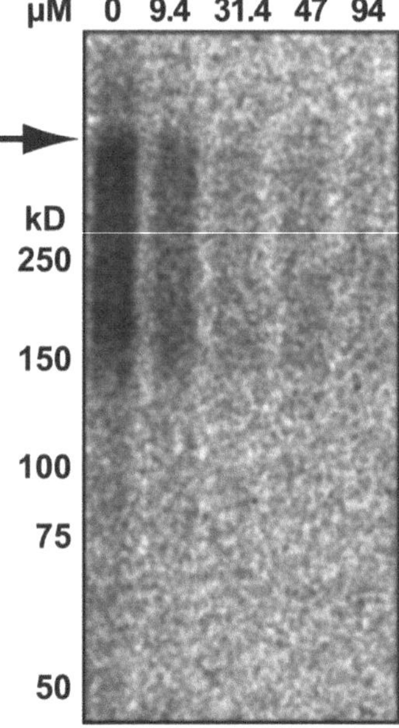

Fig. 21.1. The in vitro ubiquitination assay. The assay was performed essentially as described in **Section 3.3** using wild-type ER-derived microsomes (prepared as in **Section 3.1.1**) from yeast expressing Ste6p*-3HA. The microsomes were incubated with a final concentration of 6.5 mg/mL cytosol, the ATP-regenerating mix, ^{125}I-labeled ubiquitin, and the indicated concentration of methylated ubiquitin at 30°C for 15 min. The *arrow* indicates the boundary between the stacking and running gel.

3. Add 2 μL ^{125}I-labeled ubiquitin (*see* **Note 7**).
4. Incubate up to 1 h at 23°C (*see* **Note 3**).
5. At the desired time point, add 80 μL of 1.25% SDS stop solution and briefly agitate (~2 s) on a vortex mixer at high speed.
6. Incubate at 37°C for 30 min.
7. Add 400 μL of 2% Triton X-100 solution or 900 μL of 1% Triton X-100 solution and place the tubes on ice.
8. Add 2 μL of anti-HA (10 μg) antibody and gently mix the solution by rotating overnight at 4°C.
9. Add 30 μL of 50% (v/v) Protein A-Sepharose and continue to rotate at 4°C for 2–3 h.
10. Harvest and then wash the immunoprecipitates with 800 μL of ice-cold IP wash buffer four times (collect the Sepharose beads by centrifugation at 2,100*g* for 10 s at room temperature using a mini-centrifuge). The samples are placed on ice between each step. After the final wash, remove as much buffer as possible from the Sepharose with a gel-loading micropipet tip.
11. Add 30 μL of either 2× SDS–PAGE sample buffer or TCA sample buffer and elute bound proteins by incubating at 37°C for 30 min.
12. Spin down the Sepharose by centrifugation at 7,000 *g* for 10 s at room temperature.
13. Transfer the supernatant (~28 μL) to a new Eppendorf tube and analyze via 6% SDS–PAGE (*see* **Section 3.5**).

*3.4. Membrane Extraction of Ubiquitinated Ste6p**

The membrane extraction assay is similar to the ubiquitination assay except that each sample is separated into membrane and cytosolic fractions by centrifugation after the ubiquitination reaction.

1. Set up 25-μL reactions using the procedure outlined in **Section 3.3**, steps 1–4 (we typically set up 25-μL reactions to obtain a 20-μL supernatant, which does not disturb the pellet). To observe Cdc48p/p97-dependent extraction of ubiquitinated Ste6p*, use *cdc48-3* mutant cytosol and its isogenic wild-type cytosol prepared as in **Section 3.2**. Optimal Cdc48 dependence is observed when the membranes are incubated for 5 min on ice and washed 2 times with B88.
2. Following the incubation, pellet the microsomes in a refrigerated microcentrifuge at 18,000 *g* for 10 min at 4°C.
3. Return the reaction tube to ice and quickly transfer 20 μL of supernatant (containing extracted/ubiquitinated Ste6p*) to a new microcentrifuge tube on ice. Be careful not to disturb the pellet.

4. Remove the remaining supernatant completely and resuspend the pelleted microsomes in 25 μL of ice-cold B88. Transfer 20 μL of suspension to a new tube.
5. Add 80 μL of 1.25% SDS stop solution to the supernatant and resupspended microsomes and briefly agitate the mixture on a vortex mixer. Process the samples as in **Section 3.3**, steps 6–13.

3.5. Data Collection and Analysis

3.5.1. Autoradiography

1. Half of the final sample (~12 μL, *see* **Section 3.3**, step 13) is analyzed by SDS–PAGE. Typically, 6 cm 6% gels are used to resolve the ubiquitinated species and are run at 20 mA (constant current) until the bromophenol blue dye front is at the bottom of the gel (*see* **Note 6**). Unmodified Ste6p* will reside approximately in the center of the separating phase of the gel and the "smear" of ubiquitinated Ste6p* will reside in the upper half of the gel.
2. The gel is gently placed in SDS–PAGE fixative for 15 min to 2 h and gently shaken at room temperature. Gels are dried on filter paper for approximately 45 min on a vacuum drier with heating (~80°C) and are then cooled to room temperature before the vacuum is broken. Typically, the resulting autoradiograph requires 1 day of exposure on a phosphoimage screen although the use of old label may require significantly longer exposure times (e.g., ~1 month).

3.5.2. Western Blotting

1. The other half of the sample (~12 μL) is used to detect unmodified Ste6p* by western blotting in the same manner as described in **Section 3.5.1**, step 1.
2. Proteins are transferred from gels to a nitrocellulose membrane, which is then blotted with anti-HA antibody followed by decoration with horseradish peroxidase-conjugated secondary antiserum. The bound secondary antibody is detected with the SuperSignal West Pico Chemiluminescent Substrate (Thermo Scientific) according to the manufacture's instructions.

4. Notes

1. We typically store ^{125}I-labeled ubiquitin at –80°C in aliquots of 20 μL. Although repeated freeze and thaw cycles (~3 times) do not seem to be detrimental to activity, best results are seen when the reagent is used within 2 months (half-life of ^{125}I is ~60 days) after preparation. Non-labeled ubiquitin is also stored at –80°C in aliquots of 20 μL.

2. Protein concentration of the cytosol is measured by the Bradford method with the protein assay kit (Bio-Rad). Bovine serum albumin (BSA) is used as the standard. No detectable loss of the activity was seen when cytosol was stored at –80°C for up to approximately 12 months.
3. The in vitro ubiquitination of Ste6p* requires physiological temperature and does not occur on ice. The optimal temperature is 23°C and the extent of ubiquitination becomes inefficient at higher temperatures (e.g., 37°C), possibly because the misfolded substrate protein aggregates. However, the phenotype of some temperature-sensitive mutant alleles (e.g., *ssa1-45*), which is most evident at 37°C in vivo, is exhibited at 23°C in the in vitro reaction.
4. Microsomes prepared from homogenates after spheroplast formation (*see* **Section 3.1.1**) or after glass beads disruption in a medium scale (*see* **Section 3.1.3**) are more E3 ligase enzyme-dependent than microsomes prepared from homogenates after a small-scale glass bead disruption (*see* **Section 3.1.2**).
5. The addition of an inhibitor for deubiquitination (ubiquitin aldehyde) or a proteasome inhibitor (MG132 "n-cbz-leu-leu-leu-al") does not result in increased ubiquitin chain extension. In addition, higher concentrations of cytosol (>~8 mg/mL) decrease the signal intensity.
6. Cytosol at a final concentration of 1–2 mg/mL results in a low-molecular weight ubiquitinated species, but addition of more cytosol (at a final concentration of 4–6 mg/mL) "shifts" the ubiquitinated species to a higher molecular weight. The use of a 6% gel is critical to differentiate these two species, and this molecular weight shift is due to Ufd2p in the cytosol (12).
7. All reaction samples containing ^{125}I-labeled ubiquitin should be shielded by an approximately 1-mm lead plate to prevent excess exposure. The aliquots of ^{125}I-labeled ubiquitin should be stored in a substantially more shielded lead container. Radioactivity at each step of this protocol should be surveyed with a γ-detecting monitor, and all items that contact ^{125}I should be properly disposed.

References

1. Tsai, B., Ye, Y., and Rapoport, T.A. (2002) Retro-translocation of proteins from the endoplasmic reticulum into the cytosol. *Nat. Rev. Mol. Cell Biol.* **3,** 246–255.
2. Kostova, Z., and Wolf, D.H. (2003) For whom the bell tolls: protein quality control of the endoplasmic reticulum and the ubiquitin-proteasome connection. *EMBO J.* **22,** 2309–2317.
3. Meusser, B., Hirsch, C., Jarosch, E., and Sommer, T. (2005) ERAD: the long road to destruction. *Nat. Cell Biol.* **7,** 766–772.

4. Römisch, K. (2005) Endoplasmic reticulum-associated degradation. *Annu. Rev. Cell Dev. Biol.* **21,** 435–456.
5. Nakatsukasa, K., and Brodsky, J.L. (2008) The recognition and retrotranslocation of misfolded proteins from the endoplasmic reticulum. *Traffic.* **9**, 861–870.
6. Vashist, S., and Ng, D.T. (2004) Misfolded proteins are sorted by a sequential checkpoint mechanism of ER quality control. *J. Cell Biol.* **165,** 41–52.
7. Nishikawa, S., Brodsky, J.L., and Nakatsukasa, K. (2005) Roles of molecular chaperones in endoplasmic reticulum (ER) quality control and ER-associated degradation (ERAD). *J. Biochem.* **137,** 551–555.
8. Denic, V., Quan, E.M., and Weissman, J.S. (2006) A luminal surveillance complex that selects misfolded glycoproteins for ER-associated degradation. *Cell.* **126,** 349–359.
9. Carvalho, P., Goder, V., and Rapoport, T.A. (2006) Distinct ubiquitin-ligase complexes define convergent pathways for the degradation of ER proteins. *Cell.* **126,** 361–373.
10. Loayza, D., Tam, A., Schmidt, W.K., and Michaelis, S. (1998) Ste6p mutants defective in exit from the endoplasmic reticulum (ER) reveal aspects of an ER quality control pathway in *Saccharomyces cerevisiae. Mol. Biol. Cell.* **9,** 2767–2784.
11. Huyer, G., Piluek, W.F., Fansler, Z., Kreft, S.G., Hochstrasser, M., Brodsky, J.L., and Michaelis, S. (2004) Distinct machinery is required in Saccharomyces cerevisiae for the endoplasmic reticulum-associated degradation of a multispanning membrane protein and a soluble luminal protein. *J. Biol. Chem.* **279,** 38369–38378.
12. Nakatsukasa, K., Huyer, G., Michaelis, S., and Brodsky, J.L. (2008) Dissecting the ER-associated degradation of a misfolded polytopic membrane protein. *Cell.* **132,** 101–112.
13. Zhang, Y., Nijbroek, G., Sullivan, M.L., McCracken, A.A., Watkins, S.C., Michaelis, S., and Brodsky, J.L. (2001) Hsp70 molecular chaperone facilitates endoplasmic reticulum-associated protein degradation of cystic fibrosis transmembrane conductance regulator in yeast. *Mol. Biol. Cell.* **12,** 1303–1314.
14. Kaiser, C.A., Michaelis, S., and Mitchell, A. (1994) Methods in Yeast Genetics., Cold Spring Harbor Laboratory Press, Cold Spring Harbor, NY.
15. Nossal, N.G., and Heppel, L.A. (1966) The release of enzymes by osmotic shock from Escherichia coli in exponential phase. *J. Biol. Chem.* **241,** 3055–3062.
16. Shen, S.H., Chrétien, P., Bastien, L., and Slilaty, S.N. (1991) Primary sequence of the glucanase gene from Oerskovia xanthineolytica. Expression and purification of the enzyme from Escherichia coli. *J. Biol. Chem.* **266,** 1058–1063.
17. McFarlane, A.S. (1958) Efficient trace-labelling of proteins with iodine. *Nature* **182,** 53.
18. Helmkamp, R.W., Goodland, R.L., Bale, W.F., Spar, I.L., and Mutschler, L.E. (1960) *Cancer Res.* **20,** 1495–1500.
19. Rothblatt, J.A., and Meyer, D.I. (1986) Secretion in yeast: reconstitution of the translocation and glycosylation of alpha-factor and invertase in a homologous cell-free system. *Cell* **44,** 619–628.
20. Deshaies, R.J., and Schekman, R. (1989) SEC62 encodes a putative membrane protein required for protein translocation into the yeast endoplasmic reticulum. *J. Cell Biol.* **109,** 2653–2664.
21. Brodsky, J.L., and Schekman, R. (1993) A Sec63p-BiP complex from yeast is required for protein translocation in a reconstituted proteoliposome. *J. Cell Biol.* **123,** 1355–1363.
22. Latterich, M., Fröhlich, K.U., and Schekman, R. (1995) Membrane fusion and the cell cycle: Cdc48p participates in the fusion of ER. *Cell* **82,** 885–893.

Chapter 22

Studying the ArfGAP-Dependent Conformational Changes in SNAREs

Fernanda Rodriguez and Anne Spang

Abstract

Vesicle SNAREs (v-SNAREs) are included with high fidelity into each transport vesicle generated in the cell. These SNAREs determine the fate of vesicles, as they are the key factors deciding with which compartment a particular vesicle will fuse. The mechanism of high fidelity inclusion of SNAREs into transport vesicles is very difficult to study in vivo. Therefore, we use in vitro assays aiming to recapitulate SNARE uptake into vesicles. One of the key assays is a pull-down with SNARE tails fused to GST in the presence or absence of ArfGAPs and coat components such as the small GTPase Arf1 and coatomer. This in vitro assay allowed us to show that the ArfGAPs Glo3 and Gcs1 can induce a conformational change in SNAREs. Protease protection assays were used to confirm the conformational change and can also be used to address the question about the nature of the different conformations in SNARE proteins.

Key words: SNAREs, Glo3p, Arf1p, conformational changes, SNARE binding assay, protease protection assays.

1. Introduction

Communication between different organelles in the cell is mostly mediated by vesicles that are formed at a donor compartment and are consumed by a specific target compartment (1, 2). Vesicle formation is regulated by small GTPases of the Arf superfamily. These GTPases rely on guanine-nucleotide exchange factors (GEFs) for their activation and GTPase-activating proteins (GAPs) to control the time of GTP hydrolysis. Therefore, the spatial and temporal control of GTPase activity requires the concerted action of GEFs and GAPs. ArfGAPs are not only important to stimulate GTP hydrolysis on Arf1p but also at other steps in

A. Economou (ed.), *Protein Secretion*, Methods in Molecular Biology 619,
DOI 10.1007/978-1-60327-412-8_22,

vesicle biogenesis, for example, for initiation of vesicle formation (3, 4). ArfGAPs positively influence vesicle generation and play an essential role for the incorporation of SNARE proteins into transport carriers (5, 6).

Six ArfGAP proteins are present in *Saccharomyces cerevisiae*, of which Gcs1p and Glo3p are the best characterized. ArfGAPs are divergent proteins: the N-terminal GAP domain is well conserved, whereas the remainder of the proteins share low sequence identity. None of the six ArfGAPs is essential. However, stimulation of GTPase hydrolysis is a vital process, indicating that they have at least partially overlapping functions. Such overlapping function has been demonstrated for Glo3p and Gcs1p, because both of them can act as ArfGAP in retrograde transport from the Golgi to the ER. Glo3p is the preferred GAP while Gcs1p can take over in the absence of Glo3p (7). As expected the simultaneous loss of Glo3p and Gcs1p is lethal. Furthermore, Glo3p and Gcs1p promote the interaction of the SNAREs with Arf1p and coatomer (5, 6). This can be demonstrated by using a SNARE binding assay. In these experiments, the trans-membrane domain of SNAREs was replaced by a glutathione-S-transferase-tag (GST-tag). These GST-tagged SNAREs were immobilized on glutathione beads and then incubated together with Glo3p or Gcs1p and Arf1p and coatomer. The unbound material was subsequently washed away, and Arf1p or coatomer recruitment to the immobilized SNAREs could be observed on Coomassie-stained SDS–PAGE gels. The SNAREs have at least two different conformations: one competent and the other incompetent for interaction with Arf1p and coatomer. Glo3p and Gcs1p act catalytically to introduce conformational changes in SNAREs, and in this alternative conformation SNAREs can bind to Arf1p and coatomer. Another ArfGAP Age2 does not possess this activity that would induce a conformational change (6). The conformational change can be monitored by protease protection assays and CD spectroscopy. Using protease-protection assays, we showed that SNAREs are more resistant to protease treatment when pre-incubated with Glo3p. An increase of the α-helical content of SNAREs upon treatment with Glo3 was observed using CD spectroscopy. In this chapter, we describe the purification protocols of Glo3p, Arf1p and SNAREs, the SNARE binding assay, and the protease-protection assay for monitoring these conformational changes.

2. Materials

2.1. Media for Purification

1. 2xYT medium: dissolve 16 g bacto-tryptone, 10 g bacto-yeast extract, and 5 g NaCl in 900 mL of water. Adjust the pH to 7.0 with NaOH and the volume to 1 L with water. Autoclave at 121°C for 15 min.

2. 1000x ampicillin stock (100 mg/mL): dissolve 5 mg of ampicillin (Sigma) in 50 mL of deionized water. Sterilize by filtration. Aliquot and store at –20°C.
3. 150x Isopropyl β-D-1-thiogalactopyranoside (IPTG) stock: dissolve 0.36 g of IPTG in 10 mL of deionized water. Sterilize by filtration. Aliquot and store at –20°C.

2.2. Purification of Glo3p

1. Lysis buffer: 20 mM Tris–HCl (pH 8), 8 M urea, 0.1 M NaH_2PO_4, 20 mM imidazole.
2. Buffer A: 20 mM Tris–HCl (pH 6.3), 8 M urea, 0.1 M NaH_2PO_4, 20 mM imidazole.
3. Buffer B: 20 mM Tris–HCl (pH 6.3), 8 M urea, 0.1 M NaH_2PO_4, 250 mM imidazole.
4. Buffer C: 25 mM HEPES–KOH (pH 7.2), 150 mM KAc, 0.1 mM $ZnCl_2$, 1 mM DTT, 20% glycerol.
5. Buffer D: 25 mM HEPES–KOH pH 7.2, 150 mM KAc, 0.01 mM $ZnCl_2$, 0.5 mM DTT, 20% glycerol.
6. Buffer E: 25 mM HEPES–KOH pH 7.2, 150 mM KAc, 0.01 mM $ZnCl_2$, 0.1 mM DTT, 20% glycerol.
7. Ni^{2+}-NTA (Qiagen).
8. Tween 20 stock solution (20 %): dissolve 20 mL of Tween 20 in 100 mL final volume of deionized water.
9. Econo-Pac column (Biorad).
10. Amicon Ultra centrifugal filters Ultra-15, MWCO 10 kDa (Millipore).
11. Bradford solution (Biorad).

2.3. Purification of His_6-Arf1ΔN17p

1. STE buffer: 50 mM Tris–HCl (pH 8.0), 25% (w/v) sucrose, 40 mM EDTA.
2. Lysozyme stock solution (50 mg/mL): dissolve 1 g of lysozyme (Sigma) in 20 mL of water. Make 400 μL aliquots and store at –20°C.
3. Triton buffer: 50 mM Tris–HCl (pH 8.0), 0.2 % Triton X-100, 100 mM $MgCl_2$.
4. Buffer F: 20 mM HEPES–KOH (pH 7.4), 1 mM $MgCl_2$, 1 mM DTT, 200 mM KCl, 20 mM imidazole.
5. Buffer G: 20 mM HEPES–KOH (pH 7.4), 1 mM $MgCl_2$, 1 mM DTT, 200 mM KCl, 250 mM imidazole.
6. Buffer H: 20 mM HEPES–KOH (pH 7.4), 1 mM EDTA, 100 mM NaCl, 1 mM DTT, 2 mM $MgCl_2$.

2.4. Purification of SNARE–GST

1. Buffer I: 20 mM HEPES–KOH (pH 7.2), 150 mM KAc, 0.05 % Tween 20, protease inhibitor Cocktail tablet (Roche).

2. Glutathione–agarose beads (Sigma, *see* **Note 1**).
3. Buffer K: 20 mM HEPES–KOH (pH 7.2), 50 mM glutathione, 150 mM KAc, 0.05 % Tween 20, one tablet of protease inhibitor Cocktail (Roche).
4. PBS buffer: 8 g of NaCl, 0.2 g of KCl, 1.44 g of Na_2HPO_4, 0.24 g of KH_2PO_4. Take to a final volume of 1 L.
5. PBS buffer with 15 % glycerol: 8 g of NaCl, 0.2 g of KCl, 1.44 g of Na_2HPO_4, 0.24 g of KH_2PO_4, 150 mL of glycerol. Take to a final volume of 1 L.

2.5. SNARE–Arf1 Binding

1. BBP buffer: 25 mM HEPES–KOH (pH 6.8), 1 mM DTT, 0.5 mM $MgCl_2$, 300 mM KAc, 0.2 % Triton X-100.
2. Glutathione–agarose beads (Sigma, *see* **Note 1**).
3. Hepes buffer: 20 mM HEPES–KOH (pH 6.8).

2.6. Protease Protection Assay

1. B88 buffer: 20 mM HEPES–KOH (pH 6.8), 250 mM sorbitol, 150 mM KAc, 5 mM $MgAc_2$.
2. V8 stock solution (5 mg/mL): dissolve 50 mg in 10 mL of B88 buffer. Aliquot and store at –20°C.
3. Trypsin stock solution (5 mg/mL): dissolve 50 mg in 10 mL of B88 buffer. Aliquot and store at –20°C.

2.7. Gel Electrophoresis

1. 5x sample buffer: 312.5 mM Tris–HCl (pH 6.8), 10% SDS, 50% glycerol, 25% β-mercaptoethanol, 0.025% bromophenol blue.
2. 30% Acrylamide, 0.8 % *N,N*'-methylene-bis-acrylamide solution.
3. 4x separating gel buffer: 1.5 M Tris–HCl (pH 8.8).
4. 4x stacking gel buffer: 500 mM Tris–HCl (pH 6.8).
5. 20% SDS.
6. 10% ammonium persulfate (APS). Aliquot and store at –20°C.
7. 12.5% separation gel: 25 mL acryl-bisacryl solution, 15 mL 4x separating buffer, 300 μL 20% SDS, 19.5 mL deionized water, 40 μL TEMED, 400 μL APS.
8. Stacking gel: 5 mL acryl-bisacryl solution, 7.5 mL 4× stacking buffer, 150 μL 20% SDS, 17.1 mL deionized water, 36 μL TEMED, 240 μL APS.
9. 5x SDS-running buffer: 125 mM Tris, 960 mM glycine, 0.5% SDS.
10. Coomasie blue stain: dissolve 2.5 g of Coomasie Brilliant Blue R 250 in 454 mL methanol and 400 mL water. Add 92 mL glacial acetic acid and bring to a final volume of 1 L with water. Filter the solution.

11. De-staining solution: 25% methanol, 10% glacial acetic acid (650 mL H_2O + 250 mL MeOH + 100 mL HAc).

2.8. Others

1. 100% Trichloroacetic acid (TCA, 100 g/100 mL).
2. 1 M Tris–HCl pH 8.0.

2.9. Equipment

1. Sonifier (Branson).
2. Centrifuges: Eppendorf microcentrifuge 5417R, Sorvall RC5C.

3. Methods

3.1. Purification of Glo3p

Glo3p is expressed in *Escherichia coli* with a COOH-terminal fusion His tag, Glo3p-His_6 (*see* **Note 2**). *GLO3* is cloned into the pET21b (Novagen) vector that encodes a hexahistidine tag (7). This plasmid contains a drug-resistant marker for ampicillin. pET21b-*GLO3*-His_6 is transformed into BL21(DE3). Glo3p is prone to aggregation when it is expressed in *E. coli*. Therefore, we purified Glo3 under denaturing conditions and then refolded after the purification.

1. Inoculate 30 mL of 2xYT medium supplemented with ampicillin (100 μg/mL) with one colony of BL21(DE3) transformed with plasmid pET21b-*GLO3*-His_6. Grow culture overnight at 37°C.
2. Take the overnight culture and inoculate 3 L of 2×YT medium containing ampicillin. Grow culture for 3 h at 37°C with vigorous shaking to OD of 0.7. Induce protein expression by adding 1 mM of IPTG. Induce protein production for 3 h at 37°C (*see* **Notes 14** and **15**).
3. Spin cells for 10 min at 4,000 *g* at 4°C, discard the supernatant, and freeze the cells in liquid nitrogen.
4. Thaw the cell pellet on ice. Add 50 mL of lysis buffer and incubate for 60 min at room temperature rotating end over end (*see* **Note 3**).
5. Remove cell debris by centrifugation at 10,000 *g* at room temperature for 15 min.
6. During this time, wash 2 mL of 50 % Ni^{2+}-NTA-Agarose slurry (Qiagen) first with water and then with lysis buffer. Add the slurry to the lysate and incubate for 1 h at room temperature in a rotary shaker (*see* **Note 4**).
7. Remove the supernatant by centrifugation for 5 min at room temperature and 700 *g*. Wash three times with lysis buffer and three times with buffer A.

8. Load the resin with the last wash into an Econo-Pac column.
9. Elute with buffer B and collect 1 mL fractions (*see* **Note 5**). Assess the presence of protein in a microtiter plate by mixing 1 μL of each fraction with 200 μL Bradford reagent (*see* **Note 6**). Continue elution until the Bradford reagent does not turn blue anymore. Analyze protein-containing fractions by SDS–PAGE and Coomassie blue stain (*see* **Note 7**).
10. Pool protein-containing fractions and dilute to 0.01 mg/mL. Add Tween 20 to a final concentration of 0.05% and dialyse against buffer C at room temperature for 3 h. Change the dialysis buffer to buffer D and dialyse overnight at 4°C. Finally, dialyse 3 h in dialysis buffer E at 4°C (*see* **Note 5**).
11. Concentrate the dialysed protein solution 100 times with an Amicon 30 filter. Aliquot and freeze in liquid nitrogen. Store at –80°C.

3.2. Purification of His_6-Arf1ΔN17p

The recruitment of Arf1ΔN17p to SNARE tails does not depend on its nucleotide state. Arf1ΔN17p-Q71L (8), a guanidine nucleotide hydrolysis-deficient Arf1p mutant, and Arf1ΔN17p-T31N (8), which is in GDP-bound form, are both recruited to the SNAREs. We describe here the purification of wild-type N-terminal hexahistidine-tagged Arf1ΔN17p but any of the mutants can be purified with the same protocol (5).

The N-terminus of Arf1p contains a hydrophobic helix and a myristoylation site that are both involved in membrane anchoring. To increase protein solubility, to avoid protein aggregation, and to reduce unspecific binding, a truncated version of Arf1p that lacks the first 17 N-terminal amino acids (Arf1ΔN17p) was purified (9). Arf1ΔN17 was cloned into pQE30 vector (ampicillin resistance).

1. Inoculate 50 mL of 2xYT medium supplemented with ampicillin (100 μg/mL) with one colony of BL21-pCodonPlus *E. coli* cells (Invitrogen) carrying the *ARF1* expressing plasmid. Grow culture overnight at 37°C.
2. Take the overnight culture and inoculate 1.5 L of 2xYT medium containing ampicillin. Grow the culture at 37°C to an OD of 0.5. Add 1 mM of IPTG to induce protein expression. Allow protein production for 3 h at 37°C (*see* **Notes 14** and **15**).
3. Pellet cells by centrifugation at 4,000 *g* at 4°C for 10 min, remove supernatant, and freeze cells in liquid nitrogen.
4. Thaw cells on ice. Resuspend cells in 20 mL of pre-chilled STE buffer. Add 0.4 mL of lysozyme (final concentration

1 mg/mL) and incubate for 15 min at room temperature mixing gently. Add 8 mL of pre-chilled Triton buffer. Disrupt the cells by sonification with four cycles at 15 s/cycle. Check lysis under the microscope. Spin for 15 min at 16,000 *g* to remove cell debris.

5. Add 2.5 mL of 50% NiNTA-agarose previously washed first with water and then with buffer F. Incubate for 1 h at 4°C rotating end over end.
6. Wash the resin four times with cold buffer F. After each wash remove the buffer by centrifugation for 5 min at 700 *g* and 4°C.
7. Pour the resin into an Econo-Pac column and elute the protein with buffer G. Collect 1-mL fractions until no protein elutes anymore. Check by quick Bradford (*see* **Note 6**).
8. Run and SDS–PAGE to check the purity of the protein and pool the Arf1p-containing fractions (*see* **Note 9**).
9. Dialyse twice against buffer H at 4°C for 3 h each.
10. Aliquot and freeze in liquid nitrogen. Store at –80°C.

3.3. Purification of SNARE–GST

The SNAREs Bet1p and Sec22p are purified as GST C-terminal fusion in which the GST replaces the transmembrane domain of the SNAREs. *BET1* and *SEC22* were cloned in pGEX-2T vectors (pGEX-2T-BET1-GST and pGEX-2T-SEC22-GST) that contain the *bla* gene conferring resistance to ampicillin (10). The purification protocol for both proteins is identical. Bet1p-GST is prone to degradation while Sec22p-GST is more stable. All steps are performed on ice or at 4°C. All buffers should be stored at 4°C.

1. Inoculate 30 mL of 2xYT supplemented with ampicillin (100 μg/mL) with BL21-pCodonPlus cells (Invitrogen) with pGEX-2T-BET1-GST or pGEX-2T-SEC22-GST. Grow it overnight at 37°C.
2. Inoculate 1.5 L of 2xYT containing ampicillin with 30 mL of the overnight culture. Grow cells at 37°C for 2 h (OD_{600}= 0.5) with vigorous shaking. Induce protein production with 1 mM IPTG for 3 h at 37°C (*see* **Notes 14** and **15**).
3. Collect the cells by centrifugation at 4,000 *g* for 10 min at 4°C. Discard the supernatant. Keep cells always on ice prior to freezing in liquid nitrogen.
4. Add to the cells 30 mL of pre-chilled buffer I (*see* **Note 8**). Lyse the cells by sonification thrice for 30 s with pauses on ice to avoid warming up of the sample. Check lysis under the microscope.
5. Centrifuge the lysate at 16,000 *g* for 30 min at 4°C.

6. Add 2 mL of 50% glutathione-agarose to the supernatant. Incubate for 1 h at 4°C rotating end over end.
7. Spin down at 700 *g* for 5 min and remove the supernatant. Wash three times with buffer I.
8. Pour the beads into an Econo-Pac column. Work in a cold room.
9. Add buffer K and collect 1 mL fractions until the presence of protein cannot be detected by quick Bradford (*see* **Note 6**). Analyze the protein-containing fractions by SDS–PAGE and Coomassie blue stain (*see* **Note 10**).
10. Pool all the protein-containing fractions and dialyse twice against PBS buffer containing 15% glycerol at 4°C for 3 h each.
11. Aliquot and freeze in liquid nitrogen. Store aliquots at –80°C.

3.4. SNARE-Arf1 Binding Assay

1. Mix 25 μL of 50% glutathione-agarose slurry with 5 μg of SNARE-GST in BBP (final volume: 50 μL) in a low-binding microfuge tube (to reduce background binding). Incubate for 30 min at 4°C in a rotator. To remove the unbound material, wash the resin three times with BBP buffer (*see* **Note 11**).
2. Add 20 nM Glo3p and 1.2 μM His_6-Arf1ΔN17p and BBP buffer to a final volume of 100 μL. Incubate for 2 h at 4°C in a rotator.
3. Wash thrice with BBP and once with HEPES buffer (*see* **Note 12**). Remove the wash buffer by centrifugation at 4°C at 1,000 *g*.
4. Add 10 μL of 2x SDS sample buffer and incubate for 10 min at 65°C. Spin down the samples at maximum speed in a tabletop centrifuge and analyze the samples by SDS gel electrophoresis. Stain the gel with Coomasie Blue stain. Arf1ΔN17p binds to SNAREs only in the presence of Glo3p or when the SNARE is pre-incubated with Glo3p.

3.5. Protease Protection Assay

3.5.1. Bet1p-GST

1. Immobilize Bet1p-GST on glutathione-agarose beads as described in **Section 3.4**. Take 160 μL of 50% GSH-agarose slurry per each reaction.
2. Split evenly the washed beads into two tubes. Incubate each half either with or without 20 nM Glo3p-His_6 for 1 h at 4°C.
3. Wash beads three times with BBP buffer and once with B88 buffer to remove Glo3p-His_6 (*see* **Note 11**). Add B88 buffer to a final volume of 80 μL.

4. Transfer 10 μL of the beads to a new microfuge tube containing 6 μL of 5x SDS-sample buffer. To the rest of the beads add 0.5 μg/mL V8 protease and incubate at 37°C under agitation. Take 10 μL after 5, 10, 15, 20, 30, and 60 min and pipet into a microfuge tube containing 6 μL of 5x SDS-sample buffer. Heat samples immediately for 5 min at 95°C (*see* **Note 13**).
5. Analyze eluted proteins by SDS gel electrophoresis and Coomasie Blue stain.

3.5.2. Sec22p-GST

1. Immobilize Sec22p-GST on glutathione-agarose beads as described in **Section 3.4**. Take 160 μL of 50% glutathione-agarose slurry per each reaction.
2. Divide the washed beads into two tubes. Incubate each half either with or without 20 nM Glo3p-His_6 for 1 h at 4°C.
3. Wash beads three times with BBP buffer and once with B88 buffer to remove Glo3p-His_6 (*see* **Note 11**). Add B88 buffer to a final volume of 80 μL.
4. Transfer 10 μL of the beads to a new microfuge tube containing 6 μL of 5x SDS-sample buffer. Add to the rest of the beads 2.5 μg/mL trypsin protease and incubate at 37°C under agitation. Take 10 μL after 5, 10, 15, 20, 30, and 60 min and pipet into a microfuge tube containing 6 μL of 5× SDS-sample buffer. Heat samples immediately for 5 min at 95°C (*see* **Note 13**).
5. Analyze eluted proteins by SDS gel electrophoresis and Coomasie Blue stain.

4. Notes

1. Sigma glutathione-agarose is provided as a lyophilized powder stabilized with lactose. To prepare the glutathione-agarose gel, swell the lyophilized powder in water to a final concentration of 200 mL/g. Incubate for 30 min at room temperature. Alternatively, it can be incubated over-night at 2–8°C. After swelling, wash the agarose beads thoroughly with 10 volumes of PBS buffer to remove the lactose present in the lyophilized product. The resin can be stored in 2 M NaCl (or 20 mM HEPES–KOH, pH 6.8) containing 1 mM sodium azide at 2–8°C.
2. The GAP activity of Glo3p can be assessed by a single-round hydrolysis of Arf1-bound GTP (11). Using this approach, it has been has shown that N-terminal hexahistidine-tagged Glo3p is not active (12).

3. All the urea containing buffers have to be prepared freshly before use. Urea is not stable in solution because it is hydrolysed to ammonia and carbon dioxide. To avoid salt precipitation keep the solution at room temperature.
4. Purification is done in batch in a 50-mL falcon tube. For washing, add buffer, mix by inversion, and centrifuge at 700 *g* for 5 min. Remove supernatant carefully.
5. DTT and imidazole must be added fresh just before use as they are both relatively unstable.
6. One fast way to check the presence of protein after purification is by using Bradford reagent in a microtiter plate (quick Bradford). Mix 1 μL of each fraction with 200 μL Bradford reagent. Protein-containing fractions turn blue.
7. The molecular mass of Glo3p-His$_6$ is approximately 56 kDa but it runs as 60-kDa band.
8. Always add protease inhibitor to buffers right before use.
9. The molecular mass of His$_6$-Arf1ΔN17p is approximately 20 kDa.
10. The molecular mass of Sed5-GST and Bet1-GST are approximately 66 and 43 kDa.
11. To wash the beads to which proteins were immobilized, spin down the sample in a tabletop centrifuge at 1,000 *g* for 3 min. Then, remove the supernatant carefully avoiding sucking up the beads. Use a gel loading tip either connected to a vacuum device or a pipetman.
12. The wash with HEPES buffer is necessary to remove the KAc that is present on the BBP buffer. Potassium forms with dodecylsulfate a precipitate and proteins may be trapped by precipitation.
13. TCA precipitation: add to all samples 1/10 volume of 100% TCA. Incubate for 30 min on ice. Spin at 13,000 *g* at 4°C for 15 min. Discard supernatant and add 1 mL of acetone at –20°C. Incubate for 30 min on ice. Spin at 13,000 *g* at 4°C for 15 min. Discard supernatant and let pellet dry for 2 h at room temperature or 10 min at 65°C. Dissolve the pellet in 10 μL of 1 M Tris–HCl pH 8 and add Lämmli buffer. To prepare the 100% TCA solution: add 20 mL of Milli-Q water in a 100-mL graded cylinder. Add 100 g TCA and mix. Adjust the volume to 100 mL with Milli-Q water. Keep it at 4°C in the dark.
14. If expression is low, either use freshly transformed cells (after recovery, transfer cells to 25 ml LB-Amp medium without plating them out first) or harvest cells of the pre-culture by centrifugation and resuspend them in fresh

medium. In the latter case the concentration of potential Lac repressors is reduced.

15. To check the induction efficiency, take samples prior and after induction and run an SDS–PAGE.

Acknowledgments

The work on ArfGAPs and SNAREs in the Spang lab is funded by the University of Basel and the Swiss National Science Foundation.

References

1. Kirchhausen, T. (2002) Clathrin adaptors really adapt. *Cell* **109**(4), 413–416.
2. Spang, A. (2002) ARF1 regulatory factors and COPI vesicle formation. *Curr Opin Cell Biol* **14**(4), 423–427.
3. Lewis, S.M., Poon, P.P., Singer, R.A., Johnston, G.C., Spang, A. (2004) The ArfGAP Glo3 is required for the generation of COPI vesicles. *Mol Biol Cell* **15**(9), 4064–4072.
4. Lee, S.Y., Yang, J.S., Hong, W., Premont, R.T., Hsu, V.W. (2005) ARFGAP1 plays a central role in coupling COPI cargo sorting with vesicle formation. *J Cell Biol* **168**(2), 281–290.
5. Rein, U., Andag, U., Duden, R., Schmitt, H.D., Spang, A. (2002) ARF-GAP-mediated interaction between the ER-Golgi v-SNAREs and the COPI coat. *J Cell Biol* **157**(3), 395–404.
6. Schindler, C., Spang, A. (2007) Interaction of SNAREs with ArfGAPs precedes recruitment of Sec18p/NSF. *Mol Biol Cell* **18**(8), 2852–2863.
7. Poon, P.P., Cassel, D., Spang, A., Rotman, M., Pick, E., Singer, R.A., Johnston, G.C. (1999) Retrograde transport from the yeast Golgi is mediated by two ARF GAP proteins with overlapping function. *EMBO J* **18**(3), 555–564.
8. Kahn, R.A., Clark, J., Rulka, C., Stearns, T., Zhang, C.J., Randazzo, P.A., Terui, T., Cavenagh, M. (1995) Mutational analysis of Saccharomyces cerevisiae ARF1. *J Biol Chem* **270**(1), 143–150.
9. Paris, S., Béraud-Dufour, S., Robineau, S., Bigay, J., Antonny, B., Chabre, M., Chardin, P. (1997) Role of protein-phospholipid interactions in the activation of ARF1 by the guanine nucleotide exchange factor Arno. *J Biol Chem* **272**(35), 22221–22226.
10. Springer, S., Schekman, R. (1998) Nucleation of COPII vesicular coat complex by endoplasmic reticulum to Golgi vesicle SNAREs. *Science* **281**(5377), 698–700.
11. Albert, S., Gallwitz, D. (1999) Two new members of a family of Ypt/Rab GTPase activating proteins. Promiscuity of substrate recognition. *J Biol Chem* **274**(47), 33186–33189.
12. Poon, P.P., Cassel, D., Huber, I., Singer, R.A., Johnston, G.C. (2001) Expression, analysis, and properties of yeast ADP-ribosylation factor (ARF) GTPase activating proteins (GAPs) Gcs1 and Glo3. *Methods Enzymol* **329**, 317–324.

Chapter 23

Studying Endoplasmic Reticulum Function In Vitro Using siRNA

Cornelia M. Wilson and Stephen High

Abstract

In eukaryotic cells, N-glycosylation is typically the most common protein modification that occurs in the endoplasmic reticulum (ER) lumen. N-glycosylation is facilitated by a large heterologous protein complex called the oligosaccharyltransferase (OST) that allows the attachment of a high mannose oligosaccharide from a dolichol pyrophosphate donor *en bloc* onto suitable asparagine residues of newly synthesized nascent chains during translocation into the ER lumen (1). While the complexity of the OST is highly conserved in eukaryotes, the role of its different subunits is poorly defined. We have investigated the function of three OST subunits, the ER translocon-associated component ribophorin I, and two isoforms of the presumptive catalytic subunit, STT3. We use a combination of siRNA-mediated knockdown of individual proteins combined with a semi-permeabilized mammalian cell system to provide a robust read out for OST subunit function during N-glycosylation of model substrates in vitro. This approach is equally applicable to the study of other cellular components.

Key words: Endoplasmic reticulum, N-glycosylation, oligosaccharyltransferase, semi-permeabilized cells, siRNA.

1. Introduction

The endoplasmic reticulum (ER) is the site where most secretory proteins acquire their native conformation and gain access to the secretory pathway, and the ER provides the necessary environment and components to facilitate the folding and onward transport of newly synthesized membrane and secretory proteins (2). During translocation, growing nascent chains can be glycosylated by an enzyme complex called the oligosaccharyltransferase (OST). The OST is a large heterologous complex localized in the ER membrane close to the ER translocation channel

A. Economou (ed.), *Protein Secretion*, Methods in Molecular Biology 619,
DOI 10.1007/978-1-60327-412-8_23,

that allows the passage of newly synthesized proteins into the ER lumen (3). The majority of mammalian OST subunits have obvious yeast equivalents, namely ribophorin I (Ost1p), ribophorin II (Swp1p), OST48 (Wbp1p), STT3A and STT3B (Stt3p), N33 and IAP (Ost3p and Ost6p), and DAD1 (Ost2p) (4–8). In addition, there are two putative mammalian OST subunits, DC2 and KCP2, that are less well defined (7).

In order to investigate the function of individual subunits of the mammalian OST, we have developed an in vitro translation assay that is supplemented with semi-permeabilized cells (SP cells). This basic approach is well established and involves the preparation of digitonin-solubilized cultured mammalian cells as a source of ER membranes for the N-glycosylation of proteins synthesized in a cell-free translation system (9). We have now established that SP cells can be prepared following the siRNA-mediated knockdown of three OST subunits (*see* **Fig. 23.1**) and have used these SP cells to assay the effect of subunit depletion upon the N-glycosylation efficiency of various precursors (10, 11).

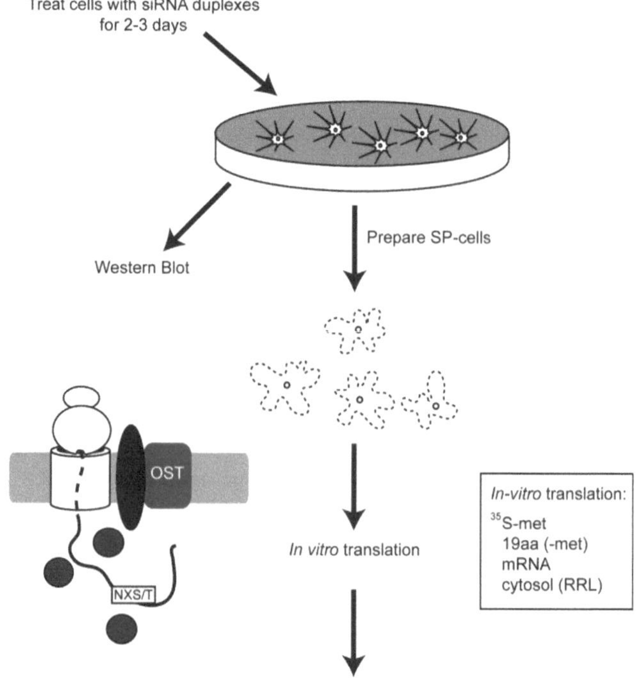

Fig. 23.1. Assessing ER function in vitro using siRNA. Experimental approach to assess the N-glycosylation efficiency of the OST complex in vitro. HeLa cells are transfected with siRNA duplexes, incubated for 2 days, semi-permeabilized (SP), and used for in vitro biogenesis of glycosylated precursors. Following synthesis, membrane-associated translation products can be isolated and analyzed by SDS–PAGE. The efficiency of subunit knockdown in SP cells is assessed by Western blotting.

2. Materials

2.1. RNA Interference

1. 21 nucleotide duplexes corresponding to:
   - human ribophorin I (aagcgcacagtggacctaagc)
   - human STT3A (gcagtaggatcatatttgatt)
   - human STT3B (tatcaacgatgaaagagtatt)
   - siControl Risc-free (Dharmacon)
2. Dulbecco's modified Eagle Media (DMEM) with 4.8 g/L glucose supplemented with 5 mL non-essential amino acids (NEAA) (Lonza, UK) and 5 mL glutamine (200 mM stock) per 500 mL (Lonza).
3. DMEM supplemented with 5 mL NEAA, 5 mL glutamine, and 50 mL foetal calf serum (Lonza) per 500 mL (complete media).
4. HeLa cells (60% confluent) grown in 10 cm^2 dishes seeded 24 h prior to treatment.
5. Oligofectamine (Invitrogen).
6. Tunicamycin (prepared in water, stock solution at 2 mg/mL and stored at –20°C).

2.2. Preparation of SP Cells

1. siRNA-treated HeLa cells (10 cm^2 dish of subconfluent cells), incubated for 2–3 days.
2. Solution of Trypsin (0.25%) and ethylenediamine tetraacetic acid (EDTA) (1 mM) (Lonza).
3. Phosphate-buffered saline (Lonza).
4. DMEM with 4.8 g/L glucose supplemented with 5 mL NEAA (10 mM), 5 mL glutamine (200 mM), and 10% foetal calf serum (Lonza) per 500 mL.
5. Soyabean trypsin inhibitor (SBTi) (50 mg/mL) (Sigma).
6. KHM buffer: 20 mM HEPES, pH 7.2, 110 mM KOAc, 2 mM MgOAc.
7. HEPES buffer: 50 mM HEPES, pH 7.2, KOAc 50 mM.
8. Digitonin (40 mg/mL in DMSO, stored at –80°C) (Calbiochem).
9. Trypan Blue solution (0.4%) (Sigma).
10. $CaCl_2$ (0.1 M, stored at –20°C).
11. Micrococcal nuclease (15,000 U at 1 mg/mL in sterile water, stored at –20°C in small aliquots, avoid repeated refreezing).
12. EGTA (0.4 M, stored at –20°C).

2.3. mRNA Transcription In Vitro

1. PCR purification kit (Qiagen)
2. 10 μg linerized plasmid DNA (from restriction digest or PCR), containing gene of interest downstream from a SP6, T7, or T3 polymerase promoter, in RNase-free water
3. 5x Transcription buffer (Promega), thawed and at RT (store at –20°C)
4. Ribonucleotide triphosphates (ATP, UTP, CTP, and GTP) (25 mM each) (Roche Diagnostics), keep on ice (store at –20°C)
5. 100 mM DTT (Sigma), keep at RT (store at –20°C)
6. SP6 RNA polymerase (50 U/μL) (New England Biolabs), keep on ice (store at –20°C)
7. RNase inhibitor (100 U/μL) (Promega), keep on ice (store at –20°C)
8. DEPC water (RNase free)

2.4. Translation In Vitro

1. Rabbit reticulocyte lysate (RRL), nuclease treated (Promega) and thawed on ice (stored at –80°C).
2. 19 amino acid mix, 1 mM each (minus methionine) (Promega), stored at –20°C
3. EasyTag^{TM} L [^{35}S] methionine (15 μCi/μL) (Perkin Elmer), stored at 4°C
4. Aurintricarboxylic acid (ATCA) made up in 100 mM HEPES–KOH (pH 7.9) at stock solution of 2.5 mM (Sigma), stored at –20°C

2.5. Western Blotting

1. Transfer buffer: 20 mM Tris, 150 mM glycine, 20% (v/v) methanol, 0.05% SDS. Store at RT.
2. Supported nitrocellulose membrane (Hybond-C, GE Healthcare) and 3 MM chromatography paper (Whatman).
3. Tris-buffered saline with Tween-20 (TBS-T): Prepare 10x stock with 500 mM Tris–HCl, pH 7.4, 1.4 M NaCl, 20 mM KCl, and 5% Tween-20. Dilute 100 mL of 10x stock in 900 mL water before use and store at 4°C.
4. Blocking buffer: prepare fresh 5% (w/v) dried skimmed milk powder in TBS-T.
5. Primary antibodies diluted in TBS-T buffer supplemented with 3% (w/v) Fraction V bovine serum albumin (BSA) (Sigma).
6. Rabbit polyclonal antisera recognizing Ribophorin I and STT3A were made to order by Invitrogen. Antisera specific for STT3B and α-tubulin (TAT-1) were obtained from Reid Gilmore (University of Massachusetts Medical School, USA) and Keith Gull (University of Oxford, UK), respectively.

7. Secondary antisera – anti-rabbit/anti-mouse IgG conjugated to horse radish peroxidase (Sigma).
8. Enhanced chemiluminescent, Western Lightning (ECL) reagents (Perkin Elmer) and BioMax MR film (Kodak) – (*see* **Note 1**)

3. Methods

The methods described below outline (1) the siRNA-mediated knockdown of individual proteins and (2) the expression and characterization of in vitro synthesized glycoprotein precursors in SP cells.

3.1. RNA Interference

RNA interference is a natural mechanism occurring in mammalian cells that interferes with gene expression by hindering the mRNA production from specific genes. This mechanism is initiated by double-stranded RNA (dsRNA) which is homologous in sequence to the silenced gene. The dsRNA is processed to short interfering RNAs (siRNAs) between 21 and 30 nucleotides in length which then function as a component of a 'silencing complex' to repress the expression of the target gene. Tuschl and colleagues were able to introduce artificially prepared 21 nucleotide siRNAs into cultured mammalian cells and demonstrate that target genes could be efficiently repressed after 2 days of incubation (12).

1. HeLa cells (60% confluent) grown in 10-cm^2 dishes seeded 24 h prior to treatment (*see* **Note 2**).
2. Set up in two separate 1.5-mL microcentrifuge tubes the following mixes:
   (a) 60 μL 20 μM siRNA duplex + 400 μL DMEM (minus foetal calf serum).
   (b) 56 μL oligofectamine + 400 μL DMEM (minus foetal calf serum).
   Incubate mix (b) for 5 min at RT (*see* **Note 3**).
3. Combine tubes (a) and (b) together and mix thoroughly by pipetting up and down with 1-mL pipette and incubate for 20–25 min at RT.
4. Remove the media from the cells and wash twice with PBS.
5. Replace with 4.1 mL of DMEM (minus serum) and add the oligofectamine–siRNA mix gently dropwise to the cells.
6. Incubate the cells with transfection mix at 37°C/5% CO_2 for 5 h (*see* **Note 4**).
7. After incubation, wash the cells twice with PBS and replace with 10 ml of complete media.

8. RNAi-treated cells are incubated for 48–72 h (*see* **Note 5**). As a control, tunicamycin (2 μg/mL) is added to untreated cells 12 h prior to the preparation of SP cells as a source of ER membranes for in vitro translation (*see* **Note 6**).

3.2. Protein Expression in SP Cells

SP cells allow the researcher to reconstitute the initial stages in the assembly and modification of proteins entering the secretory pathway, as it would occur in an intact cell. The procedure involves treating cells grown in culture with the detergent digitonin and isolating the resulting cells that are depleted of their cytosolic components. In subsequent experiments, specifically generated mRNA transcripts can be translated in a cell-free translation supplemented with the SP cells prepared as outlined below.

There are a number of distinct advantages to using this system. As this is an in vitro approach, individual components can be easily manipulated, allowing cellular processes to be studied under a variety of conditions. Furthermore, ER-mediated processes such as N-glycosylation and glucose trimming often occur more rapidly than comparable reactions in isolated canine pancreatic microsomes. In addition, membrane permeable chemical cross-linking reagents can be added and can readily access proteins within the ER lumen.

3.2.1. Preparation of SP Cells

This procedure uses a modified protocol based on those of Plutner et al. (13) and Wilson et al. (9) that has been adapted for quantifying N-glycosylation efficiency (10) (*see* **Note 7**).

1. Harvest HeLa cells grown to subconfluency for 48–72 h post-transfection by gently washing the cell monolayer twice with PBS and add 2 mL trypsin. Incubate the cells until all cells have been removed from the dish by gently tapping the dish.
2. Resuspend cell suspension in 4 mL KHM buffer + 8 μL SBTi (*see* **Note 8**) and transfer to a 15-mL Falcon tube.
3. Centrifuge at 280 *g* for 3 min at 4°C. Aspirate the supernatant from the cell pellet.
4. Resuspend pelleted cells in 6 mL of ice-cold KHM and permeabilize by the addition of 6 μL digitonin (from 40 mg/mL stock) to a final concentration of 40 μg/mL, mix immediately by inversion and incubate on ice for 5 min (*see* **Note 9**).
5. Adjust the volume to 14 mL with ice-cold KHM and pellet cells by centrifugation as step 3.
6. Discard supernatant and resuspend cells in 14 mL ice-cold HEPES buffer. Incubate on ice for 10 min and pellet by centrifugation as step 3.

7. Discard the supernatant and resuspend cells carefully in 1 mL ice-cold KHM (using a manual pipette with P1000 tip gently up and down). Place on ice.
8. Transfer a 10 μL aliquot to a separate 1.5 mL microcentrifuge tube and add 10 μL of Trypan blue.
9. At this stage, count the cells in a haemocytometer and check for permeabilization by trypan blue staining. >99% of cells should appear blue under the microscope.
10. Transfer cells to a 1.5-mL microcentrifuge tube and spin for 30 s at 12,000 *g*. Discard supernatant and resuspend the cells in 80 μL KHM using a pipette (approximately 1×10^6 cells).
11. Remove endogenous mRNA from the cells by treating with a calcium-dependent nuclease. Add 1 μL of 0.1 M $CaCl_2$ and 1 μL of Micrococcal nuclease (15 U) and incubate at room temperature for 12 min.
12. Add 1 μL of 0.4 M EGTA to chelate the calcium and inactivate the nuclease. Pellet the cells by centrifuging at 12,000 *g* for 30 s in a microcentrifuge and resupend in 80 μL KHM.
13. Use approximately 10^5 cells per 25 μL translation reaction (~4 μL of the 80 μL obtained).

3.3. Western Blotting

This procedure is used to assess the efficiency of knockdown of a protein after siRNA treatment (*see* **Note 10**, *see* **Fig. 23.2**).

1. To normalize loading, make a note of the cell count for each SP-cell preparation and load approximately 10^5 cells per lane (i.e., ~4 μL of the 80 μL obtained in 20 μL SDS–PAGE sample buffer).
2. After proteins have been resolved by SDS–PAGE they are transferred to supported nitrocellulose membrane. The following instructions are for the use of a Bio-rad semi-dry transfer cell. Remove the stacking gel and equilibrate in the separating gel transfer buffer for 10–15 min at RT (*see* **Note 11**). Six pieces of 3-MM paper and a sheet of nitrocellulose cut slightly larger than the size of the gel are presoaked in transfer buffer.
3. A transfer sandwich consisting of Whatman 3 M filter paper and nitrocellulose membrane is assembled.
4. First place three pieces of pre-soaked paper onto the platinum anode and roll out any air bubbles using a pipette (*see* **Note 12**).
5. Next place the nitrocellulose membrane on top and roll out any air bubbles as above.

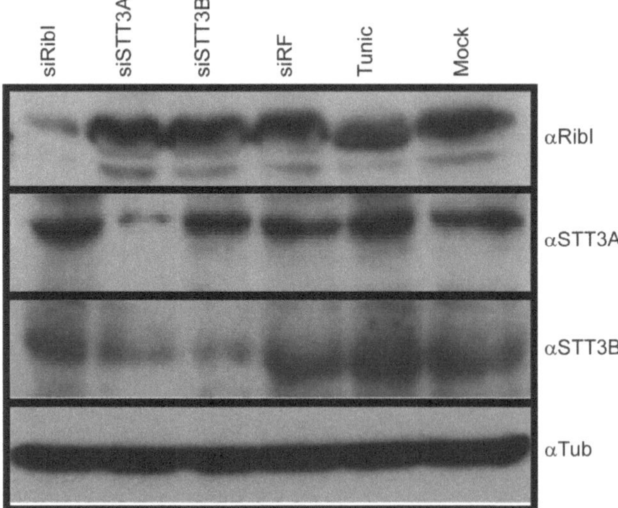

Fig. 23.2. Determining the efficiency of siRNA-mediated knockdown. Rib1, STT3A, and STT3B siRNA-mediated knockdowns. Lysates of cells 48 h after transfection with ribophorin I, STT3A, STT3B siRNA duplexes (*lanes 1* and *3*), non-functional siRisc-free control (siRF) (*lane 4*), or mock transfected cells (*lane 6*) were probed with antibodies specific for ribophorin I, STT3A, STT3B, or α-tubulin as indicated. The knockdown of STT3A causes a reduction in cellular levels of both STT3A and STT3B (10).

6. Gently layer the gel on top of the membrane, making sure that it is in the centre of the membrane and avoiding any of the gel being outside the perimeter of the nitrocellulose.
7. Finish the sandwich by placing three more pre-soaked papers on top of the gel and again roll out any air bubbles as above.
8. Gently place the cathode onto the stack trying not to disturb the filter paper stack.
9. Transfer samples onto the nitrocellulose membrane for approximately 30 min at 15 mA. After the transfer, carefully remove the membrane, cut the top left-hand corner for orientation, and place in water. To check for efficient transfer, the nitrocellulose blot can be stained briefly with Ponceau S and washed in distilled water.
10. Then block the blots with TBS-T (50 mM Tris–HCl, pH 7.6, 140 mM NaCl, 2 mM KCl, 0.5% Tween-20) containing 5% non-fat skimmed milk powder for 1 h at room temperature or overnight at 4°C (*see* **Note 13**).
11. Add primary antibodies at 1:1000 dilution (Ribophorin I), 1:500 dilution (STT3A and STT3B), and 1:200 dilution (α-tubulin) to TBS-T and incubate for 1 h at room temperature (or overnight at 4°C).

12. The blots are washed three times in TBS-T and the secondary anti-rabbit/mouse goat peroxidase antibody is added at 1:2000 dilution in TBS-T for 1 h at room temperature.
13. The blots are washed three times in TBS-T for 30 min. Blots are developed using the ECL^{TM} system according to manufacturer's instructions.

3.4. Transcription In Vitro

The cDNA encoding a glycoprotein of interest is subcloned into an in vitro expression vector such as pSPUTK downstream of a suitable promoter (typically T7 or SP6) containing an RNA polymerase binding site from which in vitro transcription can be initiated. The plasmid may be linearized by restriction endonuclease digestion to generate a template suitable for mRNA synthesis (*see* **Note 14**). Alternatively, a transcription template can be generated by PCR (*see* **Note 15**). The transcription method outlined is a modification of a method previously described elsewhere (14).

1. Prior to the preparation of the transcription reaction, the DNA obtained by either PCR or restriction digest must be 'cleaned' using a PCR purification kit (Qiagen) and eluted with 30 μL H_2O giving a ~28 μL final eluted volume (*see* **Note 16**).
2. Prepare a 100 μL transcription mixture containing 20 μL transcription buffer (5x), 4 μL 0.1 M DTT, 28 μL linearized DNA (5–10 μg), 1 μL RNase inhibitor (40 units), 3 μL of each nucleotide, and 42 μL H_2O.
3. Add 2 μL of the appropriate RNA polymerase (T7 or SP6) (160 units) and incubate for 2 h at 37°C (*see* **Note 17**).
4. The RNA transcript can be most simply purified using a RNA cleanup kit (Qiagen) (*see* **Note 18**). Elute the purified RNA pellet in 100 μL RNase-free H_2O containing 1 mM DTT and 1 μL RNase inhibitor. Store the RNA at –80°C.
5. To assess the yield of RNA, remove 1 μL and analyze on a 1% agarose gel (*see* **Note 19**).

3.5. Translation In Vitro

SP cells appear most compatible with rabbit reticulocyte lysate-based translation systems that are typically supplemented with tRNA and an ATP-regeneration system.

1. Prepare a 25-μL translation mixture on ice containing 17.5 μL rabbit reticulocyte lysate, 0.5 μL 19 amino acids mix, 0.5 μL KCl, 1 μL EasyTagTM ^{35}S-methionine (0.75 mCi/ml), 1 μL mRNA, and 4 μL SP cells (*see* **Note 20**). Incubate the translation reaction at 30°C for 30–60 min and then place on ice.
2. Add 2.5 μL ATCA (2.5 mM stock) to the reaction and incubate at 30°C for a further 10 min (*see* **Note 21**).

3. Isolate the membrane fraction by centrifuging at 12,000 *g* for 1 min.
4. Remove the supernatant and wash the cell pellet with 100 μL KHM buffer and re-centrifuge at 12,000 *g* for 1 min.
5. To prepare the samples for SDS–PAGE, resuspend the cell pellet in 10 μL KHM buffer and add 50 μL SDS–PAGE sample buffer (0.0625 M Tris–HCl, pH6.8, SDS (2%, w/v), glycerol (10%, v/v), and Bromophenol Blue plus 2 μL DTT (1 M) and heat for 5 min at 95°C (*see* **Note 22**). Pellet the samples by centrifugation at 12,000 *g* for 30 s and load half of the sample, keep the remainder at –20°C for further analysis as necessary.
6. Separate samples using SDS–PAGE and choose a separating gel with a composition appropriate for the expected molecular weight of the protein of interest (*see* **Note 23**). This is typically between 8 and 14% acrylamide. After electrophoresis, the SDS–PAGE gel should be dried and exposed to autoradiography film (Kodak X-Omat AR film) or a phosphorimage screen (Fuji) (*see* **Fig. 23.3**).

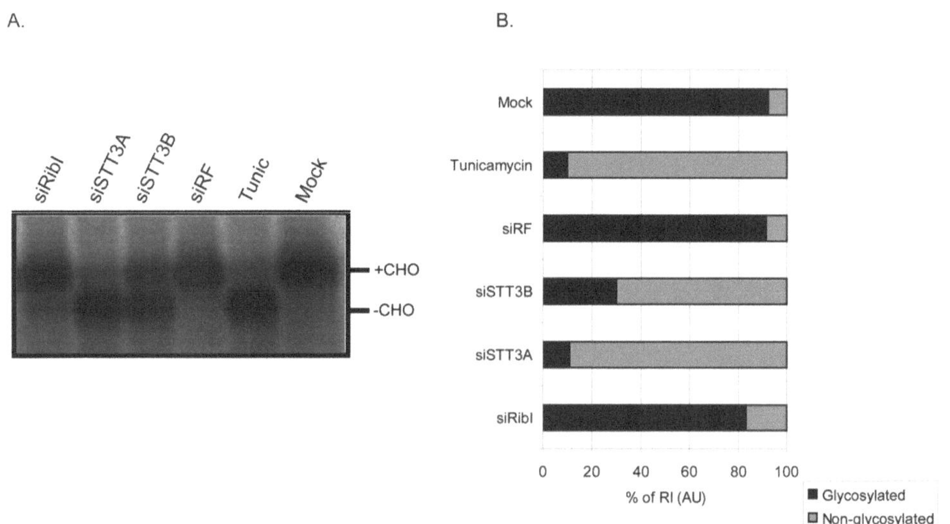

Fig. 23.3. Effect of OST subunit knockdown on the N-glycosylation of glycophorin C. (**A**) Glycophorin C was synthesized as a radiolabeled polypeptide using a rabbit reticulocyte lysate system supplemented with semi-permeabilized HeLa cells prepared 48 h after transfection with siRNAs specific for the mRNAs encoding ribophorin I (*lane 1*), STT3A (*lane 2*) or STT3B (*lane 3*), a non-functional control siRNA (siRF) (*lane 4*), or following mock transfection (*lane 6*). As a positive control for loss of N-glycosylation, HeLa cells were incubated with 2 mg/mL tunicamycin for 12 h prior to isolation on day 2 (*lane 5*). The resulting glycosylated (+CHO) and non-glycosylated (–CHO) polypeptides are shown. (**B**) The relative proportion of glycosylated (*black bar*) and non-glycosylated (*grey bar*) products for each sample was calculated and expressed as a percentage of relative intensity (% of RI).

4. Notes

1. We find that the Kodak BioMax MR film provides very good resolution for visualizing western blots. In the case of low protein expression or low affinity antibodies more sensitive films such as Kodak XAR maybe advantageous.
2. It is crucial to optimize the number of cells required for transfection as this can affect transfection efficiency.
3. All procedures are performed in a sterile laminar flow-hood. All plastic ware including tips, stripettes, and tubes are either sterilized by the supplier or must be autoclaved before use
4. Cells can be incubated for longer than 5 h but be careful to include serum if leaving overnight.
5. The length of siRNA treatment required will depend upon a number of factors including (i) transfection efficiency; (ii) protein stability (generally short-lived proteins ($t^1/_2$ > 24 h) will require 2 days while long-lived proteins ($t^1/_2$ < 24 h) will require at least 3 days); (iii) efficiency of siRNA sequence chosen; (iv) transcriptional rate of the gene of interest; and (v) growth characteristics of the cell line chosen.
6. Tunicamycin treatment can be reduced to shorter incubations of 5 h by using a concentration of 20 μg/mL.
7. The procedure takes approximately 1 h and should be carried out immediately prior to using the SP cells for translation in vitro. SP cells can be stored at –80°C in 100 μL of sterile sucrose (0.2 M) and 1 μL of PMSF (0.1 M). Resuspend in 100 μL KHM and test for translocation efficiency to ensure that is there is no loss of activity. It is advisable to use a minimum of one 10 cm^2 dish of cells to prepare SP cells as it proves difficult to work with a smaller quantity. Note that the size of the cell pellet will decrease during the procedure since the loss of cytosol is accompanied by a decrease in cell volume.
8. It is important to add SBTi; otherwise any carry over of trypsin will result in the degradation of in vitro translation products.
9. The digitonin concentration has been optimized for permeabilization of HeLa cells. If a different cell line is used, the concentration of digitonin required for permeabilization should be assessed by titration and trypan blue staining.

10. If antibodies are not available against the protein of interest, quantitative real-time qPCR (RT-qPCR) maybe used to assess the efficiency of knockdown.
11. Time for equilibration of the gel will vary according to its thickness and the methanol in the transfer buffer will shrink the gel by dehydration. Generally, 10–15 min of incubation is sufficient for 1.5 mm thickness gels.
12. If the air bubbles are not removed by rolling with a pipette or a test tube, this will lead to inefficient/or uneven transfer of proteins onto the nitrocellulose membrane
13. Depending on the size of the gel, we routinely use either a 50-mL falcon tube or a square-shaped Petri dish as the vessel to perform all the subsequent steps in the procedure.
14. In vitro transcription using a linearized template having a 3′ overhang may inhibit translation or produce aberrant translation products. Restriction endonucleases that generate a 5′ overhang or blunt end are favoured for use prior to the transcription reaction. To avoid low mRNA yields or partially degraded mRNA always use sterile pipette tips, microcentrifuge tubes, and RNase-free water.
15. The 5′ oligonucleotide used for priming the PCR reaction should ideally be at least 150 bp upstream of the promoter site.
16. Alternatively, the DNA template obtained either by restriction digest or PCR can be extracted with phenol/chloroform (*see* **Note 18** for details).
17. The yield of mRNA can be increased by a further addition of RNA polymerase (1 μL) after 1 h.
18. Alternatively, RNA products can be purified by adding an equal volume of phenol/chloroform (1:1), then twice with chloroform and precipitate by adding 1/10 volume of 3 M NaOAc (pH 5.2) and 3 volumes of ethanol. The RNA pellet is resuspended in 100 μL RNase-free H_2O containing 1 mM DTT and 1 μL RNase inhibitor. Store the RNA at –80°C.
19. Run 1 μL of RNA sample on a 1% agarose gel containing 2% ethidium bromide. For running RNA, use either electrophoresis equipment that is for RNA only or run samples for a limited time, i.e. 30 min at 100 v using ordinary DNA gel electrophoresis equipment and visualize immediately.
20. To evaluate the translation efficiency of a new RNA sample, set up a single 25-μL reaction including 4 μL of sterile water instead of SP cells.
21. ATCA treatment prevents any further initiation of translation but allows the elongation of already initiated chains.

22. The temperature of denaturation may be varied according to the characteristics of the protein of interest. For example, polytopic membrane proteins such as opsin are usually denatured at 37°C for 30 min while single spanning or secretory proteins can be denatured either at 70°C for 10 min or 95°C for 5 min.
23. If there are no products from the translation reaction, the RNA may need to be heated to 60°C for 10 min prior to translation in order to denature any secondary structure that inhibits efficient translation. Additional products with molecular weights smaller than the major translation product may be observed due to alternative initiation at downstream methionine residues or premature chain termination.

Acknowledgements

We are eternally grateful to all our colleagues who have contributed their time and materials towards this chapter. The work described in this chapter was supported by grant funding from the Biotechnology and Biological Sciences Research Council.

References

1. Knauer, R., and Lehle, L. (1999). The oligosaccharyltransferase complex from yeast. *Biochim Biophys Acta* **1426,** 259–273.
2. Helenius, A., and Aebi, M. (2004). Roles of N-linked glycans in the endoplasmic reticulum. *Annu Rev Biochem* 73, 1019–1049.
3. Nikonov, A.V., Snapp, E., Lippincott-Schwartz, J., and Kreibich, G. (2002). Active translocon complexes labeled with GFP-Dad1 diffuse slowly as large polysome arrays in the endoplasmic reticulum. *J Cell Biol 158*, 497–506.
4. Kelleher, D.J., and Gilmore, R. (1997). DAD1, the defender against apoptotic cell death, is a subunit of the mammalian oligosaccharyltransferase. *Proceedings of the National Academy of Sciences of the United States of America* **94,** 4994–4999.
5. Kelleher, D.J., and Gilmore, R. (2006). An evolving view of the eukaryotic oligosaccharyltransferase. *Glycobiology* **16,** 47R–62R.
6. Kelleher, D.J., Karaoglu, D., Mandon, E.C., and Gilmore, R. (2003). Oligosaccharyltransferase isoforms that contain different catalytic STT3 subunits have distinct enzymatic properties. *Mol Cell* **12,** 101–111.
7. Shibatani, T., David, L.L., McCormack, A.L., Frueh, K., and Skach, W.R. (2005). Proteomic analysis of mammalian oligosaccharyltransferase reveals multiple subcomplexes that contain Sec61, TRAP, and two potential new subunits. *Biochemistry* **44,** 5982–5992.
8. Silberstein, S., Kelleher, D.J., and Gilmore, R. (1992). The 48-kDa subunit of the mammalian oligosaccharyltransferase complex is homologous to the essential yeast protein WBP1. *J Biol Chem* **267,** 23658–23663.
9. Wilson, R., Allen, A.J., Oliver, J., Brookman, J.L., High, S., and Bulleid, N.J. (1995). The translocation, folding, assembly and redox-dependent degradation of secretory and membrane proteins in semi-permeabilized mammalian cells. *Biochem J* **307** (Pt 3), 679–687.
10. Wilson, C.M., and High, S. (2007). Ribophorin I acts as a substrate-specific facilitator of N-glycosylation. *J Cell Sci* **120,** 648–657.

11. Wilson, C.M., Roebuck, Q., and High, S. (2008). Ribophorin I regulates substrate delivery to the oligosaccharyltransferase core. *Proceedings of the National Academy of Sciences of the United States of America* **105,** 9534–9539.
12. Elbashir, S.M., Harborth, J., Lendeckel, W., Yalcin, A., Weber, K., and Tuschl, T. (2001). Duplexes of 21-nucleotide RNAs mediate RNA interference in cultured mammalian cells. *Nature* **411,** 494–498.
13. Plutner, H., Davidson, H.W., Saraste, J., and Balch, W.E. (1992). Morphological analysis of protein transport from the ER to Golgi membranes in digitonin-permeabilized cells: role of the P58 containing compartment. *J Cell Biol* **119,** 1097–1116.
14. Gurevich, V.V., Pokrovskaya, I.D., Obukhova, T.A., and Zozulya, S.A. (1991). Preparative in vitro mRNA synthesis using SP6 and T7 RNA polymerases. *Analy Biochem* **195,** 207–213.

Chapter 24

High-Quality Immunofluorescence of Cultured Cells

Dibyendu Bhattacharyya, Adam T. Hammond, and Benjamin S. Glick

Abstract

Immunofluorescence microscopy of cultured cells often gives poor preservation of delicate structures. We have obtained dramatically improved results with a simple modification of a standard protocol. Cells growing on a coverslip are rapidly dehydrated in a cold organic solvent and then are rehydrated in a solution containing a homobifunctional crosslinker. The crosslinking reaction stabilizes cellular structures during subsequent incubation and wash steps, usually without compromising antigenicity. This method reproducibly yields high-quality images of endomembrane compartments and cytoskeletal elements.

Key words: Immunofluorescence, formaldehyde, paraformaldehyde, methanol, acetone, organic solvents, crosslinking, transitional ER, ER exit sites, ER export sites.

1. Introduction

Studies of the secretory pathway rely on methods to determine the intracellular locations of secretory cargo proteins and of trafficking machinery components. A common approach is to overexpress fluorescently tagged proteins, but both overexpression and tagging can cause aberrant localization. Immunofluorescence microscopy of cultured cells avoids this limitation by detecting endogenous proteins at their normal expression levels (1).

Unfortunately, the quality of immunofluorescence images is often poor. This problem can be ascribed to the gentle fixation procedures that are used to retain antigenicity. The most commonly used fixative is formaldehyde (1). However, when we fixed cultured mammalian cells with formaldehyde to visualize transitional endoplasmic reticulum (ER) sites (tER sites; also known as

A. Economou (ed.), *Protein Secretion*, Methods in Molecular Biology 619,
DOI 10.1007/978-1-60327-412-8_24, © Springer Science+Business Media, LLC 2010

ER exit sites), much of the information was lost. As judged by GFP tagging, discrete tER sites were present in the cell periphery and concentrated in the juxtanuclear Golgi region (2, 3); yet when tER sites were viewed by immunofluorescence microscopy, the peripheral sites were often invisible while the juxtanuclear sites appeared as a diffuse blob. If the cells were fixed with organic solvents rather than formaldehyde (4), the results were better but still not satisfactory. Our troubleshooting suggested that cellular architecture was being disrupted by liquid flow during the wash steps. We therefore developed an improved immunofluorescence method that incorporates a chemical crosslinking reaction. This method reliably preserves tER sites as well as other cellular structures, including Golgi compartments and microtubules (3).

Our approach is to dehydrate cultured cells and then rehydrate them in a fixative solution. Cells grown on a coverslip are transferred to acetone or methanol at –20°C. This treatment

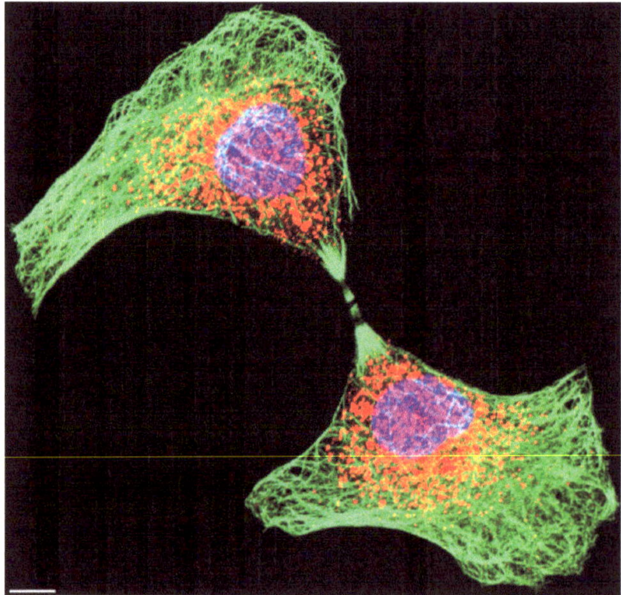

Fig. 24.1. Immunofluorescence image of a dividing normal rat kidney cell. Microtubules (*green*) were stained with a monoclonal anti-β-tubulin antibody (clone KMX-1, Roche, Indianapolis, IN) followed by Cy2-conjugated donkey anti-mouse antibody. tER sites (*red*) were stained with an affinity-purified polyclonal anti-Sec13 antibody (6) followed by Rhodamine Red-X-conjugated donkey anti-rabbit antibody. Both primary antibodies were diluted in the ratio of 1:100 and both secondary antibodies (from Jackson Immunoresearch, West Grove, PA) were diluted in the ratio of 1:200. DNA (*magenta*) was stained by supplementing the mounting medium with 4 mM TOTO-3 (Molecular Probes, Eugene, OR). Separate Z-stacks in three fluorescence channels were collected with a Zeiss (Thornwood, NY) LSM 510 confocal microscope equipped with a 100X 1.4-NA Plan-Apo objective and with standard filters for visualizing FITC/Cy2, Rhodamine Red-X, and Cy5/TOTO-3. These images were then projected and combined using the Zeiss software. The background staining outside of the cells was removed using Adobe Photoshop. Scale bar, 10 μm.

appears to extract the lipids almost instantaneously while precipitating the proteins in place (4). The dehydrated cells are then rehydrated in the presence of a homobifunctional amine-reactive crosslinker. At this point, the sample can be processed using standard procedures for antibody labeling and DNA staining. The crosslinking reaction strongly stabilizes cellular architecture, typically without compromising antibody binding (2, 3). A representative image is shown in **Fig. 24.1**, which displays tER sites, microtubules, and nuclear DNA in a dividing cell.

2. Materials

2.1. Coverslips and Wells

1. No. 1.5 thickness glass coverslips.
2. Materials for creating wells on a coverslip. One option is to punch holes in a laminating film sheet such as Cleer Aheer® from C-Line Products (Mt. Prospect, IL) and to attach a piece of the perforated sheet to a standard coverslip. A second option is to use coverslips with wells that are created by a silicone backing, such as SecureSlip™ coverslips from Grace Bio-Labs (Bend, OR) (*see* **Note 1**).

2.2. Instruments for Processing Coverslips

1. Fine-point forceps, such as jeweler's microforceps.
2. Rubber hose attached to a vacuum trap.

2.3. Solutions Made in Advance

1. Phosphate-buffered saline (10X PBS): dissolve 80 g NaCl, 2.0 g KCl, 14.4 g Na_2PO_4, and 2.4 g KH_2PO_4 in deionized water at a final volume of 1 L. Sterilize by autoclaving and store at room temperature. Upon 10-fold dilution, the pH should be approximately 7.4.
2. BS^3 solution (100X): dissolve bis(sulfosuccinimidyl)suberate (Pierce, Rockford, IL) to 10 mM in deionized water. Store at –80°C in 10-μL aliquots.
3. Ethylenediamine is prepared as a 100 mM solution. Add 669 μL pure ethylenediamine to 90 mL deionized water. Add 2 mL 6 M HCl to bring the pH to approximately 8. Then adjust the pH to 7.5 with additional 6 M HCl. Adjust to 100 mL with deionized water. Filter sterilize and store at 4°C protected from light.
4. Mounting solution can be purchased commercially or prepared in the laboratory (*see* **Note 2**). We use the following custom mixture. To 90 mL glycerol, add 10 mL of 10X PBS that had been adjusted to pH 9 with 0.5 M Na_2CO_3.

Dissolve *n*-propyl gallate in this solution to 5% (w/v) using bath sonication. Store at –80°C in 200-μL aliquots.

5. VALAP: combine an equal weight of paraffin (m.p. 51–53°C), lanolin, and vaseline. Heat and mix until homogeneous. Store in a beaker at room temperature.
6. Normal horse serum (Vector Laboratories, Burlingame, CA) is stored at 4°C.
7. Gelatin solution: dissolve cell culture grade porcine skin gelatin (Sigma-Aldrich, St. Louis, MO) to 0.1% (w/v) in deionized water. Filter sterilize and store at room temperature.
8. Hoechst solution (1%): dissolve Hoechst 33258 (Molecular Probes, Eugene, OR) to 1% (w/v) in deionized water. Store at 4°C protected from light.

2.4. Solutions Made Fresh

1. PBS+: dilute 10X PBS 10-fold in deionized water and add 0.1% *n*-octyl-β-D-glucopyranoside (Sigma-Aldrich, St. Louis, MO) (also known as octyl glucoside—*see* **Note 3**).
2. Blocking buffer: to 10 mL PBS+ add 0.1 g non-fat powdered milk, 222 μL 45% cold water fish skin gelatin (Sigma-Aldrich, St. Louis, MO), and 0.1 mL normal horse serum (*see* **Note 4**).
3. Antibody solutions: dilute the desired primary and secondary antibodies in blocking buffer. Dilutions typically range from 1:50 to 1:5000 and must be determined empirically for each antibody. Spin 5 min at maximum speed in a microcentrifuge and retain the supernatant. If DNA staining is desired, supplement the secondary antibody mixture with a 1:5000 dilution of 1% Hoechst solution.

3. Methods

3.1. Growth of Cells on Coverslips

1. If standard glass coverslips are being used, place the sterile coverslips in suitable culture dishes. It may be helpful to etch an asymmetric mark on the top of each coverslip. If SecureSlip™ coverslips are being used, the coverslips should be pre-coated with gelatin as follows. Aseptically remove SecureSlip™ coverslips from the package and place them in a culture dish with the wells facing up. Add 20 μL 0.1% gelatin to each well. Cover the culture dish and let it sit for 10 min in the hood. Then aspirate the excess gelatin completely using a Pasteur pipet attached to a vacuum trap. Fill the culture dish with culture medium. Make sure that the SecureSlip™ coverslips are completely submerged and not floating. If a SecureSlip™ coverslip does float, push it down

to the bottom of the culture dish using a sterile pipette tip or a sterile forceps.

2. Plate the cells at a density that will yield about 60% confluency on the day of the experiment. Grow the cells in normal culture medium under standard conditions (*see* **Note 5**).

3.2. Preparations for Immunofluorescence Processing

1. On the day of the experiment, for each coverslip, fill a 50-mL conical plastic tube with 25 mL of either acetone or methanol (depending on the antigen—*see* **Note 6**). Cool these tubes of organic solvent to –20°C.
2. Prepare fresh PBS+ and blocking buffer (*see* **Subheading 2.4**).
3. Perform any desired experimental manipulations of the cultured cells.
4. Prepare a humidified chamber for the incubations. A suitable chamber can be created by placing a moist paper towel in a standard Petri dish.

3.3. Organic Solvent Treatment and Crosslinking

1. Remove a culture dish from the incubator. Working quickly, lift a coverslip out of the culture dish using forceps (*see* **Note** 7).
2. Remove as much culture medium as possible (this point is especially important—*see* **Note 8**). If a standard coverslip is being used, wick away the liquid by touching an edge of the coverslip to a paper towel. If a SecureSlip™ coverslip is being used, the sides of the well protect the cells, so the culture medium should be removed by inverting the coverlip on a paper towel and pressing gently on the bottom of the coverslip with forceps.
3. Immediately drop the coverslip into a tube of cold acetone or methanol. The goal during these manipulations is to transfer the coverlip from 37 to –20°C as quickly as possible, ideally in less than 5 s.
4. Leave the tube at –20°C for 5 min (or longer—*see* **Note 9**).
5. Remove the coverslip from the organic solvent using forceps. Hold the coverslip vertically in a tissue culture hood until the solvent has completely evaporated. The best results are obtained if the solvent is dried rapidly by touching a corner of the coverslip to the air flow grating.
6. Set the coverslip down on a clean surface with the cells facing up. If desired, the dehydrated cells can be left at room temperature for up to several hours.
7. If a standard coverslip is being used, create wells as follows. Cut a piece of laminating film to about the size of the

coverslip. Use a leather punch to make 5-mm holes in the laminating film. Then carefully seal the laminating film to the coverslip by rubbing with a pipette tip. Up to four wells fit readily on a 22 × 22 mm^2 coverslip. If a SecureSlip™ coverslip is being used, wells are already present on the coverslip (*see* **Note 10**).

8. Prepare diluted BS^3 crosslinker by adding a 10-μL aliquot of 10 mM BS^3 to 990 μL PBS+. Pipette 10 μL of diluted BS^3 into each well (unless the epitope is sensitive to BS^3 treatment—*see* **Note 11**). Incubate for 30 min at room temperature in a humidified chamber.
9. During this incubation, prepare the diluted and centrifuged primary and secondary antibodies (*see* ***Subheading 2.4***). Prepare enough of each antibody solution to add 10 μL per well.
10. If a standard coverslip is being used, remove the BS^3 by aspiration using a Pasteur pipet attached to a vacuum trap. If a SecureSlip™ coverlip is being used, the preferred method is to blot away the liquid by inverting the coverslip onto a paper towel. In either case, it is important to remove the liquid gently during this and subsequent steps.
11. Wash each well three times with a drop of PBS+. For each wash, add a drop of PBS+ to the well, then gently aspirate or blot away the liquid.
12. To quench any unreacted BS^3, pipette 10 μL of ethylenediamine solution into each well and incubate for 15 min at room temperature (*see* **Note 12**).
13. During this incubation, remove particulate matter from the blocking buffer by spinning in a tabletop centrifuge for 15 min at 2000 *g* (typically about 3000 rpm). Transfer the supernatant to a fresh tube and use this centrifuged blocking buffer from now on.

3.4. Antibody Incubations

1. To block nonspecific binding sites, add a drop of centrifuged blocking buffer to each well. Incubate for 1 h at room temperature.
2. Gently aspirate or blot away the blocking buffer. Add 10 μL of primary antibody solution to each well. Incubate for 30–60 min at room temperature.
3. Gently wash each well eight times with a drop of blocking buffer.
4. Add 10 μL of secondary antibody solution to each well. Incubate for 30 min at room temperature.
5. Gently wash each well 10 times with a drop of blocking buffer.

3.5. Mounting the Samples

1. Completely aspirate the final drop of blocking buffer from each well. It is essential that no liquid remains (*see* **Note 13**).
2. Add 10 μL of mounting solution to each well (*see* **Note 2**).
3. Invert the coverslip onto a glass slide. If a standard coverslip is being used, seal the edges with VALAP. If a SecureSlip™ coverslip is being used, seal the edges with clear nail polish.
4. If possible, the samples should be viewed immediately, but good results can be obtained for several weeks if the slides are stored at room temperature in the dark.

4. Notes

1. Most of our experiments have employed single-well SecureSlip™ coverslips (cat. no. MSR12-0.5), which can be placed in individual wells of a six-well culture dish. The multi-well versions (such as cat. no. MSR-12) are suitable for placing in a Petri dish.
2. We have used traditional glycerol-based mounting media, but newer formulations that eliminate refractive index mismatch may significantly improve image quality (5).
3. Octyl glucoside is optional but reduces background labeling with some antibodies. This detergent will not solubilize transmembrane proteins at the concentration used. Octyl glucoside readily absorbs water from the air and should be stored with desiccation. If the octyl glucoside powder was refrigerated, the bottle should be warmed completely to room temperature before opening.
4. Other additives can be also used to make a blocking buffer. The formulation given here has worked well for us with a variety of antibodies.
5. The confluency and culture medium can be varied as needed for the purposes of the experiment.
6. Many antigens are visualized well with acetone, but some are visualized better with methanol, which tends to denature proteins more extensively. The choice of organic solvent should be made empirically.
7. Removing the coverslip from the culture dish can be tricky. The easiest method is to lift one edge of the coverslip with a syringe needle before grabbing the coverslip with forceps.
8. If a significant amount of culture medium is transferred with the coverslip, the dehydration will initially be

incomplete, and the mixture of water and organic solvent will disrupt cellular architecture. It is therefore essential to remove as much of the culture medium as possible before dropping the coverslip into the organic solvent. However, the cells cannot be allowed to dry completely.

9. If desired, coverslips can be left for days or even weeks in organic solvent at –20°C.
10. After prolonged storage of a SecureSlip™ coverslip in organic solvent, the silicone backing may detach. In this case, dry the coverslip and silicone backing separately, and then reattach the silicone backing, taking care not to perturb the cells.
11. Some epitopes might be blocked by treatment with BS^3, which reacts mainly with lysine side chains. If the staining is weak after BS^3 treatment, rehydration can be done with PBS+ lacking BS^3. In this case, the washes should be exceptionally gentle to avoid disrupting cellular architecture.
12. Ethylenediamine is a very potent quencher of amine-reactive crosslinkers. More common quenchers such as glycine would presumably also be effective, but the incubation period might need to be prolonged.
13. If any droplets of liquid remain, they may not mix completely with the mounting solution, and the resulting refractive index mismatch will distort the images.

Acknowledgments

This work was supported by NIH grant GM-61156. The anti-Sec13 antibody was a kind gift of Bor Luen Tang and Wanjin Hong (National University of Singapore).

References

1. Donaldson, J.G. (1998) Immunofluorescence Staining, in *Current Protocols in Cell Biology*. John Wiley & Sons. pp. 4.3.1–4.3.6.
2. Bhattacharyya, D. and Glick, B.S. (2007) Two mammalian Sec16 homologs have nonredundant functions in ER export and transitional ER organization. *Mol. Biol. Cell.* **18**, 839–849.
3. Hammond, A.T. and Glick, B.S. (2000) Dynamics of transitional endoplasmic reticulum sites in vertebrate cells. *Mol. Biol. Cell.* **11**, 3013–3030.
4. Melan, M.A. and Sluder, G. (1992) Redistribution and differential extraction of soluble proteins in permeabilized cultured cells. Implications for immunofluorescence microscopy. *J. Cell Sci.* **101**, 731–743.
5. Staudt, T., Lang, M. C., Medda, R., Engelhardt, J., and Hell, S.W. (2007) 2,2′-Thiodiethanol: a new water soluble mounting medium for high resolution optical microscopy. *Microsc. Res. Tech.* **70**, 1–9.
6. Tang, B.L., et al. (1997) The mammalian homolog of yeast Sec13p is enriched in the intermediate compartment and is essential for protein transport from the endoplasmic reticulum to the Golgi apparatus. *Mol. Cell. Biol.* **17**, 256–266.

Chapter 25

Trapping Oxidative Folding Intermediates During Translocation to the Intermembrane Space of Mitochondria: In Vivo and In Vitro Studies

Dionisia P. Sideris and Kostas Tokatlidis

Abstract

The MIA40 pathway is a novel import pathway in mitochondria specific for cysteine-rich proteins of the intermembrane space (IMS). The newly synthesised precursors are trapped in the IMS by a disulfide relay mechanism that involves introduction of disulfides from the sulfhydryl oxidase Erv1 to the redox-regulated import receptor Mia40 and then on to the substrate. This thiol–disulfide exchange mechanism is essential for the import and oxidative folding of the incoming cysteine-rich substrate proteins. In this chapter we will describe the experimental methods that have been developed in order to study and characterise disulfide-trapped intermediates in yeast mitochondria.

Key words: Mitochondrial protein import, mitochondrial intermembrane space, MIA40 pathway, Blue Native PAGE, thiol-disulfide exchange, mixed disulfide intermediates.

1. Introduction

When thinking of thiol–disulfide exchange reactions, two cellular compartments are immediately brought to our minds; the periplasm of bacteria and the endoplasmic reticulum (ER) of eukaryotes. Both of these compartments harbour dedicated oxidative systems that shuttle disulfide bonds in substrate proteins that are ultimately required for their proper folding and activity (1–6). The mitochondrial intermembrane space (IMS) has been recently identified as an additional cellular compartment that is able to accommodate sulfhydryl oxidation reactions (7–10). It has long been thought that the IMS of mitochondria is inhospitable

A. Economou (ed.), *Protein Secretion*, Methods in Molecular Biology 619,
DOI 10.1007/978-1-60327-412-8_25, © Springer Science+Business Media, LLC 2010

to thiol exchange reactions due to the pores on the outer membrane that allow free diffusion of reduced glutathione molecules from the cytosol into the IMS (11). The surprising identification of disulfide bonds in many cysteine-rich proteins in the IMS was pivotal in the search of a system that mediates the acquisition of disulfide bonds in mitochondria.

Mia40 is an essential protein that was recently identified to act as an IMS receptor which selectively recognises and oxidises substrate proteins (12–17). Together with the essential sulfhydryl oxidase Erv1, which recycles Mia40 from its reduced to its oxidised state, Erv1 and Mia40 make the MIA40 pathway (13, 16, 18–20). This pathway constitutes a disulfide relay mechanism that ultimately transfers structural and/or catalytic disulfide bonds to the substrate concomitantly with the transfer of electrons to the respiratory chain (21–23).

The substrates of Mia40 that have been identified so far contain a twin CX3C or CX9C motif (9, 20, 24, 25). Twin CX3C substrate proteins that have been extensively used and served as model substrates in many studies include members from the small Tim family. Tim proteins reside in the IMS where they act as chaperone complexes for the transport of hydrophobic precursors to the inner membrane (26–29). Intramolecular oxidation of the twin CX3C motif mediated by the MIA40 pathway secures the protein in a folded conformation, which is a prerequisite for the assembly of the protein in the chaperone complex (8, 27, 30–32).

Mixed disulfide intermediates in the ER and the bacterial periplasm are especially difficult to detect probably due to the very fast kinetics of the interaction. However, in the IMS of mitochondria, substrate proteins have been efficiently trapped covalently with Mia40. This unique characteristic is particularly advantageous in identifying new substrates and characterising the mechanistic parameters that define these interactions. In this chapter we will analyse the in vivo and in vitro methods that have been developed and used in order to trap oxidative folding intermediates during translocation to the IMS of yeast mitochondria.

2. Materials

2.1. Protein Import into Isolated Yeast Mitochondria Followed by Trapping of Mixed Disulfide Intermediates

Prepare all solutions in distilled H_20 unless otherwise stated.

1. TNT-coupled reticulocyte lysate system (Promega) and 10 μCi/μL ^{35}S-methionine (Perkin Elmer).
2. Import buffer (2X): 1.2 M sorbitol, 4 mM KH_2PO_4, 100 mM KCl, 100 mM HEPES, 20 mM $MgCl_2$, 0.5 M Na_2EDTA, 10 mM L-methionine, 2 mg/mL fatty-acid-free BSA; adjust pH to 7.2 with KOH, solution stable for months at –20°C.

3. 0.1 M ATP (adjust with KOH to pH 7.2; store at –20°C; protect from hydration by keeping tightly sealed in the presence of silica gel).
4. 0.5 M NADH can be stored in small aliquots at –80°C.
5. Proteinase K: 10 mg/mL in distilled water; freshly made; keep on ice prior to use.
6. Make 200 mM phenylmethanesulfonyl fluoride (PMSF) in isopropanol, as it is unstable in water (caution: very toxic) make fresh.
7. Wash buffer: 0.6 M sorbitol, 20 mM HEPES–KOH (pH 7.4). Store at 4°C.
8. 100 mM *N*-ethyl maleimide (NEM), prepare 1 M stock solution in absolute ethanol and then dilute with water. Store at –20°C in small aliquots, protect from photoxidation by wrapping in aluminium foil and use once.

2.2. Analysis of Protein Complexes by Blue Native Electrophoresis After Import into Isolated Yeast Mitochondria

1. Solubilisation buffer: 0.16% *n*-dodecylmaltoside (n-DDM), 50 mM NaCl, 20 mM HEPES–KOH (pH 7.4), 2.5 mM $MgCl_2$, 1 mM EDTA, 10% glycerol, 1 mM PMSF (add just before use) (*see* **Note 1**).

2.3. Blue Native PAGE

1. Sample buffer (10X): 5% Coomassie Brilliant Blue G-250, 0.5 M 6-aminocaproic acid, 100 mM BisTris–HCl (pH 7.0), 1 mM PMSF. Store at –20°C.
2. Anode buffer: 50 mM BisTris–HCl (pH 7.0). Store at 4°C.
3. Cathode buffer: 50 mM Tricine, 15 mM BisTris–HCl (pH 7.0), 0.02% Coomassie Blue G-250. Store at 4°C.
4. Colourless cathode buffer: as above without the Coomassie. Store at 4°C (*see* **Note 2**).
5. Gel buffer: 100 mM BisTris–HCl (pH 7.0), 0.1 M 6-aminocaproic acid.
6. Acrylamide solution: 48% acrylamide, 1.5% bisacrylamide. Filter through 0.44-μm acrodisc. Keep the solution in the dark at room temperature for months.
7. High Molecular Weight Electrophoresis Markers.

2.4. In Vitro Trapping Assay of Mixed Disulfide Intermediates with Mia40

1 Beads saturated (50 μL) with immobilised Mia40 as a 6 × His or GST tag (*see* **Note 3**).

2 Wash solution: 150 mM NaCl, 50 mM Tris–HCl (pH 7.4), 0.1% BSA, 0.1% Triton X-100.

3. Methods

3.1. In Vitro Synthesis and Radioactive Labelling of Precursor Proteins

A precursor protein can be synthesised in vitro by cloning the respective gene in a plasmid containing a RNA polymerase promoter like SP6, T3 or T7 (**Note 4**). A commercially available reticulocyte lysate system (Promega) allows the synthesis of the protein by coupling transcription and translation. The reaction is carried out in the presence of a radioactive amino acid, like [^{35}S] methionine, in order to label the newly synthesised protein for detection. The rabbit reticulocyte lysate system has been used almost exclusively for protein synthesis and import into mitochondria. Wheat germ lysate is also available but is not an optimal system for mitochondrial import as it lacks the hsp70 chaperone. This chaperone has been shown to interact with some mitochondrial precursors and is needed so that the precursor protein is presented unfolded on the mitochondrial surface. The protocol described here is according to the manufacturer's instructions for the TNT-coupled rabbit reticulocyte lysate system (Promega) and previous publications (33, 34).

1. Rapidly thaw all components by hand and place on ice.
2. Place an Eppendorf on ice and add in the following order 25 μL TNT rabbit reticulocyte lysate, 2 μL TNT reaction buffer, 2 μL [^{35}S] methionine (1000 Ci/mmol at 10 mCi/mL), 1 μL TNT RNA polymerase (SP6, T3 or T7), 1 μL amino acid mix minus methionine, 1 μL RNasin® ribonuclease inhibitor (40 u/μL).
3. Add 1 μg of plasmid DNA.
4. Add nuclease-free water to a final volume of 50 μL.
5. Shield the mixture from light to prevent photo-oxidation and disulfide bond formation.
6. Incubate the reaction at 30°C for 90 min without shaking.
7. Centrifuge the reaction mixture at 100,000 *g* for 15 min at 4°C in order to remove ribosomes.
8. Transfer the supernatant to a new Eppendorf and store at –20°C (*see* **Note 5**).

3.2. Protein Import into Isolated Yeast Mitochondria and Trapping of Mixed Disulfide Intermediates

Mitochondrial protein import can be reconstituted in an in vitro experiment by incubating a radiolabelled precursor with isolated yeast mitochondria and analysing the reaction products by gel electrophoresis and autoradiography. When it comes to import of IMS precursors this event is described as translocation across the outer membrane. In this sense it is important after an import experiment to check the protection of the precursor against externally added protease.

A thiol–disulfide exchange is a fast and very transient reaction that occurs between an oxidase and a substrate. Therefore, it is not easy to trap and monitor a mixed disulfide intermediate between these two components. A few things should be kept in mind when trying to create conditions that will enhance the efficiency of trapped intermediates. Before importing a cysteine-rich precursor it is important that the precursor is unfolded and reduced. The radioactive precursor can be incubated with 10 mM DTT on ice for 10 min prior to import to ensure reduction of the substrate, but it is important not to exceed a concentration of 0.5 mM DTT in the final import reaction. It has been shown that a concentration of 0.5 mM DTT will enhance the amount of imported material but a concentration over 2 mM DTT will reduce Mia40 (13) and decrease the import of some precursors by 50% (9). Furthermore, long import reactions should be avoided and shorter time points should be taken when trying to trap a mixed disulfide intermediate because redox reactions are fast events. Import time points ranging from 30 s to 20 min have been used in many publications. Metal binding may be involved in the folding of a precursor protein after being imported in mitochondria. The use of a metal chelator such as EDTA and orthophenanthroline during an import reaction can discriminate between precursors that require metal binding for their final oxidation and those that do not, using the metal-dependent activity of the matrix processing peptidase MPP as control (35). Finally, the use of an alkylating agent, such as NEM or iodoacetamide, at the end of the import reaction has been shown to further boost the amount trapped disulfide adducts.

Typically, for every import reaction 30–50 μg of isolated mitochondria is used and 5 20% (v/v) of radioactive precursor. The protocol described here is for one time point import (*see* **Note 6**) by using 50 μg of mitochondria and 5 μL of radioactive precursor.

1. Thaw 50 μg of isolated yeast mitochondria quickly at 30°C.
2. Place an Eppendorf on ice and add 50 μL 2 import buffer, 2μL 0.1 M ATP, 0.5 μL of 0.5 M NADH, 50 μg mitochondria, and make the volume to 100 μL with H_2O.
3. Start the import reaction by adding 5 μL of radioactive precursor, mix gently and place in a 30°C waterbath.
4. To arrest the import reactions add 25 μL of 100 mM NEM and place the tube on ice for 2 min.
5. Spin at 14000 *g* for 5 min at 4°C to collect mitochondria.
6. Resuspend mitochondria gently in 100 μL of wash buffer that contains 0.1 mg/mL proteinase K and incubate on ice for 20 min.

7. Add 1 μL of 0.2 M PMSF to inactivate proteinase K and leave on ice for further 10 min.
8. Spin at 14000 *g* for 5 min at 4°C to recollect mitochondria.
9. Resuspend the mitochondrial pellet with laemmli sample buffer that does not contain a reducing agent, heat for 5 min at 95°C and keep on ice or store at –20°C until ready to load a SDS–PAGE.

In the example of **Fig. 25.1**, radioactive yeast Tim10 has been imported into isolated wild-type yeast mitochondria for time points ranging from 1 to 20 min (*see* **Note 7**). It is clear from this example that a mixed disulfide intermediate with Mia40 is formed immediately upon import into mitochondria and that accumulation of the monomeric Tim10 in the inter-membrane space is increasing with time. The graph on panel B that shows the ratio of the intermediate to the monomeric Tim10 for every time point is indicative of the fast nature of the covalent interaction between the oxidase and the substrate. In some cases the precursor has different migration patterns depending on its redox state. This

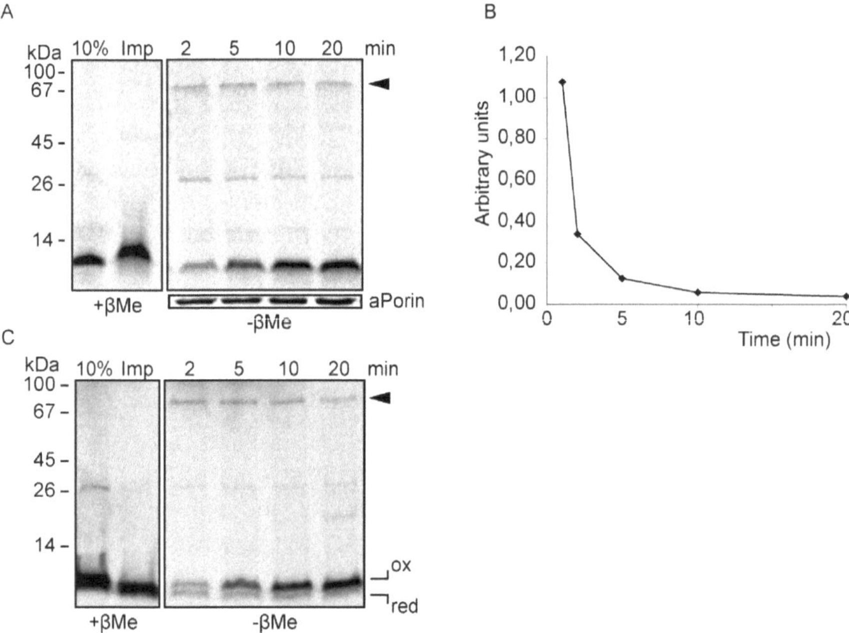

Fig. 25.1. (**A**) Import of radioactive yeast Tim10 into isolated wild-type yeast mitochondria. The precursor is imported in a time-dependent manner in the intermembrane space where it forms a mixed disulfide intermediate with Mia40 (indicated by an arrowhead). This intermediate is sensitive to reducing agent (in this case + βMe, left panel). In the right panel the imported material has been analysed in the absence of a reducing agent. The samples were analysed by SDS–PAGE and autoradiography. (**B**) The graph shows the ratio of the mixed disulfide intermediate to the monomeric Tim10 accumulating in the intermembrane space. (**C**) Import kinetics of radioactive Cox17 into isolated wild-type yeast mitochondria. The covalent mixed disulfide intermediate is shown with an arrowhead while the annotations "ox" and "red" denote the oxidised and reduced forms of the precursor.

intrinsic property of some proteins can provide significant information regarding the time-dependent oxidation of the substrate. In **Fig. 25.1** panel C radioactive Cox17 has been imported in wt mitochondria where the different redox species are analysed. From this example it is evident that while the mixed disulfide intermediate is immediately formed and decreasing in time, the amount of the oxidised form of the precursor is increasing to the detriment of the reduced form.

3.3. Analysis of Protein Complexes by Blue Native Electrophoresis After Import into Isolated Yeast Mitochondria

Blue native electrophoresis is a technique that is used to dissect precursor import intermediates (*see* **Note 8**). The main advantage of BN–PAGE is the fact that the electrophoretic mobility of the proteins is dependent solely on their molecular weight (36, 37). This is achieved by inducing a charge shift on the proteins when the anionic dye Coomassie blue G250 (*see* **Note 9**) binds to the hydrophobic regions of the solubilised proteins. The basic principle involves import of the radioactive precursor into mitochondria followed by solubilisation of the membranes with a mild detergent that while extracting the proteins leaves the complexes intact. The choice of the detergent is crucial because the protein complexes may dissociate by treating too harshly, and on the other hand too mild conditions may not be enough to extract the proteins (*see* **Note 1** and **10**).

The protocol described here is an immediate continuation from step 8 after import of radioactive precursor into isolated mitochondria.

1. Resuspend the mitochondrial pellet in 50 μL solubilisation buffer and leave on ice for 30 min.
2. Spin at 16000*g* at 4°C for 20 min to pellet insoluble aggregates.
3. Transfer the supernatant that contains the soluble material to another Eppendorf, add 5 μL 10X BN sample buffer and load on the BN (*see* **Note 11**).

In **Fig. 25.2** the import stages of Tim10 after import in mitochondria have been determined by BN–PAGE. Even in the early time points the Mia40 covalent complex is formed, which as time progresses gives rise to the assembly of the soluble TIM10 complex, followed by its association with the TIM22 complex. In order to see if any of the complexes formed are via disulfide bonds, the samples can be treated with 20 mM DTT during the solubilisation (*see* **Note 12**). In this case the only complex that is reduced upon incubation with DTT is the covalent MIA40 complex (lane 5). Furthermore it should be noted that some complexes are extremely sensitive to protease treatments. It is crucial that the protease is well inactivated after the import reaction because any residual protease will degrade proteins during the solubilisation step and therefore jeopardise the detection of some

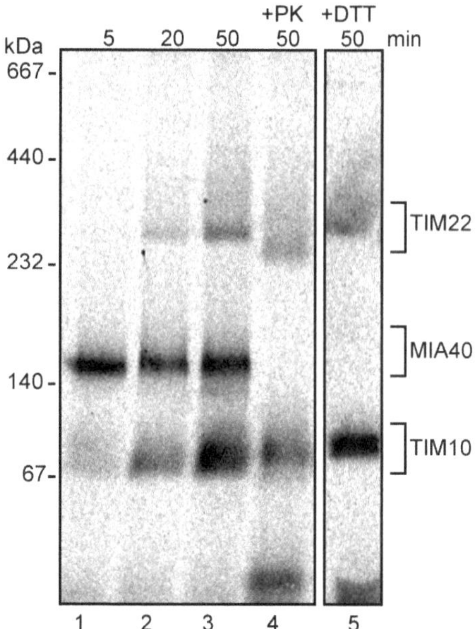

Fig. 25.2. Import of radioactive Tim10 into isolated wild-type yeast mitochondria for the respective time points. After solubilisation in DDM buffer the samples were separated on BN-PAGE and visualised by autoradiography. The sample in lane 4 was incubated with 0.1 mg/mL proteinase K during the solubilisation and the sample in lane 5 was with 20 mM DTT during the solubilisation. The complexes have been annotated on the right of the figure.

complexes. In the example of **Fig. 25.2** (lane 4) the sample was treated with proteinase K, which resulted in complete degradation of the MIA40 complex and partial degradation of the TIM22 complex.

3.4. In Vitro Trapping Assay of Mixed Disulfide Intermediates with Mia40

It is possible to reconstitute a covalent interaction between two components in vitro. This can be done by directly incubating a purified protein immobilised on affinity beads with a radioactive substrate. The basic principle of this assay is that the substrate in a reduced state should covalently interact with pure, oxidised Mia40 that is prebound on beads (*see* **Notes 13** and **14**). To arrest the reaction NEM is added to irreversible alkylate and therefore block the mixed disulfide intermediate. The following protocol can be similarly done with the protein containing a histidinetag.

1. Keep the glutathione beads containing Mia40 in a buffer with 150 mM NaCl and 50 mM Tris–HCl (pH 8.0) in a 50% slurry.
2. Take 50 μL of glutathione beads saturated with Mia40 and place on ice (14).
3. Add 5 μL of radioactive protein in the tube and incubate for the desired time points at 4°C with mild shaking.

4. Arrest the reaction with the addition of 12.5 μL of 100 mM NEM.
5. Spin at 5000 *g* for 2 min at 4°C and remove the supernatant which contains the unbound material.
6. Wash the beads with 250 μL wash buffer for 5 min on a rotating wheel to remove unspecific binding.
7. Spin at 5000 *g* for 2 min at 4°C and repeat washing for two more times (three in total).
8. At the last spin resuspend the beads in laemmli sample buffer without a reducing agent, vortex, heat for 5 min at 95°C, vortex again, quickly spin down the beads and load on SDS–PAGE the supernatant.

In **Fig. 25.3** the interaction between Mia40 and radioactive Tim10 is shown. The presence of NEM is important in monitoring the reaction as it presumably generates a more homogeneous population of the covalent intermediate. This assay can also be used to test mutant forms of either the oxidase or the substrate and hence understand the mechanistic parameters that define this interaction. Specifically this assay was used by two independent studies when the interaction between Mia40 and Tim10 was found to be dependent on the first cysteine of the substrate (24, 25).

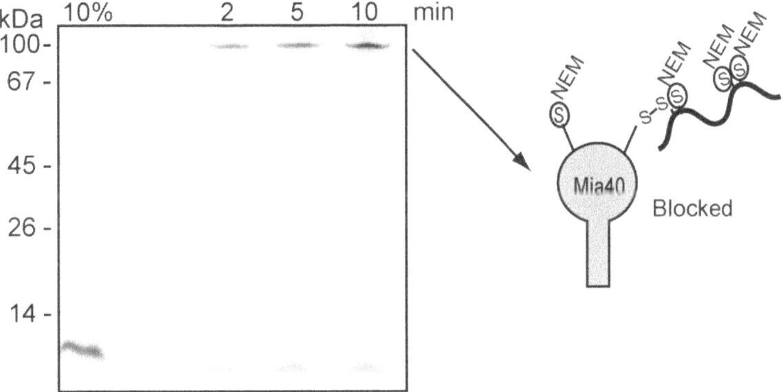

Fig. 25.3. (**A**) Interaction of radioactive Tim10 with pure bead immobilised Mia40 for the indicated time points. The samples were analysed by SDS–PAGE and visualised by autoradiography. On the right a schematic representation of the interaction between Mia40 and the substrate is depicted.

4. Notes

1. The choice of the detergent is crucial for best analysis of protein complexes. For yeast mitochondria 1% digitonin

and 0.16% DDM are commonly used as they give good lysis while keeping most protein complexes intact. Triton X-100 can also be used in a range from 0.2 to 1%. When choosing a detergent it is important that it extracts the proteins, particularly membrane-associated/embedded proteins, while preserving the protein complexes.

2. Anode and cathode buffers for BN–PAGE can be reused up to five times.
3. It is important not to keep the protein immobilised for more than 3 days as it can get easily degraded by proteases.
4. For T3- or T7-based transcription a linear DNA template has been shown to give better results while for SP6 a circular plasmid is more efficiently transcribed.
5. The in vitro synthesised radioactive precursor should not be stored for a long period of time at –20°C. It should be used as quickly as possible (ideally freshly made) as it can get oxidised with time, which can impinge on its import efficiency into mitochondria.
6. When performing an import reaction for several time points make one import master mix and take aliquots at desired time points, terminate by adding NEM and by placing on ice.
7. To make sure that the amount of mitochondria used for every time point is equal a western analysis can be performed against an abundant mitochondrial protein. In the case of **Fig. 25.1A** the antibody against porin, an outer membrane protein has been used as a loading marker.
8. A more detailed description on how to cast the gel and use Blue Native PAGE as a method to identify protein complexes in mitochondria is also given in the chapter of Vögtle et al., in this issue.
9. Use Coomassie blue G250 only and not R250.
10. Pilot experiments with different detergents at different concentrations should be performed by checking whether the protein of interest is in the supernatant or in the pellet fraction on a SDS gel.
11. The BN should be run at 4°C with all the buffers pre-chilled.
12. Reducing agents can diffuse in the gel so it is advisable not to load these samples right next to the non-reducing ones.
13. It is not necessary to have the protein immobilised on beads in order to perform the experiment. The assay can also be done with as little as 0.1 μg of pure oxidase and

1 μL of radioactive substrate, albeit from our experience at lower efficiencies. The fact that the protein is immobilised on beads probably creates a more packed environment for the interaction to occur, which would explain why the efficiency of trapping a mixed disulfide intermediate is higher.

14. Consider the amount of beads used for the assay as 50% slurry. The amount that is used is not crucial as the oxidase is in large excess over the substrate. However, it is important that the bed volume of the beads is maintained throughout the experiment and that beads are not lost during sample handling.

Acknowledgments

This work was supported by intramural funds from IMBB-FORTH, the University of Crete and the European Social Fund and national resources (to KT). DS was supported by a PENED grant. We are grateful to members of our lab for comments.

References

1. Bardwell, J. C. (1994) Building bridges: disulphide bond formation in the cell. *Mol Microbiol* **14,** 199–205.
2. Kadokura, H., Katzen, F., and Beckwith, J. (2003) Protein disulfide bond formation in prokaryotes. *Annu Rev Biochem* **72,** 111–135.
3. Nakamoto, H., and Bardwell, J. C. (2004) Catalysis of disulfide bond formation and isomerization in the Escherichia coli periplasm. *Biochim Biophys Acta* **1694,** 111–119.
4. Ritz, D., and Beckwith, J. (2001) Roles of thiol-redox pathways in bacteria. *Annu Rev Microbiol* **55,** 21–48.
5. Sevier, C. S., and Kaiser, C. A. (2002) Formation and transfer of disulphide bonds in living cells. *Nat Rev Mol Cell Biol* **3,** 836–847.
6. Sevier, C. S., and Kaiser, C. A. (2006) Conservation and diversity of the cellular disulfide bond formation pathways. *Antioxid Redox Signal* **8,** 797–811.
7. Allen, S., Lu, H., Thornton, D., and Tokatlidis, K. (2003) Juxtaposition of the two distal CX3C motifs via intrachain disulfide bonding is essential for the folding of Tim10. *J Biol Chem* **278,** 38505–38513.
8. Lu, H., Allen, S., Wardleworth, L., Savory, P., and Tokatlidis, K. (2004) Functional TIM10 chaperone assembly is redox-regulated in vivo. *J Biol Chem* **279,** 18952–18958.
9. Mesecke, N., Terziyska, N., Kozany, C., Baumann, F., Neupert, W., Hell, K., and Herrmann, J. M. (2005) A disulfide relay system in the intermembrane space of mitochondria that mediates protein import. *Cell* **121,** 1059–1069
10. Webb, C. T., Gorman, M. A., Lazarou, M., Ryan, M. T., and Gulbis, J. M. (2006) Crystal structure of the mitochondrial chaperone TIM9.10 reveals a six-bladed alpha-propeller. *Mol Cell* **21,** 123–133.
11. Benz, R. (1994) Permeation of hydrophilic solutes through mitochondrial outer membranes: review on mitochondrial porins. *Biochim Biophys Acta* **1197,** 167–196.
12. Chacinska, A., Pfannschmidt, S., Wiedemann, N., Kozjak, V., Sanjuan Szklarz, L. K., Schulze-Specking, A., Truscott, K. N., Guiard, B., Meisinger, C., and Pfanner, N. (2004) Essential role of Mia40 in import and assembly of mitochondrial intermembrane space proteins. *EMBO J* **23,** 3735–3746.
13. Grumbt, B., Stroobant, V., Terziyska, N., Israel, L., and Hell, K. (2007) Functional characterization of Mia40p, the central component of the disulfide relay system of the

mitochondrial intermembrane space. *J Biol Chem* **282,** 37461–37470.
14. Hofmann, S., Rothbauer, U., Muhlenbein, N., Baiker, K., Hell, K., and Bauer, M. F. (2005) Functional and mutational characterization of human MIA40 acting during import into the mitochondrial intermembrane space. *J Mol Biol* **353,** 517–528.
15. Naoe, M., Ohwa, Y., Ishikawa, D., Ohshima, C., Nishikawa, S., Yamamoto, H., and Endo, T. (2004) Identification of Tim40 that mediates protein sorting to the mitochondrial intermembrane space. *J Biol Chem* **279,** 47815–47821.
16. Rissler, M., Wiedemann, N., Pfannschmidt, S., Gabriel, K., Guiard, B., Pfanner, N., and Chacinska, A. (2005) The essential mitochondrial protein Erv1 cooperates with Mia40 in biogenesis of intermembrane space proteins. *J Mol Biol* **353,** 485–492.
17. Terziyska, N., Lutz, T., Kozany, C., Mokranjac, D., Mesecke, N., Neupert, W., Herrmann, J. M., and Hell, K. (2005) Mia40, a novel factor for protein import into the intermembrane space of mitochondria is able to bind metal ions. *FEBS Lett* **579,** 179–184.
18. Hell, K. (2008) The Erv1-Mia40 disulfide relay system in the intermembrane space of mitochondria. *Biochim Biophys Acta* **1783,** 601–609.
19. Lee, J., Hofhaus, G., and Lisowsky, T. (2000) Erv1p from Saccharomyces cerevisiae is a FAD-linked sulfhydryl oxidase. *FEBS Lett* **477,** 62–66.
20. Muller, J. M., Milenkovic, D., Guiard, B., Pfanner, N., and Chacinska, A. (2008) Precursor oxidation by Mia40 and Erv1 promotes vectorial transport of proteins into the mitochondrial intermembrane space. *Mol Biol Cell* **19,** 226–236.
21. Allen, S., Balabanidou, V., Sideris, D. P., Lisowsky, T., and Tokatlidis, K. (2005) Erv1 mediates the Mia40-dependent protein import pathway and provides a functional link to the respiratory chain by shuttling electrons to cytochrome c. *J Mol Biol* **353,** 937–944.
22. Bihlmaier, K., Mesecke, N., Terziyska, N., Bien, M., Hell, K., and Herrmann, J. M. (2007) The disulfide relay system of mitochondria is connected to the respiratory chain. *J Cell Biol* **179,** 389–395.
23. Dabir, D. V., Leverich, E. P., Kim, S. K., Tsai, F. D., Hirasawa, M., Knaff, D. B., and Koehler, C. M. (2007) A role for cytochrome c and cytochrome c peroxidase in electron shuttling from Erv1. *EMBO J* **26,** 4801–4811.
24. Milenkovic, D., Gabriel, K., Guiard, B., Schulze-Specking, A., Pfanner, N., and Chacinska, A. (2007) Biogenesis of the essential Tim9-Tim10 chaperone complex of mitochondria: site-specific recognition of cysteine residues by the intermembrane space receptor Mia40. *J Biol Chem* **282,** 22472–22480.
25. Sideris, D. P., and Tokatlidis, K. (2007) Oxidative folding of small Tims is mediated by site-specific docking onto Mia40 in the mitochondrial intermembrane space. *Mol Microbiol* **65,** 1360–1373.
26. Curran, S. P., Leuenberger, D., Oppliger, W., and Koehler, C. M. (2002) The Tim9p-Tim10p complex binds to the transmembrane domains of the ADP/ATP carrier. *Embo J* **21,** 942–953.
27. Curran, S. P., Leuenberger, D., Schmidt, E., and Koehler, C. M. (2002) The role of the Tim8p-Tim13p complex in a conserved import pathway for mitochondrial polytopic inner membrane proteins. *J Cell Biol* **158,** 1017–1027.
28. Luciano, P., Vial, S., Vergnolle, M. A., Dyall, S. D., Robinson, D. R., and Tokatlidis, K. (2001) Functional reconstitution of the import of the yeast ADP/ATP carrier mediated by the TIM10 complex. *EMBO J* **20,** 4099–4106.
29. Vergnolle, M. A., Baud, C., Golovanov, A. P., Alcock, F., Luciano, P., Lian, L. Y., and Tokatlidis, K. (2005) Distinct domains of small Tims involved in subunit interaction and substrate recognition. *J Mol Biol* **351,** 839–849.
30. Lu, H., Golovanov, A. P., Alcock, F., Grossmann, J. G., Allen, S., Lian, L. Y., and Tokatlidis, K. (2004) The structural basis of the TIM10 chaperone assembly. *J Biol Chem* **279,** 18959–18966.
31. Vergnolle, M. A., Alcock, F. H., Petrakis, N., and Tokatlidis, K. (2007) Mutation of conserved charged residues in mitochondrial TIM10 subunits precludes TIM10 complex assembly, but does not abolish growth of yeast cells. *J Mol Biol* **371,** 1315–1324.
32. Vial, S., Lu, H., Allen, S., Savory, P., Thornton, D., Sheehan, J., and Tokatlidis, K. (2002) Assembly of Tim9 and Tim10 into a functional chaperone. *J Biol Chem* **277,** 36100–36108.
33. Glick, B. S. (1991) Protein import into isolated yeast mitochondria. *Methods Cell Biol* **34,** 389–399.
34. Tokatlidis, K. (2000) Directing proteins to mitochondria by fusion to mitochon-

drial targeting signals. *Methods Enzymol* **327,** 305–317.

35. Luciano, P., Tokatlidis, K., Chambre, I., Germanique, J. C., and Geli, V. (1998) The mitochondrial processing peptidase behaves as a zinc-metallopeptidase. *J Mol Biol* **280,** 193–199.
36. Schagger, H., Cramer, W. A., and von Jagow, G. (1994) Analysis of molecular masses and oligomeric states of protein complexes by blue native electrophoresis and isolation of membrane protein complexes by two-dimensional native electrophoresis. *Anal Biochem* **217,** 220–230.
37. Schagger, H., and von Jagow, G. (1991) Blue native electrophoresis for isolation of membrane protein complexes in enzymatically active form. *Anal Biochem* **199,** 223–231.

Chapter 26

Native Techniques for Analysis of Mitochondrial Protein Import

F.-Nora Vögtle, Oliver Schmidt, Agnieszka Chacinska, Nikolaus Pfanner, and Chris Meisinger

Abstract

During the evolution of eukaryotic cells, the majority of mitochondrial genes have been transferred to the nuclear genome with the consequence that most mitochondrial proteins have to be imported into the organelle. This process occurs usually in a post-translational manner. In order to accomplish this task elaborate protein machineries have evolved that import precursor proteins in a concerted fashion and sort them into the four mitochondrial subcompartments. Native techniques such as Blue Native Electrophoresis allow to analyze the composition of the import machineries and to characterize the cooperation of import components. The analysis has led to the discovery of new components and import pathways of mitochondria.

Key words: Mitochondria, protein import, translocation intermediates, Blue Native PAGE, protein complex analysis.

1. Introduction

The mitochondrial intermembrane space (IMS) and the innermost compartment, the mitochondrial matrix, are confined by two lipid bilayers, the outer and inner mitochondrial membranes. Mitochondrial proteins have to cross at least one of these two barriers to get to their final location within the organelle. Given that 99% of about 1,000 different mitochondrial proteins are encoded by the nuclear genome and synthesized in the cytosol this imposes a complex task on the mitochondrial protein import machinery.

A. Economou (ed.), *Protein Secretion*, Methods in Molecular Biology 619,
DOI 10.1007/978-1-60327-412-8_26,

These machineries do not only have to tackle the problem of importing hundreds of different precursors, but also face the task of assembling various proteins in a step-wise manner into sophisticated oligomeric machineries. This is especially critical in the mitochondrial membranes, where various large protein complexes like the complexes of the respiratory chain have evolved. Recent developments in the identification of novel protein import pathways have shed light on how mitochondria resolve this problem, and several studies have focused on analyzing translocation intermediates of different precursor proteins.

Blue Native Electrophoresis (Blue Native Polyacrylamide Gel Electrophoresis, BN–PAGE), first introduced by Schägger and von Jagow in 1991 (1, 2), is a leading technique for the analysis of mitochondrial preprotein import intermediates. The basic principle includes first the solubilization of the membranes with a mild non-ionic detergent that leaves protein–protein interaction intact (e.g., Digitonin, Dodecylmaltoside, Triton X-100). Following solubilization, the detergent is replaced during the electrophoretic run by the dye Coomassie blue G 250 that binds to hydrophobic regions of the extracted membrane proteins, thus preventing aggregation and adding negative charges to the protein complex. As an alternative possibility, protein complexes can be separated by sucrose gradient centrifugation and analyzed by SDS–PAGE and immunodecoration (3, 4).

We use mitochondria from the yeast *Saccharomyces cerevisiae* to investigate the interactions between the different import complexes and to dissect the intermediate stages of precursor proteins, i.e. their transient association with distinct import complexes, on their way from the cytosol to their final destination.

2. Materials

2.1. Blue Native Electrophoresis

1. Gel buffer (3X): 220 mM ε-amino-*n*-caproic acid, 150 mM Bis–Tris–HCl, pH 7.0. Store at 4°C.
2. Acrylamide (stock solution): 49.5% (w/v) acrylamide, 3% (w/v) Bis-acrylamide. Store in the dark at room temperature.
3. Anode buffer (10X): 500 mM Bis–Tris–HCl, pH 7.0. Store at 4°C.
4. Cathode buffer (10X): 500 mM Tricine, 150 mM Bis–Tris–HCl, pH 7.0, 0.2% Coomassie blue G 250. Store at 4°C.
5. Loading dye (10X): 5% Coomassie blue G 250, 500 mM ε-amino-*n*-caproic acid, 100 mM Bis–Tris–HCl, pH 7.0. Store at 4°C.

6. Solubilization buffer: 0.5–1.5% Digitonin (High Purity, Calbiochem), 20 mM Tris–HCl, pH 7.4, 0.1 mM EDTA, 50 mM NaCl, 10% glycerol, 1 mM PMSF (prepare 0.2 M stock solution in isopropanol, store at room temperature). Store at 4°C (*see* **Note 1**).
7. Ammonium persulfate (APS): prepare 10% (w/v) solution in water. Store at 4°C.
8. *N,N,N,N′*-Tetramethyl-ethylenediamine (TEMED). Store at 4°C.
9. Gradient mixer with peristaltic pump.
10. Molecular weight marker.
11. Isopropanol.
12. Destainer: 20% (v/v) ethanol, 10% (v/v) acetic acid. Store at room temperature.

2.2. In Vitro Import into Isolated Mitochondria

1. BSA buffer: 3% (w/v) BSA (bovine serum albumin, fatty acid free), 250 mM sucrose, 80 mM KCl, 5 mM $MgCl_2$, 5 mM KH_2PO_4, 5 mM methionine, 10 mM MOPS–KOH, pH 7.2. Store in aliquots at –20°C. Each import reaction is supplemented with 2 mM NADH (make fresh), 2–4 mM ATP–KOH, pH 7.0 (store in aliquots at –20°C), 10 mM creatine phosphate (store in aliquots at –20°C), 0.1 mg/mL creatine kinase (make fresh), which are all solubilized in water.
2. SEM buffer: 250 mM sucrose, 10 mM MOPS–KOH, pH 7.2, 1 mM EDTA. Store at 4°C.
3. ^{35}S-labeled precursor proteins synthesized in rabbit reticulocyte lysate (5). Store in aliquots at –80°C.

2.3. TOM–TIM–Preprotein Supercomplex Formation

1. Recombinant-purified $b_2\Delta$-dihydrofolate reductase (DH-FR) protein (3). Store in aliquots at –80°C.
2. Methotrexate (MTX): 10 mM in 100 mM MOPS–KOH, pH 7.2. Store in aliquots at –20°C. To reach the final concentration of 500 μM dilute the MTX shortly before the start of the import reaction 1/20 in water.
3. AVO mix (100X): 100 μL valinomycin (1 mM stock), 200 μL oligomycin (10 mM stock), 100 μL antimycin A (8 mM stock), 600 μL ethanol. All stocks are prepared in ethanol and are highly toxic. Dilute 1/100 in import reaction. Store at –20°C (*see* **Note 2**).

2.4. Sucrose Gradient Preparation and Centrifugation

1. Sucrose solution: 25% (w/v) sucrose, 0.4% (w/v) Digitonin, 20 mM Tris–HCl, pH 7.4, 50 mM NaCl, 1 mM PMSF, protease inhibitor mix without EDTA.
2. Ultracentrifugation tubes (e.g., from Beckman Coulter, Ultra-Clear™, 14 × 89 mm, 13.2 mL volume; *see* **Note 3**).

3. 72% (w/v) Trichloroacetic acid (TCA). Store at room temperature.
4. 1.25% (w/v) Sodium desoxycholate. Store at room temperature.

3. Methods

As an example for the resolution of import intermediates we have chosen the maturation pathway of the Tom40 protein, which can be monitored using BN–PAGE. Tom40 is the main component of the translocase of the outer mitochondrial membrane (TOM complex) and forms two assembly intermediates (I, II) before assembling into the mature complex of 400 kDa (6). After translocation through the pre-existing TOM pore (7), the hydrophobic precursor is chaperoned by the small Tim proteins through the aqueous IMS to the sorting and assembly machinery (SAM) of the outer membrane (8). This interaction between the Tom40 precursor and the SAM complex can be visualized on BN–PAGE, where it forms the assembly intermediate I of approximately 250 kDa. In a later assembly step a 100-kDa complex (assembly intermediate II) is formed that is comprised of membrane-integrated Tom40 in a complex with the small Tom5 protein (9). Following the precursor's association with the remaining TOM components, the mature TOM complex is generated (**Fig. 26.1**).

The TOM complex also cooperates with the Translocase of the Inner mitochondrial Membrane (TIM23) complex in order to transport precursor proteins into the matrix. For this collaboration, the two complexes form an interaction induced

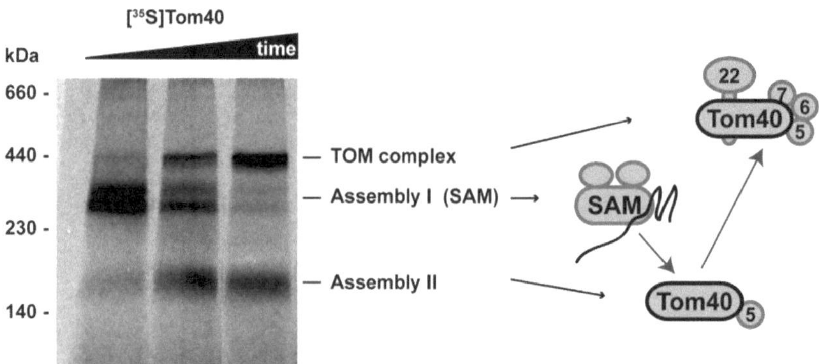

Fig. 26.1. The Tom40 precursor protein forms two assembly intermediates before its maturation into the TOM complex. The radiolabeled precursor protein was imported for increasing periods of time into isolated yeast mitochondria. After lysis in Digitonin buffer, the samples were analyzed by BN-PAGE and autoradiography. The cartoon on the right side depicts the different assembly intermediates of the Tom40 precursor protein before it assembles into the mature TOM complex.

by a precursor protein in transit (3). The interaction is transient and can only be resolved on BN–PAGE by arresting translocation of the precursor protein in a two-membrane spanning fashion. An excellent tool for this accumulation is the preprotein $b_2\Delta$-DHFR, consisting of the N-terminal matrix-targeting signal of cytochrome b_2 fused to murine DHFR. Upon addition of the substrate analogue MTX, the C-terminal DHFR domain is stably folded and cannot be translocated through the TOM channel. This leads to an arrest, in which the preprotein simultaneously spans the outer and the inner mitochondrial membranes. This so-called TOM–TIM–preprotein supercomplex can be analyzed either on BN–PAGE (**Fig. 26.2**) or by sucrose gradient centrifugation.

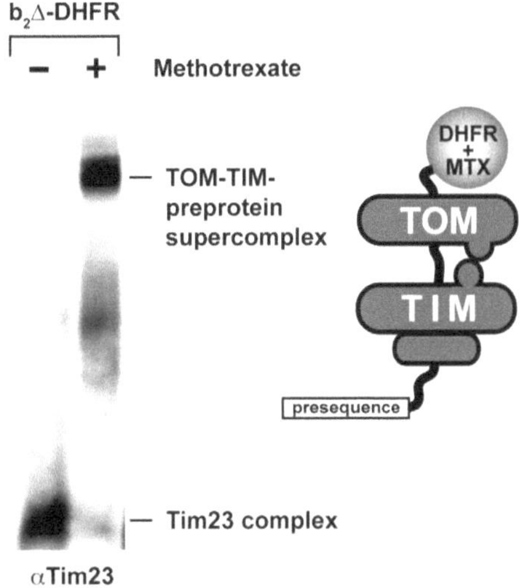

Fig. 26.2. Arrest of the $b_2\Delta$-DHFR preprotein with Methotrexate (MTX) leads to formation of a TOM-TIM-preprotein supercomplex. An import reaction with wild-type mitochondria and the recombinant protein $b_2\Delta$-DHFR was performed in the absence (–) or presence (+) of 5 μM MTX. Samples were lysed in Digitonin buffer and separated by BN-PAGE followed by immunodecoration with antisera against Tim23.

3.1. Casting the Blue Native gel

1. Assembly and casting of the Blue Native gel is based on the use of the Hoefer gel system (SE 600 series, GE Healthcare). Also required are a gradient mixer and a peristaltic pump.
2. Scrub the glass plates with water and clean them as well as the spacers and the comb carefully and thoroughly with ethanol. If you use a detergent for washing make sure all residual detergent is washed away as it can result in the dissociation of protein complexes.
3. Prepare the gel solutions. Gradient Blue native gels are made from two different solutions that differ in their acrylamide concentration and are obtained by shifting the water to acrylamide ratio. Pouring is carried out by a gradient maker

and peristaltic pump and results in a linear gradient. The choice in composition of acrylamide solutions depends on the molecular weight of the protein complexes to be analyzed. For separation of the assembly pathway of the outer membrane protein Tom40 typically a gradient of 6–16.5% is used. The standard volume for the separation gel is 18 mL in total, composed of equal amounts of both acrylamide concentrations. The 6% mix consists of 3 mL of 3X gel buffer, 1.07 mL of acrylamide (stock solution), 4.88 mL of water. The 16.5% mix is made up of 3 mL of 3X gel buffer, 2.35 mL of acrylamide (stock solution), 1.8 mL of glycerol and 1.12 mL of water (*see* **Note 1**). Add 35 μL of 10% APS and 3.5 μL of TEMED to both solutions directly before pouring (*see* **Note 4**).

4. After pouring overlay the resolving gel gently with isopropanol. This will ensure a smooth surface. Let the gel polymerize for approximately 20–30 min.
5. Remove the isopropanol layer and rinse the gel thoroughly several times with water. Residual water can be removed with the help of Whatman paper.
6. For the stacking gel 7.5 mL of a 4% acrylamide solution is used that is composed of 2.5 mL of 3X gel buffer, 0.6 mL of acrylamide and 4.37 mL of water. Add 50 μL of APS and 5 μL of TEMED to the prepared gel solution and mix thoroughly. Overlay the resolving gel with the help of a 1 mL tip. Immediately insert the comb and let the gel polymerize for another 15–20 min.
7. If the gel is not used immediately store it at 4°C (*see* **Note 5**).

3.2. In Vitro Protein Import into Isolated Mitochondria for Analysis by BN–PAGE

1. Thaw the BSA buffer, ATP and creatine phosphate and combine them with the freshly prepared NADH and creatine kinase in the indicated concentrations (*see* **Section 2.2**, *see* **Note 6**). The total volume of this import mix should be 85 μL.
2. Thaw mitochondria isolated from *Saccharomyces cerevisiae* and the ^{35}S-labeled Tom40 reticulocyte lysate on ice.
3. Add 50 μg of mitochondria (our mitochondria are adjusted to a final protein concentration of 10 mg/mL in SEM buffer, so that 50 μg correspond to 5 μL). Gently vortex the sample and incubate it at 25°C for 3 min (*see* **Note 7**).
4. Add 10 μL of the Tom40 lysate, gently vortex and incubate at 25°C for 5–45 min. The total volume of the import mix is now 100 μL (*see* **Note 8**).
5. Stop the import reaction by putting the samples on ice.

6. Re-isolate the mitochondria by centrifugation at 12,000 *g* at 4°C for 10 min. The following steps are carried out on ice. Discard the supernatant and wash the pellet with 100 μL of ice-cold SEM buffer (*see* **Note 9**).
7. Centrifuge the mitochondria according to step 6. The pellet can now be solubilized in Digitonin buffer (*see* **Subheading 3.5.1**) and further analyzed by Blue Native PAGE.

3.3. Chase of Intermediates to Their Mature Form

1. To analyze sequential steps in protein import and assembly, a pulse-chase reaction can provide important information. In principle a standard import reaction is performed for a short period of time, then the mitochondria are re-isolated and resuspended, followed by an additional incubation.
2. Follow the protocol depicted in 3.2 until step 4. After adding the reticulocyte lysate place the import reaction at 25° for 5 min ("pulse").
3. Re-isolate the mitochondria by centrifugation at 12,000 *g* at 4°C for 10 min. Discard the supernatant (*see* **Note 10**).
4. Resuspend the mitochondrial pellet in BSA buffer, which can be supplemented with various chemicals (ATP, NADH, CK, CP, etc.) (*see* **Note 11**), the total volume of the reaction should be 100 μL. Incubate the samples for the desired period of time (approximately 5–45 min) at 25°C ("chase").
5. Stop the reaction by putting the samples on ice, all subsequent steps are identical to steps 6 and 7 of the in vitro import protocol (*see* **Section 3.2**).

3.4. Formation of a TOM–TIM–Preprotein Supercomplex

1. Supplement the BSA buffer with ATP, NADH and the ATP regenerating system CK and CP (*see* **Section 2.1**). The details below are given for one reaction, which will be analyzed by BN–PAGE. The final volume of the import reaction is 100 μL. For the control reaction add 1 μL of AVO mix (100X) to dissipate the membrane potential and thereby preclude supercomplex formation.
2. Thaw the mitochondria and the recombinant purified $b_2\Delta$-DHFR precursor protein on ice.
3. Add the 20X diluted MTX to a final concentration of 5 μM and 3 μg of $b_2\Delta$-DHFR preprotein. After gentle vortexing the samples are incubated at 25°C for 2 min.
4. Add 70 μg of mitochondria (protein amount), vortex gently and incubate for a further 15 min.
5. Stop the import reaction by dissipation of the membrane potential by adding 1 μL of the AVO mix.
6. Mitochondria are re-isolated by centrifugation at 12,000 *g* at 4°C for 10 min. All following steps are carried out on ice.

Discard the supernatant and wash the pellet with 100 μL of cold SEM buffer (*see* **Note 9**).

7. Centrifuge the mitochondria according to step 4. The pellet can now be solubilized in Digitonin buffer and further analyzed by Blue Native PAGE or a sucrose gradient.

3.5. Analysis of Protein Complex Formation by BN–PAGE

1. Mitochondria (50–100 μg protein) are centrifuged at 12,000*g* at 4°C for 10 min. The pellet is resuspended in 50 μL of ice-cold solubilization buffer with Digitonin. Dissolve the pellet by pipetting up and down 10 times. It is essential that each sample is treated equally as differences in solubilization could impair the reproducibility (*see* **Note 12**).
2. Incubate the samples on ice for 10–15 min.
3. Centrifuge the samples for 10 min at 4°C at 12,000 *g* (*see* **Note 13**). The supernatant is transferred to another reaction tube and mixed with 5 μL of 10X loading dye by pipetting.
4. Apply the sample to the precast blue native gel.
5. Prepare the 1X cathode buffer by diluting 50 mL of its 10× stock with 450 mL of water. Very gently overlay the gel with the pre-cooled 1X cathode buffer, assemble the gel into the cooling chamber that contains 4 L of pre-cooled 1X anode buffer. Start the electrophoresis at 600 V and 15 mA per gel at 4°C. The gel run will take approximately 3 h. For running the gel overnight the voltage should be decreased to 70–80 V.
6. After completion of the run, the gel is destained in the destaining solution for approximately 30 min and subsequently vacuum-dried for 1–2 h at 80°C. For detection of the radiolabeled proteins, the dried gels are exposed to a radiation-sensitive storage phosphor screen.

In case proteins will be detected by immunodetection, the cathode buffer must be exchanged when the electrophoresis has run for approximately 30 min. Stop the electrophoresis, discard the cathode buffer and replace it with 500 mL of cooled 1X cathode buffer without Coomassie Blue G 250 (*see* **Note 14**), proceed with the electrophoresis. When the run has been completed the gel can undergo standard Western blotting onto a polyvinylidene fluoride (PVDF) membrane.

3.6. Separation of the TOM–TIM–Preprotein Supercomplex by Sucrose Gradient Centrifugation

1. Fill the ultracentrifugation tubes with the sucrose solution and place them in an upright position at –20°C for at least 1 h. During this freezing a sucrose and salt gradient of approximately 5–45% is obtained (*see* **Note 15**).
2. Thaw the gradient for 1 h at room temperature before loading the sample.

3. After the clarifying spin the solubilized mitochondria (1 mg of mitochondrial protein in 500 μL of solubilization buffer) are loaded on top of the gradient and centrifuged at 210,000 *g* for 19 h at 4°C (e.g. in the SW 41 Ti rotor from Beckman Coulter; *see* **Note 16**).
4. After centrifugation the fractions of the gradient are collected starting from the top by using a 1000-μL tip. Each fraction will range between 0.3 and 0.6 mL, resulting in a total of 20–40 fractions.
5. Precipitate the proteins by addition of TCA (final concentration 15%) and sodium desoxycholate (final concentration 0.0125%). Mix well and incubate the samples on ice for 30 min, alternatively they can be stored overnight at –20°C. Centrifuge for 30 min at 4°C and 12,000 *g*. Discard the supernatant carefully and wash the pellet by addition of 500 μL of ice-cold acetone. Spin again, remove the supernatant and dry the pellet at 30°C for 10–15 min. Pellets are solubilized in sample buffer for SDS–PAGE (e.g. Laemmli) by incubation at 60°C for 15 min under vigorous shaking.
6. Further analysis of fractions by SDS–PAGE and Western Blotting. For detection of the supercomplex, antibodies against Tim23, Tom40 or an antiserum directed against DHFR can be used. However, it is recommended to utilize affinity-purified antibodies.

4. Notes

1. Every solution for BN gels should be stored at 4°C as protein complexes are temperature labile. The gel electrophoresis should also be performed at 4°C.
2. The AVO mixture efficiently inhibits the membrane-potential ($\Delta\psi$)-dependent protein import by dissipation of the membrane potential and inhibition of its regeneration. It is commonly used to abolish protein import that requires a membrane potential and to stop the import reaction.
3. The ultracentrifugation tubes should be filled up almost completely after addition of the sucrose solution and the sample volume.
4. Gel solutions without APS and TEMED can be prepared in advance (e.g. mixtures for up to 4 gels) and stored at 4°C up to 2 weeks.
5. Gels not used the same day can be wrapped in wet paper towels and a plastic disposal bag and can be stored overnight at 4°C.

6. Addition of the ATP-regenerating system creatine kinase and creatine phosphate to the import reaction has a beneficial effect only when the import reaction time exceeds 30 min.
7. The crucial factor for a successful Tom40 import on BN–PAGE is the integrity of the outer mitochondrial membrane. It can be scrutinized by titration of the mitochondria with increasing concentrations of Proteinase K followed by separation on SDS–PAGE and immunoblotting to test for the integrity of IMS-localized marker proteins. If the import quality is poor despite intact mitochondria you can raise the amount of mitochondria loaded per lane to 100 μg (protein amount) or the amount of reticulocyte lysate to 10% (v/v) of the import reaction.
8. Compared to in vitro import on SDS–PAGE more reticulocyte lysate is needed to detect the signal on BN–PAGE; however, the ratio of reticulocyte lysate to the total reaction volume should not exceed 10% as this may lead to rupture of the outer mitochondrial membrane.
9. Washing the mitochondrial pellet: Since every resuspension step inflicts additional damage on the fragile mitochondrial membranes, it is sufficient to simply add the SEM buffer and to centrifuge the tubes in the reverse position so that the pellet is formed on the opposite side of the tube.
10. Re-isolation of mitochondria harbors the advantage that non-imported precursor proteins are removed. By additional incubation of the sample only the already imported proteins can be further sorted. This is of particular advantage for the analysis of mutant precursors or strains.
11. Mitochondria can be resuspended in different buffers for performing the chase reaction. The only obligation is its isotonic character (i.e. 250 *m*M sucrose for yeast mitochondria). By addition or removal of different supplements (ATP, NADH, CK, CP, etc.) the chase conditions can be easily varied.
12. The quality and purity of Digitonin is vital for an effective solubilization and extraction of membrane proteins. Different batches can contain various amounts of impurities. Therefore it is essential to test every freshly prepared stock of Digitonin (5% in water) for efficient and reproducible solubilization under established standard conditions. The amount of Digitonin needed for solubilization is individual for each protein complex. It ranges from 0.5–1.5% and has to be experimentally adjusted with each new batch of Digitonin and for each protein complex.

13. This so-called clarifying spin ensures that only the extracted proteins and protein complexes are loaded onto the gel, while aggregates and non-solubilized proteins remain in the pellet fraction.
14. The dye might interfere with blotting efficiency on PVDF membranes.
15. The sucrose gradient can also be poured from two different sucrose concentrations. However, the gradients obtained by freeze-thawing result in a higher reproducibility and can be stored at –20°C for a longer period of time but they yield a difference in salt concentrations in addition to sucrose concentrations.
16. The amount of mitochondria loaded depends on the desired number of gels for further analysis (we routinely load 1 mg of mitochondrial protein on the sucrose gradient, which will allow the run of around six SDS gels for analyzing fractions by Western blotting). However, one should keep in mind that due to the loss of mitochondrial proteins in the pellet and the distribution of proteins in the different gradient fractions, the overall quantity of proteins from the starting material will decrease. In general the gradient can be loaded with 0.5–3 mg of mitochondria (protein amount).

Acknowledgments

This work was supported by the Trinational Research Training Group GRK 1478, the Sonderforschungsbereich 746, the Excellence Initiative of the German Federal and State Government (EXC 294) and the Gottfried Wilhelm Leibniz Program.

References

1. Schägger, H., and von Jagow, G. (1991) Blue native electrophoresis for isolation of membrane protein complexes in enzymatically active form. *Anal Biochem.* **199,** 223–231.
2. Schägger, H., Cramer, W.A., and von Jagow, G. (1994) Analysis of molecular masses and oligomeric states of protein complexes by blue native electrophoresis and isolation of membrane protein complexes by two-dimensional native electrophoresis. *Anal Biochem.* **217,** 220–230.
3. Chacinska, A., Rehling, P., Guiard, B., Frazier, A.E., Schulze-Specking, A., Pfanner, N., Voos, W., and Meisinger, C. (2003) Mitochondrial translocation contact sites: separation of dynamic and stabilizing elements in formation of a TOM-TIM-preprotein supercomplex. *EMBO J.* **22,** 5370–5381.
4. Yamamoto, H., Esaki, M., Kanamori, T., Tamura, Y., Nishikawa, S., and Endo, T. (2002) Tim50 is a subunit of the TIM23 complex that links protein translocation across the outer and inner mitochondrial membranes. *Cell* **15,** 519–528.5.
5. Wiedemann, N., Pfanner, N., and Rehling, P. (2006) Import of precursor proteins into isolated yeast mitochondria. *Methods Mol Biol.* **313,** 373–383.

6. Model, K., Meisinger, C., Prinz, T., Wiedemann, N., Truscott, K.N., Pfanner, N., and Ryan, M.T. (2001) Multistep assembly of the protein import channel of the mitochondrial outer membrane. *Nat Struct Biol.* **8,** 361–370.
7. Hill, K., Model, K., Ryan, M.T., Dietmeier, K., Martin, F., Wagner, R., and Pfanner, N. (1998) Tom40 forms the hydrophilic channel of the mitochondrial import pore for preproteins. *Nature* **395,** 516–521.
8. Becker, T., Vögtle, F.N., Stojanovski, D., and Meisinger, C. (2008) Sorting and assembly of mitochondrial outer membrane proteins. *Biochim Biophys Acta.* **1777,** 557–563.
9. Wiedemann, N., Kozjak, V., Chacinska, A., Schönfisch, B., Rospert, S., Ryan, M.T., Pfanner, N., and Meisinger, C. (2003) Machinery for protein sorting and assembly in the mitochondrial outer membrane. *Nature* **424,** 565–571.

Subject Index

A. Economou (ed.), *Protein Secretion*, Methods in Molecular Biology 619,
DOI 10.1007/978-1-60327-412-8, © Springer Science+Business Media, LLC 2010

MIX
Papier aus verantwortungsvollen Quellen
Paper from responsible sources
FSC® C105338

If you have any concerns about our products,
you can contact us on
ProductSafety@springernature.com

In case Publisher is established outside the EU,
the EU authorized representative is:
Springer Nature Customer Service Center GmbH
Europaplatz 3, 69115 Heidelberg, Germany

Printed by Libri Plureos GmbH
in Hamburg, Germany